The
Ascent of
Mathematics

The Ascent of Mathematics

Raymond Coughlin

David E. Zitarelli

Temple University

McGraw-Hill Book Company

New York
St. Louis
San Francisco
Auckland
Bogotá
Hamburg
Johannesburg
London
Madrid
Mexico

Montreal
New Delhi
Panama
Paris
São Paulo
Singapore
Sydney
Tokyo
Toronto

The Ascent of Mathematics

1 2 3 4 5 6 7 8 9 0 DOCDOC 8 9 8 7 6 5 4

ISBN 0-07-013215-1

This book was set in Times Roman by University Graphics, Inc. The editors were Peter R. Devine and James W. Bradley; the designer was Joseph Gillians; the production supervisor was Leroy A. Young. The photo editor was Lorinda Morris (Photoquest). The drawings were done by J & R Services, Inc.
R. R. Donnelley & Sons Company was printer and binder.

Library of Congress Cataloging in Publication Data

Coughlin, Raymond F.
    The ascent of mathematics.

    Includes index.
    1. Mathematics—1961–   I. Zitarelli, David E.
II. Title.
QA39.2.C69 1984   510   83-17548
ISBN 0-07-013215-1

This book is dedicated
with sincere appreciation
to Morris Kline

Contents

Preface   ix

Chapter 1   **Section 1** Early Systems   2
Natural Numbers 1 **Section 2** The Hindu-Arabic System   19
                       Section 3 The Binary System   33
                       Section 4 Divisibility and Prime Numbers   44
                       Section 5 Pythagorean Mathematics and
                                      Pattern Recognition   56
                       Section 6 Order Amid Chaos: Conjecture versus Theorem   69

Chapter 2   **Section 1** Rational Numbers   82
Real Numbers 81   **Section 2** Infinite Sums and Radioactivity   94
                       Section 3 Decimals   105
                       Section 4 Irrational Numbers   119
                       Section 5 Sets   130
                       Section 6 A Digression on Proof   140

Chapter 3   **Section 1** The Inception of Geometry   152
Geometry   **Section 2** Euclidean Geometry   164
and Logic 151   **Section 3** Conditional Statements   174
                       Section 4 Arguments   186
                       Section 5 The Theory of Parallels   203
                       Section 6 Aftermath   213

Chapter 4   **Section 1** Rebirth of Mathematics   230
The Power   **Section 2** Saccheri and Euclid's Flaw   240
of Logic 229
                       Section 3 Spherical Geometry   250

Chapter 5   **Section 1** Descartes and His Analytic Geometry   266
Linear   **Section 2** Linear Equations and Straight Lines   277
Programming 265   **Section 3** Systems of Linear Equations   294

Section 4	Linear Inequalities	305
Section 5	Linear Programming Problems	318
Section 6	Graphical Solutions of Linear Programming Problems	332

Chapter 6 Probability 345

Section 1	A Genetic Approach to Probability	346
Section 2	Theoretical Probability Experiments	358
Section 3	Sample Spaces and Addition Rules	369
Section 4	Conditional Probability	390
Section 5	Independent Events	409
Section 6	Counting Techniques	424
Section 7	Solving Probability Problems	437

Chapter 7 Transition from Probability to Statistics 445

Section 1	The Monte Carlo Method	446
Section 2	Empirical Probability	455
Section 3	Design of an Experiment	467
Section 4	Descriptive and Inferential Statistics	477

Chapter 8 Statistics 489

Section 1	Histograms, or a Picture Is Worth 1000 Words	490
Section 2	Measures of Central Tendency	505
Section 3	Standard Deviation	519
Section 4	The Normal Curve	533
Section 5	The Normal Approximation for Data	551
Section 6	Sample Surveys	561

Answers to Odd-Numbered Exercises	573
Index	601

Preface

The main theme of this book is that mathematics is a vibrant, evolutionary discipline. Our premise is that every mathematical technique was created by someone to solve a human problem. Jacob Bronowski's classic work "The Ascent of Man" (Little, Brown and Company) showed that science is intimately intertwined with human evolution. The following quotation from the foreword of Bronowski's book captures the spirit of our text when we take the liberty of twice replacing the word "science" with "mathematics."

> Knowledge in general, and mathematics in particular, does not consist of abstract but of manmade ideas, all the way from its beginnings to its modern and idiosyncratic models. Therefore the underlying concepts that unlock nature must be shown to arise early and in the simplest cultures of man from his basic and specific faculties. And the development of mathematics which joins them in more and more complex conjunctions must be seen to be equally human: discoveries are made by men, not merely by minds, so they are alive and charged with individuality.

Our text, "The Ascent of Mathematics," was written to extend Bronowski's theme. We have tried to dispel the often preconceived notion that mathematics is an endless procession of apparently meaningless manipulations. We have countered that view by presenting traditional mathematical skills in a very untraditional setting. Throughout the text, we immerse the mathematics in a specific episode of its evolution. We try to capture the spirit of *who* the creators of the mathematics were, *why* they were confronted with the particular societal problem, and *how* they solved the problem. We have highlighted cultural settings, personalities, and the influence of the mathematical inventions.

There is one important distinction between Bronowski's "Ascent" and ours. His is aimed at the general reader. Ours is written for students. Thus this book contains examples that demonstrate the material and exercises that reinforce and extend it. The emphasis is on skills and techniques, because one cannot learn *about* mathematics without *doing* mathematics. The vehicle used is a series of ascent themes that place the material in a cultural, historical, or human setting. These themes also serve as a common thread that unifies the presentation and ties together the various parts of the subject. However, they are not intended to present a precise historical devlopment of mathematics. Limitations of space have forced us to choose the vignettes that best illustrate our main theme. We have been guided by what we think is of most interest to students and by what we hope will convince them that mathematics is a living discipline.

Format

One of the most innovative features of the text is the separation of the ascent themes from the skills portions of the text. Rather than fuse the techniques with the thematic material, as most texts do, we have developed a method that keeps them separate and distinct, both in content and format. This means that it is easy for a teacher to cover a particular skill while skipping the theme or substituting other motivating material. Hence the text is very flexible.

We accomplish this by using a different type style to distinguish the themes from the skills. The themes, written in a more literary style, are set in a reduced type. This makes it very easy to emphasize different themes. Thus two instructors could cover the same topics but, by highlighting different ascent themes, teach two entirely different courses that reflect their individual expertise and enthusiasm.

Each section has been partitioned into subsections. The first such subsection is the ascent theme. The remaining subsections have titles that are set off like the title "Format" above. This division will help the instructor to prepare lectures and the students to organize the material.

We have tried to keep the length of each section to what can be covered in a typical 50-minute class. Sometimes, however, the material itself has dictated more extensive coverage.

Writing Styles

We have employed two distinct writing styles—one for the themes and another for the skills. The themes are written in a more literary style—neither as formal as a history text nor as light as a magazine or newspaper article. On the other hand, we have tried to remain as rigorous as possible in the skills portions. The examples, which are the backbone of the text, are set off by a distinct type style and a vertical color rule in order to easily distinguish them from more general discussions.

The one common thread that runs through both styles—the literary and the more rigorous—is our challenge to the reader to think, to probe, to guess, and, in short, to *get involved*. It is the only way to understand mathematics.

This is especially apparent in the two formats used in the examples. When we encounter a particular problem for the first time, we proceed directly to its solution. But when we illustrate a skill already introduced, we distinguish the problem from its solution by using this format:

Example
Problem . . . .

Solution . . . .

The student should be able to make a good attempt at solving the problem in this kind of example before reading its solution.

Course Organization

The CUPM Panel on Mathematics Appreciation Courses published its official report in the *American Mathematical Monthly,* January 1983 (pp. 44–51). Reprints of this timely, informative report may be purchased for $1.00 from the MAA Publications Department, 1529 Eighteenth Street, NW, Washington, D.C. 20036. It discusses several types of courses whose "main goal is to get students to appreciate the significant role that mathematics plays in society, both past and present." The report notes that there are nearly as many ways to teach such courses as there are teachers of them. The flexible format in this book allows for numerous approaches. Here are three suggestions.

1 A course that emphasizes skills applied in the social sciences and humanities. It would include Chapter 1 (Sections 1-1 to 1-4); Chapter 2 (Sections 2-1 to 2-4); and Chapters 3, 4, 6, and 8. Parts of Chapter 7 would be assigned as reading material.

2 A course that includes an emphasis on proof and logic. It would consist of Chapters 1, 2, 3, 4, 6, and 8.

3 A course that emphasizes skills generally found in a finite mathematics course. It would include Chapter 1 (Sections 1-1 to 1-4); Chapter 2 (Sections 2-1 to 2-4); and Chapters 5, 6, 7, and 8.

Chapter Dependence

The text can naturally be divided into three parts: Numbers (Chapters 1 and 2), Geometry (Chapters 3, 4, and 5), and Probability and Statistics (Chapters 6, 7, and 8). All the chapters except 4 and 7 are approximately the same length. We have found that five or six chapters can be covered in a semester, presuming that about half the ascent themes are discussed in class. Of course, if fewer themes are covered, then more chapters can be included, and vice versa.

Several sections in Chapters 1 and 2 can easily be omitted in order to cover more material in later chapters. Much of Chapter 7, which is included primarily to demonstrate the connection between probability and statistics, either can be omitted or assigned as reading material.

Chapter 8 can be covered without Chapter 7. Here is a schematic display of the dependence of the chapters:

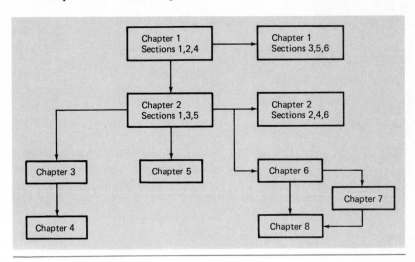

Exercises

We have taken great care to build the exercises around the examples in the text. For each example or set of examples there is a group of routine exercises that develop that skill. Thus the student gets to practice each technique presented in the text through a series of exercises modeled after the examples. The answers to the odd-numbered exercises appear in the back of the book. Each odd-numbered exercise is followed by a very similar even-numbered exercise. Thus an instructor can assign any proportion of homework problems that have answers as opposed to those that do not.

Once the skills presented in the section have been adequately developed in the exercises, there are more challenging problems designed to motivate, entice, or humor. They require a greater degree of reasoning than the routine problems.

In assigning homework, instructors will find it easy to choose problems from the routine exercises because they follow the order and level of difficulty of the examples. However, care should be taken when problems are selected from the more challenging exercises. Since they are generally more difficult than the others, instructors should assign only those that they care to spend time on in class.

Supplements

There is a supplement to the text, called the "Instructor's Resource Guide." It gives an outline of each section and lists its objectives. It also contains an annotated bibliography of the sources used, as well as some suggestions of material for updating or extending particular subjects. It also provides the answers to the even-numbered problems.

Acknowledgments

It is a pleasure to acknowledge the help we received on this book from many people. Our biggest support came from our wives, Judy and Anita, who tolerated with patience and understanding our total immersion in this project over several years. They deserve special praise for ignoring our boorish behavior at social affairs and for raising our children in our absence.

Several colleagues at Temple University played vital roles in the evolution of this text. We thank Leon Steinberg for suggesting that we put our classnotes into print and David Elesh (sociology), Judith Goode (anthropology), Richard Joslyn (political science), and James Hilty (history) for the advice they offered in their respective specialties.

Our reviewers played an important part in transforming classnotes into a book. Two of them deserve special thanks for their unstinting, constructive criticism of the entire manuscript: Louis F. Hoelzle, Bucks County Community College, and John Novosel, Richard J. Daley College. We are very appreciative, too, for the advice we received from the following reviewers: William D. Blair, Northern Illinois University; Sebastian S. Koh, West Chester State College; Peter H. Maserick, Pennsylvania State University; John Mathews, California State University, Fullerton; Carl S. Myhill, Western Connecticut State College; Frederick Rickey, Bowling Green State University; James Stasheff, University of North Carolina at Chapel Hill; and William R. Weller, Shippensburg University.

We extend our gratitude to four of our former undergraduate students who provided numerous helpful comments and conducted research in the library: Mary Lynne Aspinall, Lisa Duckworth, Gerhart Keller, Jr., and Elaine Carpey.

The staff at McGraw-Hill has been particularly supportive. We especially thank editor Peter Devine for the interest and enthusiasm he has given the book. We also thank Donald Chatham for his astute guidance, James W. Bradley for his editorial supervision, Joseph Gillians for his design, Lorinda Morris for her photo research, and Kathleen Civetta for developmental editing. Finally, we thank the typists who converted our own brand of hieroglyphics into print: Mittie Davis, Patricia Libbey, Gerry Sizemore, Donna Sowell, and Anita Zitarelli.

Raymond Coughlin
David E. Zitarelli

1 Natural Numbers

This chapter is concerned with what are sometimes called whole numbers. It shows their evolution from early times to the present. The chapter may be divided into two parts, each of which treats a different aspect of natural numbers. The first part, Sections 1-1 to 1-3, is concerned with various systems of numeration. The second, Sections 1-4 to 1-6, investigates various properties of natural numbers as seen through the eyes of the classical Greeks.

Section 1-1 begins with the Mary Leakey expedition to Tanzania, which sets the stage for the emergence of the concept of number. It traces this development from the earliest times until the classical Egyptian and Mayan civilizations, and examines the number systems of these two great cultures. The transmission of early number systems to the west is discussed in Section 1-2. It is motivated by a discussion of the Thor Heyerdahl expedition aboard a replica of a 5000-year-old boat and its implications for how some early civilizations might have communicated. Section 1-3 examines a useful modern number system, the binary system. It leads to a presentation on exponents and ends with the question, "Which system is best?" In Section 1-4 a "silent lecture" from 1903 serves as a connection between the so-called fundamental theorem of arithmetic, first proved 25 centuries ago, and a result in abstract algebra resolved in 1981. Section 1-5 provides a glimpse into other types of numbers studied in ancient Greece, bearing such names as "square," "triangular," "perfect," and "friendly." The importance of patterns is stressed here. This theme is developed further in Section 1-6, where it is shown how mathematicians distinguish between order and chaos in problems concerning prime numbers. Important distinctions are drawn between the meanings of three basic terms: theorem, counterexample, and conjecture.

1-1 EARLY SYSTEMS

There they were, a group of supposedly mature people, playing a child's game. They were part of the Mary Leakey expedition in Tanzania seeking clues to humanity's origins, when the monotony of the daily chores and the inevitable heat and wind took their toll. So they relaxed by pelting each other with dry elephant dung, just as children do at the seashore. Except that children throw sand.

To avoid one such missile Dr. Andrew Hill ducked low. It missed. Suddenly he stopped laughing. He bent over slowly, crouching to the ground. The entire group's merriment ceased as his colleagues rushed to his side.

What could be wrong? Was he ill?

Their fears were allayed immediately when they found that he was actually studying the ground in front of him. He was examining a series of punctures in the volcanic tuff, and a closer examination revealed that they were animal footprints.

The game ended. The work began anew. What significance would this stroke of luck have in the quest to find the origins of humanity? And what meaning would it have for the origin of numbers?

Gathering Information about Early Humans

The scene above provides the background setting for this section, in which we will be concerned with the early evolution of the number concept. There is an interesting parallel between our knowledge of early humans and the development of our number system. The ways in which we discover information about humankind's use of numbers in prehistoric times are the same means used by scholars to unlock the mysteries of the origins of humanity. We must rely on the research being done in a variety of fields outside of mathematics. For example, anthropologists study societies in the cultural mainstream and compare them with some that have developed in relative isolation. By studying the differences in their number systems we get a measure of the impact of societal and cultural demands on the development of numbers.

Of course, the work in the field of archeology has been just as important. Artifacts unearthed in Africa, the Middle East, and Central America give us a hint as to how early humans counted and kept records. As indicated in our opening story, new finds continually add to our store of knowledge, corroborating, or sometimes questioning, existing theories.

Educational psychology provides a useful barometer against which we can measure the evolution of mathematics as compared with the way in which our mathematical education develops. In other words, does the development of our mathematical knowledge, from the elementary concepts that you were exposed to in your preschool years up to the present, follow the same evolutionary process as mathematics itself? More simply, did your mathematical education imitate the historical development of the subject? In these first two chapters we will supply you with enough detail of the ascent of our number system for you to be able to answer this question confidently.

Artifacts

Many elementary questions confront anyone who sets out to study the initial development of the number concept. Some of them include:

When did humans first feel the need to count and what means were used?

Did humans always count in the same way that we do now?

Were early number systems radically different from ours?

What were some of the difficulties encountered by early humans that necessitated the invention of a more complex number system?

Why would a particular culture be compelled to improve its number system?

To answer these questions we will outline some of the recent history of anthropology. In 1959 Louis and Mary Leakey startled the scientific world with their announcement of the unearthing in northern Tanzania of the remains of *Homo habilis*, an erect-walking hominid estimated to be 1,750,000 years old. Later they found a human skull in Kenya, estimated to be 2,600,000 years old. However, in 1973 Donald Johanson proposed a conflicting theory about humanity's origins, based on his findings in Ethiopia. Then the Mary Leakey expedition experienced their stroke of luck. First the animal footprints were found, and then, in 1976, a trail of hominid footprints was discovered nearby. Further investigation revealed that these tracks were left by hominids walking upright on the hardened volcanic ash at least 3,600,000 years ago. Mary Leakey reported on these findings in April 1979, but rather than clear up the matter it actually led to an academic brouhaha involving Mary's son Richard and Donald Johanson, who clung to his own theory that humans had evolved from his own *Australopithecus* and not from Leakey's hominids.

The question of when *Homo sapiens* came into existence is another matter of disagreement. The estimates range from more than a million years ago to as little as 50,000. Once again Mary Leakey is at the forefront. In a recent article in the English journal of natural science, *Nature*, she and her associates indicated that the modern human being began to evolve from its precursor *Homo erectus* between 150,000 and 90,000 years ago. They warned, however, that further anatomical analysis on the specimen, Laetoli Hominid No. 18, would have to be carried out before their conclusion could be considered final.

These numbers, dealing in tens of thousands and millions, are somewhat beyond our comprehension. A number that is easier to deal with is 8000 B.C., when the world's first town, Jericho, was founded. In fact, it was not until about 4000 B.C. that the first cities and urban sprawls appeared. Then, starting around 3600 B.C., a profound change occurred along the Nile River, a development which we will pursue subsequently.

The issue of when it all began, and precisely where, remains unsettled. But what is agreed upon unequivocally is that humankind's origins lie in east Africa, probably in Tanzania or Ethiopia. Thus it is natural to focus our initial investigations regarding the origins of numbers upon the African continent. Surprisingly, though, such work was begun only recently, and the pioneering book "Africa Counts," by Claudia Zaslavsky, was published in 1973 (by Prindle, Weber & Schmidt, Boston) "as a preliminary survey of a vast field awaiting investigation."

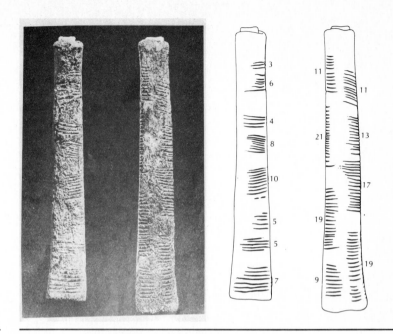

FIGURE 1 Ishango tally bone.

What is also surprising is that the oldest known mathematical artifact comes not from Africa but from Europe. It is a 7-inch bone of a wolf discovered in Czechoslovakia in 1937; it is estimated to be 30,000 years old. There are 55 notches on the bone, and they are arranged into two groups, with 25 in one and 30 in the other. Within each group the notches are grouped by 5s. Of course, we can only conjecture as to what the purpose of this arrangement might have been.

The second-oldest artifact was excavated from Africa. It was found in 1960 along the shore of Lake Edward in Zaire. Dating back to the period between 9000 and 6500 B.C., it is a small bone handle which has come to be known as the Ishango bone (see Figure 1). It contains groups of notches arranged in three distinct columns:

11	11	3
13	21	6
17	19	4
19	9	8
		10
		5
		5
		7

Several possible meanings have been suggested for these groupings, and there is no agreement on any one of them. The anthropologists who found this bone believe the first column is related to prime numbers, the second to the base of a system, and the third to doubling. Another anthropologist claims the numbers are related to a lunar calendar. One mathematician has questioned all such theories, asserting "there is always the chance that the tallies may simply record several days' catch of fish!"

Tallying

The answer to the question of when humans first had the need to use numbers is sketchy. However, it is important to note that both the artifacts described above are examples of tallying, where *tally* refers to a succession of strokes on wood, bone, or paper, where each stroke stands for one of a series of objects or events. (Webster informs us that the term "tally" is derived from the Latin *talea*, meaning cutting, from which we also get our word "tailor.") We frequently use such a system for recording ballots in small elections even today when we group the results into 5s by ‖‖‖ .

Tally sticks have been used throughout history, and one curious incident concerning them centers in England. From the twelfth until the early nineteenth century the British Exchequer used tally sticks to record financial transactions officially. Each stick had two sets of identical notches carved into it (see Figure 2). It was then split into two, and each party was given a piece, the theory being that fraudulent methods could thus be prevented. It was not until 1826 that an act of Parliament ended this obsolete practice. Eight years later the tally sticks were ceremoniously burned in a huge bonfire. Unfortunately the fire raged out of control and burned down the House of Commons.

Let us return to our discussion of tallying. It is not surprising that the earliest known artifacts are tally bones. The very existence of these remains raises the question of what types of tallying processes could have been used earlier. For example if a herder wanted to check whether any livestock had strayed overnight, how would this have been done? One way was to point to a series of parts of the human body in a prescribed order. For instance, the Bugilai of British New Guinea used their right index finger to touch the following sequence of body parts: the fingers on the left hand, from the little finger to the thumb, followed by the left wrist, elbow, shoulder, and breast, and then the right breast. This accounted for the numbers from 1 to 10. The trouble with this method, naturally, is that it has very limited scope and duration. Later, piles of shells and stones were used instead. For example, the immense gold supply of the Asante kingdom was balanced every 20 days by means of cowrie shells, and many societies carried out everyday market calculations with the aid of piles of pebbles. In fact, the term "calculator" is derived from the Latin word *calculus*, which means pebble.

Not one of these tallying procedures required spoken words. Such "number words" did evolve though, but even this development does not imply that the concept of a number was understood, for at the first stage different words were used to express different sets of objects. For example, distinct words were used for two eyes, two nostrils, and two ears. (This explains the existence of such diverse terms as "flock," "gaggle," "school," "pride," and "bray" to describe sets of various kinds of living creatures.)

The method of finger gesturing is sometimes related directly to the num-

FIGURE 2   Tally sticks.

ber words themselves. For instance, the Arusha Maasai of northern Tanzania rarely say a number without the accompaniment of a finger sign. In order to see the finger gestures for the numbers 1 to 9 you can consult the photographs on pages 249 to 251 of the book "Africa Counts." It should be pointed out, however, that they do not do justice to such reckoning unless they are accompanied by motion pictures and sound. Incidentally, we still use the word "digit" to refer to numbers as well as to fingers and toes.

It was indeed an important step in the evolution of number when this common property of number, or "two-ness" in our example above, was first understood. It is impossible to determine when it evolved, and whether it evolved in one area of the world and diffused into others or evolved independently in various parts of the world. But one point should be made clear: the concept of number evolved, it did not suddenly spring up in the hands of some mathematical Moses. In fact, if we examine the earliest written documents that were produced in the potamic regions bordering the Nile, Tigris and Euphrates, Indus, and Yangtze rivers, we see that the number concept had already achieved a rather advanced state, including the use of various operations on the numbers themselves.

We have seen that although we have gained a great deal of knowledge about the origins of early humans, there is no universal agreement concerning exact dates. It is not surprising then that there is little consensus over exactly when and how humans began to count.

Numeration Systems

It took quite a long time before humans experienced the need for a written means of recording numbers. We discussed some of these tallying methods previously, but what happens when a culture is compelled to use large numbers? The next step in the ascent of number was to replace simple tallying by a numeration system, or number system. A *numeration system* is an organized way of representing numbers by various combinations of symbols. These symbols are called *numerals*. Two distinct number systems will have different numerals representing the same number.

By the way, do you know the name of our number system? Most likely, you have heard of it but cannot recall it. It is called the *Hindu-Arabic* system, and it is based on the 10 familiar numerals

0, 1, 2, 3, 4, 5, 6, 7, 8, 9

Sometimes we will just refer to it as "our number system." This system did not evolve directly from any one simple tallying system but from many other systems, as its very name suggests.

It is instructive at this juncture to study in some detail some ancient number systems. This will provide us with some concrete information from which we can confidently make conjectures about the origins of our number system and will also allow us to graphically see how number systems evolved.

We will consider the Egyptian and Mayan (mý-ahn) number systems. There are many such ancient systems to choose from, but we feel that the comparison of these two in particular will provide us with ample similarities to and differences from each other and with our own number system.

Egyptian Numeration System

We mentioned earlier that the world's first town was founded about 8000 B.C., that cities began appearing by about 4000 B.C., and that a profound change occurred some 400 to 500 years later along the Nile River. This profound change was the rise of the ancient Egyptian civilization.

By 3500 B.C. the Egyptian civilization had developed a numeration system which was capable of handling very large numbers. A summary of the spoils taken by Menes during his plunderings can be seen on a royal mace now on exhibit at Oxford University. It boasts of having taken 400,000 cows, 1,422,000 goats, and 120,000 captives. But the symbols used to express these numbers are quite different from ours. In other words, on the royal mace you will not find the symbol 4 followed by five 0s. A completely different set of symbols was used. Our first task will be to learn how to translate from the Egyptian system to our own.

The Egyptian type of picture writing is known as *hieroglyphics*, or "sacred carvings." It was used chiefly in inscriptions on stone. Later it evolved into a sort of script called *hieratic*, which was used by the priests when writing on papyrus. Because of Egypt's excessively dry climate, many of these papyrus scrolls have been preserved, and it is from them that we have derived most of our knowledge of the ancient Egyptians and, in particular, of their mathematics.

An example of hieroglyphics is given in Figure 3 on page 8. It is taken from a sandstone stele and is part of an account of the expedition of Amenhotep III in 1450 B.C. Note that the marks on the left side of the third row from the bottom are

Each vertical stroke represents a unit, each heel bone 10, and each remaining symbol 100. So these markings, when read from right to left,

FIGURE 3   Amenhotep III.

represent seven 100s, four 10s, and three 1s (units). This amounts to 743, as can be seen by the following translation of the Egyptian symbols into our own Hindu-Arabic symbols:

	Ones	Tens	Hundreds
Egyptian	III	∩∩ ∩∩	9999 999

Hindu-Arabic $3 \cdot 1 + 4 \cdot 10 + 7 \cdot 100 = 3 + 40 + 700 = 743$

Notice that not only did the Egyptians use different symbols than ours, but they combined them in a different way to express the specific numbers. They repeated a symbol many times in small groups to express multiples of the symbol. Thus 30 would be written as three 10s, or ∩∩ ∩, and 400 would be written as four 100s, or 99 99 .

Note that unlike our numeration system the smaller units are placed on the left. The Egyptians did not maintain any consistency, however, and the number 743 could just as well have been written

9999 ΠΠ |||
999 ΠΠ

This representation differs from our system in two other ways: the symbols are placed vertically as well as horizontally, and each symbol is written repeatedly (e.g., there are four heel bones).

Symbols for larger numbers included a lotus flower for 1000, a bent finger for 10,000, a tadpole for 100,000, a kneeling figure for 1,000,000, and a figure thought to be a rising sun for 10,000,000. The Egyptians had to create a new symbol for every power of 10. Compare this with our system, in which only 10 symbols are needed to describe any number, no matter how large it is. Table 1 gives a correspondence between the symbols of the two number systems.

TABLE 1

**Comparison of Egyptian (Hieroglyphics) Number Symbols
and Our (Hindu-Arabic) Number Symbols**

Egyptian							
	ǀ	Π	9	ℊ	⌒	⌒	𝔜
Ours	1	10	100	1000	10,000	100,000	1,000,000

The method that is used to translate Egyptian numerical symbols into our own is straightforward. First separate the number into subgroups consisting of repeated symbols, translate each symbol into its corresponding Hindu-Arabic symbol (always a power of 10), multiply the latter symbol by the number of times the symbol appears, and then add the numbers. For example, a number similar to the one on the sandstone stele in Figure 3 can be translated as follows:

Egyptian
|| ΠΠΠ 9999
||| ΠΠΠ 9999
↓   ↓   ↓

Hindu-Arabic $5 \cdot 1 + 6 \cdot 10 + 8 \cdot 100 = 5 + 60 + 800 = 865$

Let us consider another example. It would be instructive for you to read the problem and then try to solve each part of the example on your own before you read the solution.

Example 1

Here are four numbers written in Egyptian hieroglyphics.

(a) ||∩   (b) |||∩∩ 9   (c) ∩∩ 99   (d) ||| ∩∩∩∩ 9999
        || ∩∩   ∩∩ |||| ∩∩∩∩ 99999

Problem   Express each number in Hindu-Arabic notation.

Solution   (a) There are two 1s and one 10 so the number is $2 + 10 = 12$.

(b) We perform the translation as follows, noting that there are five 1s, four 10s, and one 100:

||| ∩∩
|| ∩∩ 9
↓   ↓   ↓

$5 \cdot 1 + 4 \cdot 10 + 1 \cdot 100 = 5 + 40 + 100 = 145$

(c) Note that there are no 1s. Since there are four 10s and two 100s, we get

∩∩
∩∩ 99
↓   ↓

$4 \cdot 10 + 2 \cdot 100 = 40 + 200 = 240$

(d) This problem demonstrates explicitly which number system is more efficient in expressing numbers:

||| ∩∩∩∩ 9999
|||| ∩∩∩∩ 99999
↓   ↓   ↓

$7 \cdot 1 + 8 \cdot 10 + 9 \cdot 100 = 987$

Remember that all the numbers expressed in hieroglyphics could be written in the reverse order and sometimes were. We will continue to use the order specified in Figure 3 just to emphasize that the order of the symbols is unimportant.

When you study a foreign language, it is usually more difficult to translate from your own into the foreign language than vice versa. This usually holds true for translating numbers written in other languages, but you will probably find that the Egyptian system is so picturesque and similar to Hindu-Arabic numeration that the task will be fairly easy. Let us try an example.

Example 2

Problem   Translate the following numbers into Egyptian numeration symbols:

(a) 22   (b) 719   (c) 503   (d) 967

Solution (*a*) There are two 10s and two 1s in 22; so the Egyptian form of 22 is ‖∩∩.

(*b*) There are seven 100s, one 10, and nine 1s; so the Egyptian form is

‖‖ 999
‖‖‖ ∩ 9999

(*c*) There are five 100s and three 1s; so the Egyptian form is

‖ 99
999 . How does the 0 get translated?

(*d*) Here is another lengthy one. It is ‖‖ ∩∩∩ 9999 / ‖‖‖ ∩∩∩ 99999 .

We hope that you are finding that translating between our system and hieroglyphics is an easy exercise. Can you see why it is readily mastered? It is because both number systems are based on the number 10. We will explain this in more detail in the next section. This is also one of the reasons why you will probably find the next number system more difficult to understand.

Mayan Numeration System

The Mayan civilization produced one of the most advanced cultures in history. For six centuries, beginning about A.D. 150, they built scores of architecturally fabulous cities throughout what is now Mexico, Guatemala, and El Salvador. They also developed an advanced astronomy which included a 365-day calendar and the orbital periods of Mars and Venus. Suddenly, around A.D. 900, the Mayan culture collapsed. Nobody knows why. Of course the people did not vanish; more than 2 million Mayan-speaking Indians still populate the Yucatán peninsula today. But the highly advanced civilization, as we know it, seems to have just disappeared.

Our knowledge of this early Mayan way of life continues to expand today. For instance, one basic assumption in anthropology has been that the change from a hunting-gathering existence to that of a settled village was spurred by the intermediate development of agricultural crops. But this assumption did not seem to apply to the Mayas, because their soil was difficult for growing crops and because game was scarce. So how did they support their population, estimated at between 2 and 3 million people? The answer was supplied in 1980, when an extensive system of canals covering 11,000 square miles was discovered, thus indicating that adequate supplies of water and game were available.

Our knowledge of Mayan mathematics is rather complete, however. You might be surprised to learn, though, that it was not until the 1880s that the Mayan number system was deciphered. The reason for this late date is that the Spaniards had systematically destroyed the native writ-

TABLE 2

Mayan Numerals

1	•	6	•̲	11	≝•̲	16	≣•̲
2	••	7	••̲	12	≝••̲	17	≣••̲
3	•••	8	•••̲	13	≝•••̲	18	≣•••̲
4	••••	9	••••̲	14	≝••••̲	19	≣••••̲
5	̲	10	≝	15	≣		

ings of the meso-Americans after their conquests, and this practice has deprived us of many of the original documents.

It has been said that the complexity of a civilization is mirrored in the complexity of its numbers. If this is true, then the Mayan civilization must have been complex indeed, as we shall soon see. We should point out, however, that the Maya developed two elaborate systems of numeration beginning about 300 B.C., mainly for constructing a calendar and studying astronomy. The more commonly used system, which like Morse code uses dots (pebbles) and dashes (sticks), is shown in Table 2.

Notice that a dot is the symbol for 1 unit and a dash the symbol for 5 units. Notice also that the dots are grouped horizontally above the dashes, and that the dashes are stacked vertically. This leads to a vertical numeration system that we will contrast with the Egyptian and Hindu-Arabic systems, both of which are horizontal.

For the numbers 1 to 19 the Mayan system resembles the Egyptian system in the sense that numbers are formed by grouping like symbols together. However, there are some contrasts, too, as the accompanying chart shows.

Property	Egyptian	Mayan
Placement of groups	Horizontal	Vertical
First symbol (unit)	|	•
Second symbol	∩	
Value of second symbol	10	5

The Mayan system begins to assume a distinct character of its own with numbers greater than 20. In general, these numbers are composed of vertical groups, where each group consists of one of the numbers in

Table 2. Let us look at an example to illustrate what we mean. Here is a typical Mayan number:

$$\begin{array}{c} \bullet\;\bullet \\ \hline \bullet\;\bullet\;\bullet \end{array}$$

This number consists of two groups of dots and dashes. We will use brackets to single out each group, and use Table 2 to read off the numerals within each group:

$$\underline{\bullet\;\bullet}\;\}\;7$$

$$\underline{\bullet\;\bullet\;\bullet}\;\}\;13$$

The numeral in the bottom group represents the number of 1s, or units, in the number. So this group represents 13 units, or $13 \times 1 = 13$. The numeral in the top group denotes 20s, so this group represents seven 20s, or $7 \times 20 = 140$. The Mayan number $\begin{array}{c} \bullet\bullet \\ \hline \bullet\bullet\bullet \end{array}$ then represents 153, which we indicate in the following way:

$$\underline{\bullet\;\bullet}\;\}\;\;7 \times 20 = 140$$

$$\underline{\bullet\;\bullet\;\bullet}\;\}\;\;13 \times 1 = \underline{+13}$$
$$153$$

The next example supplies two more illustrations of this procedure. The second part introduces one of the shortcomings of the Mayan system, and we will return to it in Section 1-2.

Example 3

Problem What number does each of these Mayan numbers represent?

$(a)\;\begin{array}{c} \bullet\;\bullet\;\bullet \\ \hline \bullet\;\bullet \\ \hline \end{array}\;\;(b)\;\underline{\bullet\;\bullet}^{\bullet}$

Solution (a) We follow the method described previously. First we form groups of dots and dashes. Each numeral in the bottom group denotes 1s and each numeral in the next group above denotes 20s. The number in (a) becomes

$$\underline{\bullet\;\bullet\;\bullet}\;\}\;13 \times 20 = 260$$

$$\begin{array}{c}\underline{\bullet\;\bullet}\\\hline\end{array}\;\}\;17 \times\;\;1 = \underline{+17}$$
$$277$$

Thus $\begin{array}{c} \bullet\bullet\bullet \\ \hline \bullet\bullet \\ \hline \end{array}$ represents 277.

(b) You might be tempted to say that the number $\underline{\bullet\;\bullet}^{\bullet}$ represents 8 because it consists of three 1s and one 5. If so, compare this numeral with the one for 8 that is given in Table 2. It is $\underline{\bullet\;\bullet\;\bullet}$, and it is very important to note that the three dots are arranged horizontally. Thus the arrangement •ᐧ• is different from • • • . Now let us see what number •ᐧ• represents:

$$
\begin{array}{ll}
\bullet & \}\ 1 \times 20 = 20 \\
\bullet\ \bullet & \}\ 7 \times\ 1 = \underline{+7} \\
& \qquad\qquad\ 27
\end{array}
$$

Thus our Hindu-Arabic numeral for •ᐧ• is 27.

The three numbers considered so far are sufficient for showing how to translate Mayan numbers which consist of two groups of dots and dashes, but what if there are more groups? Consider, for example,

. Again the first step is to separate the number into groups, and

the second is to read off the numeral in each group from Table 2:

$$
\begin{array}{l}
\bullet\ \bullet\ \}\ 7 \\
\underline{\bullet}\ \}\ 11 \\
\bullet\ \bullet\ \bullet\ \}\ 8
\end{array}
$$

The third step is to assign each one of these groups a value. We have already seen that the bottom group denotes 1s and the middle group 20s. The top group denotes 360s. Thus we obtain

$$
\begin{array}{ll}
\bullet\ \bullet & \}\ 7 \times 360 = 2520 \\
\underline{\bullet} & \}11 \times\ 20 =\ 220 \\
\bullet\ \bullet\ \bullet & \}\ 8 \times\ 1 = \underline{+8} \\
& \qquad\qquad\quad 2748
\end{array}
$$

The last step is to add the numbers that each group denotes. The result is that the Mayan numeral represents 2748.

This example shows that the values of the bottom three groups are 1, 20, and 360. You obtain the value of each additional group by multiplying the value of the preceding group by 20. This leads to the numbers listed in Table 3, which should be read from bottom to top. The Mayan system was used primarily for counting days within a calendar year of 360 days; there were also 5 holy days. This explains how the number 360 was derived; otherwise, following the rule of multiplying by 20, it would have been $400 = 20 \times 20$.

The next example presents some of the complexities involved in translating numbers from the Mayan system. Keep in mind that groups are formed only by a string of dots over one or more dashes. A group cannot consist of dots over dots or of dashes over dots. Moreover, a string

TABLE   3

Values of Each Group

$7200 \times 20 =$	144,000
$360 \times 20 =$	7,200
$18 \times 20 =$	360
$1 \times 20 =$	20
$1 \times 1 =$	1

of dots cannot have more than four, and a pile of dashes cannot have more than three.

Example 4
Consider the Mayan numerals

(a) ☰ ••• (b) • • (c) •• ☰ —

Problem What number does each one represent?
Solution

(a)
$11 \times 7200 = 79,200$
$8 \times 360 = 2,880$
$2 \times 20 = 40$
$1 \times 1 = +1$
$\overline{82,121}$

(b)
$6 \times 360 = 2160$
$6 \times 20 = 120$
$1 \times 1 = +1$
$\overline{2281}$

(c) In general, four bars are never stacked together. The spacing in this number indicates that another group is intended. This number becomes

$1 \times 360 = 360$
$17 \times 20 = 340$
$5 \times 1 = +5$
$\overline{705}$

The examples so far have all concentrated on translating Mayan numbers into our system. Now we will proceed in the opposite direction and convert our numerals into the Mayan system. The procedure for this conversion is to divide the given number by the basic Mayan numbers listed in Table 3 that are less than the given number. The next example will demonstrate the procedure.

Example 5
Problem Express each of the following numbers in the Mayan system.
   (a) 27   (b) 101   (c) 3723
Solution (a) The numbers in Table 3 that are less than 27 are 1 and 20. Divide 20 into 27. The quotient is 1 and the remainder 7, so $27 = 1 \cdot 20 + 7$. Thus

$$27 = \begin{cases} 20 = 1 \times 20 \longrightarrow \bullet \\ +7 = 7 \times 1 \longrightarrow \bullet\bullet \end{cases}$$

Another way of obtaining this result is as follows:

$$20\overline{)27} \begin{array}{l} \frac{1}{} \rightarrow \bullet \\ \frac{20}{7} \rightarrow \underline{\bullet\bullet} \end{array} \Bigg\} \quad \underline{\overset{\bullet}{\bullet\bullet}}$$

(*b*) First divide 20 into 101 and get $101 = 5 \cdot 20 + 1$. Hence the bottom group contains $\bullet$ and the top contains _____ . The Mayan number for 101 is $\overline{\bullet}$. You might observe that this number is quite different from $\underline{\bullet}$, which is 6. The alternate method of obtaining the result is

$$20\overline{)101} \begin{array}{l} \frac{5}{} \rightarrow \underline{} \\ \frac{100}{1} \rightarrow \bullet \end{array} \Bigg\} \quad \underline{\overset{}{\bullet}}$$

(*c*) The largest number in Table 3 that is less than 3723 is 360. The procedure is to divide 360 into 3723, obtain the remainder, then divide the remainder by 20. We get

$$3723 = 10 \cdot 360 + 6 \cdot 20 + 3$$

so the groups are $\underline{\underline{}}$, $\underline{\bullet}$, and $\bullet\bullet\bullet$. We group them from top to bottom to obtain the Mayan number $\underline{\overset{\underline{\underline{}}}{\underset{\bullet\bullet\bullet}{\bullet}}}$. Alternately,

$$360\overline{)3723} \begin{array}{l} \frac{10}{} \rightarrow \underline{\underline{}} \\ \frac{360}{123} \\ \frac{0}{123} \end{array}$$

$$20\overline{)123} \begin{array}{l} \frac{6}{} \rightarrow \bullet \\ \frac{120}{3} \rightarrow \bullet\bullet\bullet \end{array} \Bigg\}\quad \underline{\overset{\underline{\underline{}}}{\underset{\bullet\bullet\bullet}{\bullet}}}$$

Examples 3 to 5 have illustrated how to convert from and into the Mayan number system. They have shown that several complexities can arise, like the distinction between $\overset{\bullet}{\bullet\bullet}$ and $\bullet\bullet\bullet$, and the meaning of $\underline{\underline{}}$. Can you see how this system cries out for a symbol for zero, a concept that the Egyptian system had no need for? The Mayans did indeed develop such a symbol; they were the first civilization to do so. We will discuss this matter fully in Section 1-2. For now it is better for you to gain some familiarity with these systems by turning to the exercises.

Summary

Humans have walked the face of the earth for a long time, but it was not until the recent past that they had any need for a system of numbers. Some scattered artifacts provide clues to the use of early methods of tallying. All these are based on primitive tally methods.

We studied two numeration systems in detail, the Egyptian and the Mayan. We saw that the Egyptian system requires many different symbols, and that the same symbol can be repeated in a group up to nine separate times. However, the system is very easy to understand and implement. The Mayan system, on the other hand, uses only two different symbols, a dot • and a dash ——. The dots are strung out horizontally, with up to four in a string; the dashes are stacked vertically, with up to three in a stack. Numerals are composed of vertical groups of dots and dashes. This system is more compact and efficient than the Egyptian. However, it is more cumbersome to work with at first.

EXERCISES

In Problems 1 to 12 translate the numbers into our Hindu-Arabic system.

1 

2 

3 

4 

5 

6 

7 

8 

9 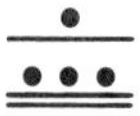

10 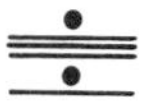

11 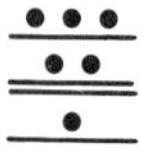

12 

In Problems 13 to 18 translate the numbers into the Egyptian system.

13 36

14 571

15 759

16 1984

17 405

18 450

In Problems 19 to 26 translate the numbers into the Mayan system.

19	17	21	72	23	435	25	1984
20	19	22	93	24	502	26	1492

27 (*a*) Multiply ∩∩ ||| by ∩ and write the product in the Egyptian system.
   (*b*) Can you invent a rule for multiplying by ∩ ?
   (*c*) Apply your method to 999∩∩| .

28 What type of multiplicative rule can be used for describing the Mayan system? (Model your answer after Problem 27.)

29 What is the rule for multiplying by 10 in the Hindu-Arabic system?

30 Consider the three systems: Egyptian, Mayan, Hindu-Arabic.
   (*a*) In which one is it easiest to multiply by 10?
   (*b*) In which one is it hardest to multiply by 10?

31 (*a*) Write the Mayan numeral for 25.
   (*b*) How does this numeral compare to the numeral for 6?
   (*c*) How do you think the Mayans distinguished between the two?

32 Are •̄ and __•__ the same number?

33 (*a*) Do you think the Mayan system includes the numeral ⋯≣ ?
   Why?
   (*b*) Convert 366 to the Mayan system.

34 The Mayan numeral for 20 is not ≣ .

   (*a*) What is the numeral for 20?
   (*b*) What is the numeral for 360?

35 Roman numerals are used on clocks, chapters of books, and dates of movies (and perhaps in some other obsolete instances as well). This section was written in MCMLXXXIII. What year is this Roman numeral?

36 What are *quipu* and who used them? (Use any source you want. Most dictionaries are sufficient.)

37 Which of the following time divisions were prehistoric humans likely to notice: a year, a month, a week, a day, an hour?

38 Which do you think came first, number names or number symbols? Why?

39 Ishango is a (choose one):
   (*a*) Suburb of Dallas
   (*b*) Modern dance
   (*c*) Bone with tally marks

40 A dash is (choose one):
   (*a*) The amount of bitters placed in a Manhattan
   (*b*) A 100-meter run
   (*c*) A stroke of a pen
   (*d*) The symbol for 5 in the Mayan system

1-2   THE HINDU-ARABIC SYSTEM

Some people feel that the Garden of Eden was a lush, verdant strip of land located between the Tigris and Euphrates rivers in what is now lower Iraq. They may be right. But the famed sailor Thor Heyerdahl (Tor HÍ-er-dahl) had plenty of reason to think otherwise. Within days after he and his international crew set sail from that area on November 23, 1977, aboard a small craft that was modeled after those built some 5000 years earlier, they encountered violent storms that threatened to thrash the boat and risk their lives. The Garden of Eden seemed more like a curse than a promise to them.

What possible reason could induce an international team of explorers to risk their lives aboard a 5000-year-old craft? Why would Heyerdahl risk fame and fortune to set sail from such a place?

If they had been mountain climbers they would answer such questions with the stock phrase "Because it is there." But Heyerdahl had a much more serious purpose in mind.

Consider first the craft, which was named the *Tigris*. Its first true test came when that violent gale mentioned above thrashed it with all the fury it could muster. A modern-day hull made of wooden planks would surely have ripped apart under such constant pounding. But the *Tigris'* hulls were constructed of large bundles of reed, called *bardi*, which acted like giant sieves, straining the waves and allowing them to pass through harmlessly. The Sumerian sailors who had designed and built them long ago had certainly mastered their trade.

The *Tigris* earned a good grade on its first test. But could it survive the rigors of a voyage that was to take it down the Persian Gulf, east to the Indus Valley, west across the Arabian Sea, and finally up the Red Sea to Egypt? Heyerdahl was confident that it could. He had studied the Sumerian scrolls that explained how such crafts were constructed, and he had spent 2 years in planning and building the *Tigris*.

The question remains: Why would Heyerdahl want to risk life and limb on such a project? The answer lies in his insatiable quest to know how ancient civilizations might have communicated. There were three such great civilizations which had attained a high degree of culture by about 3000 B.C.: the Sumerians in Mesopotamia, the Hindus in the Indus Valley of India, and the Egyptians in north Africa. Recent evidence had suggested that the three of them communicated and traded together. Skeptics countered, however, that it was virtually impossible for these civilizations to interact closely because of the vast distance separating them. Heyerdahl, not content with mere speculation, set out to prove that their proposed contact was along trade routes by sea instead of land.

The entire project ended in bittersweet success. After 5 months and 4200 miles of catastrophe and narrow escape, after battling many of the perils that seafaring Sumerians must have faced, and after dodging sea goliaths plying Middle East oil through the most congested route in open seas, the *Tigris* reached Djibouti in Somalia at the head of the Gulf of Aden. From there it was just a short jaunt up the Red Sea to their final destination. Total success was in sight. But then twentieth-century politics reared its ugly head. A local war in Eritrea was raging. The *Tigris* was forbidden to continue its journey.

Heyerdahl and his crew were angry and frustrated. They had proved their point—the three great ancient civilizations could have carried on constant trade. There was but one last step to take. In an "appeal to men of reason to resume the cause of peace in a corner of the world where civilization first dawned," they set the *Tigris* ablaze at sea and sullenly watched as it sank into the waters below.

This episode in recent history is interesting, but what relation does it bear to mathematics? A little phrase in our account of the Heyerdahl saga provides a clue: "recent evidence had suggested that the three of them communicated and traded together." The evidence that we had in mind was the similarity of their number systems. Knowledge of the Egyptian system had existed since the Rosetta stone was deciphered in the early 1800s. We have already studied this system. But the number system from Mesopotamia, referred to as Babylonian, only came to light with the publication of Otto Neugebauer's discoveries in the 1940s. We did not present this system because its complexity would cause us to digress too far from our main theme. This leaves the number system of the Hindus, and we shall turn to its history now to explain how our own system is a direct descendant of it.

Our number system is called the Hindu-Arabic system because it was born in India and reared by the Arabs. The details of how this number system spread from the Hindus to the Arabs and thence to Egypt are fragmented. This is where the Heyerdahl expedition is so helpful. It was known that the system originated in India, but all theories about how the Arabs learned of the system were mere conjecture. Heyerdahl showed that it was possible, indeed probable, that the great civilizations of Sumer, the Indus Valley, and Egypt traded regularly and exchanged ideas. It was possible for the Sumerians to sail vast distances in their bardi craft.

The initial development of the system in India followed the same pattern as the Egyptian and Mayan systems. Archeological digs have unearthed inscriptions containing vertical strokes which were arranged in groups. These inscriptions date to about 3000 B.C., the same era as that of the Egyptian pyramid builders. Some time after the third century B.C. this was replaced by a system in which Brahmi letters (see Figure 1) were used to denote numbers. It is not known when the Hindus recognized that their positional system required only ten letters, but it seems likely that the transition was made only gradually. Also, the source of the inspiration for the change is uncertain. Most scholars feel that it was the result of internal developments within India itself, but others argue that the newer system was borrowed from the rod numerals used in China. In any event, a mathematical work of about A.D. 500 contains the phrase "from place to place each is ten times the preceding," indicating that a base 10 positional system was in use at the time. In this section we will explain what is meant by a "base 10 positional system," although you can probably make a good guess right now. All that was needed to complete the system was the introduction of zero, and this occurred by the end of the seventh century.

How did the system spread from India to the Arab world? Heyerdahl's voyage suggests that it was a gradual change brought about by the recognition by Arab merchants of the superior number system of the Hindus. The most widely accepted hypothesis before Heyerdahl's voyage was that the

Arabs adopted the Hindu system as one of their spoils after they had conquered large sections of western Asia after the birth of Islam in 622.

In any case, a piece of concrete evidence of the Hindu-Arabic connection is a book written ca. A.D. 825 by the Arabic mathematician al-Khowarizmi. It begins with a description of the system and demonstrates how to compute with it. His examples were taken from "cases of inheritance, legacies, partition, lawsuits, and trade."

Next the Arabs turned their interest to the west, and as their conquests moved from Persia to Egypt, along north Africa, and into Spain and Italy, their newly acquired number system accompanied them. It was in this way that the system we now use spread into the rest of Europe. This spread was slow and gradual, and the numerals changed as they traveled from one culture to another. When al-Khowarizmi's book was translated into Latin in the twelfth century, the numerals were incorrectly assumed to be of Arabic origin, though the Arabs themselves never laid claim to the invention and always recognized their indebtedness to the Hindus. This is the reason why our system is now called the Hindu-Arabic number system; it reflects our debt to both these civilizations.

In 1202 the mathematician known as Fibonacci wrote a book whose aim was to explain this new number system "in order that the Latin race might no longer be deficient in that knowledge." It begins:

> These are the nine figures of the Indians: 9 8 7 6 5 4 3 2 1. With these nine figures and with this sign 0 . . . any number may be written, as will be demonstrated below.

Fibonacci included a wealth of practical examples and showed in every case that the new system was superior to the Roman numerals. This undoubtedly explains why his book became so influential and why Europe soon adopted this system.

In spite of Fibonacci's advocacy of the Hindu-Arabic number system, it encountered strong cultural resistance throughout Europe. An interesting example of this is a statute of 1299 which forbade the bankers of Florence from using Arabic numerals instead of Roman numerals. Eventually, however, the commercial interests of the merchants of Florence, Pisa, and Venice led them to adopt the system because it enhanced trade with the Arabic-speaking people, who by that time encircled the entire Mediterranean Sea.

Figure 1 displays the genealogy of our numerals. The main part of the figure, consisting of the period from the Brahmi numerals in the second row to the sixteenth-century numerals in the next-to-last row, is the commonly accepted version. The symbols in the next-to-last row became standardized after Gutenberg's invention, in the fifteenth century, of a printing press with movable type. As irony would have it, the numerals used in the modern Arabic world are different from those used throughout the western world.

The top row in Figure 1 reflects the theory that the Hindu numerals had their origins in China, beginning about 1300 B.C. This theory is supported by the recent work of Frank Swetz, an American authority on Sino mathematics. If it becomes accepted, this would mean that the numeration system com-

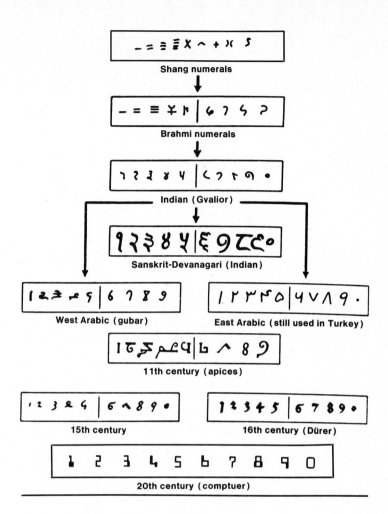

FIGURE 1

monly used throughout the modern world had its origins in China 34 centuries ago!

The bottom row in Figure 1 obviously reflects the most recent change in the shape of our numerals, and is due to the influence of computers. Undoubtedly the computer will have as much influence on our lives as did Gutenberg's printing press some 500 years earlier.

We have mentioned only three ancient number systems, Hindu-Arabic, Egyptian, and Mayan, but there were many others that were developed, each with its own characteristics. What were the fundamental properties of the Hindu-Arabic system that rendered it important enough to be universally adopted throughout the world today? To explain these properties let us briefly return to the Egyptian and Mayan systems to point out their primary advantages and disadvantages.

Fundamental Questions

There are three fundamental questions that are pertinent when you are describing a number system. They are:

1   How many distinct symbols are used?

2   Is there a prescribed order that is used to combine the symbols?

3   Is there a need for a symbol representing zero?

We are about to answer these questions for the Egyptian and the Mayan numeration systems, but perhaps you would like to answer them yourself before reading our discussion. Can you answer them for our own Hindu-Arabic system?

WARNING!

> *Mathematics is not a spectator sport.* If you have not attacked the exercises in the previous section yet, you should do so now before proceeding further, because we are going to make use of some of them below. Moreover, your ability to follow the main lines of our development will be enhanced greatly if you have already devoted independent thought to these exercises.

Base

Stanley Kubrick's film "2001: A Space Odyssey" has become a classic. The number in the title is usually pronounced "two thousand one." It raises two issues. First, there are no definite guidelines for pronouncing future dates. The year 1900 is pronounced "nineteen hundred." Will the year 2000 be pronounced "twenty hundred"? Would you prefer to say "two thousand" instead? Perhaps your preference is to pronounce all dates after that in terms of "twenty," so that 2001 will become "twenty-oh-one." This is an intriguing matter, and it will not be too long before we see whose preferences turn out to be adopted.

The second issue to emerge from the number 2001 is that its pronunciation should not include the word "and." We are not just nitpicking here either. When you write a check for the amount $2001, you must spell out "two thousand one dollars." The expression "two thousand and one dollars" is incorrect because the word "and" is used only to signify the fractional part of a unit. With this in mind, let us consider what is meant by the base of a system.

> **Example 1**
> Suppose a check is made out for the sum of four thousand two hundred three dollars.
> **Problem** Write the Hindu-Arabic and Egyptian numerals for this amount.

Solution The Hindu-Arabic numeral is 4203. It consists of four 1000s, two 100s, no 10s, and three 1s. We will write this in the form

$$(4 \times 1000) + (2 \times 100) + (0 \times 10) + (3)$$

Now it is easy to go directly from this form to the Egyptian system. Recall that the numerals for 1000 and 100 are ⌐ and 𝟅, respectively. Then the Egyptian numeral for 4203 becomes ⌐ ⌐ ⌐ ⌐ 𝟅𝟅 ||| .

Notice that the number 4203 is composed of 1s, 10s, 100s, and 1000s. The Hindu-Arabic representation requires a symbol for zero but the Egyptian numeral does not. In general, any number can be written in both the Egyptian and Hindu-Arabic number systems in terms of so many units, so many 10s, so many 100s, so many 1000s, etc. For this reason we say that these systems have *base 10*. This explains why it is easy for us to translate between them.

Recall the first question raised earlier:

How many distinct symbols are used?

The Egyptian system would require an endless number of symbols, but the Hindu-Arabic requires only 10. Do you see why the Egyptians would eventually be compelled to adopt a different system? Notice, however, that both systems have base 10. Thus the question above is not the same thing as asking what the base of a given number system is.

You can easily guess why the Egyptians chose the number 10 for the base of their number system. It is very natural because human beings have 10 fingers. When you are counting and run out of fingers, you must make a mark to recall that you reached 10 once and are starting over again tallying on your fingers.

Is there another natural base? How about 5? As we mentioned in the previous section, even today when we are counting things like ballots we make a new mark or symbol to represent groups of 5. For instance, which group of marks is easiest to read, the first with no groupings or the second?

||||||||||||||||||

卌 卌 卌 |||

Obviously the second represents 18 votes. The first group also represents 18, but it is not nearly so clear.

Now take this tally procedure two steps further. Suppose it seemed more natural to use a dot instead of a vertical line, and more economical to place the dots above the horizontal slash and to align the slashes vertically. Then how would the tally of 18 votes appear? It would be

• • •
══

Does this look familiar? Of course; it is the Mayan system.

Initially the Mayan system appears to be base 5. Look back at the numbers in Table 2 in Section 1-1. You will see that each number is composed of so many units and so many 5s. But numbers greater than 19 do not follow this pattern. For instance, 23 is *not* composed of four 5s and three 1s. Instead it is written as one 20 and three 1s, that is, $\overset{\bullet}{\underset{\bullet\bullet\bullet}{}}$.

Why did 20 enter the picture? We again suggest anatomical reasons. The Egyptians used base 10 because human beings have 10 fingers, so it was just as natural for the Mayans to adopt 5, for the fingers on one hand. Extend this reasoning to include your feet and you number 20 digits. This seems very natural, but it is only a conjecture, of course.

We call the Mayan system a *mixed-base* system because it consists of more than one base. Historically this system has been called *vigesimal*, meaning base 20, but this designation is misleading. Instead it should be referred to as quinary-vigesimal, or 5–20. In fact, it is worse than that. The Mayan numeration system is a mixed 5–20 base with a 360 exception.

In spite of its complexity the Mayan system requires only two distinct symbols. Clearly what was needed was to blend the single-base property of the Egyptians with the economy of symbols of the Mayans. Do you see how our own system could have evolved from such a mixture?

Most of us have had little or no experience in dealing with mixed-base systems. As a result, we find them very cumbersome. In addition, they contain several inherent shortcomings that would have to be overcome if they were going to be used widely. For instance, Problem 31 in Section 1-1 shows the difficulty in determining whether ____•____ represents 6 or 25. There are several other difficulties too that show why number systems eventually had to evolve from mixed-base to single-base systems.

You might think that the Mayan system represents a descent of numbers rather than an ascent, because it is much more complicated and very different from our own. However, we will turn now to a discussion of some other properties of the Mayan system that show that it represents a definite ascent.

Position

The Mayan system is much more complicated than the Egyptian system. It is clear that the latter developed quite naturally from a simple tallying method. As larger numbers were needed, a new symbol was created. The Mayans, however, must have felt compelled to design a system in which economy of space was a key ingredient. Perhaps writing utensils were more scarce in North America than in the fertile Nile Valley. Let us compare the space used by each number system with a concrete example.

Example 2
Problem Express 999 in the Egyptian and Mayan systems.

Solution   The Egyptian form is

$$99999 \quad \cap\cap\cap\cap\cap \quad |||||$$
$$9999 \quad \cap\cap\cap\cap \quad ||||$$

To express 999 in the Mayan system first determine how many 360s and 20s are in the number. We get $999 = 2 \cdot 360 + 279 = 2 \cdot 360 + 13 \cdot 20 + 19$. Hence the Mayan form is

The Egyptian numeral requires far more space than the Mayan.

The Mayan form is much more compact. As larger numbers are considered, the difference in the economy of space between the systems is even more dramatic.

Do you see what property the Mayans invented for their system in order to achieve this compact representation of numbers? It is the use of the position of the symbols. If two dots are written next to each other, the number represented, • •, is 2 since they are both in the units, or bottom, position. If one dot is written on top of the other, as in •̣, the bottom dot is in the units position and refers to 1 unit, whereas the upper dot is in the 20s position and it represents one 20. Hence •̣ is the number 21.

We say that the Mayan system is *positional* because the same symbol can represent different values depending upon where it is placed in the number. The Egyptian system is not positional since a symbol can refer to only one number no matter where it appears. Of course, for convenience and aesthetic appearance the separate symbols were invariably grouped together. Even so, the number $\cap | \cap$ would always represent 21, as would $| \cap \cap$ and $\cap \cap |$. Is our number system positional? Compare the two numerals 21 and 12.

Thus the first fundamental property of a number system is base. The second is *position* or, equivalently, *place value*.

Can you relate question number 2 to the matter of whether a system is positional? The positional property reflects one way in which the Mayan system represents an ascent beyond the Egyptian system. Another example of the evolution comes from the concept of zero.

Zero

It is difficult for us today to conceive of a number system which did not use a symbol for zero. Yet the role that zero has played in history illustrates vividly the theme that number systems evolve.

The Maya were the first people to use a symbol for zero. It looked like a small shell, ⬭ , and was used to denote the absence of 20s, or 360s.

Example 3

Problem   What number does ‾⬭̲ represent?

Solution   There are three groups of symbols. The bottom one is ‾•••‾ , the middle one is ⬭ , and the top one is ——— . Hence the number is:

———} 5 × 360 = 1800

⬭ } 0 × 20 =   0

•••}13 × 1 = +13

                  ————

                  1813

The next example will demonstrate how to convert from numbers that require the use of ⬭ in the Mayan system. One part will show how this symbol removes possible ambiguity; the other will show that some ambiguity still remains, however.

Example 4

Problem   Write each of these numbers in the Mayan numeration system.
   (*a*) 1085   (*b*) 720

Solution   (*a*) We first must find how many 360s, 20s, and 1s are in each number. We have $1085 = 3 \cdot 360 + 0 \cdot 20 + 5$. Since there are no 20s we must use the symbol for zero in the twenties position. Therefore the Mayan numeral for 1085 is $\overset{\bullet\bullet\bullet}{\underset{\rule{1em}{0.4pt}}{⬭}}$. Notice how the ⬭ removes the ambiguity between this number and ‾•••‾ .

   (*b*) There are two 360s in 720. That is, $720 = 2 \cdot 360$. Thus • • is the Mayan numeral for 720.

Examine part (*b*) in Example 4. It dramatizes a significant inadequacy of the Mayan system. Do you recognize it? The Mayan representation of 720 is • •. How does this differ from the Mayan numeral for 2? Note that no zero symbols are appended at the bottom to give this number the clarity it requires; that is, the Mayans did not write

In fact, • • represents many other numbers, for example, 40. It seems incredible to us today that such an advanced civilization, one which made use of numerous astronomical and calendrical computations, could miss this seemingly obvious step. Ah, the power of hindsight!

This is a very serious flaw in the Mayan system. How was the reader supposed to know which meaning the writer had in mind? First, the context made it clear. Often there would be a series of designs placed next to the number to indicate the separate groupings, one design per group. But second, and more important, these ancient civilizations were closed societies and only the scribes and priests were literate. They knew each other and communicated among themselves so there was no need for a more elaborate system. This made it possible for them to know what number was meant, though it complicates the task for us. You can see, then, that as a society's needs progressed its numeration system would have to change also.

Let us point out that ⬭ was not used to denote an absence of • or _____ . It was used to denote the omission of an entire group. Thus ⬭̇ was not used to denote

$$
\begin{array}{r}
\bullet \quad \} 1 \times 20 = 20 \\
⬭ \quad \} 5 \times 1 = +5 \\
\hline
25
\end{array}
$$

Instead __•__ was written. But how was the reader to know that 6 was not the intended meaning? Apparently, again, both the context and a knowledge of the writer made it possible.

The concept of zero had a curious and complicated evolution. It proved to be a very elusive and difficult abstraction for ancient civilizations. Even today many people still find the three expressions, 0/1, 1/0, and 0/0 mysterious if not downright diabolical. In the exercises we will ask you to grapple with them and then we will give you a thorough explanation of the meaning of each.

Figure 2 displays some of the ancient symbols for zero. The Hindus initially used a dot for this placeholder—their word for it, *sunya*, meaning empty, blank, or void. *Sunya* was translated in Arabic as *sifr* and then into Latin as *zephirum*, from which our "zero" is derived. By about A.D. 500 some Greek writers used the letter omicron to denote zero. (It is the first letter in *ouden*, which means "nothing.") The Greeks, however, never recognized zero as a number. They used it only as a place-

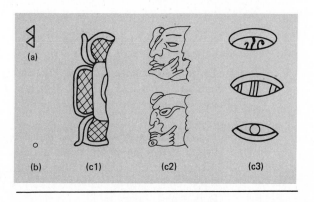

FIGURE 2   Ancient symbols for zero. (*a*) Babylonian cuneiform; (*b*) India; (*c*) Maya: (*c*1) monumental inscription style, (*c*2) face inscriptions, (*c*3) written in books.

holder. The Hindus were the first to accept zero as an actual number, though they initially committed the frequently made error of dividing by zero. (Of course, *you* know enough never to do that! Right?) Mahavira wrote about A.D. 850: "A number multiplied by zero is zero, and that number remains unchanged which is divided by, added to, or diminished by zero." Some 300 years later the famous Hindu mathematician Bhaskhara corrected this by saying that "fractions [whose] denominator is zero are termed infinite quantities." Today we say that such fractions are meaningless.

Hindu-Arabic System: Its Properties

Let us list the properties that our Hindu-Arabic number system possesses.

1 It is a base 10 system because the 10 symbols, 0 to 9, are used to represent all the numbers, and any number can be written as a combination of powers of 10. We have used this fact throughout these first two sections. Thus you have used the concept of our number system being base 10 without necessarily understanding the significance of the term.

2 The position that the symbols are placed in the expression of the number is significant; so the Hindu Arabic number system is positional. For example, the two combinations of the symbols 1 and 9 yield two different numbers, 91 and 19.

3 The zero in the Hindu-Arabic system has full placeholder ability. This allows us to write numbers like 50 with no ambiguity. Also zero is recognized as a number.

The Hindu-Arabic numeration system presents a simple and concise way of recording numbers, while the notation simultaneously assists in performing calculations with these numbers. Of course, it has the further advantage of being uniform throughout the world. A traveler to Europe might hear the words *neun, neuf,* or *neuve,* but he or she will see only the numeral 9.

These are the major properties of our number system that we will build upon in the ensuing sections. We have tried to indicate how number systems evolved. You might want to make some conjectures now as to why we use the system that we do. For example, why is it base 10 instead of some other number? Is it more economical as far as the space used to write numbers than the other systems discussed? Is this "economy of space" a useful property and, if so, why? Why not use a base 100 system that would allow much more economy of space? We have only touched the surface with these types of questions. We hope we have at least touched the surface of your intellectual curiosity for the subject.

EXERCISES

In Problems 1 to 4 translate the numbers expressed in Mayan into our Hindu-Arabic system.

1

3

2

4

In Problems 5 to 10 express the given number in the Mayan system.

5   366

6   722

7   3601

8   1805

9   1920

10   120

11   What numbers do these Mayan numerals represent?

(*a*) ═══   (*b*) ──── (*c*)

12   (*a*)   Write the Mayan numeral for 365.
     (*b*)   How does this numeral compare with the numeral for 6?

In Problems 13 to 16 write the words for checks that are drawn in the amounts.

13   $203

14   $1909

15   $2001

16   $4005

In Problems 17 to 20 write the Egyptian and Hindu-Arabic numerals for the numbers.

17   Ten thousand eighteen

18   Fourteen thousand six

19   One million two thousand three hundred four

20   One hundred thousand eighty eight

Water-meter readings present us with an everyday application of the base 10 system. In Problems 21 to 24 determine the number of cubic feet in each of the types of representative readings.

21

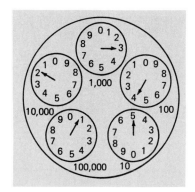

23

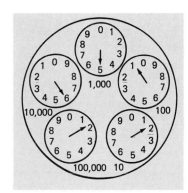

22

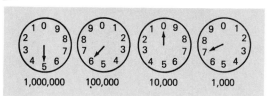

24

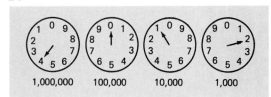

In Problems 25 and 26 write the numbers in our Hindu-Arabic system. How does each one illustrate the fact that this system is positional?

25 One hundred one

26 Two hundred twenty two

27 (*a*) Write 102 in the Egyptian system.
(*b*) Did the Egyptian system have a need for a symbol like ⬯ ?

28 Is there a need for a zero symbol in a nonpositional number system?

29 Evaluate:
(*a*) $5 + 0$
(*b*) $5 - 0$
(*c*) 5×0
(*d*) $0 \div 5$
(*e*) (Be careful!) $5 \div 0$

30 Evaluate
(*a*) $0 + 4$
(*b*) $0 - 4$
(*c*) 4×0
(*d*) $0/4$
(*e*) $4/0$

31 What number base is indicated in the following expressions?
(a) "Four score and seven years ago" (A. Lincoln)
(b) "The number of the beast ... is six hundred three score and six" (Revelations 12:18)

32 What number base is indicated within the Roman numeral system by the numerals V, L, and D?

33 What number base is conjectured to be exhibited on the middle column of the Ishango tally bone in Figure 1 of Section 1-1?

34 What number base is indicated by the French words *soixante-dix, quatre-vingt,* and *quatre-vingt dix*?

35 What fundamental property of number systems explains why when adding numbers it is important to "keep the columns straight"?

36 Find the roots of the words "eleven" and "twelve." What number base is indicated? (A dictionary will be helpful.)

37 Pronounce aloud the words for 19 and 91.
(a) Is pronunciation positional? That is, are number words always read from left to right or right to left?
(b) The number 19 is read from right to left. What other numbers under 100 have this property?
(c) 91 is read from left to right. What other numbers under 100 have this property?
(d) What base is suggested by parts (b) and (c)?

38 The word "million" is derived from the French *mille* and the Italian *on*. What does it mean literally?

39 Base is (choose one):
(a) Part of the infield
(b) Where people in the Navy live
(c) A property of a number system

40 A system is positional if it:
(a) Reads from top to bottom
(b) Is supine
(c) Depends on the placement of the digits

1-3   THE BINARY SYSTEM

It seems that we are now in the midst of a second industrial revolution, and that integrated circuits will be the key to the revolution. Made of such materials as silicon and germanium, integrated circuits are so called because they can amplify, switch, store, and control electric signals. Their size is incredibly small—about four can fit on a hard contact lens—so they are usually simply called *chips*. As a result we will call it the "chip revolution." Such chips have made possible innovations like hand-held calculators (even programmable ones) and digital watches, but these advances will pale by comparison with the changes that will surely take place by the turn of the century. Let us consider four areas in which all of our lives will be affected.

1   *Home computers.* Connected to a central storage bank via telephone, the user will be able to obtain information about the stock market, legal advice, recipes, and even TV schedules. In fact, the use of books as we know it today may become a thing of the past, for the home viewer will be able to read a book on a TV screen. (One chip alone can store the contents of a 50,000-word book. What do you think future libraries will look like?)

2   *Television.* We mentioned a TV screen above but you should not visualize the usual screen. In fact, chips will undoubtedly be developed to replace the one remaining tube, the picture tube, so that the resulting screen will be very thin and entirely flat.

3   *Mail.* Forget writing letters. Just send out a radio wave on the computer attached to the TV set. This service will function "through rain, snow, and gloom of night," though it might experience difficulties in electrical storms.

4   *Telephone.* Both Motorola, Inc., and Bell Telephone Laboratories, Inc., are feverishly working to complete the technology needed for portable pocket telephones. So the next time you see a chauffeured limousine with someone talking on the phone, just smile inwardly at this primitive means of communication.

How is all this circuitry possible? The answer lies in the newest (and yet the oldest) numeration system devised by humans: the binary system. On an integrated circuit the number 1 is represented by the presence of an electric impulse, and 0 is represented by the absence of such a pulse. Thus every message to the computer, that is, every number, can be represented in a form just like the decimal number system, except that the only allowable digits are 0 and 1. There are two conversions which are possible, from decimal to binary, and conversely. In order to perform conversions from one system to another, as well as to understand in more detail the role of the base of a system, it is necessary to introduce the concept of exponents.

Exponents

In the preceding section the thirteenth-century mathematician Fibonacci was quoted as saying that every number could be represented in terms of the 10 basic digits. Such a representation depends upon powers of 10. These numbers can be expressed in two ways: as a 1 followed by 0s, or in exponential form. There are occasions when each is appropriate.

The headline for an editorial described the 1980s as "The Age of 1,000,000,000,000" because the national debt was expected to reach $1 trillion during the decade. Writing "The Age of a Trillion" would not have achieved nearly the same impact.

Take another example. Which of these statements is more dramatic?

J. Paul Getty is a billionaire.

or

J. Paul Getty is worth $1,000,000,000.

There is another side to this issue, as is illustrated by an embarrassing incident that happened in April 1982 to the president of the New York Public Library. After mistaking the size of a gift to the library, he exclaimed, "I couldn't believe it. I thought at first it was $10,000, because you don't pay attention to zeros." He missed two 0s. The check was worth more than a million dollars—that is, more than $1,000,000.

The other way of writing these numbers is in exponential form. In general, if a and b are numbers, then numbers in the form

$$b^a$$

are called *exponentials*. The term b^a is read "b raised to the a power" or simply "b to the ath." The number b is called the *base,* while a is called the *exponent* or the *power*.

Numbers in which the exponent is a natural number have an easy explanation. They can be viewed as a string of multiplications. For instance, 10^6 means 10 multiplied by itself 6 times. We will write this as

$$10^6 = \underbrace{10 \times 10 \times 10 \times 10 \times 10 \times 10}_{6}$$

After performing these multiplications you obtain:

$$10^6 = 1,000,000$$

This shows that another way of writing the number 1 million is 10^6. This way does not achieve the dramatic effect of 1,000,000 mentioned above, but it is far more economical. In fact, exponential notation was developed

TABLE 1

Exponential Form of Some Numbers

Number	Numeral	Exponential Form
One	1	10^0
Ten	10	10^1
Hundred	100	10^2
Thousand	1,000	10^3
Million	1,000,000	10^6
Billion	1,000,000,000	10^9
Trillion	1,000,000,000,000	10^{12}

in the late sixteenth century because it is both precise and economical at the same time.

Table 1 lists the exponential form of some common numbers. Do you see a relationship between the number of zeros in the numeral and the exponent in the exponential form? We will let you answer this question in the exercises.

The top row in Table 1 contains an important fact:

$$10^0 = 1$$

This is a specific case of the convention that

$$b^0 = 1$$

for every number b (except $b = 0$). This notation becomes crucial when we write a number in its expanded version, as illustrated by Example 1 below. At first it might appear that we are making a mountain out of a molehill in this example, but your mastery of the binary system (or any other base, for that matter) will be greatly enhanced if you understand it. The expanded version of a number focuses attention on the vital role played by exponential notation.

Example 1
In order to write the expanded version of the number 21,034, proceed as follows:

$$21,034 = 20,000 + 1000 + 30 + 4$$
$$= (2 \times 10,000) + (1 \times 1000) + (3 \times 10) + 4$$
$$= (2 \times 10^4) + (1 \times 10^3) + (3 \times 10^1) + (4 \times 10^0)$$

Notice that the exponent 2 is missing in the last line. By adding it, that is, by inserting the placeholder 0, we obtain the expanded version

$$(2 \times 10^4) + (1 \times 10^3) + (0 \times 10^2) + (3 \times 10^1) + (4 \times 10^0)$$

Recall, by definition $2^2 = 2 \times 2 = 4$ and $2^3 = 2 \times 2 \times 2 = 8$. As we indicated above, 2^a is usually read "2 to the ath," but two cases are read differently. If $a = 2$ then 2^2 is read "2 squared," while if $a = 3$ then 2^3 is read "2 cubed." However, all other cases use the "-th" form, so that 2^4 is read "2 to the fourth" and 2^5 is "2 to the fifth." The next example shows that the order in which exponentials are formed is very important.

Example 2

Problem   Evaluate (a) 2^5 and (b) 5^2.

Solution   The direction "to evaluate" means "to find the number to which it is equal." In part (a) we have

$$2^5 = \underbrace{2 \times 2 \times 2 \times 2 \times 2}_{5} = 32$$

In part (b) we have

$$5^2 = \underbrace{5 \times 5}_{2} = 25$$

Note that 2^5 is *not* equal to 5^2.

Example 3

Problem   If you know that $2^3 = 8$ and $2^4 = 16$, then what is 2^7?

Solution   The direct way to compute 2^7 is to multiply 2 by itself 7 times. However, by grouping the string of seven 2s in the following way you can make use of the information given.

$$2^7 = \underbrace{(2 \times 2 \times 2 \times 2)}_{4} \times \underbrace{(2 \times 2 \times 2)}_{3} = 2^4 \times 2^3$$

Thus $2^7 = 2^4 \times 2^3 = 16 \times 8 = 128$.

Example 3 demonstrates one of the fundamental rules of exponentiation:

$$b^m \times b^n = b^{m+n}$$

Two additional rules will be discovered in the exercises. The three of them will help to decrease computation time considerably.

Suppose you put $b = 2$, $m = 3$, and $n = 0$ into the equality above. This leads to the equality

$$2^3 \times 2^0 = 2^3$$

Do you see why we define 2^0 to be equal to 1?

Examples 2 and 3 showed how to evaluate exponential forms. The next example will illustrate how to proceed in the opposite direction.

Example 4
Problem Write 1024 in exponential form, using base (a) 4 and (b) 2.
Solution (a) Since $1024 = 4 \times 4 \times 4 \times 4 \times 4$, we can write $1024 = 4^5$. Thus the exponent is 5 when the base is 4.

(b) The direct way to proceed is to write 1024 as a string of 2s. A quicker way, however, is to make use of part (a) and an exponentiation rule you will discover in the exercises.

$$1024 = 4^5 = (2^2)^5 = (2)^{2\times 5} = 2^{10}$$

Thus the exponent is 10 when the base is 2.

Binary System

The term "binary" refers to the number 2. Thus the term "binary system" is synonymous with base 2. There is nothing sacred about 2. We could have considered number systems of base 5 or 12, but we chose base 2 because its role in modern computer technology makes it a relevant base to consider.

The following table of powers of 2 will be handy.

TABLE 2

Some Powers of 2

$2^0 = 1$	$2^6 = 64$
$2^1 = 2$	$2^7 = 128$
$2^2 = 4$	$2^8 = 256$
$2^3 = 8$	$2^9 = 512$
$2^4 = 16$	$2^{10} = 1024$
$2^5 = 32$	$2^{11} = 2048$

We will now illustrate how to convert from our base 10 system (Hindu-Arabic) to a base 2 system (binary). In order to express 42 in the binary system you find those numbers in Table 2 which add up to 42. Notice that you need only consider numbers in the first column since those in the second are greater than 42. In other words, you must first find the largest power of 2 less than 42. It is 32. So $42 = 32 + 10$. Next find the largest power of 2 less than 10. It is 8. Then

$$42 = 32 + 8 + 2$$

so

$$42 = 2^5 + 2^3 + 2^1$$

TABLE 3

Hindu-Arabic Numeral	Binary Numeral
0	0
1	1
2	10
3	11
4	100
5	101
6	110
7	111
8	1000
9	1001
10	1010

Now write the expanded version of 42 in base 2, using all exponents from 5 to 0, including those with a multiple of 0:

$$42 = (1 \times 2^5) + (0 \times 2^4) + (1 \times 2^3)$$
$$+ (0 + 2^2) + (1 \times 2^1) + (0 \times 2^0)$$

The binary number is formed from the multiples of the powers of 2,

101010

This may seem foreign to you but it is exactly the same procedure that is used to form our familiar Hindu-Arabic numbers. Compare it with Example 1.

Notice that it takes many more digits to express 42 in base 2 than in the Hindu-Arabic system. Of course, this is one of the primary advantages of our number system. The advantage of the binary system is that it requires only two symbols, 0 and 1, and that arithmetic operations are very easy to perform. The Mayan system also used only two symbols, but its mixed base made computations difficult to carry out.

In this section it must always be clear in which numeration system a number is expressed. As we just saw, the numeral 101010 in base 2 represents the base 10 numeral 42. For this reason we read the numeral 101010 in base 2 as "one-zero-one-zero-one-zero" and in base 10 as "one hundred one thousand ten." Table 3 lists the binary equivalent for our Hindu-Arabic numbers from 0 to 10. The next two examples show how to translate from one system to the other.

Example 5

Express the following numbers in the binary system: (*a*) 87 and (*b*) 523.

Solution (*a*) The largest power of 2 less than 87 is 64; so we write $87 = 64 + 23$. The largest power of 2 less than 23 is 16; so we write $87 = 64 + 16 + 7$. Since $7 = 4 + 2 + 1$, we write $87 = 64 + 16 + 4 + 2 + 1$. The expanded version of 87 in base 2 is

$$87 = (1 \times 2^6) + (0 \times 2^5) + (1 \times 2^4) + (0 \times 2^3)$$
$$+ (1 \times 2^2) + (1 \times 2^1) + (1 \times 2^0)$$

Hence 87 is equal to 1010111 in binary notation.

(*b*) The largest power of 2 less than 523 is 512; so we write $523 = 512 + 11$. Then $523 = 512 + 8 + 2 + 1$. Hence

$$523 = 1 \times 2^9 + 0 \times 2^8 + 0 \times 2^7 + 0 \times 2^6 + 0 \times 2^5 + 0 \times 2^4$$
$$+ 1 \times 2^3 + 0 \times 2^2 + 1 \times 2^1 + 1 \times 2^0$$

Hence 523 is equal to 1000001011 in binary notation.

Example 6

Find the equivalent Hindu-Arabic numerals for the binary numerals (*a*) 11001 and (*b*) 11001001.

Solution It is easiest to work from right to left.

(*a*) Write

$$
\begin{array}{ccccc}
1 & 1 & 0 & 0 & 1 \\
2^4 & 2^3 & 2^2 & 2^1 & 2^0
\end{array}
$$

Thus 11001 in binary notation is equal to $2^4 + 2^3 + 2^0 = 16 + 8 + 1 = 25$ in our system.

(*b*) We write

$$
\begin{array}{cccccccc}
1 & 1 & 0 & 0 & 1 & 0 & 0 & 1 \\
2^7 & 2^6 & 2^5 & 2^4 & 2^3 & 2^2 & 2^1 & 2^0
\end{array}
$$

Thus 11001001 in binary notation is equal to $2^7 + 2^6 + 2^3 + 2^0 = 128 + 64 + 8 + 1 = 201$ in our system.

Arithmetic Operations in the Binary System

We have pointed out one reason why the binary numeration system is useful for computers—because every number can be represented by a series of storage units with each unit either possessing an electric impulse or not, corresponding to a series of 0s or 1s. An alternate way to visualize how numbers are stored in a computer is to picture the series of storage units as a sequence of switches that are either on or off. If the switch is off then it represents 0, and if it is on it represents 1. Therefore any number expressed in base 2, say, 1001011, can be represented in the computer by a sequence of switches that are in the positions, from left to right: on-off-off-on-off-on-on.

The other reason why the binary system is the numeration system best fitted for the computer is that arithmetic operations performed in base 2 are extremely simple, so much so that addition becomes almost mechanical. It can be done by simply passing a current of electricity through a sequence of storage cells. Because electricity flows so fast the speed of the calculations is tremendous.

The primary reason why addition in base 2 is so simple is that there are only four rules that must be considered:

$$
\begin{array}{cccc}
0 & 0 & 1 & 1 \\
+0 & +1 & +0 & +1 \\
\hline
0 & 1 & 1 & 10
\end{array}
$$

Remember that all of these symbols are in base 2 so that 1 and 0 are still read "one" and "zero," respectively, but 10 is read "one-zero" and not "ten" since it represents the number 2. Let us look at an example of how we apply these rules to addition problems.

Example 7

Problem Solve the following addition problems in base 2.

(a)   10   (b)   101
     + 11   (b) + 110

Solution (a) We first add the two right-most digits, $0 + 1 = 1$. We then add $1 + 1 = 10$. Hence

$$
\begin{array}{r}
10 \\
+ \ 11 \\
\hline
101
\end{array}
$$

You should translate each number into our Hindu-Arabic notation to check your answer. For this problem, 10 in base 2 is 2 and 11 in base 2 is 3, while 101 in base 2 is 5. It checks.

(b) This is very similar to (a). We have

$$
\begin{array}{r}
101 \\
+ \ 110 \\
\hline
1011
\end{array}
$$

Neither of the problems in Example 7 required us to carry from one place to the next. Let us look at this type of problem in detail. In order to do the addition

$$
\begin{array}{r}
11 \\
+ \ 11
\end{array}
$$

we have to learn how to carry because the sum in the right-most column is $1 + 1 = 10$. Just as in addition problems in our Hindu-Arabic number system, we place the 0 in the right-most column of the sum and carry the 1 into the next column. Most people either keep the 1 in their heads or write it at the top of the next column. The latter method is akin to what the computer does. When it performs $1 + 1$ it inserts a 0 in the sum and a 1 in the next column. Schematically, we describe this by putting a 1 below the second column and using an arrow to show that it is being carried:

$$
\begin{array}{r}
11 \\
+ \ 11
\end{array}
$$

Carry	1
	↖
Sum	0

The second column consists of the sum $1 + 1 + 1 = 11$ and we simply place it in the bottom row:

$$
\begin{array}{r}
11 \\
+ 11 \\
\end{array}
$$

Carry	1
	↖
Sum	110

The final sum is 110. Note that 11 in base 2 is 3 so the problem in Hindu-Arabic is $3 + 3 = 6$. The answer checks since 6 is 110 in base 2.

Example 8

Problem What is $111 + 111$ in the binary system?

Solution The start of the problem is the same as the one above:

$$
\begin{array}{r}
111 \\
+ 111 \\
\end{array}
$$

Carry	1
	↖
Sum	0

Since $1 + 1 + 1 = 11$, we must place 11 in the proper place. Write the right-most 1 in the sum and carry the left-most 1 to the next column. The sum becomes

$$
\begin{array}{r}
111 \\
+ 111 \\
\end{array}
$$

Carry	11
	↖↖
Sum	1110

Example 9

Problem Perform the following addition in the binary system.

$$
\begin{array}{r}
1011011101 \\
+ 1011110100 \\
\end{array}
$$

Solution

$$
\begin{array}{r}
1011011101 \\
+ 1011110100 \\
\end{array}
$$

Carry	111111
	↖↖↖↖↖
Sum	10111010001

The Best System

We have examined four number systems while studying the evolution of the concept of number: the Egyptian, Mayan, Hindu-Arabic, and binary. Hundreds of other systems are known to have been used, most having a mixed base such as 5-10, 5-20, or 10-60, but some having single base 8, 6, or 2. Anthropologists define a "cultural universal" to be any property that every known culture has possessed. One of the primary cultural universals is number. This explains why the radio signal transmitted from earth in search of other advanced civilizations in the universe consists of a numeric message. In fact, the first part of the message establishes a binary counting code.

Which system is the best? It is tempting just to avoid this question because different systems arose to meet the needs of different cultures. Yet it is a fact that today the Hindu-Arabic system is used throughout most of the world. It has evolved over a very long period of time and has incorporated many improvements, such as the introduction of fractions, negative numbers, decimals, and scientific notation, without its foundation being altered at all.

Despite the many advantages of the Hindu-Arabic system, there are reasons for preferring other systems. One organization, the Dozenal Society of America, lobbies for base 12. Some machines make use of a base 8 system, called an octonary system, or octal notation, and such a system would be eminently convenient for working with stock market quotations. The genetic code is a base 4 system in which every feature of every living being is conveniently described. Finally, without the binary system the computer revolution would be only food for the fantasies of Jules Verne and Isaac Asimov.

EXERCISES

In Problems 1 and 2 write the expanded version of the numbers.

1 80,421   2 105,605

In Problems 3 to 8 evaluate the exponential forms.

3 10^6 6 4^5

4 10^8 7 3^6

5 3^4 8 4^3

9 Use Problems 5 and 7 to evaluate 3^{10}.

10 Use Problems 6 and 8 to evaluate 4^8.

11 Use Problems 5 and 7 to evaluate $3^6/3^4$.

12 Use Problems 6 and 8 to evaluate $4^5/4^3$.

13 Problem 11 suggests a rule for $3^m/3^n$. What is it?

14 Write a^n/a^m as a power of the number a.

15 Use Problem 5 to evaluate 3^{12}. Note that $3^{12} = 3^4 \cdot 3^4 \cdot 3^4$.

16 Write $(a^m)^n$ as a power of the number a.

In Problems 17 and 18 write the numbers in exponential form with base 3.

17 243   18 729

In Problems 19 to 24 numbers are given in Hindu-Arabic form. Express them in binary form.

19 14   22 101

20 38   23 180

21 58   24 1213

In Problems 25 to 30 numbers are given in binary form. Express them in Hindu-Arabic form.

25 1010   28 1100111

26 11101   29 100010101

27 1001010   30 110001001

In Problems 31 to 36 perform the indicated addition in base 2. Check your answer by translating the numbers into Hindu-Arabic form.

31 $\begin{array}{r} 10110 \\ + 11001 \\ \hline \end{array}$ 34 $\begin{array}{r} 10111 \\ + 11101 \\ \hline \end{array}$

32 $\begin{array}{r} 11011 \\ + 10100 \\ \hline \end{array}$ 35 $\begin{array}{r} 100110111 \\ + 110101101 \\ \hline \end{array}$

33 $\begin{array}{r} 11011 \\ + 10110 \\ \hline \end{array}$ 36 $\begin{array}{r} 1001011101 \\ + 1110010111 \\ \hline \end{array}$

37 You might expect a mathematician's favorite novel to have a number in the title. One of our favorites is Joseph Heller's "Catch-22." In it, Yossarian refers to the "fighting 256th squadron" as "that's two to the fighting eighth power." Is Heller's arithmetic correct?

38 Robert Heinlein, one of the creators of modern science fiction, wrote in his book, "The Number of the Beast," that the number of universes besides our own is "six raised to its sixth power, and the result in turn raised to its sixth power." Write this number in exponential form with base 6.

39 Write each of the following numbers in exponential form with base 10.
    (a) 10,000   (b) 100,000

40 Edward Kasner introduced the names "googol" for 10^{100} and "googolplex" for 10^{googol}. How many zeros does each term contain? (Incidentally, Kasner claims that the number of grains of sand on the beach at Coney Island is much less than a googol.)

41 What is the relationship in Table 1 between the number of zeros in the "Numeral" column and the exponent in the "Exponential Form" column?

42 How many times richer is a billionaire than a millionaire?
(a) twice
(b) 10
(c) 100
(d) 1000
(e) million

43 An exponent is (choose one):
(a) A former ponent
(b) Proponent's brother
(c) A power

44 Binary refers to (choose one):
(a) 2
(b) Being next to a nary
(c) Buying nothing

1-4 DIVISIBILITY AND PRIME NUMBERS

The American Mathematical Society (AMS) is typical of the many professional organizations which are devoted to research and advancement in their field. Most of them hold annual meetings at which the members present the fruits of their recent labors. In October 1903, the AMS held a meeting in New York City. One of the papers on the program bore the modest title, "On the factorization of large numbers." The speaker was F. N. Cole.

We use the word "speaker" loosely, not just because Cole was a man of few words, but because in this "talk" Cole was a man of no words. He walked to the board and worked out in longhand the arithmetic for raising 2 to the sixty-seventh power. Then he subtracted 1. The number $2^{67} - 1$ has 20 digits. He moved silently to the other side of the board and multiplied

$$193,707,721 \times 761,838,257,287$$

The two calculations were equal. For the first time, and probably the only time since, an audience of the AMS vigorously applauded the speaker presenting a paper. Everyone in the room had immediately realized the significance of Cole's result and appreciated the startling amount of effort he must have applied to unearth it. No one questioned him at the time, but later the noted writer E. T. Bell asked him how long it took him to crack the mystery, and Cole responded, "Three years of Sundays."

This incident lies at the heart of this section and the following two sections. It serves to illustrate the fact that mathematicians today are just as interested in properties of numbers as were the classical Greeks of some 2500 years ago. It also reflects two very important aspects of mathematics: factorization and pattern recognition. Undoubtedly the response by Cole's audience is a mystery to you. Why would sane individuals (presuming mathematicians *are* sane, of course) vigorously applaud the fact that one of their peers discovered that two mammothly large numbers are equal? It will take us a while to explain, but we are confident that at the end of this chapter you too will express amazement at Cole's achievement.

When summarizing the properties of the Hindu-Arabic system at the end of Section 1-2, we said that this system afforded us two advantages: (1) numbers can be represented easily, and (2) the form of the representation enhances the ability to perform arithmetic operations. In this section we will be concerned with the latter advantage. In particular we will investigate properties of the set of *natural numbers*. We will denote this set by *N,* so that

$$N = \{1, 2, 3, \ldots\}$$

where the ellipsis, or three dots, means that this sequence of numbers continues without end. Sometimes N is called the set of counting numbers. Throughout this section whenever we say "number" we mean "natural number."

The Classical Greeks

Our study of the set of natural numbers begins with another great step in the evolution of our number system. It was taken by the classical Greeks, who lived in the period from about 600 to 300 B.C. The ancient Greeks, for the first time in mathematics, as similarly in many other diverse fields such as philosophy, physics, and literature, asked the question, "Why?" *Why* does the diameter of a circle bisect the circle? *Why* is the sum of two even numbers always even? For the first time the idea of proof, of certainty, of demonstrative methods and logical reasoning, began to appear. In this section and the two sections that follow, we will indicate how this type of reasoning enabled the Greeks to discover subtle and important properties of our number system.

Once we study the methods of the Greeks we will be able to understand why F. N. Cole's seemingly unassuming presentation before the staid audience of the AMS provoked such unprecedented admiration and a round of applause.

At the heart of these three sections is the idea of observing patterns from experience, generalizing from them, and then determining whether the pattern is *always* true. Of course, this type of mental process is at the core of not only mathematics but most other intellectual disciplines as well. Let us now consider how one facet of the Greek philosophy affected the study of natural numbers.

The term "atomism" refers to the doctrine that the universe is composed of simple, indivisible minute particles. The Greek philosopher Plato believed that there were four such primary constituents of all matter: earth, air, fire, and water. As the field of chemistry evolved, it was found that there are many more, diverse building blocks, called the *elements*. Modern physics has carried atomism to an extreme. At the turn of the century it was found that the elements are made up of atoms, which themselves are composed of a nucleus, formed by protons and neutrons, and revolving electrons. The aim of researchers today in molecular physics is to establish a single, comprehensive theory, called a "grand unification," according to which the basic, indivisible particles are "quarks."

Generally, mathematicians are atomists too; they attempt to study structures by decomposing them into simpler structures whose properties are more easily established. But they have been no more successful in establishing the primary building blocks of mathematics than have the chemists and physicists. Early in this century it was thought that set theory would provide this base, but such a hope was dashed by the Austrian mathematician Kurt Gödel (1906–1978) in 1931. It is now agreed that a more fundamental notion is needed.

Elementary Building Blocks

The mathematicians of classical Greece determined the irreducible parts of the natural numbers. They correctly singled out the operation as multiplication and the primitive elements as the prime numbers. We say that a number is a *prime number* if no number except 1 and itself divides the number.

Let us be a bit more specific about the term "divides." The number 3 divides 24 because there is a natural number, namely 8, such that $24 = 8 \cdot 3$. In general, if x and y are natural numbers, we say that x *divides* y if there is a natural number q such that $y = qx$. Our notation for this is $x \mid y$, which is read "x divides y." Thus $3 \mid 24$ because $24 = 8 \cdot 3$ (here $q = 8$). Also $13 \mid 221$ because $221 = 17 \cdot 13$ ($q = 17$). Sometimes the phrase "13 divides 221 evenly" is used, but in mathematical jargon the word "evenly" is superfluous. When $x \mid y$ we also say that x is a *divisor* of y, y is *divisible by* x, and y is a *multiple* of x.

Let n represent a number greater than 1. Then it is always true that $1 \mid n$ and $n \mid n$. Hence a prime number is a number n such that 1 and n are the only divisors of n. Let us list the first few primes:

2, 3, 5, 7, 11, 13, 17, 19, 23, 29, 31, 37, 41, . . .

A number which is not prime is called a *composite number*. The first few composite numbers are:

4, 6, 8, 9, 10, 12, 14, 15, 16, 18, 20, 21, 22, 24, 25, 26, 27, 28, . . .

Notice that the number 1 does not appear in either list; it is neither prime nor composite.

In Section 1-6 we will discuss the lore of prime numbers. For now we are more interested in seeing how primes form the building blocks for all natural numbers. Let us look at an example of how we can write a composite number as a product of prime numbers.

Consider the number 72. We can write $72 = 8 \cdot 9$. We can also write $72 = 6 \cdot 12$. Thus 72 can be factored into smaller numbers in two distinct ways. Surely you can come up with additional ways. Note however that the factors 8, 9, 6, and 12 are all composite. If we express each one of these factorizations in terms of prime factors we obtain

$$72 = 8 \cdot 9 = 2^3 \cdot 3^2$$

and

$$72 = 6 \cdot 12 = 2 \cdot 3 \cdot 2^2 \cdot 3$$

If we rearrange the numbers in the latter factorization, it becomes

$$72 = 2 \cdot 2^2 \cdot 3 \cdot 3 = 2^3 \cdot 3^2$$

which coincides with the first factorization.

This example provides one instance of the unique factorization theorem. It says that the only way that 72 can be factored into primes is $2 \cdot 2 \cdot 2 \cdot 3 \cdot 3$, where we do not distinguish between this factorization and any permutation of it, say, $2 \cdot 2 \cdot 3 \cdot 2 \cdot 3$. We always write the factors as powers of primes, and put the smaller primes on the left. The more traditional name of the unique factorization theorem is the *fundamental theorem of arithmetic*. It was stated and proved in Euclid's *Elements* around 300 B.C.

FUNDAMENTAL THEOREM OF ARITHMETIC: Every composite natural number can be factored into primes in one and only one way.

How do you determine whether a number is prime or composite? The direct approach is to determine whether it has any factors other than 1 and itself. If one such factor is found, the fundamental theorem assures us that it is possible to factor the number uniquely into prime numbers.

Divisibility Tests

Sometimes it is easy to determine if a number has a factor other than 1 or itself. For example, if the last digit is a 0, you know it is divisible by 10, and since $10 = 2 \cdot 5$, you know that the two primes 2 and 5 divide the number. Are there any other numbers that are divisible by 5? You probably remember that if the last digit is 5 the number is divisible by 5. Thus if a number's last digit is 0 or 5 then it is divisible by 5. Are there any other numbers that have 5 as a divisor? No; so we can also say that if a number is divisible by 5 then its last digit is 0 or 5. We abbreviate these last two statements by writing:

> DIVISIBILITY BY 5: A number n is divisible by 5 if and only if the last digit of n is 0 or 5.

Similarly, if a number is divisible by 2, then it is an even number, and so its last digit is 0, 2, 4, 6, or 8. Conversely, if the last digit of a number is 0, 2, 4, 6, or 8, then it is divisible by 2. We record this in the following rule:

> DIVISIBILITY BY 2: A number n is divisible by 2 if and only if the last digit of n is 0, 2, 4, 6, or 8.

Therefore, to determine whether a number is divisible by 2 or 5, we need only inspect the last digit. Thus the numbers 10, 40, 85, 635, and 1225 are divisible by 5, while 12, 44, 76, 198, and 40 are divisible by 2.

The tests for divisibility not only tell us when one number is divisible by another but also when it is not. That is, if a number fails to satisfy the criterion, then it is not divisible by that number. Hence 5 does not divide 221. Rather than write the entire expression "5 does not divide 221" we write simply $5 \nmid 221$. This manner of imprinting a diagonal slash over the symbol for some relation to denote the negation of that relation is customary practice. Thus, for example, $\neq$ means "is not equal to" and $\nless$ "is not less than."

The question of what numbers divide a given number is not always easy to answer. Take, for example, the number 1001. What numbers divide it? Clearly neither 2 nor 5 does since 1001 is neither even nor ends in 5. But could 3 be a divisor? There is a rule which applies to this case:

> DIVISIBILITY BY 3: A number n is divisible by 3 if and only if the number formed by adding the digits of n is itself divisible by 3.

Since the sum of the digits of 1001 is 2 ($= 1 + 0 + 0 + 1$) and since $3 \nmid 2$, it follows from our divisibility test that $3 \nmid 1001$. The expression "sum of the digits of n" will occur repeatedly, so we will shorten it to the *digital sum of n.*

A related term is *digital root,* which is the number between 0 and 9, inclusive, that you get when you find successive digital sums. An example will illustrate these terms.

Example 1
Problem Compute the digital root of the number 1,234,567,890.
Solution Let $n = 1,234,567,890$. The digital sum of n is 45 since

$$1 + 2 + 3 + 4 + 5 + 6 + 7 + 8 + 9 + 0 = 45$$

The digital sum of 45 is 9. Since 9 lies between 0 and 9, this is the digital root of n. Note that $3 \mid n$. We will see that $9 \mid n$ too.

Now we can restate the test above in terms of digital root.

DIVISIBILITY BY 3: A number n is divisible by 3 if and only if its digital root is 0, 3, 6, or 9.

Example 2
Problem What numbers in the 280s are divisible by 3?
Solution The set in question is $\{280, 281, 282, \ldots, 289\}$. The digital sum of 280 is $2 + 8 = 10$, so the digital root of 280 is 1. Thus $3 \nmid 280$. Similarly $3 \nmid 281$. However $3 \mid 282$ since its digital root is 3. Once we know that 3 divides a number it follows that 3 divides every three numbers greater than and less than that number. Thus the fact that $3 \mid 282$ implies that $3 \mid 285$ and $3 \mid 288$ also.

Earlier, when we discussed divisibility by 2 and 5, we indicated that you can tell whether a given number is divisible by one of these numbers by inspection. By this we meant that the digit in the units place alone determines whether that number is divisible by either 2 or 5. This is not the case with 3, as can be seen by listing some multiples of 3:

3, 6, 9, 12, 15, 18, 21, 24, 27, 30

Notice that each of the 10 digits occurs in the units place of one of the multiples. The situation is similar for the number 9, but there is another divisibility test which is similar to that for 3. It is sometimes called "casting out nines."

DIVISIBILITY BY 9: A number n is divisible by 9 if and only if the digital root of n is 0 or 9.

Example 3

The Springfield Little League had 288 boys and girls come to its first day of practice. Head coach Abner Tripleday wanted the kids to be broken up into teams of 9. Could this be accomplished? Since the digital sum of 288 is $2 + 8 + 8 = 18$ and the digital root is 9, it follows that $9 \mid 288$. Thus it is possible to divide up the kids evenly. (How many teams are there? This is where the q used in the definition of divisibility enters in.)

The test for divisibility by 4 is similar to the one for 2 except that you must consider not just the last digit of the number but the last two digits. Thus we have

---

DIVISIBILITY BY 4: A number n is divisible by 4 if and only if 4 divides the number formed by the two right-hand digits of n.

Since $4 \mid 48$, it follows that $4 \mid 748$ and $4 \mid 7648$, but $4 \nmid 78$, and so $4 \nmid 278$.

An Interesting Pattern

We have presented some of the more elementary divisibility tests. There are many more. Let us investigate a few additional tests in order to test your ability at pattern recognition.

Let us restate the tests for 2 and 4 in a more compact form (we shorten "if and only if" to "iff"*):

$2 \mid n$ iff 2 divides the number formed by the last digit of n.

$4 \mid n$ iff 4 divides the number formed by the last two digits of n.

The tests for 8 and 16 are very similar to these. Take a guess before we give them to you. They are:

$8 \mid n$ iff 8 divides the number formed by the last three digits of n.

$16 \mid n$ iff 16 divides the number formed by the last four digits of n.

This pattern can be generalized into a test for divisibility by the powers of 2, namely 2^m. Recall the exponential form of the powers of 2:

$$1 = 2^0 \quad 2 = 2^1 \quad 4 = 2^2 \quad 8 = 2^3 \quad 16 = 2^4$$

If you rewrite each of the previous four tests, using this exponential form, you can probably guess that the pattern yields the following test:

---

*We have no idea how "iff" should be pronounced.

$2^m \mid n$ iff 2^m divides the number formed by the last m digits of n.

We will have more to say about patterns like this one in the next section.

Applying the Fundamental Theorem

Now that we have the use of some divisibility tests the application of the fundamental theorem will be much easier. The method to find the prime factorization of a number entails searching for specific divisors. There is no set procedure, but you will often find it convenient to use the divisibility tests to determine if a number is a divisor and then do the division. Then apply the same technique to the quotient.

Example 4

Problem Apply the fundamental theorem of arithmetic to the number 2700.

Solution Since 2700 ends in 0 we see that $10 \mid 2700$ and $2700 = 270 \cdot 10$. The quotient, 270, also ends in 10, and $270 = 27 \cdot 10$. Thus $2700 = 27 \cdot 10 \cdot 10$. Since $10 = 2 \cdot 5$ and we know that $10 \cdot 10 \mid 2700$ we have that $2 \cdot 5 \cdot 2 \cdot 5 \mid 2700$. The last quotient was 27. It is not divisible by 2; so we proceed to 3. The digital root of 27 is 9, so $3 \mid 27$ and $27 = 3 \cdot 9 = 3 \cdot 3 \cdot 3 = 3^3$. Hence $2700 = 2 \cdot 5 \cdot 2 \cdot 5 \cdot 3^3$. Now we rearrange the primes in standard form, consisting of powers of primes with the smaller primes to the left. Hence

$$2700 = 2^2 \cdot 3^3 \cdot 5^2$$

This is the only way that 2700 can be factored into powers of primes.

Example 5

Problem Apply the fundamental theorem of arithmetic to the number 1134.

Solution We arrange the work in two columns. The left-hand column will give the explanation of why a particular divisor is chosen and the right-hand column will contain the division.

1 1134 ends in 4 so it is divisible by 2 (but not by 4 since $4 \nmid 34$). $\qquad 1134 = 567 \cdot 2$

2 The digital root of 567 is 9 so $9 \mid 567$. $\qquad 1134 = 63 \cdot 9 \cdot 2$

3 The digital root of 63 is 9 so $9 \mid 63$. $\qquad 1134 = 7 \cdot 9 \cdot 9 \cdot 2$

Hence

$$1134 = 2 \cdot 3^4 \cdot 7$$

Example 6
Problem Apply the fundamental theorem of arithmetic to the number 27,720.
Solution

1 27,720 ends in 0, so 10 | 27,720. $27{,}720 = 2772 \cdot 10$

2 4 | 72 so 4 | 2772. $27{,}720 = 693 \cdot 4 \cdot 10$

3 The digital root of 693 is 9, $27{,}720 = 77 \cdot 9 \cdot 4 \cdot 10$
so 9 | 693.

4 $77 = 7 \cdot 11$. $27{,}720 = 7 \cdot 11 \cdot 9 \cdot 4 \cdot 10$

Hence

$$27{,}720 = 7 \cdot 11 \cdot 3^2 \cdot 2^2 \cdot 2 \cdot 5$$
$$= 2^3 \cdot 3^2 \cdot 5 \cdot 7 \cdot 11$$

The primitive elements here are the prime numbers and the operation is multiplication. Do you think that natural numbers have unique factorizations into primes if we use addition instead of multiplication? See Problem 47.

**A Recent Step
in the Ascent**

The fundamental theorem of arithmetic is over 2200 years old. Its discovery by the Greeks marked a significant advance in the evolution of number because it showed that the primes are the primitive building blocks of the natural numbers. The theorem has been applied continuously since its discovery. In fact, it was one of the key tools in a recent exciting development in mathematics. In 1981 a very old problem, worked on by scores of scientists, was finally solved. In the exercises we present a more complete account of this very recent step in the ascent of mathematics.

EXERCISES

In Problems 1 to 10 determine whether the number is divisible by 2, 5, and/or 10.

1 24   **6** 720

2 482   **7** 321

3 30   **8** 727

4 85   **9** 3860

5 615   **10** 10,321

In Problems 11 to 16 determine whether the number is divisible by 3 or 9.

11 321

12 162

13 462

14 729

15 810

16 1212

In Problems 17 to 22 determine whether the number is divisible by 4 and/or 8.

17 688

18 1088

19 612

20 1812

21 1872

22 6842

In Problems 23 to 30 apply the fundamental theorem of arithmetic to the numbers.

23 2480

24 6318

25 40,050

26 62,730

27 62,370

28 24,948

29 14,872

30 14,800

Problems 31 to 34 deal with proper divisors, where a *proper* divisor of a number x is any divisor except x. Thus 1 is always a proper divisor.

31 Calculate the sum of all of the proper divisors of 28.

32 Find a number less than 28 which has the property indicated in Problem 31.

33 Find the sum of the proper divisors of 220.

34 Find the sum of the proper divisors of 284.

35 Which column in the Ishango tally bone consists entirely of primes? (See Figure 1 in Section 1-1.)

36 Is 1001 a prime number?

37 There are 5280 feet in a mile. Can you tell whether there is an exact number of yards in a mile without performing division? How?

38 Can you construct a divisibility test for 25?

39 Why do you think we used the letter q in our definition of divisibility?

40 Find a number that is less than 100 and has exactly five distinct divisors.

41 What is the smallest number that can be divided by each of the first 10 numbers: 1, 2, 3, . . . , 10?

42 Find a number that has exactly eight distinct factors.

43 Write 18 as a sum of two primes in two distinct ways.

44 In 1874 W. Stanley Jevons asserted in his book "Principles of Science," "Can the reader say what two numbers multiplied together will produce the number 8,616,460,799? I think it unlikely that anyone but myself will ever know; for they are two large prime numbers." Do you think the ensuing century has been good to Jevons?

Group Theory

This set of exercises relates the fundamental theorem of arithmetic to one of the most exciting recent developments in mathematics. The word "group" has a very special meaning to mathematicians; loosely it is a bunch of things with a way of combining them, subject to four restrictions. The origins of group theory go back to the 1820s, when a teenager, Evariste Galois, was trying to determine when certain equations in algebra could be factored. As with natural numbers, the aim was to factor a given group into its irreducible elements. In the case of natural numbers the irreducible elements turned out to be prime numbers. For groups they turned out to be "simple groups," and for over 100 years group theorists attempted to classify all of them. By 1979 the end seemed in sight. An announcement for a 6-week conference held that summer bubbled with enthusiasm.

> The tremendous progress in finite group theory in recent years promises to change the entire field. After more than a quarter of a century of intensive effort by several hundred people, the classification of all finite simple groups appears to be rapidly nearing conclusion. One can safely predict that by 1980 the general theory of simple groups will have been fully developed.

The prediction turned out to be correct. Almost. A total of 179 mathematicians from 14 countries attended the conference, including two from Russia and two from mainland China, but the finishing touches were not applied until 1981. All the simple groups fit nicely into separate categories except for an elusive bunch of 26 mavericks, which have come to be called *sporadic groups.* Table 1 lists these groups according to their *order,* where the order of a group is the number of elements in it. The table is taken from a research article by one of the pioneers, Richard Brauer. The exercises refer to Table 1.

45 The order of group 6, called the Janko group, is given as 175,600. This is not correct. Multiply the given factors and determine the correct order.

46 Group 4, the fourth Mathieu group, has order 10,200,960, but its prime factorization in the table contains an error. Determine the correct factorization.

How many digits are contained in each of the following orders:

47 Group 23, the Thompson group

48 Group 19, the Lyons-Sims group

49 Group 25, the "monster" or "friendly giant"

50 Group 22, the "baby monster"

TABLE 1

The 26 Sporadic Groups

	Group			Order
1–5	The five Mathieu groups	M_{11}	$2^4 \cdot 3^2 \cdot 5 \cdot 11$	$= 7{,}920$
		M_{12}	$2^6 \cdot 3^3 \cdot 5 \cdot 11$	$= 95{,}040$
		M_{22}	$2^7 \cdot 3^2 \cdot 5 \cdot 7 \cdot 11$	$= 443{,}520$
		M_{23}	$2^7 \cdot 3^2 \cdot 5 \cdot 5 \cdot 11 \cdot 23$	$= 10{,}200{,}960$
		M_{24}	$2^{10} \cdot 3^3 \cdot 5 \cdot 7 \cdot 11 \cdot 23$	$= 244{,}823{,}040$
6	The Janko group	J	$2^3 \cdot 3 \cdot 5 \cdot 7 \cdot 11 \cdot 19$	$= 175{,}600$
7	The Janko-Hall group	J_1	$2^7 \cdot 3^3 \cdot 5^2 \cdot 7$	$= 604{,}800$
8	The Janko-Higman and McKay group	J_2	$2^7 \cdot 3^5 \cdot 17 \cdot 19$	$= 50{,}232{,}960$
9	The Held-Higman and McKay group		$2^{10} \cdot 3^3 \cdot 5^2 \cdot 7^3 \cdot 17$	$= 4{,}030{,}387{,}200$
10	The Higman-Sims group		$2^9 \cdot 3^2 \cdot 5^3 \cdot 7 \cdot 11$	$= 44{,}352{,}000$
11	The McLaughlin group		$2^7 \cdot 3^6 \cdot 5^3 \cdot 7 \cdot 11$	$= 898{,}128{,}000$
12	The Suzuki sporadic group		$2^{13} \cdot 3^7 \cdot 5^2 \cdot 7 \cdot 11 \cdot 13$	$= 448{,}345{,}497{,}600$
13–15	The three Conway groups	1	$2^{21} \cdot 3^9 \cdot 5^4 \cdot 7^2 \cdot 11 \cdot 13 \cdot 23$	$\sim 4 \times 10^{18}$
		2	$2^{18} \cdot 3^6 \cdot 5^3 \cdot 7 \cdot 11 \cdot 23$	$\sim 4 \times 10^{12}$
		3	$2^{10} \cdot 3^7 \cdot 5^3 \cdot 7 \cdot 11 \cdot 23$	$\sim 5 \times 10^{11}$
16–18	The three Fischer groups	F_{24}	$2^{21} \cdot 3^{16} \cdot 5^2 \cdot 7^3 \cdot 11 \cdot 13 \cdot 17 \cdot 23 \cdot 29$	$\sim 1.3 \times 10^{24}$
		F_{23}	$2^{18} \cdot 3^{13} \cdot 5^2 \cdot 7 \cdot 11 \cdot 13 \cdot 17 \cdot 23$	$\sim 4 \times 10^{18}$
		F_{22}	$2^{17} \cdot 3^9 \cdot 5^2 \cdot 7 \cdot 11 \cdot 13$	$\sim 6 \times 10^{13}$
19	The Lyons-Sims group		$2^8 \cdot 3^7 \cdot 5^6 \cdot 7 \cdot 11 \cdot 31 \cdot 37 \cdot 67$	$\sim 5 \times 10^{16}$
20	The O'Nan-Sims group		$2^9 \cdot 3^4 \cdot 7^3 \cdot 5 \cdot 11 \cdot 19 \cdot 31$	$\sim 1.5 \times 10^{17}$
21	The Rudvalis-Conway and Wales group		$2^{14} \cdot 3^3 \cdot 5^3 \cdot 7 \cdot 13 \cdot 29$	$= 145{,}926{,}144{,}000$
22	The "Baby Monster"		$2^{41} \cdot 3^{13} \cdot 5^6 \cdot 7^2 \cdot 11 \cdot 13 \cdot 17 \cdot 19 \cdot 23 \cdot 31 \cdot 47$	$\sim 4.25 \times 10^{33}$
23	The Thompson group		$2^{15} \cdot 3^{10} \cdot 5^3 \cdot 7^2 \cdot 13 \cdot 19 \cdot 31$	$\sim 9 \times 10^{16}$
24	The Harada-F. S. Norton and P. E. Smith group		$2^{14} \cdot 3^6 \cdot 5^6 \cdot 7 \cdot 11 \cdot 19$	$\sim 2.73 \times 10^{14}$
25	The "Monster"		$2^{46} \cdot 3^{20} \cdot 5^9 \cdot 7^6 \cdot 11^2 \cdot 13^3 \cdot 17 \cdot 19 \cdot 23 \cdot 29 \cdot 31 \cdot$ $\cdot 41 \cdot 47 \cdot 59 \cdot 71$	$\sim 8 \times 10^{53}$
26	The Janko "group"	J_4	$2^{24} \cdot 3^3 \cdot 5 \cdot 7 \cdot 11^3 \cdot 23 \cdot 29 \cdot 31 \cdot 37 \cdot 43$ $= 86{,}775{,}571{,}046{,}077{,}562{,}800$	$\sim 8.68 \times 10^{19}$

Determine whether each of the following statements is true or false for all the sporadic groups.

51 The order is even.

52 There is one group whose order lies between 1000 and 10,000; one between 10,000 and 100,000; one between 1,000,000 and 10,000,000; etc.

53 The order is divisible by 16.

54 The order is divisible by 8.

55 The order is divisible by 10.

56 The order has a factor whose exponent is greater than 1.

57 The order has a factor whose exponent is equal to 1.

58 The order is divisible by 7, 11, or 13.

1-5 PYTHAGOREAN MATHEMATICS AND PATTERN RECOGNITION

Before the time of the classical Greeks, numbers were used only in a very practical sense. Each new improvement in a number system related to the utility of the system; that is, each made it possible to operate with numbers more efficiently. This probably parallels your own mathematical education in that you were told that you should study a new branch of mathematics because it would allow you to solve some type of practical problems. But now our story of the evolution of number takes a curious twist, for the Greeks themselves were a different brand of people from any that preceded them. They looked at life from wholly new perspectives, as is clearly manifested in their approach to mathematics.

The Greeks studied numbers for their own sake, because they felt that a number system had an innate beauty unto itself and not because a more thorough knowledge of it would help them solve more practical problems. In fact, they distinguished between the mundane art of mere computing with numbers, which they referred to as *logistica,* and the more lofty branch of study concerning abstract relationships between numbers, called *arithmetic.* The practice of logistica was delegated to slaves, while sages and philosophers studied arithmetic. Of course, today we refer to both studies as arithmetic, but in this section we will center our attention on many of the ideas that the Greeks discovered when they practiced their brand of arithmetic.

Perhaps at times in this section you will say to yourself, "Why am I studying this; what is the use of it all?" If so, you do not have the same approach to the subject as the Greeks. They looked for patterns and relationships in the set of natural numbers, and when they were successful at isolating an interesting pattern and were able to show that it always holds, they considered this mental accomplishment a thing of beauty. But how does this affect you? We hope on the one hand that you will find at least some of their findings intriguing. On the other hand, by studying this type of mathematics you will be getting an introduction to pattern recognition, a mental exercise that you will be called upon to use in a vast array of disciplines, both throughout your academic career and beyond. The exercises will help you test your ability at recognizing patterns.

Numerology

We will start our discussion on the way that the ancient Greeks thought of and worked with numbers by explaining the very curious way in which they identified numbers with letters in their alphabet. The Greek numeration system, called the *alphabetic system,* was not discussed in Section 1-1 because we were concentrating on the practical aspects of the evolution of number systems. The Greek alphabetic system was hardly practical. In this system a number is associated with each letter in the alphabet. Later it developed into a form of numerology called *gematria,* in which words are represented by numbers that are obtained by adding up the values of each of the letters. In Problem 1 in the exercises we describe the method the Greeks used to identify numbers with letters. For example, the letters of the Greek word for "amen" add up to 99, and in certain Christian manuscripts the number 99 is written at the end of a prayer to signify "amen."*

The most famous example of gematria is the number 666. In Revelations 13:18 one finds:

> Here is wisdom. Let him that hath understanding count the number of the beast; for it is the number of a man, and his number is Six hundred three score and six.

Since the time that verse was written, some 2000 years ago, numerologists have been trying to fit names to the number of the beast, 666. One of the first was "Nero Caesar." A Renaissance algebraist, Michel Stifel, came up with "Pope Leo X." Later John Napier, who is known for his invention of logarithms, showed that 666 could stand for "the Pope of Rome," but one of his contemporaries hastened to add that it also stood for "Martin Luther." Even in the twentieth century it has been pointed out that the numerical value of Kaiser Wilhelm's name added up to 666. In Problem 1 we ask you to compute your own value.

Pythagoras and His Band of Merry Mathematicians

In the previous section we encountered various properties of natural numbers centering around the concept of divisibility. Since that kind of material has a modern abstract flavor to it, you might think at first that it is a product of the twentieth century. On the contrary, it is, in fact, almost 2500 years old. So let us turn the clock back to about 500 B.C. and transport ourselves to Croton, a small town in southern Italy. There we meet the mythical figure Pythagoras,

---

*In certain mathematics texts the letters QED are placed at the end of the proof to signify "that which was to be proved." The QED comes from the Latin *quod erat demonstrandum* and bears no numerological connotation whatsoever. However, some students have been heard to mutter "amen" when finally arriving at that part of the proof.

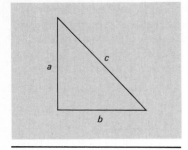

truly one of the most interesting yet puzzling persons in history. You probably know his most famous result, which is

THE PYTHAGOREAN THEOREM: In a right triangle, if a and b are the lengths of the legs and c is the length of the hypotenuse, then $a^2 + b^2 = c^2$.

Pythagoras founded a society of disciples, the Pythagoreans, and the society attributed all results to the master. This commune was founded upon certain religious tenets which are peculiarly interesting, and to give you an idea of their beliefs we will list some of the rules of the order.

1 To abstain from beans.

2 Not to pick up what has fallen.

3 Not to touch a white rooster.

4 Not to break bread.

5 Not to step over a crossbar.

6 Not to stir the fire with iron.

7 Not to eat from a whole loaf.

8 Not to pluck a garland.

9 Not to sit on a quart measure.

10 Not to eat the heart.

11 Not to walk on highways.

12 Not to let swallows share one's roof.

13 When the pot is taken off the fire, not to leave the mark of it in the ashes, but to stir them together.

14 Do not look in a mirror beside a light.

15 When you rise from the bedclothes, roll them together and smooth out the impress of the body.

You might find some of these rules strange, and so stigmatize the Pythagorean brotherhood as being composed of a bunch of societal misfits. In point of fact such a conception would be partly true, for the Pythagoreans did experience considerable friction with the Croton populace. Yet in a scientific sense the Pythagoreans were centuries, in fact millennia, ahead of their time. Their motto was

All is number.

By this they meant that mathematical relationships are the essence of all phenomena and that they alone should be used to express seemingly unrelated events. Jacob Bronowski's "The Ascent of Man" devotes one chapter to this topic. It is entitled "The Music of the Spheres" in recognition of the way in which the Pythagoreans reduced both music and the motions of planets to simple relationships among numbers. It was not until the time of Galileo in the early seventeenth century that humans again began to look for the essence of things in number, and today this philosophy applies to almost all modern science and to a considerable portion of the social sciences as well.

We mentioned gematria above because it is indicative of one facet of pythagorean mathematics, number mysticism. One other aspect of this mysticism associated sex with numbers; even numbers were said to have one sex and odd another. (See Problem 2.) Some numbers were identified with human properties, as shown in the accompanying table. The association of a square, the number 4, with the property of justice is reflected in our modern interpretation of a "square shooter" as being one who acts justly.

Number	Meaning	Reasoning
One	Reason	Consistent whole
Two	Opinion	There is another side to the issue
Four	Justice	Product of equals (2 × 2)
Five	Marriage	Union of 2 (female) and 3 (male)
Seven	Health	
Eight	Love and friendship	

There is a very interesting parallel between the development of mathematics in classical Greece and in China. Shortly after the time of Pythagoras there developed in China the dualistic theory of the yin and the yang, the two primal forces in the universe. Of interest to us is the association of odd numbers with yang, which is the male force, and even numbers with yin, the female force.

Although the Pythagoreans' main contributions were in the field of geometry, we will consider here their study of arithmetic, though keep in mind that they did not separate geometry from arithmetic as we tend to do today. Actually the type of arithmetic we will examine is called "the theory of numbers" in the United States and "higher arithmetic" in Great Britain.

Square Numbers

Since the Greek view of mathematics emphasized geometry, it is no wonder that numbers were described by geometric patterns. The circle was regarded as the most nearly perfect of all plane figures. The next most important was the square, and square arrays of pebbles were used to represent the "perfect squares."

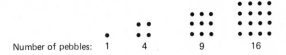

Number of pebbles: 1   4   9   16

Today, of course, we write these numbers as 1^2, 2^2, 3^2, and 4^2 and use the term "squared" to describe them.

Suppose you add two square numbers. Is the sum always another square number? Obviously not, since $1 + 4 = 5$ and 5 is not a square number. (Do you see why 5 is not a square number?) However, some sums of squares do form square numbers, such as $3^2 + 4^2 = 5^2$. In such a case we write these numbers in the form (3, 4, 5) and refer to this expression as a *pythagorean triple*. Thus a pythagorean triple (x, y, z) consists of three natural numbers x, y, and z such that

$$x^2 + y^2 = z^2$$

Note that according to the pythagorean theorem the set of all such triples coincides with the set of all right triangles whose sides have lengths which are natural numbers. Notice also that if (x, y, z) is a pythagorean triple then so is (y, x, z) since $y^2 + x^2 = z^2$, but (y, z, x) is not since $y^2 + z^2 \neq x^2$. Thus the order in which you write the three numbers is very important. For example, since (3, 4, 5) is a pythagorean triple so is (4, 3, 5), but (4, 5, 3) is not.

> **Example 1**
> **Problem** Which of the following triples of numbers are pythagorean triples?
> (a) (1, 2, 3)   (b) (6, 8, 10)   (c) (10, 8, 6)
> (d) (8, 6, 10)   (e) (0, 1, 1)
> **Solution** Both (b) and (d) are pythagorean triples since $6^2 + 8^2 = 10^2$ and $8^2 + 6^2 = 10^2$. However, (a) is not because $1^2 + 2^2 \neq 3^2$, (c) is not because $10^2 + 8^2 \neq 6^2$, and (e) is not because 0 is not a natural number.

Triangular Numbers

In any square, the line joining the point in the upper left-hand corner to the one in the lower right-hand corner is called the *main diagonal*. Another class of numbers that the Pythagoreans investigated consisted of those numbers obtained by removing the pebbles above the main diagonal of a square number.

The numbers that result have geometric forms that are triangles:

Number of pebbles: 1   3   6   10

As a result they are called *triangular numbers*. The Pythagoreans developed several properties of these numbers, and we shall consider some of them now and others in the exercises.

We will let T_n stand for a triangular number so that T_1 is the first one, T_2 the second, T_3 the third, etc. From the triangles above we see that $T_1 = 1$, $T_2 = 3$, $T_3 = 6$, and $T_4 = 10$. Consider the following table of triangular numbers:

n	1	2	3	4	5	6	7	8
T_n	1	3	6	10	15	21	28	?

What is the next number in the table; that is, what is T_8? The clue is to see that there is a pattern between successive triangular numbers. For example, the difference between $T_6 = 21$ and $T_5 = 15$ is 6 (= $21 - 15$) and the difference between T_7 and T_6 is 7. If this pattern is to continue, then we should get $T_8 - T_7 = 8$, which means that T_8 should equal $T_7 + 8$. At this point we do not know if $T_8 = T_7 + 8$, we are simply guessing that it does because previous experience shows that the pattern works up until now. So we compute T_8 independently, that is, by actually counting pebbles, and obtain $T_8 = 36$. But we also have $T_7 + 8 = 36$. Hooray, our guess is confirmed. But then a nagging doubt arises: will this pattern *always* hold? Will it hold for T_9; that is, is $T_8 + 9 = T_9$? Will it work for T_{10}, T_{100}, etc.? If we express our conjecture by using general symbols, we are asking whether the formula

$$T_{n-1} + n = T_n$$

will always be true.

The formula states that if you add n to the previous triangular number, T_{n-1}, the result is the next triangular number, T_n. Table 1 shows that the formula works for the first few values of n.

How would the Pythagoreans have showed that the formula is valid for *every* value of n, not merely the first few? They most likely expressed a proof of the formula geometrically because they considered numbers as geometric objects. Here is how their argument probably went.

TABLE 1

A Pattern to Triangular Numbers

n	$T_{n-1} + n$	$= T_n$
2	$T_1 + 2 = 1 + 2 = 3$	$= T_2$
3	$T_2 + 3 = 3 + 3 = 6$	$= T_3$
4	$T_3 + 4 = 6 + 4 = 10$	$= T_4$
5	$T_4 + 5 = 10 + 5 = 15$	$= T_5$
6	$T_5 + 6 = 15 + 6 = 21$	$= T_6$
7	$T_6 + 7 = 21 + 7 = 28$	$= T_7$
8	$T_7 + 8$	$= T_8$

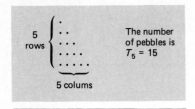

FIGURE 1

First we will proceed with the argument for the specific case where $n = 5$. Consider the geometric representation of T_5 in Figure 1.

Now remove the bottom row of pebbles in Figure 1. What is left? It is the geometrical description of T_4, which contains 10 pebbles. In other words, if we start with the pebble triangle representing T_4 with 10 pebbles, then add another row on the bottom consisting of 5 pebbles, we get the pebble triangle representing T_5 with 15 pebbles. The number fact that $T_4 + 5 = T_5$ is geometrically depicted in Figure 2.

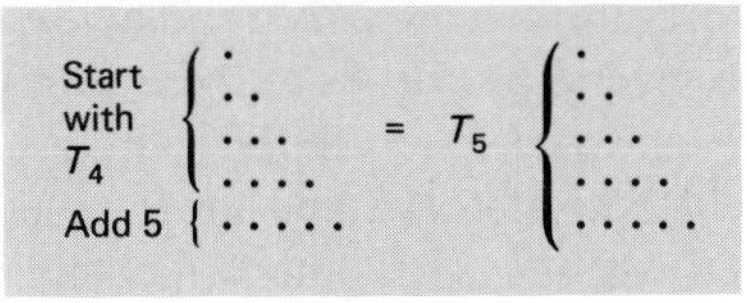

FIGURE 2

In order to prove the result for *any* value of n, we proceed in the same way except that we refer only to the general value n, not mentioning specific numbers such as 5 and 4.

Here goes. Consider the geometric depiction of T_{n-1} in Figure 3. It has $n - 1$ rows and $n - 1$ columns. Now add another row on the bottom consisting of n pebbles. The resulting triangle has n rows and n columns and it is the representation of T_n. Hence $T_{n-1} + n = T_n$.

We are not sure that the Pythagoreans actually carried out their proofs in this manner although it seems plausible. The word "calculate" comes from the Latin *calculus,* meaning a small stone, and this in turn is derived from the Greek *khalix,* which means pebble. But one point is certain: The Pythagoreans were the first people to make general statements about numbers. And it is just this characteristic which separates Greek mathematics from the mathematics of every civilization which preceded it. It presents another instance of the ascent of number.

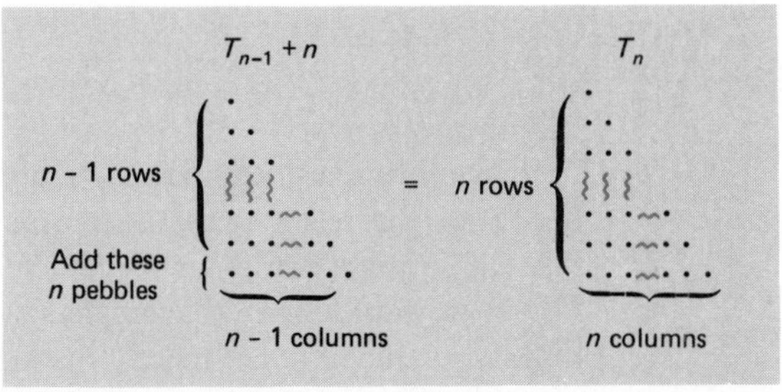

FIGURE 3

Perfect Numbers

In Problem 31 of Section 1-4 you were asked to show that the sum of the proper divisor of 28 is 28. The same is true for 6, because its proper divisors are 1, 2, 3 and $1 + 2 + 3 = 6$. The Greeks called them *perfect numbers,* and such a name is particularly apt. After all, was not the world created in 6 days and does not the moon circle the earth in 28? In general, a number n will be perfect if the sum of the proper divisors of n is equal to n.

Euclid's book the "Elements" is customarily but erroneously reputed to contain only geometry. We have already seen that it contains the fundamental theorem of arithmetic. Another arithmetic result is a method for producing perfect numbers. See if you can make anything of it.

If as many numbers as we please beginning from a unit be set out continuously in double proportion, until the sum of all become prime, and if the sum multiplied into the last make some number, the product will be perfect.

We doubt that many of you will be able to follow Euclid's prescription, since reading it is akin to reading an IRS form, so we will supply some details. First, construct the powers of 2 (Euclid's "double proportion"):

$$1, 2, 4, 8, 16, 32, \ldots$$

Second, add these numbers in succession until the sum is a prime; for instance, $1 + 2 = 3$ and $1 + 2 + 4 = 7$ are primes, but $1 + 2 + 4 + 8 = 15$ is composite. Finally, if the sum is prime, then multiply it by the last number used in the sequence of powers of 2; thus $2 \times 3 = 6$ and $4 \times 7 = 28$. The number obtained will always be perfect. In Problem 11 you will be asked to compute another perfect number following this prescription. The modern way of expressing it is as follows:

If p is prime and $2^p - 1$ is prime, then $(2^p - 1)2^{p-1}$ is perfect.

Euclid gave a proof of this proposition, so we know that it always holds. His proof is too complicated for our purposes, so we will simply make use of the result itself and take for granted that it is true. We will encounter numbers of the form $2^p - 1$ in the next section. They are called *Mersenne numbers*.

On the basis of tables of the perfect numbers that were known at that time, two Greek mathematicians concluded the following:

1 They alternately end in a 6 or an 8.

2 There is one perfect number in each interval: 1 to 10, 10 to 100, 100 to 1000, etc.

These conclusions were never proved to be true. There is a very good reason why no proof was found, because both conclusions turn out to be false, as can be seen from Table 2, which lists the first eight perfect numbers. We will have many occasions to consider patterns of numbers, and this instance indicates that you have to be very careful about drawing general conclusions based upon a few specific cases. In other words, just because the pattern holds for a few (or even many) cases, it can never be assumed to always hold unless a general proof has been given.

TABLE 2

The Smallest 8 of the 27 Known Perfect Numbers

Number	Number of Digits
6	1
28	2
496	3
8,128	4
33,550,336	8
8,589,869,056	10
137,438,691,328	12
2,305,843,008,139,952,128	19

There were 28 perfect numbers known as of September 1983. (See Problem 21 in Section 1-6.) All of them are even. It is known that Euclid's procedure will produce all of the even ones. But two questions remain unanswered.

1 Are there any odd perfect numbers?

2 Is there an infinite number of perfect numbers?

Many mathematicians are presently trying to answer these questions. The largest perfect number known today is

$$2^{44,496}(2^{44,497} - 1)$$

which is a rather large number indeed. The discovery of an earlier perfect number, then the largest known, was the cause of some celebration. The University of Illinois printed a postage-meter stamp to commemorate the occasion because the discovery was made on their computer. (See Figure 4.) To give you a feeling for the size of this number, which contains 6751 digits, we have reproduced a computer printout of it in Figure 5.

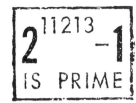

FIGURE 4 A postage-meter stamp honoring the largest prime.

```
                          2 81411 20136 97373 31529 75842 58419 18186 62382 01360 07878 92419 34934 55151 76682
27631 38107 15094 74563 32570 74198 78930 85350 71537 35644 50164 18881 80178 93905 48709 41439 18572 57571 56575 87064
78418 35674 70706 74633 49718 80530 50875 41682 16243 25680 55582 60711 11691 94660 74608 73056 96536 08305 71590 24277
49342 26866 18396 63091 85433 46251 45374 84258 65598 23862 35046 02922 75078 01410 90716 33484 39547 78109 33972 60096
90967 70918 43944 55575 42211 15477 34376 02069 79650 05708 78849 93478 01297 72778 78532 80743 22345 54020 93517 18023
10429 92316 75884 32457 03610 41108 50960 43976 90384 50365 51402 23496 25383 66575 12071 69661 69715 27322 36111 92684
64547 51701 73452 70113 79148 17510 78208 21297 62894 67956 31098 96076 74922 50494 83425 40733 34414 12192 78339 30461
53921 25289 32010 72613 66892 93688 81566 54916 71395 17471 04526 63709 17575 36037 74156 85576 65153 13827 61372 72816
96692 63352 96663 63787 28653 97699 41609 10777 71835 93336 00268 01245 17633 45149 04395 9A324 823A3 64572 51219 40639
14326 35639 22560 45560 42396 00430 77993 61927 37990 05864 00420 76309 23208 13392 26249 29420 76312 93326 80338 1A471
55525 58206 39308 88994 86655 70202 40381 58563 13578 94977 97670 46261 84532 79567 29767 28920 52623 11792 01478 62478
13331 83401 50844 75386 76052 66122 17340 57972 12374 144A5 80372 53554 63022 00953 63010 08145 86752 47046 0461A 86203
90935 55206 19532 82409 51895 10704 07932 84825 09546 25301 51872 82399 71717 64140 66331 58043 09008 61194 25783 80931
06474 89915 94407 47632 84377 e5848 82542 39211 70614 93829 40294 83257 16297 92993 88940 69587 73754 48948 0A110 83452
93394 32780 84527 29789 83413 51401 93912 41966 17994 88795 21032 82381 12742 21870 06345 41149 74345 72A72 32843 42636
93488 04878 99347 19624 03393 96785 76761 50371 60019 66502 52168 25011 77931 784A8 01200 05054 22821 36255 05205 09209
72445 98958 52366 82747 78516 19190 50325 48531 15029 40313 21789 89005 19575 11943 01340 27728 27343 90683 65112 05878
95060 19875 31218 82187 78865 70240 07291 78418 65185 89977 78851 03067 43945 89610 86452 5A766 41569 2A256 64174 47061
61533 05144 85227 38845 49635 05925 54106 06458 42732 34641 09506 68763 63144 47514 26909 49329 53219 92421 25946 95157
65500 91585 21173 42092 32758 82063 32762 54086 17963 03296 20335 72563 55360 40560 97832 11154 75399 08908 43381 69197
47615 81716 16066 20557 30700 03771 94730 01343 18155 60750 15902 78421 64901 42254 45712 24646 93679 32349 70894 95466
84254 36412 34778 53761 94310 03013 90805 68383 42077 26286 18722 64610 97075 06566 92810 28000 33941 7n434 39919 62002
05979 45655 27774 91388 32377 56792 72006 55437 68640 79217 74415 59278 27235 08230 92843 6A353 43966 7915n 22967 61018
34243 78782 04200 87274 02861 72126 84576 38873 36057 69491 22410 98265 92577 36066 62414 67280 15898 86n55 23486 34588
08822 27855 50570 63092 76349 41503 45476 77180 61829 63528 66263 00550 92222 54318 45976 81941 26777 60304 74603 44175
58102 92983 20171 22635 52344 39676 81630 99191 27574 20633 48077 19021 87541 38915 80871 52904 91878 29308 41213 34009
10419 75631 30215 40478 43660 41784 46757 73899 86320 83586 20799 22340 85162 63437 54067 71169 70732 32139 8A2A4 94377
91221 71985 95360 58979 02291 78176 82865 48287 87818 04150 60635 66004 71641 04095 4A377 72017 37448 87332 46685 5n43n
69582 62103 04316 33638 53113 84093 49002 13323 72463 44337 39774 274n5 89667 38275 44203 12857 48745 8196n 33523 2nn56
37229 31959 23692 88171 37527 67022 60450 91173 50695 04025 01666 77552 14932 07364 36541 99488 47701 03639 09372 00575
78999 89580 77577 51266 21113 05790 57174 49417 22201 60705 30243 91611 67059 90451 30425 62063 18240 15898 86n55 23486 34588
43054 97722 39514 96482 16018 38628 86144 63019 36017 71054 67775 0319n 26303 09947 47397 61857 62073 73447 72544 14271
35362 42636 08636 69327 15763 59830 45447 97181 67188 01639 86954 75251 46305 65557 18437 17916 87566 91403 20724 97856
85867 18527 58660 24396 02335 28351 39449 80064 32703 027A1 04224 14497 18836 80541 68978 47962 67391 47608 76963 92191
```

FIGURE 5 The twenty-third Mersenne prime, $2^{11,213} - 1$.

EXERCISES

1 Let the first nine letters of the alphabet assume values 1 to 9, the next nine (j to r) multiples of 10 (10 to 90), and the last eight (s to z) multiples of 100 (100 to 800). What numerical values do your first name, surname, and the way you officially sign your name have?

2 What sex does the phrase "even number" conjure up? What about "odd number"?

3 What number is referred to as "the number of the beast"?

4 Define *triskaidekaphobia*. (You will need a good dictionary.)

Problems 5 and 6 deal with pythagorean triples.

5   Which of the following triples are pythagorean triples?
   (*a*)   (15, 20, 25)   (*c*)   (3, 5, 4)
   (*b*)   (4, 5, 6)   (*d*)   (9, 40, 41)

6   Find the missing entry which makes the following triples pythagorean.
   (*a*)   (6, 8, __)
   (*b*)   (5, __, 13)
   (*c*)   (__, 40, 41)

Problems 7 to 10 deal with triangular numbers.

7   A *menorah* is a Jewish candelabra used at Hanukkah. It has places for nine candles, one of which, the *shammos,* is special and is placed in the center. Hanukkah is celebrated for eight nights. On the first night one regular candle is lit, on the second two are lit, on the third three, etc. What is the total number of regular candles lit during the Hanukkah holiday? Which triangular number is this?

8   Which triangular number represents the number of bowling pins used in a standard game?

9   On the twelfth day of Christmas a young man gave a young woman 1 partridge, 2 turtle doves, 3 French hens, . . . , up to 12 drummers drumming. How many distinct items did he give her? Which triangular number is this?

10   Which triangular number represents the number of pool balls used in a standard game?

Problems 11 to 13 deal with perfect numbers.

11   In the text we used the procedure from Euclid's "Elements" to produce the perfect numbers 6 and 28. Use it to produce the next perfect number, 496.

12   In Heller's "Catch-22" the age of the protagonist, Yossarian, is 28. In what way is this the perfect choice for an age?

13   Calculate the digital root of the perfect numbers in Table 2. Do you see a pattern forming?

Problems 14 to 22 are concerned with pattern recognition. They involve the types of numbers encountered in this section.

14   Express the first three perfect numbers in the binary system. Do you see a pattern forming?

15   Perform the following additions:

$$T_1 + T_2 = \qquad\qquad T_4 + T_5 =$$
$$T_2 + T_3 = \qquad\qquad T_5 + T_6 =$$
$$T_3 + T_4 =$$

Do you see a pattern? Use T_{n-1} and T_n to express your guess as to the makeup of the pattern. Use pythagorean pebbles to try to prove your guess.

16 Substitute some pairs of natural numbers into the "law"

$$(a + b)^2 = a^2 + b^2$$

Does equality hold? (This is the first occurrence of the so-called universal law of linearity. Two other cases will appear in subsequent exercises. Altogether these three instances comprise the most frequent occurrences of arithmetical errors which seem to occur under stressful situations, such as tests.)

17 Here is another method to arrive at T_8. Consider the square of pythagorean pebbles with 8 pebbles on each side, so the square has 64 in all. Remove the 8 pebbles on the main diagonal. There are 56 ($= 64 - 8$) pebbles remaining, with half below the main diagonal and half above it. Hence there are $\frac{1}{2} \cdot 56 = 28$ below the main diagonal. (This is T_7.) If we add the main diagonal to the pebbles below it, we get the triangle of pebbles for T_8. Thus $T_8 = 28 + 8 = 36$. Is this process indigenous to T_8 or does it work for all T_n?

18 Use the pattern described in Problem 17 to find a formula for T_n involving just n.

19 In this problem we will look at the relationship between the number of points on a circle and the regions into which the circle is divided by lines connecting the points. For example; if we mark just one point on the circle, there is no line to be drawn so there is just one region inside the circle. If we draw two points on the circle and connect them, the line divides the circle into two regions. With three points there are three lines and four regions. Fill in the following table.

Number of points	1	2	3	4	5	6
Number of regions	1	2	4	8		

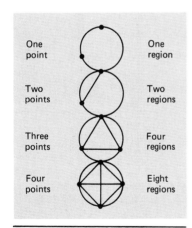

One point — One region

Two points — Two regions

Three points — Four regions

Four points — Eight regions

20 Did Problem 19 seem very simple to you? Did you see the pattern "immediately"? Did you fill in 16 and 32 under 5 and 6, respectively, without actually drawing a circle, inserting the points and lines, and then counting the regions? If you did you got the wrong answer. Notice in Problem 19 we did not ask you to prove that your assertion was correct; in fact, we did not ask you to find a pattern. It looks like the pattern should be the "doubling" pattern where you start with 1 and get each succeeding number by doubling the previous one. That is a nice pattern but it is not the one described in Problem 19 even though they start out identically. What does this exercise say about the relationship between pattern recognition and proof?

Find counterexamples to these conjectures about perfect numbers.

21 They alternately end in 6 and 8.

22 There is one in each interval: 1 to 10, 10 to 100, 100 to 1000, etc.

Amicable Pairs

We have seen that the Greeks studied properties of numbers, such as perfect numbers, and triples of numbers, such as pythagorean triples. The exercises below will examine a property of pairs of numbers. Two numbers form an *amicable pair* if the sum of the proper divisors of each one is equal to the other. They are sometimes called *friendly* numbers.

Example

The sum of the proper divisors of 220 is:

$$1 + 2 + 4 + 5 + 10 + 11 + 20 + 22 + 44 + 55 + 110 = 284$$

The sum of the proper divisors of 284 is

$$1 + 2 + 4 + 71 + 142 = 220$$

Thus 220 and 284 form an amicable pair, so they are friendly numbers. Pythagoras said, "a friend is one who is the other I, such as are 220 and 284."

Do 12 and 16 form an amicable pair? First, the sum of the proper divisors of 12 is

$$1 + 2 + 3 + 4 + 6 = 16$$

However, the sum of the proper divisors of 16 is

$$1 + 2 + 4 + 8 = 15$$

Since the sum does not equal 12, the pair 12 and 16 is not amicable.

Determine which of the following pairs of numbers form amicable pairs.

23 2620, 2924

24 945, 975

25 18, 21

26 12,285; 14,595

27 5020, 5564

28 144, 259

In each of Problems 29 to 34, one number of an amicable pair is given. Find the other number.

29 1184

32 10,744

30 6232

33 6368

31 63,020

34 1210

1-6   ORDER AMID CHAOS: CONJECTURE VERSUS THEOREM

One of life's minor joys is sitting in front of an open hearth and watching the dancing flames flicker. Another is just sitting on a beach and watching the undulating waves, or even just lying there and listening to the pounding surf. A less exotic, though just as pleasant, joy is watching the swirls and arabesques of cream in fresh-poured coffee. All three activities produce graceful relaxation and reflection. Why? Is it because there is a definite order amid all the seeming chaos? Could it be that there are well-formed patterns noticeable only to our subconscious senses?

The search for order and pattern is one of the basic focuses of mathematics. Most people have the impression that advances are hampered by lack of information; yet more frequently just the opposite occurs. You are presented with a superabundance of information. The task confronting you is to winnow out and evaluate what is crucial and simple amid the complexities of the apparent facts.

Consider, for example, Figure 1, which lists the first 1024 natural numbers via a 32 × 32 matrix. The primes are indicated by lighter print. As your eye scans over the primes it detects only chaos. But is there an order underneath? Is there perhaps some pattern which our visual senses are unable to recognize? Are the primes distributed according to some law?

This last question is typical of the kind that mathematicians have posed since the time of Pythagoras. It is very similar to the earlier question about flames, waves, and swirls. The latter question would be posed by a physicist. In fact, it is just this type of question that confounds scientists, whether they be mathematicians, physicists, chemists, political scientists, sociologists, or philosophers, when they engage in their pursuit of knowledge. It is always a question like this that stimulates the search for truth. The next step is the search for a pattern, for order amid the mass of facts drawn from experience.

This section deals with the method that is used by scientists, and in particular mathematicians, to conduct their search for order. We will consider a story about a small note left in the margin of a book by a seventeenth-century lawyer, a pattern formed by a seventeenth-century monk, and Euclid's result concerning how many prime numbers there are. Each of these stories involves a search for a general result concerning natural numbers. In each case a particular set of numbers was studied and a general pattern was derived. Then the next step was to determine if the pattern always holds.

PRIME SERIES

1	**2**	**3**	4	**5**	6	**7**	8	9	10	**11**	12	**13**	14	15	16	**17**	18	**19**	20	21	22	**23**	24	25	26	27	28	**29**	30	**31**	32
33	34	35	36	**37**	38	39	40	**41**	42	**43**	44	45	46	**47**	48	**49**	50	51	52	**53**	54	55	56	57	**58**	**59**	60	61	62	**63**	64
65	**66**	67	**68**	69	70	71	72	**73**	**74**	**75**	**76**	**77**	**78**	79	80	**81**	**82**	83	**84**	**85**	**86**	87	**88**	89	90	91	92	**93**	94	95	**96**
97	**98**	**99**	100	101	102	103	104	**105**	**106**	107	**108**	109	110	**111**	112	113	114	**115**	**116**	117	118	119	120	121	122	123	124	125	126	127	**128**
129	130	131	132	133	134	135	**136**	137	**138**	139	**140**	141	142	143	144	145	146	147	**148**	149	**150**	151	152	153	154	155	**156**	157	**158**	**159**	160
161	162	163	164	165	166	167	**168**	169	170	171	172	173	174	175	176	177	178	179	180	181	**182**	183	**184**	185	**186**	187	**188**	**189**	190	191	**192**
193	194	195	196	197	198	199	200	201	**202**	203	204	205	206	207	208	209	210	211	212	213	214	215	216	217	218	219	220	221	222	223	224
225	226	227	228	229	230	231	232	233	**234**	**235**	**236**	237	**238**	239	240	241	242	243	244	245	246	247	**248**	249	250	**251**	**252**	253	254	**255**	**256**
257	258	**259**	260	261	262	263	264	265	**266**	267	**268**	269	270	271	272	273	274	275	276	277	**278**	279	**280**	281	**282**	283	**284**	**285**	**286**	**287**	**288**
289	**290**	291	292	293	294	295	296	297	**298**	299	300	301	302	303	304	305	306	307	**308**	309	310	311	312	313	314	315	316	317	**318**	319	320
321	322	323	324	325	326	327	**328**	329	330	331	332	333	334	335	336	337	338	339	340	341	342	343	344	345	346	347	**348**	349	350	351	352
353	354	355	356	357	358	359	360	361	362	363	364	365	366	367	368	369	370	371	372	373	374	375	376	377	**378**	379	**380**	**381**	382	383	**384**
385	386	387	388	389	390	391	392	393	394	395	396	397	**398**	399	400	401	402	403	404	405	406	407	408	409	410	411	412	413	414	415	416
417	418	419	420	421	422	423	424	425	426	427	428	429	430	431	432	433	434	435	436	437	438	439	440	441	442	443	444	445	446	**447**	**448**
449	450	451	452	453	454	455	456	457	**458**	**459**	460	461	462	463	464	465	466	467	**468**	469	470	471	472	473	474	475	476	477	**478**	479	**480**
481	482	483	484	485	486	487	**488**	489	490	491	492	493	494	495	496	497	**498**	499	500	501	502	503	504	505	506	507	**508**	509	510	511	512
513	514	515	516	517	518	519	520	521	**522**	523	524	525	526	527	528	529	530	531	532	533	534	535	536	537	**538**	539	540	541	542	543	544
545	546	547	548	549	550	551	552	553	554	555	556	557	**558**	559	560	561	562	563	564	565	566	567	568	569	570	571	572	573	574	575	576
577	**578**	579	580	581	582	583	584	585	**586**	587	**588**	**589**	590	591	592	593	594	595	596	597	**598**	599	600	601	602	603	604	605	606	607	**608**
609	**610**	611	612	613	614	615	616	617	**618**	619	620	621	622	623	624	625	626	627	628	629	630	631	632	633	634	635	636	637	**638**	**639**	**640**
641	642	643	644	645	646	647	648	649	**650**	651	652	653	654	655	656	657	**658**	659	660	661	662	663	664	665	666	667	**668**	669	670	671	672
673	674	675	676	677	678	679	680	681	**682**	683	684	685	686	687	**688**	689	690	691	692	693	694	695	696	697	**698**	699	700	701	702	**703**	**704**
705	706	707	**708**	709	710	711	712	713	714	715	716	717	718	719	720	721	722	723	724	725	726	727	728	729	**730**	731	732	733	734	**735**	**736**
737	**738**	739	**740**	741	742	743	744	745	**746**	747	748	749	750	751	752	753	754	755	756	757	**758**	759	760	761	762	763	764	765	766	767	**768**
769	770	771	772	773	774	775	776	777	**778**	779	**780**	**781**	782	783	784	**785**	786	787	**788**	789	790	791	792	793	794	795	796	797	**798**	799	**800**
801	**802**	**803**	**804**	**805**	**806**	**807**	**808**	809	**810**	811	**812**	**813**	**814**	815	**816**	817	**818**	**819**	**820**	821	**822**	823	**824**	**825**	**826**	827	**828**	829	**830**	**831**	**832**
833	**834**	**835**	**836**	**837**	**838**	839	**840**	**841**	**842**	**843**	**844**	**845**	**846**	**847**	**848**	**849**	**850**	**851**	**852**	853	**854**	**855**	**856**	857	**858**	859	**860**	**861**	**862**	863	**864**
865	**866**	**867**	**868**	**869**	**870**	**871**	**872**	**873**	**874**	**875**	**876**	877	**878**	**879**	**880**	881	**882**	883	**884**	**885**	**886**	887	**888**	**889**	**890**	**891**	**892**	**893**	**894**	**895**	**896**
897	**898**	**899**	**900**	**901**	**902**	**903**	**904**	**905**	**906**	907	**908**	**909**	**910**	911	**912**	**913**	**914**	**915**	**916**	917	**918**	919	**920**	921	**922**	**923**	**924**	**925**	**926**	**927**	**928**
929	**930**	931	**932**	**933**	**934**	**935**	**936**	**937**	**938**	**939**	**940**	941	**942**	**943**	**944**	**945**	**946**	947	**948**	**949**	**950**	**951**	**952**	953	**954**	**955**	**956**	957	**958**	**959**	**960**
961	**962**	**963**	**964**	**965**	**966**	967	**968**	**969**	**970**	971	**972**	**973**	**974**	**975**	**976**	977	**978**	**979**	**980**	**981**	**982**	983	**984**	**985**	**986**	**987**	**988**	**989**	**990**	991	**992**
993	**994**	**995**	**996**	997	**998**	999	1,000	1,001	1,002	1,003	1,004	1,005	1,006	1,007	1,008	1,009	1,010	1,011	1,012	1,013	1,014	1,015	1,016	1,017	1,018	1,019	1,020	1,021	1,022	1,023	1,024

FIGURE 1

Primes

Prime numbers have always held a special fascination for humans. Remember that one column of numbers on the Ishango tally bone consisted only of primes. We have seen that they are the basic building blocks for the natural numbers. Yet their nature eludes us; they have been called "exasperating, unruly numbers" which display an "enigmatic blend of order and haphazardry." This can be seen in the distribution of the primes among the first 1024 numbers as given in Figure 1.

Yet perhaps the reason that we are unable to detect order within the apparent chaos in the distribution of the primes is because we are not viewing them in the proper perspective. It will probably take a civilization that has an entirely different approach to the subject to unravel the mysteries which continue to confound us. For instance, if we arrange the numbers as in Figure 2, consider where the primes are located. From this perspective the primes seem to exhibit an order completely missing in Figure 1.

Mathematicians would like to find a formula which has the property that whenever you plug a number into it, a prime number comes out. Consider, for example, the formula

$$n^2 - n + 41$$

	2	3	4	5	6
7	8	9	10	11	12
13	14	15	16	17	18
19	20	21	22	23	24
25	26	27	28	29	30
31	32	33	34	35	36
37	38	39	40	41	42
43	44	45	46	47	48
49	50	51	52	53	54
55	56	57	58	59	60
61	62	63	64	65	66
67	68	69	70	71	72
73	74	75	76	77	78
79	80	81	82	83	84
85	86	87	88	89	90
91	92	93	94	95	96
97	98	99	100		

FIGURE 2

Each time you insert a value of n from 1 to 40 into it you obtain a prime number. This might lead you to believe that it will always produce primes. Unfortunately if $n = 41$ is inserted then

$$(41)^2 - 41 + 41 = (41)^2$$

which is a composite number, because obviously $41 \mid (41)^2$.

Formulas like $n^2 - n + 41$ are called *polynomials*. It has been proved that no polynomial formula can produce all primes. There do exist other formulas which produce only primes, but they are of little use since the sequence of primes must be known in order to operate on them.

Even though no formula has been found a number of other facts have been discovered concerning primes. One of the first questions raised about primes by the ancient Greeks was a simple one:

QUESTION: How many primes are there?

First of all, either the number of primes is finite or it is infinite. If it is finite perhaps we can count them all. For instance, how many can you count in Figure 2? There are 26. But surely there are many primes greater than 100. And how would we know when to stop looking at larger numbers? The more we look at the question the stronger our inkling is to make the following conjecture:

CONJECTURE: There is an infinite number of primes.

Yet the nagging question remains, even after we have made our guess: How can we be sure that it is true? We will end all suspense by answering it now: The conjecture is true; there is an infinite number of primes. Just like a good novel, where the ending is not so important as how the plot is developed, it is not just the final statement which is of importance here. What *is* important is the method of reasoning.

How does one show that a set of numbers is infinite? First of all, there is no counting process, no special numeration system, that will ever be able to confirm or deny our conjecture. In fact, no counting process will ever be able to determine whether any set is infinite. Another line of reasoning must be used.

Let us look at another example that is a little easier. Take, for example, the set N of natural numbers. You know that N is infinite but suppose someone asks you how you can be sure. You would probably respond along these lines:

> Well, suppose N isn't infinite; then it's finite.
>
> Now I will show you how ridiculous that statement is.
>
> If N is finite then we can count the numbers in N until we reach the largest one.
>
> Suppose the largest number in N is g.
>
> But that's silly because the number that is 1 greater than g, g + 1, is a natural number that is greater than g.
>
> Thus [would you really say "thus"?] no such g can exist and so N cannot be finite.
>
> Therefore N is infinite.   QED.

The method of reasoning you employed is called *reductio ad absurdum*. It means that in order to prove a statement you take the "backdoor approach" of first negating it, that is, of assuming it is false, and then showing that the negation leads logically to a contradiction, or an absurdity. It is a type of reasoning that you have used many times even though you did not know its formal title. It is this type of reasoning that is used to show that a set is infinite, and we will use it now to prove our conjecture about the number of primes being infinite.

We have called it "our" conjecture, but it is usually attributed to Euclid since it appears as a formal theorem in the "Elements." The proof we will present is very similar to his.

Euclid's theorem   There is an infinite number of primes.
Proof

Suppose the number of primes is not infinite.

Then the number of primes is finite.

Then we can count them all and there must be a largest one.

Say p is the largest prime.

Form the product of all the natural numbers up to p; call it P.

Hence

$$P = 1 \cdot 2 \cdot 3 \cdot 4 \cdot \cdots \cdot p$$

Notice that P is larger than p.

Now form the number $Q = P + 1 = 1 \cdot 2 \cdot 3 \cdot 4 \cdot \cdots \cdot p + 1$.

Is Q a prime or a composite?

Since Q is larger than p and p was assumed to be the largest prime, Q cannot be prime.

Thus Q is a composite.

By the fundamental theorem of arithmetic Q has a prime divisor, say q, and q must be less than or equal to p.

Thus q is one of the numbers 1, 2, 3, 4, . . . , or p.

Hence when you divide q into Q you get a remainder of 1.

Therefore q does not divide Q.

This is our contradiction because we had assumed that q divides Q.

So we have proved that there is an infinite number of primes. The proof is not easy, but, on the other hand, it is not so difficult (we hope) that you cannot understand each step, although it may take some time. The primary reason we have included the proof here is to vividly demonstrate that even a simple-looking conjecture needs to be proved before it can be accepted as true. Quite often the mental process used to arrive at the conjecture is vastly different from the reasoning used in the proof.

What have we got so far? The job of the scientist is to look for a pattern in a large set of data. If a pattern is seen, it is called a *conjecture,* which is simply a guess that the pattern always holds. If it can be proved that the conjecture is true, it is then called a *theorem.* Let us now look at some alternate possibilities.

Mersenne Numbers

In our treatment of perfect numbers we had occasion to make use of numbers of the form $2^p - 1$, where p is a prime. We will denote these numbers by M_p. Let us list the first few of them.

$$M_2 = 2^2 - 1 = 3 \qquad M_{11} = 2^{11} - 1 = 2047$$

$$M_3 = 2^3 - 1 = 7 \qquad M_{13} = 2^{13} - 1 = 8191$$

$$M_5 = 2^5 - 1 = 31 \qquad M_{17} = 2^{17} - 1 = 131{,}071$$

$$M_7 = 2^7 - 1 = 127$$

Notice that $M_4 = 2^4 - 1 = 15$ is not one of them because 4 is not prime.

Here is the basic question. Does this list produce only chaos, or is there order to it? Put in an equivalent way: Do you see a pattern forming in the sequence of numbers

3, 7, 31, 127, 2047, 8191, . . .

Asked in yet another way, the question becomes: Do you see any common property among these numbers?

A seventeenth-century Parisian friar, Marin Mersenne, discovered a fascinating property about them. On the basis of the sequence of numbers that he was able to calculate he made the guess:

MERSENNE'S CONJECTURE: If p is a prime number, then $M_p = 2^p - 1$ is also a prime number.

As a result of his work on such numbers they came to be known as *Mersenne numbers*. There is a lesson to be learned from Mersenne's experience. His guess was based on many specific examples, and each one produced a prime; so it was natural for him to feel heady with success. But observing a pattern is one thing; proving that the pattern holds for *all* cases is an entirely different matter. That is why Mersenne's statement was called a "conjecture" instead of a "theorem." In 1876, while the United States was celebrating its centennial in Philadelphia, the Frenchman Eduard Lucas was celebrating a result of an entirely different kind. He showed that M_{67} is composite. This put an end to Mersenne's conjecture. Henceforth Mersenne numbers which are prime came to be known as *Mersenne primes*. However, Lucas was unable to produce any numbers which actually divide M_{67}. Nobody else was able to factor M_{67} either until 1903 when the Columbia University mathematician Frank Nelson Cole presented his "silent lecture" which we described at the beginning of Section 1-4.

The lesson to be learned from the Mersenne conjecture is this: Observing that a property holds for several cases does not mean that it will hold for all cases. We will encounter this lesson again subsequently.

By the way, in the introduction to Section 1-4 we described the singular reaction to Cole's presentation before his peers. We mentioned that upon first reading the story you probably had no appreciation for why they reacted with such unabashed enthusiasm. We also predicted that soon you would have an understanding and appreciation for their surprise and admiration for Cole's accomplishment. Were we right?

Fermat's Theorem

In our development so far we have had several opportunities to deal with sets of numbers in which *some* of them exhibit a certain property. The question then becomes one of determining whether the property holds for *all* the numbers. This question is really just a special case of the procedure for deciding whether a given statement is true. Several possibilities can occur:

1   You can construct a proof of the statement. By a *proof* we mean a logical chain of assertions which convinces the reader that the statement is true beyond any shadow of doubt. The proof of Euclid's theo-

rem provides us with an example of this. We will have much more to say about proofs later on.

2   You are able to find one instance in which the statement does not hold. It is called a *counterexample*. When a counterexample is found, the given statement is said to be false. Thus Mersenne's statement "if p is prime then so is M_p" turned out to be false because Lucas and Cole found the counterexample M_{67}.

There is a third possibility which arises. It is the one which is responsible for the liveliness of the subject.

3   The statement remains unresolved. No counterexample has been found, nor has a proof been constructed.

We will turn to a well-known statement in the history of mathematics that not only illustrates this possibility but indicates the extreme degree of proof that a mathematician requires before accepting a statement as true.

Earlier we considered square numbers. It seems natural to consider cube numbers after square ones. Though there is no record of the Pythagoreans having considered such numbers, they could have visualized them as the number of pebbles on a cube. The first three cube numbers would then look like this:

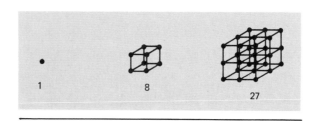

The next few cube numbers are listed below:

$$4^3 = 64 \qquad 5^3 = 125 \qquad 6^3 = 216 \qquad 7^3 = 343$$

If you add two of these numbers do you get another cube number? We saw that this sometimes happens with square numbers:

$$3^2 + 4^2 = 5^2 \qquad 5^2 + 12^2 = 13^2$$

These are examples of pythagorean triples that we considered in the previous section. We say that the triple $(3, 4, 5)$ is a *solution* of the equation

$$x^2 + y^2 = z^2$$

or that (3, 4, 5) *satisfies* the equation. As we saw before, some triples satisfy the equation (in which case they are called pythagorean triples) and others do not.

Is the situation the same for cube numbers? That is, do there exist some triples (x, y, z) that satisfy the equation $x^3 + y^3 = z^3$? Try various combinations of cube numbers and you will see that not one of them satisfies the equation. In fact, if, by chance, you do manage to find three natural numbers x, y, z satisfying

$$x^3 + y^3 = z^3 \tag{C}$$

then we can assure you that you have committed an arithmetical error. The reason for this assertion is that the eighteenth-century mathematician Leonhard Euler proved that no such solution exists. A century earlier Pierre Fermat, a French lawyer who spent his leisure time on mathematical pursuits, wrote in the margin of a Greek mathematics book he was translating that one can *never* find three natural numbers which satisfy Equation (C). In fact Fermat did not stop there; he said that one can never find any triple of natural numbers (x, y, z) such that

$$x^n + y^n = z^n \tag{F}$$

when n is greater than 2. Fermat's note read:

> To divide any power whatever into two powers of the same denomination above the second is impossible, and I have assuredly found an admirable proof of this, but the margin is too narrow to contain it.

So a narrow margin prevented Fermat from writing his proof there, and no such proof has ever been found among his papers. Ordinarily such a loss would not be damaging since eventually some other mathematician would supply the missing proof. Unfortunately, Equation (F) is not ordinary, and no proof of the impossibility of finding solutions has been found in the ensuing 300 years.

The evidence to date indicates that Fermat's assertion is indeed correct. Euler disposed of the case $n = 3$, thus showing that the sum of two cube numbers is never a cube number. Modern computers have disposed of all values of n up to 125,000 or so, and some special cases greater than that have also been decided. But the general question remains unresolved, in spite of the fact that DM 100,000 was bequeathed to the Academy of Science in Gottingen, Germany, in 1908 as a prize for the first complete proof of Fermat's so-called last theorem.

The present disposition of Fermat's "theorem" provides us with the perfect example for making two mathematical points. First, it should *not* be called a theorem because that term is reserved for statements which have been proved. True, the evidence in its favor is overwhelming, but it is circumstantial. And true, as our sociologist friends are fond of telling us, people have been hanged on far less circumstantial evidence. But until either a general proof is constructed or some triple (x, y, z) and natural number n are found to satisfy Equation (F), Fermat's assertion must remain a conjecture.

From a practical standpoint, however, we have to hedge a little, for in a certain sense Fermat's theorem is empirically true. After all, if a solution does indeed exist, then the numbers in it are so inconceivably large that we will not be able to easily deal with them. Besides, whenever Fermat asserted that he had a proof of a result it turns out that he did indeed possess one.

The second point concerns the *timelessness* of mathematical results. When Euler proved that no natural numbers satisfy Equation (C), he did not mean that none existed up to that time or that none could be found using the techniques and machines then in existence. His proof means there is no sense in trying to find triples of cube numbers because there just are not any and there never will be.

Summary

This section was our first encounter with the central mathematical concept of proof. It was concerned with establishing whether an order, detected among chaotic facts, really exists or is merely a smokescreen. Then we considered a conjecture due to Mersenne and settled it by means of a counterexample. (We supplied a proof of the fact that there are infinitely many primes.) Finally we discussed the unsolved Fermat problem, which to date has neither a proof nor a counterexample. It does, however, have a preponderance of evidence on its side, and we used this fact to illustrate the difference between a proof and the existence of circumstantial evidence.

EXERCISES

1 Examine Figure 1. What is the longest sequence of consecutive numbers which are all composite?

2 The formula $n^2 - 79n + 1601$ produces primes for every value of n from 1 to 79. But $n = 80$ produces a composite number. What is the number, and what is its factorization, when $n = 80$?

3 (a) List all divisors of 2^5.
   (b) How many divisors does 2^5 have? How many does 2^{10} have? 2^{99}?
   (c) How many divisors does 2^n have?

4 Compute the Mersenne primes M_{19} and M_{23}.

5 Call a number Z-prime if it is of the form

$$Z_p = 3^p - 1$$

and is prime. Compute Z_2, Z_3, Z_5, and Z_7. Do you see a pattern forming? How many Z-primes are there?

6 Make a conjecture relating triangular and cube numbers. Base it on the following evidence:

$$\underbrace{1}_{1^3} + \underbrace{3 + 5}_{2^3} + \underbrace{7 + 9 + 11}_{3^3} + \underbrace{13 + 15 + 17 + 19}_{4^3} = 100$$

and $1^3 + 2^3 + 3^3 + 4^3 = 10^2 = (1 + 2 + 3 + 4)^2$.

7 It has been conjectured that the cube of every natural number may be written as a difference of squares of natural numbers. Thus

$$2^3 = 3^2 - 1^2 \qquad 3^3 = 6^2 - 3^2 \qquad 4^3 = 10^2 - 6^2$$

Can you express 5^3 and 6^3 as differences of squares?

8 State whether each of the following is true or false. Be prepared to defend your position verbally.
(a) There is an infinite number of even numbers.
(b) There is an infinite number of multiples of 5.
(c) There is a smallest natural number.
(d) There is a largest prime.
(e) There is a smallest prime.
(f) There is a largest even prime.
(g) There is an infinite number of even primes.
(h) There is an infinite number of odd primes.

In Problems 9 to 18 examine each conjecture by looking at some particular examples of the general statement. Consider enough examples to convince yourself that the conjecture is either true or false. Try to prove your assertion. We will demonstrate what is meant with an example.

Example
The sum of two even numbers is even.
(i) We first consider some examples: $2 + 4 = 6$, $4 + 6 = 10$, $6 + 14 = 20$, $18 + 52 = 70$. In each case we get an even number. It looks as though it is a true statement.
(ii) We now try to prove that it always is true. An even number is of the form $2n$ for some integer n and an odd number is of the form $2n + 1$. (Try some examples if you do not follow. In other words, take 54; it is even and is in the form $2n$ where $n = 27$, i.e., $54 = 2 \cdot 27$.) Two arbitrary distinct even numbers can be expressed as $2n$ and $2m$. If we add them we get

$$2n + 2m = 2(n + m)$$

by factoring 2 from each term. Note that $2(n + m)$ is an even number. This proves that the conjecture is true.

9   The sum of two odd numbers is an even number.

10   The sum of an even number and an odd number is an odd number.

11   The product of two even numbers is an even number.

12   The product of two odd numbers is an odd number.

13   If a is an even number, then a^2 is an even number.

14   If a is an odd number, then a^2 is an odd number.

15   If a^2 is an even number, then a is an even number.

16   If a^2 is an odd number, then a is an odd number.

17   Every prime number is an odd number.

18   If a and b are both even and $a \mid b$ then $b \div a$ is an even number.

19   Find a pattern which will enable you to tell how many divisors the number 6^9 will have. (By all means do *not* compute 6^9. It is 10,077,696. Use Problem 3 as a model.)

20   A man has a 5-gallon jug and a 3-gallon jug. He wishes to purchase 4 gallons of wine from an innkeeper who has only a full 8-gallon jug. How does the man measure exactly 4 gallons without spilling a drop?

21   It has recently been conjectured that $2^{132,049} - 1$ is a prime number. Maybe by the time you read this it will have been proved. If it is prime, what is the corresponding perfect number?

Carlo Mauri/Marka

Real Numbers

This chapter is concerned with the evolution and development of various sets of numbers. It can be divided into three parts. The first part (Sections 2-1 and 2-2) deals with fractions, and the second (Sections 2-3 and 2-4) with decimals. These two types of numbers continue to be a source of frustration for many people. By examining their development we hope to show that the usual difficulties encountered with their operations can be overcome. Section 2-5 introduces set theory in the very way that it emerged in the latter part of the nineteenth century.

Section 2-1 examines the way in which fractions were handled in ancient Egypt. It will reinforce your knowledge of fractions and, along the way, introduce the solution of linear equations. Section 2-2 is optional. It extends the addition of two fractions to the addition of infinitely many fractions and leads to infinite series. The theme that binds the section together is an application of infinite series to atomic radiation.

Decimal numbers are described in the next two sections. Section 2-3 is concerned with solving various kinds of rate problems involving decimals by using one of the two types of popular hand-held calculators. It also elucidates the relationship between rational and decimal numbers. Section 2-4 extends decimal numbers to those which are not fractions. It proves that $\sqrt{2}$ is one such number. It is motivated by a tale about the person who discovered the first such number.

Section 2-5 introduces set theory as a tool for organizing the structure of the real number system and for clarifying the relationships among its principal subsets: natural numbers, integers, rational numbers, and irrational numbers. It also recapitulates the evolution of numbers from earliest times to the present. Section 2-6, which is optional, discusses what the concept of proof means to a mathematician, and how it differs from a scientist's concept of proof. The whole matter is brought to bear on the predicament facing the Food and Drug Administration about the potentially harmful affects of caffeine.

2-1 RATIONAL NUMBERS

In 1910, upon his retirement as the chancellor of Brown University, A. B. Chace and his wife took a trip to Egypt, where they came under the spell of that country's unique history. They bought a copy of an ancient papyrus scroll, and its introduction piqued their interest. It began:

> Accurate reckoning. The entrance into the knowledge of all existing things and all obscure secrets. This book was copied in the year 33, in the fourth month of the inundation season, under the majesty of the king of Upper and Lower Egypt, 'A-user-Rê', endowed with life, in likeness to writings of old made in the time of Upper and Lower Egypt, Ne-Ma'et-Rê'. It is the scribe Ahmes who copies this writing.

Any document that promises a knowledge of all existing things and a key to all obscure secrets begs to be read. There was only one problem. It was written in an ancient hieratic script. So Chace set about translating it, and his friends urged him to publish his translation. It appeared in finished form in 1929, some 19 years after his trip to Egypt.

The papyrus is often called the Rhind papyrus after A. Henry Rhind, a Scottish antique dealer who specialized in rare books. Rhind went to Egypt for his health in the winter of 1858. While there he bought a dusty old parchment which had never been translated from its ancient Egyptian script. Moreover, a big gap in it was missing. Rhind himself never did anything with the scroll, and when he died 5 years later it was purchased by the British Museum. Over 50 years later the missing portion turned up in the New York Historical Society. The two pieces were put together, resulting in a scroll 1 foot high and 18 feet long. This was the work that Chace translated. It is sometimes referred to as the Ahmes papyrus in honor of the scribe who copied it.

Many similar papyrus scrolls had been translated before Chace's time. In fact, Rhind's part of this papyrus had been translated into German in 1877. However, much of the mathematics done in ancient times has been discovered only in the relatively recent past. The clue to translating the hieroglyphics, the script in which the papyri were written, was discovered in 1799 when the Rosetta Stone was found near the town of Rashid, which Europeans called Rosetta. This famous document bore a decree in three scripts, Greek, Demotic, and hieroglyphics. A knowledge of Greek enabled Egyptologists to decipher the hieroglyphic writings, and this enabled mathematicians to translate the Rhind papyrus into modern languages.

The Rhind papyrus remains our major source of Egyptian mathematics. It is one of the oldest existing documents devoted exclusively to mathematics. Thus it is surprising that it does not deal just with the natural numbers and their properties. Instead it seems to assume a knowledge of natural numbers and unit fractions, and proceeds to show the first clumsy step in dealing with general fractions. This means that by about 1800 B.C., the period that Ahmes referred to, knowledge of the natural numbers was taken for granted and the ascent of the concept of a fraction had begun.

Unit Fractions

Recall that a *unit fraction* is one whose numerator is 1, such as $\frac{1}{4}$ or $\frac{1}{20}$. Ahmes assumed that the reader was familiar with unit fractions. The method for expressing them was to place an elongated oval sign above the denominator. Thus $\frac{1}{4}$ was written ⬭̅|||| , and $\frac{1}{20}$ as ⬭̅∩∩. The only exception was $\frac{1}{2}$, which had a special symbol. We will use our familiar Hindu-Arabic notation instead of the ancient Egyptian so that we can concentrate on the numbers themselves.

The most curious aspect of this first approach to fractions was that general fractions were not regarded as single entities. Instead they were written as sums of unit fractions. For example, the number $\frac{2}{5}$ was written as $\frac{1}{3} + \frac{1}{15}$.

The 2/n Table

The Rhind papyrus contains no introduction beyond Ahmes' statement about obscure secrets; so we know neither its purpose nor intended audience. It begins with a table of fractions whose numerators are 2. First, a unique symbol for $\frac{2}{3}$ is given; this number turns out to be special. Then it proceeds to $\frac{2}{5}$. The Egyptians had a very cumbersome interpretation of this number, as can be seen from its literal translation:

2 divided by 5

$\frac{1}{3}$ of 5 is 1 $\frac{2}{3}$, $\frac{1}{15}$ of 5 is $\frac{1}{3}$

Believe it or not, this is Ahmes' definition of the number $\frac{2}{5}$. Perhaps you can appreciate how difficult the problem was for the translators of the Rhind papyrus, because even after the symbols had been deciphered, as above, the meaning was still not clear.

Ahmes meant for the reader to focus attention on the two unit fractions $\frac{1}{3}$ and $\frac{1}{15}$. Ignoring the remainder and replacing the comma by a plus sign, we get the way in which $\frac{2}{5}$ was regarded. Namely,

2 divided by 5 "is" $\frac{1}{3} + \frac{1}{15}$

Checking to see if this statement is correct yields

$$\frac{1}{3} + \frac{1}{15} = \frac{5}{5} \cdot \frac{1}{3} + \frac{1}{15}$$

$$= \frac{5}{15} + \frac{1}{15} = \frac{6}{15} = \frac{3 \cdot 2}{3 \cdot 5} = \frac{2}{5}$$

The remaining symbols in the Egyptian interpretation of $\frac{2}{5}$ provide a sort of check on this method of expressing $\frac{2}{5}$ as the sum of two unit fractions, where the space between 1 and $\frac{2}{3}$ is regarded as a plus sign. Thus

$\frac{1}{3}$ of 5 is 1 $\frac{2}{3}$, $\frac{1}{15}$ of 5 is $\frac{1}{3}$

means

$$\frac{1}{3} \cdot 5 = 1 + \frac{2}{3}, \quad \frac{1}{15} \cdot 5 = \frac{1}{3}$$

Ahmes then intended that the reader should recognize that the number 5 mentioned twice is the denominator of the fraction and that the sum of the two latter expressions, $1 + \frac{2}{3}$ and $\frac{1}{3}$, is the numerator. In this case,

$$\left(1 + \frac{2}{3}\right) + \frac{1}{3} = 1 + \frac{3}{3} = 1 + 1 = 2$$

The important fact here is that whereas today we regard $\frac{2}{5}$ as a single irreducible fraction, the Egyptians represented it as a combination of more basic entities, the unit fractions. Of course, we sometimes think of $\frac{2}{5}$ as $\frac{1}{5} + \frac{1}{5}$, but this simplification is not at all what Ahmes had in mind.

It seems as though Ahmes was searching for a system of indecomposable unit fractions that could be regarded as the building blocks for the whole set of fractions in much the same way that the Greeks used prime numbers as the building blocks for natural numbers. As pointed out earlier, the Greeks were very successful in their quest to express all natural numbers in terms of their basic entities, prime numbers. This is manifested by the fundamental theorem of arithmetic.

An interesting question arises: Was Ahmes as successful in expressing rational numbers in terms of unit fractions? The answer will become apparent as we proceed.

In Example 1 we will consider the Egyptian form of three more fractions in Ahmes' table and pick out the unit fractions that comprised their Egyptian form.

Example 1

Problem Find our expression for the fraction and the Egyptian unit fractions from the given description in the Rhind papyrus.

(a) 2 divided by 7

$\frac{1}{4}$ of 7 is 1 $\frac{1}{2}$ $\frac{1}{4}$, $\frac{1}{28}$ of 7 is $\frac{1}{4}$

(b) 2 divided by 9

$\frac{1}{6}$ of 9 is 1 $\frac{1}{2}$, $\frac{1}{18}$ of 9 is $\frac{1}{2}$

(c) 2 divided by 13

$\frac{1}{8}$ of 13 is 1 $\frac{1}{2}$ $\frac{1}{8}$, $\frac{1}{52}$ of 13 is $\frac{1}{4}$, $\frac{1}{104}$ of 13 is $\frac{1}{8}$

Solution (a) The fraction is $\frac{2}{7}$ and the unit fractions are $\frac{1}{4}$ and $\frac{1}{28}$. Checking, we get

$$\frac{1}{4} + \frac{1}{28} = \frac{7 \cdot 1}{7 \cdot 4} + \frac{1}{28} = \frac{7}{28} + \frac{1}{28} = \frac{8}{28} = \frac{4 \cdot 2}{4 \cdot 7} = \frac{2}{7}$$

(*b*) The fraction is $\frac{2}{9}$ and the unit fractions are $\frac{1}{6}$ and $\frac{1}{18}$. Checking yields

$$\frac{1}{6} + \frac{1}{18} = \frac{3 \cdot 1}{3 \cdot 6} + \frac{1}{18} = \frac{3}{18} + \frac{1}{18} = \frac{4}{18} = \frac{2 \cdot 2}{2 \cdot 9} = \frac{2}{9}$$

(*c*) The fraction is $\frac{2}{13}$ and the unit fractions are $\frac{1}{8}$, $\frac{1}{52}$, and $\frac{1}{104}$. In order to check the Egyptian result we must find the least common multiple (lcm) of the three denominators. It is 104 since both 8 and 52 divide 104. We get

$$\frac{1}{8} + \frac{1}{52} + \frac{1}{104} = \frac{13 \cdot 1}{13 \cdot 8} + \frac{2 \cdot 1}{2 \cdot 52} + \frac{1}{104}$$

$$= \frac{13}{104} + \frac{2}{104} + \frac{1}{104} = \frac{16}{104} = \frac{8 \cdot 2}{8 \cdot 13} = \frac{2}{13}$$

A Pattern?

We have said that one of the primary ingredients in mathematics is the search for patterns. If you are in tune with us so far, you probably guessed that we are now going to ask you whether you have found a pattern in the way Ahmes chose the unit fractions. This is important, so we ask:

QUESTION 1: What method did the scribe use to factor a given fraction into unit fractions?

Let us see if we can anticipate how Ahmes generated his factorizations. Consider $\frac{2}{11}$. First we search for a unit fraction just a bit less than $\frac{2}{11}$. Since $\frac{2}{11}$ is slightly greater than $\frac{2}{12} = \frac{1}{6}$, we try $\frac{1}{6}$ as the first unit fraction and then look for a number n such that

$$\frac{2}{11} = \frac{1}{6} + \frac{1}{n}$$

For n we try the lcm of 11 and 6, namely, 66. To see if this choice works we check:

$$\frac{1}{6} + \frac{1}{66} = \frac{11 \cdot 1}{11 \cdot 6} + \frac{1}{66} = \frac{11}{66} + \frac{1}{66} = \frac{12}{66} = \frac{6 \cdot 2}{6 \cdot 11} = \frac{2}{11}$$

and thus we have the factorization $\frac{2}{11} = \frac{1}{6} + \frac{1}{66}$.

Let us see if you can apply this method to another fraction.

Example 2

Problem   Apply the method to the fraction $\frac{2}{9}$.

Solution   The first step entails noting that $\frac{2}{9}$ is a little greater than $\frac{2}{10}$ and then writing

$$\frac{2}{9} = \frac{2}{10} + \frac{1}{n} = \frac{1}{5} + \frac{1}{n}$$

We then choose the lcm of 9 and 5 for n; thus we let $n = 45$. We check to see if this choice works, and we get

$$\frac{1}{5} + \frac{1}{45} = \frac{9 \cdot 1}{9 \cdot 5} + \frac{1}{45}$$

$$= \frac{9}{45} + \frac{1}{45} = \frac{10}{45} = \frac{5 \cdot 2}{5 \cdot 9} = \frac{2}{9}$$

Hence we get the factorization is

$$\frac{2}{9} = \frac{1}{5} + \frac{1}{45}$$

We have obtained a wonderful, delightful way of factoring fractions into sums of unit fractions. Unfortunately it is not the one adopted by Ahmes, because his expression for $\frac{2}{9}$, as we saw in Example 1(*b*), is

$$\frac{2}{9} = \frac{1}{6} + \frac{1}{18}$$

Our attempt at answering Question 1 has led us to two important points. First, we have found two different ways of decomposing the fraction $\frac{2}{9}$ into unit fractions:

$$\frac{2}{9} = \frac{1}{6} + \frac{1}{18} \qquad \text{and} \qquad \frac{2}{9} = \frac{1}{5} + \frac{1}{45}$$

This did not happen when natural numbers were factored into primes. Could it be that the Greeks benefited from the Egyptians' experience?

The second point is that even though we have discovered an interesting way of factoring fractions into unit fractions, to our dismay it does not coincide with Ahmes' method. Thus we are no closer to answering the question posed than we were before.

Just what method did Ahmes employ? Nobody knows. That's right; nobody knows. A common notion seems to be that mathematics consists of a collection of procedures and facts that are well established and known. But here we have an instance where twentieth-century mathematicians have been unable to discover a process used by twentieth-century-B.C. mathematicians. Various approaches to answering Question 1 have been tried, for instance, minimize the number of terms used, or minimize the largest denominator which appears. Several possible algorithms have been constructed, but none to date corresponds to what Ahmes did. We wish that we could answer Question 1, but unfortunately, the answer is not known. In fact, this whole discussion leads us to pose another question.

> QUESTION 2: Did Ahmes employ a general method?

Perhaps he did; perhaps he did not. In any case, Question 2 has to be answered before Question 1 should even be asked. This is typical of a recurring theme in mathematics in which one should sometimes answer first the question of whether a solution exists before one should proceed to search for it. Wouldn't it be ironic if Ahmes had no general procedure in mind at all? Research into this subject continues, but it seems that the only answer lies hidden beneath our feet in yet undiscovered Egyptian ruins.

The $n/10$ Table

The next part of the papyrus is a table consisting of decompositions of the numbers of the form $n/10$ for $n = 1$ to $n = 9$. This list also points out the apparent irregularity with which the Egyptians chose the particular unit fractions to use in their factorizations. The first few entries in the table are straightforward. They read:

1 divided by 10 gives $\frac{1}{10}$

2 divided by 10 gives $\frac{1}{5}$

3 divided by 10 gives $\frac{1}{5}$ $\frac{1}{10}$

where the last line is intended to mean $\frac{3}{10} = \frac{1}{5} + \frac{1}{10}$. The table proceeds as one might expect until the expression for $\frac{7}{10}$. Then Ahmes introduces another quirk by inserting the number $\frac{2}{3}$ as one of his indivisibles. The expression for $\frac{7}{10}$ reads:

7 divided by 10 gives $\frac{2}{3}$ $\frac{1}{30}$

The expression for $\frac{8}{10}$ is even more complicated:

8 divided by 10 gives $\frac{2}{3}$ $\frac{1}{10}$ $\frac{1}{30}$

Recall that $\frac{2}{3}$ was the first entry in the $2/n$ table, and that a special symbol was reserved for it. The way in which $\frac{7}{10}$ and $\frac{8}{10}$ were written shows that $\frac{2}{3}$ was regarded as one of the important building blocks, along with the unit fractions. This second table, consisting of numbers with denominator 10, ends with $\frac{9}{10}$. You will discover in the exercises the way the Egyptians wrote $\frac{9}{10}$ in terms of the fundamental building blocks.

Fractions and Algebra

How did the Egyptians use fractions? We find a hint in the remainder of the Rhind papyrus, which consists of 85 problems and solutions, most of which are very practical, only a few being of a theoretical nature. For example, directly from the papyrus we read:

> How should 8 loaves of bread be divided amongst 10 men?

We would simply say that each man would get $\frac{8}{10}$, or $\frac{4}{5}$, of a loaf. Ahmes' solution reads:

> Each man receives $\quad \frac{2}{3} \quad \frac{1}{10} \quad \frac{1}{30}$.

Later problems involving fractions indicate that the Egyptians understood how to operate with them using some elementary algebraic techniques. For example, let us consider another problem from the papyrus:

> It is said to thee, complete $\quad \frac{2}{3} \quad \frac{1}{15} \quad$ to 1.

Ahmes intended the reader to find that number (the "complete" part) which when added to $\frac{2}{3} + \frac{1}{15}$ equals 1. Using modern notation we would express the problem as an equation in the variable x:

> Find x such that $\frac{2}{3} + \frac{1}{15} + x = 1$.

To solve for x we would first add the two fractions:

$$\frac{2}{3} + \frac{1}{15} = \frac{10}{15} + \frac{1}{15} = \frac{11}{15}$$

Then the equation becomes

$$\frac{11}{15} + x = 1$$

Next we would subtract $\frac{11}{15}$ from each side of the equation, yielding

$$x = 1 - \frac{11}{15}$$

Finally we express 1 as $\frac{15}{15}$ and perform the subtraction:

$$x = 1 - \frac{11}{15} = \frac{15}{15} - \frac{11}{15} = \frac{4}{15}$$

Ahmes' solution was written as $\frac{1}{5}$ $\frac{1}{15}$. Does our answer agree with his?

Let us look at another example of how the Egyptians used elementary algebra.

Example 3
Problem Complete $\frac{2}{3}$ $\frac{1}{30}$ to 1.
Solution We first write this in the form

$$\frac{2}{3} + \frac{1}{30} + x = 1$$

Then, multiplying $\frac{2}{3}$ by $\frac{10}{10}$ and 1 by $\frac{30}{30}$ yields

$$\frac{20}{30} + \frac{1}{30} + x = \frac{30}{30}$$

so

$$\frac{21}{30} + x = \frac{30}{30}$$

The solution is

$$x = \frac{30}{30} - \frac{21}{30} = \frac{9}{30} = \frac{3}{10}$$

Ahmes, however, obtained $\frac{1}{5}$ $\frac{1}{10}$ as a solution. Is his solution the same as ours?

We indicated above that Ahmes regarded $\frac{8}{10}$ as $\frac{2}{3}$ $\frac{1}{10}$ $\frac{1}{30}$. The next example will illustrate one way of obtaining this decomposition.

Example 4
Problem Solve $\frac{8}{10} = \frac{2}{3} + x$.
Solution We get

$$x = \frac{8}{10} - \frac{2}{3} = \frac{24}{30} - \frac{20}{30} = \frac{4}{30} = \frac{2}{15}$$

Thus we would write the answer as $\frac{2}{15}$. Of course, Ahmes had decomposed $\frac{2}{15}$ into $\frac{1}{10}$ $\frac{1}{30}$ in his $2/n$ table. That is why his final form for $\frac{8}{10}$ reads $\frac{2}{3}$ $\frac{1}{10}$ $\frac{1}{30}$.

A Rational
Historical Development

The Rhind papyrus gives us a clue as to the early attempts to understand the concept of a fraction. It seems as though the ancient Egyptians had a clear notion of a fraction only if it were in the form $1/n$, but their perception of other types of fractions was very cumbersome. Fractions of the form $2/n$ for large values of odd positive integers were expressed as the sum of as many as four unit fractions.

Not only was it difficult to comprehend such unwieldy expressions, but operating with them must have been mind-boggling. Contrast this with the fact that the Egyptians were expert engineers, erecting ageless monuments and canals, and you must be impressed with their arithmetic skill.

From these early attempts to understand fractions the progress in the evolution of rational numbers was slow. In the exercises we will see some instances of the continued use of unit fractions in Greek documents dating to A.D. 700. The idea of a general fraction of the form that we would express as n/m came about as the result of another symbiotic relationship between the Arabs and the Hindus. The Hindus invented the notation $\frac{3}{4}$ and the Arabs, around A.D. 600, introduced the bar to produce our present $\frac{3}{4}$.

The word "fraction" is derived from the Latin word *frangere,* which means "to break". Can you see why historically this word was used? Today we have other ways of expressing fractions. For example, the fraction $\frac{1}{2}$ is also equal to the decimal expression 0.5. Hence we refer to such numbers as *rational numbers,* where the word "rational" comes from the root "ratio." More formally, we adopt the following definition which will be extended in Section 2-5 to include negative numbers.

DEFINITION: A *rational number* is any number that can be expressed as the ratio a/b of two natural numbers a and b, where $b \neq 0$.

Perhaps the most significant step in the ascent of rational numbers was the development of decimal notation. Decimals have become especially important with the advent of the computer revolution and the ubiquitous hand-held calculator. We will discuss them in Section 2-3.

One Final Note
in Defense of Ahmes

There are a number of possible explanations for why the Egyptians chose their particular approach to fractions. One is notational. If you only consider unit fractions then each fraction requires only one number, the denominator, in its expression. Another advantage is that the Egyptian system of regarding nonunit fractions as the sum of units provides a convenient way to compare the relative sizes of rational numbers.

For example, their expressions for $\frac{7}{10}$ and $\frac{8}{10}$ are

$$\frac{7}{10} = \frac{2}{3} + \frac{1}{30} \qquad \frac{8}{10} = \frac{2}{3} + \frac{1}{10} + \frac{1}{30}$$

The expressions are the same except that the decomposition for $\frac{8}{10}$ contains the additional term $\frac{1}{10}$. Thus it is obvious that $\frac{8}{10}$ differs from $\frac{7}{10}$ by $\frac{1}{10}$.

Even today this type of comparison is often useful. In the next section we will show you how a modern branch of mathematics that uses fractions in a way very similar to the Egyptians describes an extremely important and timely subject, radioactivity.

In Section 2-3 we will show how the Egyptian method of using only one number to express fractions is not as backward as it seems at first glance. In fact, an idea analogous to it led to a major step in the evolution of number in the late sixteenth century.

EXERCISES

In Problems 1 to 4 find our expression for the fraction and the Egyptian unit fractions.

1   2 divided by 11

$$\frac{1}{6} \text{ of } 11 \text{ is } 1 \quad \frac{2}{3} \quad \frac{1}{6}, \quad \frac{1}{66} \text{ of } 11 \text{ is } \frac{1}{6}$$

2   2 divided by 23

$$\frac{1}{12} \text{ of } 23 \text{ is } 1 \quad \frac{2}{3} \quad \frac{1}{4}, \quad \frac{1}{276} \text{ of } 23 \text{ is } \frac{1}{12}$$

3   2 divided by 19

$$\frac{1}{12} \text{ of } 19 \text{ is } 1 \quad \frac{1}{2} \quad \frac{1}{12}, \quad \frac{1}{76} \text{ of } 19 \text{ is } \frac{1}{4}, \quad \frac{1}{114} \text{ of } 19 \text{ is } \frac{1}{6}$$

4   2 divided by 31

$$\frac{1}{20} \text{ of } 31 \text{ is } 1 \quad \frac{1}{2} \quad \frac{1}{20}, \quad \frac{1}{124} \text{ of } 31 \text{ is } \frac{1}{4}, \quad \frac{1}{155} \text{ of } 31 \text{ is } \frac{1}{5}$$

In Problems 5 to 8 apply the method shown in Example 2 to factor the fraction into two unit fractions.

5 $\dfrac{2}{7}$

6 $\dfrac{2}{5}$

7 $\dfrac{2}{13}$

8 $\dfrac{2}{15}$

9 Complete $\frac{2}{3}$ $\frac{1}{5}$ to 1.

10 Complete $\frac{1}{4}$ $\frac{1}{8}$ $\frac{1}{10}$ $\frac{1}{30}$ $\frac{1}{45}$ to $\frac{2}{3}$.

In Problems 11 to 16, solve for the unknown x.

11 $\dfrac{7}{10} = \dfrac{2}{3} + x$

14 $\dfrac{2}{5} = \dfrac{1}{3} + x$

12 $\dfrac{11}{10} = \dfrac{2}{3} + x$

15 $\dfrac{3}{4} = \dfrac{2}{3} + x$

13 $\dfrac{9}{10} = \dfrac{2}{3} + x$

16 $\dfrac{5}{7} = \dfrac{2}{3} + x$

17 The $2/n$ table includes only odd numbers for n. Why do you suppose the even numbers for n were not listed?

18 The decomposition of $\frac{2}{101}$ into fractions shows one way of decomposing each fraction in the $2/n$ table into four unit fractions:

$$\frac{2}{101} = \frac{1}{101} \quad \frac{1}{202} \quad \frac{1}{303} \quad \frac{1}{606}$$

(*a*) Use this method to decompose $\frac{2}{29}$ into four unit fractions.
(*b*) Does this method yield the same result that Ahmes gave in the $2/n$ table?

$$\frac{2}{29} = \frac{1}{24} \quad \frac{1}{58} \quad \frac{1}{174} \quad \frac{1}{232}$$

Problems 19 to 26 are taken from Greek papyri that date from the third century B.C. onward. They reveal the striking uniformity in the methods used by Egyptian scribes of 1800 B.C. and Greek scribes of the Greco-Roman period. They attest to a remarkable continuity in computational tradition and reflect the strong element of conservatism implicit in a technical tradition. In Problems 19 to 24 the decompositions of fractions are given. Determine which fraction each one is.

19 $\dfrac{1}{7}$ $\dfrac{1}{91}$

22 $\dfrac{1}{2}$ $\dfrac{1}{3}$ $\dfrac{1}{13}$

20 $\dfrac{1}{10}$ $\dfrac{1}{190}$

23 $\dfrac{1}{3}$ $\dfrac{1}{14}$ $\dfrac{1}{42}$

21 $\dfrac{1}{2}$ $\dfrac{1}{5}$

24 $\dfrac{2}{3}$ $\dfrac{1}{21}$

25 (*a*) Subtract $\frac{1}{4}$ $\frac{1}{44}$ from $\frac{1}{2}$ $\frac{1}{4}$.
(*b*) The scribe obtains $\frac{1}{3}$ $\frac{1}{11}$ $\frac{1}{33}$ $\frac{1}{44}$. Is this correct?

26 What unit fraction has the decomposition $\frac{1}{55}$ $\frac{1}{70}$ $\frac{1}{77}$?

Problems 27 to 34 deal with various aspects of unit fractions.

27 An article from the September 1964 issue of *Scientific American* read: "Today most people probably could not be trusted to add $\frac{1}{4}$ and $\frac{1}{3}$ correctly." What is the correct sum?

28 The text states that "$\frac{2}{11}$ is slightly larger than $\frac{2}{12}$." Can you show why this is true without resorting to decimals?

29 Give an example of a fraction between $\frac{1}{14}$ and $\frac{1}{15}$.

30 Find four different unit fractions whose sum is 1.

31 This problem relates Egyptian unit fractions to pythagorean perfect numbers.
   (*a*) The divisors of 6 except 1 are 2, 3, 6. What is $\frac{1}{2} + \frac{1}{3} + \frac{1}{6}$?
   (*b*) The divisors of 28 except 1 are 2, 4, 7, 14, 28. What is $\frac{1}{2} + \frac{1}{4} + \frac{1}{7} + \frac{1}{14} + \frac{1}{28}$?
   (*c*) Make a conjecture based upon parts (*a*) and (*b*).
   (*d*) Test your conjecture on 496.

32 Find four numbers *a*, *b*, *c*, and *d* satisfying $\dfrac{a}{b} + \dfrac{c}{d} = \dfrac{a + c}{b + d}$.

33 Solve the following problem, which appeared in the Rhind papyrus. "If each of 10 men receives $\frac{2}{3}$ $\frac{1}{5}$ $\frac{1}{30}$ loaves of bread, how many loaves are there altogether?"

34 What do the solutions to Exercises 13 and 33 have in common?

Problems 35 to 48 deal with a wide variety of settings in which a knowledge of rational numbers is important.

35 *(Political science)* In order for legislation to pass in the Senate it is necessary for at least $\frac{2}{3}$ of the senators to vote for it. What number is minimum for assuring passage?

36 *(Demographics)* At present, every fourth person on earth is Chinese. Of the remainder, 1 out of 5 is Indian. What fraction of the earth's population is Indian?

37 *(Pharmacy)* The following formula is used to calculate a child's dosage for a drug, where *A* is the child's age.

$$\text{Child's dosage} = \frac{A \cdot (\text{adult dosage})}{A + 12}$$

Find the dose of digoxin for an 8-year-old if the adult dose is $\frac{1}{4}$ milligram.

38 *(Farming)* If it takes 25 seconds to discharge 500 cubic centimeters from a small sprayer, how long should you spray to apply 420 cubic centimeters?

39 *(Pediatrics)* A 2-year-old male baby is said to be of average length if he is between $33\frac{3}{4}$ and $35\frac{1}{4}$ inches long. A "rule of thumb" says that one's height at maturity is approximately equal to double the length at age 2. What then is the average height of a mature male?

40 *(Nursing)* A doctor orders $\frac{1}{450}$ grain of a drug for a patient. The nurse has a vial labeled $\frac{1}{300}$ grain per 2 milliliters and a device that only measures in minims. Given that 1 milliliter = 15 minims, how many minims does the nurse administer?

41 *(Fitness)* The standard way to determine your physical-fitness category is to measure how far you can run and/or walk in 12 minutes. You are in fair shape if you can go $1\frac{1}{4}$ miles and good if $1\frac{1}{2}$ miles. For each category determine what your average pace in terms of minutes per mile must be.

42 *(Fitness)* Jogger Z runs 8 miles at a $7\frac{1}{2}$- minutes-per-mile pace, and jogger C runs $7\frac{1}{2}$ miles at an 8-minutes-per-mile pace.
(*a*)   Who runs the longest amount of time?
(*b*)   What "law" does this illustrate?

43 *(Sports)* Franklin Jacobs held the world's high-jump record at 7 feet $7\frac{1}{4}$ inches. He is 5 feet 8 inches tall. How much higher was the bar set over his head?

44 *(Sports)* Alberto Salazar runs a marathon (about 26 miles) in around 2 hours and 10 minutes. What is his fractional minutes-per-mile pace?

45 *(Geography)* A travel article describing an 18-day journey on the Colorado River through the Grand Canyon on a wooden, motorless dory said, "The river drops an average of 7 feet a mile—2200 feet over the 280-mile course we had set for ourselves." Is this the precise average or an approximation?

46 *(Sales)* Joan Didion's best seller "The White Album" is a collection of essays about the 1960s. A computer analysis of its initial sales showed that "two-thirds of its sales are in the East with another quarter concentrated on the West Coast." This is surprising because the book pays special attention to California. What fraction was sold in the rest of the country?

47 *(Education)* Professor Morris Kline wrote, "Mathematics education has been a debacle. One need only ask a typical college graduate how much two-thirds of three-quarters is." Well?

48 *(Suds)* The United States is the largest exporter of beer. However, we are only tenth in per capita consumption, averaging 16 gallons per person per year. How many 12-ounce cans is this?

2-2   INFINITE SUMS AND RADIOACTIVITY

For years nuclear energy was hailed as the answer to humanity's fuel shortage and pollution problems. Then the accident at the Three Mile Island nuclear plant occurred. It set off a bitter debate about radiation hazards and led to numerous protests at several proposed sites of nuclear power plants around the world.

    A major concern over nuclear reactors is the threat of radioactivity, both from its release into the atmosphere due to a reactor malfunction and the long-term threat of the radioactive wastes. It is the latter hazard that many people do not fully comprehend. What does it mean to say that the nuclear by-product of a reactor will "never" disappear? The answer to this question relies on an aspect of mathematics called *infinite sums*. This type of analytical skill will not only help us understand radioactive decay but it will help us to recognize patterns and think abstractly. It is also interesting to see how the ancient Egyptian view of fractions is similar to this type of mathematics in the way it uses the sum of fractions.

The Perils of Radioactivity

A radioactive element is one that continuously decays into new elements. In the process of decay high-energy radiation is given off. When emissions from a radioactive substance enter the human body, they injure cells by tearing electrons away from atoms. This process is formally called *ionization*. If the damage is slight, or takes place slowly, the body usually makes repairs. But if the damage is great, adequate repairs are impossible and the biological consequences can be severe. Certain parts of the body, such as the gonads, thyroid, and bone marrow, are especially sensitive to radiation.

Clearly, ionizing radiation is something to avoid if at all possible. Yet you are bathed in low-level radiation all the time, some of it from natural sources and some from human-made ones. For instance, cosmic rays from space subject you to about 40 millirems a year at sea level and even more at higher altitudes. (A rem is the standard unit of radiation exposure. A millirem is $\frac{1}{1000}$ rem.) Altogether the average person receives a dose of about 200 millirems per year, half from natural sources (including those in your food, water, and soil) and half from human-made ones (x-rays, TV sets, and even certain wristwatches).

How much radiation does it take to hurt you? Radiobiologists have found that a single dose, to the whole body, of 10 rems can damage the lymph nodes and spleen and decrease the bone marrow and blood cells; 100 whole-body rems can cause radiation sickness; and 600 rems is lethal to most people. However, many scientists believe it is prudent to assume there is no safe level of ionizing radiation, even though we are constantly being exposed to it.

Do Two Half-Lives Make a Whole?

Humans use various standard units of measure to organize data. For instance, geologists use ages and historians centuries. Similarly, scientists use half-life as the standard unit for measuring radioactivity. The *half-life* of a radioactive element is the time it takes for one-half of the element to decay. Take, for example, iodine 131, which is denoted by I 131. It has a half-life of about 8 days, so half of any given amount of it will decay in 8 days. Then half the remaining portion will decay in the next 8 days, and this process will continue throughout every 8-day period. We have chosen I 131 to illustrate radioactivity because it is produced in nuclear reactors and is known to cause cancer of the thyroid if it enters the human food chain.

Example 1

The half-life of I 131 is 8 days. We want to determine what fraction of it will decay in 16 days. The period of 16 days corresponds to 2 half-lives. By definition, $\frac{1}{2}$ of the I 131 will decay in the first half-life. Then $\frac{1}{2}$ of this portion will decay in the second half-life, or $\frac{1}{2} \cdot \frac{1}{2} = \frac{1}{4}$ of it. This means that altogether $\frac{1}{2} + \frac{1}{4} = \frac{3}{4}$ will decay after 2 half-lives, so $\frac{1}{4}$ ($= 1 - \frac{3}{4}$) will remain.

This example addresses the common fallacy that a radioactive element will decay entirely after 2 half-lives. It shows that in fact $\frac{1}{4}$ will remain. The next example extends this to 3 half-lives.

Example 2

Problem   What fraction of I 131 will decay in 3 half-lives? What fraction will remain?

Solution   We know from Example 1 that $\frac{3}{4}$ will decay after 2 half-lives. Now $\frac{1}{2}$ of the remaining $\frac{1}{4}$ will decay in the next half-life. That is, $\frac{1}{2} \cdot \frac{1}{4} = \frac{1}{8}$ of the original amount will decay in this third half-life. Altogether, the fraction of decay in 3 half-lives is

$$\frac{1}{2} + \frac{1}{4} + \frac{1}{8} = \frac{7}{8}$$

Thus the fraction of I 131 that remains is $\frac{1}{8}$.

The graph below might help you to see these results a little more vividly. The letters D and R stand for decay and remain, respectively.

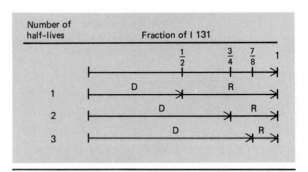

Examples 1 and 2 show that the fraction of the remaining part of a radioactive element is directly related to the number of half-lives. Can you see relationships in Table 1 between half-lives and remainder, and between half-lives and decay?

The relationships in Table 1 are easier to see if exponents are used. For instance, after 5 half-lives the fraction that remains will be 1/32 which is $1/2^5$. Here is the general form.

TABLE 1

Half-Lives of I 131

Time (Days)	Half-Lives	Remainder	Amount Decayed
0	0	1	0
8	1	$\frac{1}{2}$	$\frac{1}{2}$
16	2	$\frac{1}{4}$	$\frac{1}{2} + \frac{1}{4}$
24	3	$\frac{1}{8}$	$\frac{1}{2} + \frac{1}{4} + \frac{1}{8}$
32	4	$\frac{1}{16}$	$\frac{1}{2} + \frac{1}{4} + \frac{1}{8} + \frac{1}{16}$
40	5	$\frac{1}{32}$	$\frac{1}{2} + \frac{1}{4} + \frac{1}{8} + \frac{1}{16} + \frac{1}{32}$

> The fraction of a radioactive element remaining after n half-lives is $1/2^n$.

The fraction of decay after 5 half-lives is $1 - \frac{1}{32} = \frac{31}{32}$. This is one way of obtaining the decay, but we will describe another way because it leads to a very useful formula.

A Useful Formula

Following the pattern established in the fourth column of Table 1, the fraction of decay of I 131 after 10 half-lives will be the sum

$$F = \tfrac{1}{2} + (\tfrac{1}{2})^2 + (\tfrac{1}{2})^3 + \cdots + (\tfrac{1}{2})^{10}$$

We will describe how to evaluate this sum by forming only one power of $\frac{1}{2}$ instead of forming all of its powers up to 10.

The method involves three steps initially. First write the given formula on one line. Multiply both sides of it by $\frac{1}{2}$, and write the new formula below it. Then subtract the respective sides of the new formula from the original, and write it on the third line.

$$F = \tfrac{1}{2} + (\tfrac{1}{2})^2 + (\tfrac{1}{2})^3 + \cdots + (\tfrac{1}{2})^{10}$$
$$\tfrac{1}{2}F = \qquad (\tfrac{1}{2})^2 + (\tfrac{1}{2})^3 + \cdots + (\tfrac{1}{2})^{10} + (\tfrac{1}{2})^{11}$$
$$\overline{F - \tfrac{1}{2}F = \tfrac{1}{2} \qquad\qquad\qquad\qquad\qquad - (\tfrac{1}{2})^{11}}$$

Now factor F from the left-hand side:

$$F(1 - \tfrac{1}{2}) = \tfrac{1}{2} - (\tfrac{1}{2})^{11}$$

Dividing each side by $1 - \tfrac{1}{2}$ yields the formula

$$F = \frac{\tfrac{1}{2} - (\tfrac{1}{2})^{11}}{1 - \tfrac{1}{2}}$$

This formula for evaluating F is much easier than the original one because only one power, the eleventh, has to be determined.

The last formula for F is a special case for computing the following sum. We denote it by $S(r, n)$ because it stands for the sum of the powers of r up to n.

$$S(r, n) = r + r^2 + r^3 + \cdots + r^n$$

The formula suggests that a more effective means of computation is the formula

$$S(r, n) = \frac{r - r^{n+1}}{1 - r}$$

For the case above, we had $r = \tfrac{1}{2}$ and $n = 10$ so

$$S(\tfrac{1}{2}, 10) = \frac{\tfrac{1}{2} - (\tfrac{1}{2})^{11}}{1 - \tfrac{1}{2}}$$

The next two examples illustrate this formula. The first illustrates the fact that $S(\tfrac{1}{2}, n)$ gives the fraction of a radioactive element that decays after n half-lives.

Example 3
Problem   What fraction of I 131 will decay after 8 half-lives?
Solution   This can be obtained in two ways. One way is to make use of the fact that $S(\tfrac{1}{2}, 8)$ is the desired fraction. Then

$$S(\tfrac{1}{2}, 8) = \frac{\tfrac{1}{2} - (\tfrac{1}{2})^9}{1 - \tfrac{1}{2}} = \frac{\tfrac{1}{2} - \tfrac{1}{512}}{\tfrac{1}{2}} = \frac{\tfrac{256}{512} - \tfrac{1}{512}}{\tfrac{256}{512}} = \frac{\tfrac{255}{512}}{\tfrac{256}{512}} = \frac{255}{256}$$

The other way is to compute the fraction that remains, $(\tfrac{1}{2})^8 = \tfrac{1}{256}$. Then the fraction that decays is $1 - \tfrac{1}{256} = \tfrac{255}{256}$.

Example 4

Problem Compute $\frac{2}{3} + (\frac{2}{3})^2 + \cdots + (\frac{2}{3})^9$.

Solution The notation for this sum is $S(\frac{2}{3}, 9)$. The formula yields

$$S(\tfrac{2}{3}, 9) = \frac{\frac{2}{3} - (\frac{2}{3})^{10}}{1 - \frac{2}{3}}$$

This becomes

$$S(\tfrac{2}{3}, 9) = \frac{\frac{2}{3} - 1024/59,049}{\frac{1}{3}} = \frac{39,366/59,049 - 1024/59,049}{19,683/59,049} = \frac{38,342}{19,683}$$

which is slightly less than 2.

The next example shows that the formula can be used for values of *r* greater than 1.

Example 5

Problem Compute $5 + 25 + 125 + \cdots + 5^{10}$.

Solution The formula yields

$$S(5, 10) = \frac{5 - 5^{11}}{1 - 5} = \frac{-48,828,120}{-4} = 12,207,030$$

Infinite Sums

Consider the following question, whose answer is crucial for the disposal of atomic waste.

QUESTION: Does I 131 ever lose its radioactivity?

Let us approach this question by considering two time periods, $\frac{1}{2}$ year and 10 years. Since the half-life of I 131 is 8 days, there will be about 23 half-lives in $\frac{1}{2}$ year. Thus after $\frac{1}{2}$ year $1/2^{23}$, or about $1/67,108,864$ will remain. This is a rather small amount, whose decimal equivalent is roughly 0.0000000149. Although this is a very small portion of the original, the important point is that *some* radioactivity still exists.

What fraction will remain after 10 years? There are about 460 half-lives in 10 years, so the fraction remaining is $1/2^{460}$. This number is so small that no computer can print its decimal equivalent. Again, the portion remaining is minuscule, yet *some* does remain. We hope that this evidence is enough to convince you that the answer to the question posed above is

---

ANSWER: No, I 131 will never lose its radioactivity in any finite time period.

Notice the insertion of the word "finite" into our answer. From a practical viewpoint we can only consider finite time periods. But what about a theoretical viewpoint? What will happen after an infinite amount of time? The expression for the fraction of decay after an infinite amount of time is

$$\tfrac{1}{2} + \tfrac{1}{4} + \tfrac{1}{8} + \tfrac{1}{16} + \cdots$$

where the ellipsis indicates that there is no end to the number of fractions. From physical considerations all of the element will decay, so this sum must equal 1. This gives us our first infinite sum,

$$\tfrac{1}{2} + (\tfrac{1}{2})^2 + (\tfrac{1}{2})^3 + \cdots = 1$$

If the argument based upon radioactivity does not convince you that it is possible to add an infinite amount of numbers and yet arrive at a finite answer, then perhaps an argument from geometry will. Begin with a 1×1 square. Shade half of it. Then shade half the remaining area. Continue this process of shading the remaining half. The first three steps are shown below.

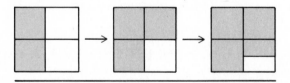

Let us consider another infinite sum,

$$\tfrac{2}{3} + (\tfrac{2}{3})^2 + (\tfrac{2}{3})^3 + \cdots$$

In Example 4 we found that the sum of its first 9 terms is $S(\tfrac{2}{3}, 9) = 38{,}342/19{,}683$. Here we must go further. We must evaluate $S(\tfrac{2}{3}, n)$ as n goes to infinity. We will use the notation $S(\tfrac{2}{3})$ to denote the sum, so our aim is to derive a formula for

$$S(\tfrac{2}{3}) = (\tfrac{2}{3}) + (\tfrac{2}{3})^2 + (\tfrac{2}{3})^3 + \cdots$$

Recall the formula for adding the first n terms.

$$S(\tfrac{2}{3}, n) = \frac{\tfrac{2}{3} - (\tfrac{2}{3})^{n+1}}{1 - \tfrac{2}{3}}$$

Notice that the number $(\tfrac{2}{3})^{n+1}$ approaches 0 as n gets increasingly large. Because it becomes negligible we set it equal to 0. Thus the infinite sum becomes

$$\frac{\tfrac{2}{3} - 0}{1 - \tfrac{2}{3}} = \frac{\tfrac{2}{3}}{\tfrac{1}{3}} = 2$$

This provides our second infinite sum

$$S(\tfrac{2}{3}) = (\tfrac{2}{3}) + (\tfrac{2}{3})^2 + (\tfrac{2}{3})^3 + \cdots = 2$$

The derivation of the formula for $S(\tfrac{2}{3})$ suggests a more general formula for

$$S(r) = r + r^2 + r^3 + \cdots$$

where r is any number that lies between 0 and 1. All you have to do is put the letter r wherever the number $\tfrac{2}{3}$ occurs in the derivation above. We will display the formula to highlight it.

$$S(r) = \frac{r}{1 - r}$$

We can use this formula to confirm the two infinite sums obtained so far:

$$S(\tfrac{1}{2}) = \frac{\tfrac{1}{2}}{1 - \tfrac{1}{2}} = \frac{\tfrac{1}{2}}{\tfrac{1}{2}} = 1$$

$$S(\tfrac{2}{3}) = \frac{\tfrac{2}{3}}{1 - \tfrac{2}{3}} = \frac{\tfrac{2}{3}}{\tfrac{1}{3}} = \tfrac{2}{3} \cdot \tfrac{3}{1} = 2$$

Example 6

Problem Compute $\tfrac{3}{4} + (\tfrac{3}{4})^2 + (\tfrac{3}{4})^3 + \cdots$.

Solution This expression is denoted by $S(\tfrac{3}{4})$. By formula,

$$S(\tfrac{3}{4}) = \frac{\tfrac{3}{4}}{1 - \tfrac{3}{4}} = \frac{\tfrac{3}{4}}{\tfrac{1}{4}} = \tfrac{3}{4} \cdot \tfrac{4}{1} = 3$$

Thus $\tfrac{3}{4} + (\tfrac{3}{4})^2 + (\tfrac{3}{4})^3 + \cdots = 3$.

This type of addition problem is perhaps the first infinite sum you have encountered. It is sometimes called an *infinite series* or a *geometric series*. An infinite series is said to *converge* if its sum is a finite number. We have already seen several examples: for instance, $\frac{1}{2} + (\frac{1}{2})^2 + \cdots$ converges to 1 and $\frac{3}{4} + (\frac{3}{4})^2 + \cdots$ converges to 3. A series which does not converge is said to *diverge*. The next example will supply an example of a series which diverges.

Example 7

Problem For each series below, determine whether it converges or diverges. We label the first one H because it is called the *harmonic series*, and the second one A because we can use algebra to evaluate it.

$$H: \frac{1}{2} + \frac{1}{3} + \frac{1}{4} + \frac{1}{5} + \cdots$$

$$A: \frac{3}{4} + \frac{3}{16} + \frac{3}{64} + \frac{3}{256} + \cdots$$

Solution The approach to series H is to group its terms as follows:

$$\frac{1}{2} + \underbrace{\frac{1}{3} + \frac{1}{4}} + \underbrace{\frac{1}{5} + \frac{1}{6} + \frac{1}{7} + \frac{1}{8}} + \underbrace{\frac{1}{9} + \cdots + \frac{1}{16}} + \cdots$$

The sum within each of these groups is greater than $\frac{1}{2}$. For instance, each of the four terms in $\frac{1}{5} + \frac{1}{6} + \frac{1}{7} + \frac{1}{8}$ is at least $\frac{1}{8}$, so the sum of the four of them is greater than $\frac{1}{2}$. Since there is an infinite number of terms, there is an infinite number of $\frac{1}{2}$s. Thus the sum becomes infinite, so series H diverges.

The approach to series A is to factor the common numerator 3. This yields

$$3(\frac{1}{4} + \frac{1}{16} + \frac{1}{64} + \frac{1}{256} + \cdots) = 3[\frac{1}{4} + (\frac{1}{4})^2 + (\frac{1}{4})^3 + (\frac{1}{4})^4 + \cdots]$$

The number inside the square brackets is

$$S(\tfrac{1}{4}) = \frac{\frac{1}{4}}{1 - \frac{1}{4}} = \frac{\frac{1}{4}}{\frac{3}{4}} = \frac{1}{3}.$$

Thus the sum of series A is $3 \cdot \frac{1}{3} = 1$, so it converges to 1.

Example 7 shows that even though the numbers in an infinite sum get smaller and smaller, you cannot immediately conclude whether the sum is finite or infinite. It is considerations like this that make this subject so hard for, yet so attractive to, many mathematicians.

EXERCISES

In Problems 1 to 6 evaluate each sum.

1 $S(\frac{1}{2}, 8)$

2 $S(\frac{1}{2}, 9)$

3 $S(\frac{1}{10}, 8)$

4 $S(\frac{1}{10}, 9)$

5 $S(3, 8)$

6 $S(3, 9)$

In Problems 7 to 10 determine the fraction of I 131 that will decay in the time periods.

7 4 half-lives

8 6 half-lives

9 48 days

10 56 days

Problems 11 to 14 deal with an isotope of uranium, U 234, which has a half-life of about 250,000 years. What fraction of it will decay in the time periods?

11 4 half-lives

12 6 half-lives

13 1 million years

14 2 million years

In Problems 15 to 20 find the sum of the infinite series.

15 $\dfrac{1}{10} + \left(\dfrac{1}{10}\right)^2 + \left(\dfrac{1}{10}\right)^3 + \cdots$

16 $\dfrac{1}{9} + \left(\dfrac{1}{9}\right)^2 + \left(\dfrac{1}{9}\right)^3 + \cdots$

17 $\dfrac{2}{5} + \left(\dfrac{2}{5}\right)^2 + \left(\dfrac{2}{5}\right)^3 + \cdots$

18 $\dfrac{3}{5} + \left(\dfrac{3}{5}\right)^2 + \left(\dfrac{3}{5}\right)^3 + \cdots$

19 $1 - \left(\dfrac{1}{2}\right) - \left(\dfrac{1}{2}\right)^2 - \left(\dfrac{1}{2}\right)^3 - \cdots$

20 $1 - \left(\dfrac{2}{3}\right) - \left(\dfrac{2}{3}\right)^2 - \left(\dfrac{2}{3}\right)^3 - \cdots$

In Problems 21 to 29 determine whether each series converges or diverges. Find the sum if the series converges.

21 $\dfrac{1}{10} + \dfrac{1}{100} + \dfrac{1}{1000} + \cdots$

22 $\dfrac{1}{7} + \dfrac{1}{7^2} + \dfrac{1}{7^3} + \cdots$

23 $\dfrac{3}{2} + \dfrac{3}{4} + \dfrac{3}{8} + \dfrac{3}{16} + \cdots$

24 $\dfrac{4}{3} + \dfrac{4}{9} + \dfrac{4}{27} + \dfrac{4}{81} + \cdots$

25 $\dfrac{1}{2} + \dfrac{1}{4} + \dfrac{1}{6} + \cdots$

26 $\dfrac{1}{5} + \dfrac{1}{10} + \dfrac{1}{15} + \cdots$

27 $\dfrac{3}{2} + \left(\dfrac{3}{2}\right)^2 + \left(\dfrac{3}{2}\right)^3 + \cdots$

28 $\dfrac{3}{4} + \left(\dfrac{3}{4}\right)^2 + \left(\dfrac{3}{4}\right)^3 + \cdots$

29 $\dfrac{1}{3} + \dfrac{1}{5} + \dfrac{1}{7} + \cdots$

30 The half-life of iodine 129 is approximately 17 million years. How much radioactivity remains after 170 million years?

31 The half-life of iodine 133 is 21 hours. How much of its radioactivity will remain after 1 week?

32 One popular account read that "after 20 half-lives, only a millionth of the original radioactive element remains." Is this precise or approximate?

33 (*a*) What is $1 + 1 + 1 + 1 + 1$?
(*b*) Evaluate $S(1, 5)$ by the formula for $S(n, r)$.
(*c*) Why are the solutions to (*a*) and (*b*) different?

34 (*a*) What is $1 + 1 + 1 + \cdots$?
(*b*) Evaluate $S(1)$ by the formula for $S(r)$.
(*c*) Why are the solutions to (*a*) and (*b*) different?

35 The price of computers has gone down by about one-half every year for the past 20 years. If a computer sold for $5 million 20 years ago, how much would it cost today?

36 Hungary's delegate to the League of Nations, Paul de Hevesy, proposed a way to achieve worldwide disarmament. His plan was to hold a plebiscite in all countries every 5 years, asking whether one-half of the existing armaments should be destroyed. If the people voted "yes" for one century, what fraction of the armaments would remain?

37 One of the problems in the Rhind papyrus reads as follows: "Sum the geometrical progression of 5 terms, of which the first term is 7 and the multiplier 7." This means to form the sum

$$7 + 49 + 343 + 2401 + 16{,}807$$

(*a*) The sum is $S(r, n)$ for some r and n. What are r and n?
(*b*) What is the sum?

38 Solve the following eighteenth-century Mother Goose rhyme:

> As I was going to St. Ives
> I met a man with 7 wives.
> Every wife had 7 sacks,
> Every sack had 7 cats,
> Every cat had 7 kits.

Kits, cats, sacks, and wives,
How many were there going to St. Ives?

39 This problem will give you some food for thought. Recall that the half-life of I 131 is 8 days. What fraction of I 131 will decay in 4 days?

The usual conception of mathematics is that it consists of a bunch of formulas which can be applied only to the natural sciences. This is just *not* the case. We offer, as counterexamples, three topics from the art world, archeology, and ecology, which might be used for oral presentations to the class or for written reports. Background material for these topics can be found in the marvelous book "Differential Equations and their Applications," by Martin Braun (Springer-Verlag, 1983).

40 The Rembrandt Society bought the beautiful painting "Disciples of Emmaus" for $170,000. What role did mathematics play in determining whether it was a modern forgery?

41 One of the most accurate ways of dating archeological finds is the method of carbon-14 dating. How did mathematics enter into determining the date of the paintings in the famous Lascaux cave in France during the time of Hammurabi's reign?

42 The Nuclear Regulatory Commission used to dispose of atomic waste by placing it in drums which were then dumped at sea. How was mathematics involved in determining whether the tightly sealed drums might crack upon impact with the ocean floor? How did the calculations affect U.S. and European governmental policy?

2-3 DECIMALS

By the year 1585 the world was ripe for a major mathematical breakthrough. The concept of rational number had evolved to an extent where it was the cornerstone of most complex computations. Scientists and business executives alike required increasingly accurate calculations, but this greater accuracy necessitated their dealing with very unwieldy fractions. An example of a common elementary problem confronting a scientist would be

$$\frac{175,193}{21,365} + \frac{50,895}{6264}$$

Carrying out this addition would be tedious and time-consuming. Even simple properties of these numbers are not immediately apparent. For instance, between what two whole numbers does each lie and which fraction is greater?

Stevin's System

Simon Stevin probably shared something with you—he hated fractions! The big difference was that Stevin could not avoid them because he was an astronomer and almost all his lengthy calculations were with fractions. Stevin decided to do something about it. He uncovered a first-rate idea to express fractions in a much more manageable way. In his book published in 1585, entitled "La Disme" ("The Tenth"), he confessed that he was not the inventor of the system, but that he appreciated its importance. He then urged his contemporaries in the science and business world to adopt the system.

The key to Stevin's new method of expressing fractions was strikingly similar to the Egyptian system of decomposing fractions into unit fractions. What renders a fraction like 21,625/3181 almost incomprehensible at first glance is that it requires two numbers in its expression. The Egyptians seemed to be willing to deal only with fractions expressed with one number, as they had distinct symbols only for unit fractions. Thus, since their numerators were always 1, each fraction could be expressed using only one number, the denominator. Stevin's system relied on holding the form of the denominator constant, requiring that all denominators be powers of 10. He called a given natural number the *unit* and introduced the sign ⓪ to denote "the end of it." For example, he said to refer to the number 364 as three hundred sixty four units and to write it in the form 364 ⓪. Then he came to the fundamental advance. In order to handle fractions between 0 and 1, he divided a unit into 10 parts and called them primes; he divided primes into 10 parts and called them seconds, etc. The primes have the sign ①, the seconds ②, etc. Today we refer to "primes" as tenths, "seconds" as hundredths, and "thirds" as thousandths.

Notice that a "prime" is nothing more than the numerator of a fraction whose denominator is 10. Hence 3 ① meant 3/10. Similarly, a "second" stands for a fraction whose denominator is 100. Hence 7 ② meant 7/100. Also 5 ③ stood for 5/1000. Note also that 3 ① 7 ② stood for 37/100 because 3 ① 7 ② meant 3/10 + 7/100, which when added yields 30/100 + 7/100. Similarly, 3 ① 7 ② 5 ③ was equal to the fraction 375/1000.

At this point Stevin urged the reader to notice that no use of fractions was necessary. Then he cautioned that the number under each sign except the unit should never exceed 9. That is, if the number 712/1000 is to be represented, the expression 7 ① 12 ③ is never written, but instead it would appear as 7 ① 1 ② 2 ③.

From here Stevin demonstrated the ease of operating with these numbers. In subtraction, for instance, he said that given the number 237 ⓪ 5 ① 7 ② 8 ③ from which the number 59 ⓪ 7 ① 4 ② 9 ③ is to be subtracted, place the numbers in this order:

$$
\begin{array}{cccccc}
\textcircled{0} & \textcircled{1} & \textcircled{2} & \textcircled{3} & & \\
2 & 3 & 7 & 5 & 7 & 8 \\
 & 5 & 9 & 7 & 4 & 9 \\
\hline
1 & 7 & 7 & 8 & 2 & 9 \\
\end{array}
$$

After subtracting in the usual manner, there remains 177829, which, as indicated by the signs above the numbers, is 177 ⓪ 8 ① 2 ② 9 ③, and this is the remainder required.

To further convince any lingering skeptics of the value of the decimal system, Stevin concluded his tract with applications to surveying, measuring tapestry, measuring casks (of wine), volume, astronomy, and computation for merchants. He ended the book with a plea that all units of weights and measures be changed to decimal multiples.

Decimal Numbers

How would you convert the fraction 175,193/21,365 into Stevin's system? By now of course you have realized that Stevin's system is the familiar decimal representation of numbers. To convert the fraction into its decimal equivalent you divide the denominator into the numerator. You get

$$
\begin{array}{r}
8.2 \\
21{,}365 \overline{\smash{\big)}\ 175{,}193.0} \\
\underline{170{,}920} \\
42{,}730 \\
\underline{42{,}730}
\end{array}
$$

Thus the fraction is equal to 8.2. Even in Stevin's time this long division was tedious, but once the fraction was expressed as a decimal, it was much easier to deal with.

Stevin's book was not widely accepted at first but its reputation grew in some significant inner circles. Some 30 years later his symbolism was improved in the logarithm tables of Napier and the astronomy tables of Kepler. It soon became clear that this new form of expressing fractions not only simplified notation for fractions but greatly facilitated operations with them. In fact, there had been incidental use of decimals in ancient China and later occasionally in Arabia, but they were not generally accepted by all until Stevin's book explained their use in complete yet elementary detail. However, the new notation sometimes involved a decimal point and sometimes a comma, and this matter has not even been settled today. For instance, $23\frac{45}{100}$ is written 23.45 in the United States, 23·45 in England and 23,45 in continental Europe.

The evolution of decimals took another major step forward as late as 1971 when the Bowmar Brain appeared. "What in the world is a Bowmar Brain?" you ask. You probably have one of its offspring close at hand and use it often. It was in 1971 that engineers at Texas Instruments, Inc., invented the hand-held calculator, the first of which was the Bowmar Brain. Humans have always used various kinds of computational aids, from the ancient abacus in the orient to the knotted rope cords, called *quipu,* of Peru, and the detailed logarithmic and trigonometric tables constructed soley to aid computations. In 1946 the first digital computer, called ENIAC, was built, and it relied almost totally on decimal numbers. These devices were used mainly by scientists.

Calculators

It was in the early 1970s, when hand-held calculators began to flood the market, that the average person had to become conversant with decimals, because calculators do not recognize fractions. If you enter the fraction 2/5 into a calculator, it reads 0.4. Similarly, if you compute 175,193/21,365, it will display the number as 8.2.

Can you tell which fraction is greater: 175,193/21,395 or 50,895/6264? We determined that the first number is equal to 8.2 by long division. We could do the same for the second but the work turns out to be much more tedious. However, using a calculator renders the problem almost trivial.

Before we solve the problem we would like to urge you to read the manual for your calculator thoroughly. If you have not purchased a calculator we would recommend you spend just a few dollars more than the bare minimum and buy one that has a memory and computes square roots, reciprocals, and exponents. The cost would be less than $25—inexpensive when compared with the original 1971 model Bowmar Brain, which retailed for $800.

There are basically two types of calculators. One way of distinguishing which type you have, or which you are about to buy, is to scan the keyboard for an "equals" key that looks like this: $\boxed{=}$. If there is one, the calculator is an algebraic entry (AE) type. Otherwise it is most likely a reverse Polish notation (RPN) calculator. The major difference is that an AE calculator allows you to enter calculations as they would be written horizontally. For example, to compute 3 + 4 you press 3, then $\boxed{+}$, then 4, and then $\boxed{=}$, and the calculator displays the result of 3 + 4 =, that is, 7.

The RPN calculator allows you to enter the calculations as they would be written vertically. Thus to calculate

$$\begin{array}{r} 3 \\ + \ 4 \\ \hline \end{array}$$

you press 3, then the key titled ENTER, then 4, then $\boxed{+}$, and the calculator displays the result of 3 $\boxed{\text{ENTER}}$ 4 $\boxed{+}$, that is, 7. The RPN type of calculator has some significant advantages over the AE type, but most calculators are of the latter type. In the next few examples we will include both key sequences.

The problem we left was to determine the decimal equivalent of the fraction 50,895/6264. Figure 1 gives the key sequences for the two types of calculators. It will serve as the model for the format that we will follow whenever we supply key sequences. We display all keys which are not

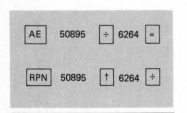

FIGURE 1 Key sequences for 50,895/6264.

numbers inside boxes and denote the $\boxed{\text{ENTER}}$ key by $\boxed{\uparrow}$. The result is 8.125. Since 8.2 is slightly larger than 8.125 we have that

$$\frac{175{,}193}{21{,}365} > \frac{50{,}895}{6264}$$

Using Your Calculator

Let us demonstrate how the calculator can ease complicated calculations.

> **Example 1**
> On August 25, Ollie bought a 2-year-old subcompact car. The salesman said that he would get about 18 miles per gallon (mpg) in the city. Ollie immediately filled the tank and noted that the odometer registered 25387.3 miles. On August 27, when he refilled it with 14.1 gallons, it registered 25541.1.
> **Problem** How many mpg did he average over this period?
> **Solution** The notation mpg for the units is both useful and suggestive. The m stands for miles driven, so m = 25541.1 − 25387.3. We are given g = 14.1, and the p stands for "per," which means "divided by." Thus to compute the mileage we evaluate
>
> $$\text{mpg} = \frac{25541.1 - 25387.3}{14.1}$$
>
> by one of the following key sequences:
>
> $\boxed{\text{AE}}$ $\quad$ 25541.1 $\boxed{-}$ 25387.3 $\boxed{\div}$ 14.1 $\boxed{=}$
>
> $\boxed{\text{RPN}}$ $\quad$ 25541.1 $\boxed{\uparrow}$ 25387.3 $\boxed{-}$ 14.1 $\boxed{\div}$
>
> The display will read 10.9 to one decimal place. Thus Ollie's car averaged 10.9 miles per gallon on its first tank of gas.

If you got 10.9 on your first try you probably feel more elated than Ollie did. In fact, Ollie most likely checked and rechecked his calculations a few times in disbelief.

This example brings up three important points: (1) intermediate calculations, (2) checking and approximating, and (3) round-off error and significant digits. Let us take them in order:

1 Notice that before pressing 14.1 we did not have to press the $\boxed{=}$ key in AE or the $\boxed{\uparrow}$ key in RPN. The calculator computed the subtraction and displayed the result, in this case, 153.8. Often you can use these intermediate results to check your calculations. For example, you might have been able to do the subtraction in your head and

so you could check to see if the display were correct, thus assuring yourself that you pressed the correct buttons. It is often the case that the calculator displays an incorrect result not because of faulty reasoning but faulty execution—you might have pressed 5 instead of 6, for instance.

2 The previous paragraph illustrates how important it is to check your computations on a calculator. Often you can inspect for errors by rounding off the numbers in the problem and doing the subsequent calculation in your head, thus giving you a "ballpark" figure to aim at. If the answer on the calculator is very different from your approximation, you probably made an error somewhere. For instance, before entering the problem in Example 1 you could have approximated the mileage by $25500 - 25300 = 200$ and "divided" by 15 to get the approximate value of 13 mpg.

3 Understanding round-off and its significance is of paramount importance when dealing with decimals, and, in particular, with calculators. There is a limit to the number of significant digits that a calculator can manipulate. This number varies from one model to another. Often eight significant digits are used, but this number is sometimes increased by using exponential notation. For example, if you enter $\frac{1}{3}$ into your calculator it will store it as 0.33333333, providing it uses eight significant digits. The latter number is a little less than $\frac{1}{3}$ since the decimal expansion of $\frac{1}{3}$ has an infinite number of 3s. Hence 0.33333333 is only an approximation of $\frac{1}{3}$ but a fairly accurate one. Note that 0.33 is another approximation of $\frac{1}{3}$, but less accurate. This demonstrates a major disadvantage of a calculator in that it must round off numbers with many significant digits. Usually this round-off process is not detrimental because most problems do not require such a large degree of accuracy. But you should be aware of when you and/or the calculator is rounding off. For instance, the calculator had to round off the calculations in Example 1 because the actual solution had more significant digits than the display of the calculator (which is usually eight). For our purposes the first two decimal places were more than enough to determine the car's mpg, so we said the answer is "about 10.9" when our calculator actually displayed 10.907801.

The next example, which shows how decimals often play an important role in everyday calculations, requires a little logical interpretation of the problem before the calculator can be used. It also demonstrates how to use the reciprocal, or $\boxed{1/x}$, key, the memory key, which we refer to as "store," or $\boxed{\text{STO}}$ (although your calculator may have another name, such as $\boxed{\text{M}}$), and the "recall" key or $\boxed{\text{RCL}}$, which retrieves the number in the memory.

Example 2

The Springfield Joggers Club conducts weekly run-for-funs on Sundays at 4:00 P.M., around a scenic 2.35-mile course. Martha participated in a recent 2-lap run which started at 4:01. She finished at 4:33:38.4. Most runners like to know their average pace, customarily in terms of minutes per mile (min/mi). We will call this the RP, for runner's pace.

Problem What was Martha's RP?

Solution The units for RP are min/mi so we must compute

$$\frac{\text{Time (min)}}{\text{Distance (mi)}} = \frac{4:33:38.4 - 4:01}{2 \times 2.35}$$

First we must convert 38.4 seconds to minutes. We get 38.4 sec = 38.4/60 min = 0.64 min. Hence we must compute

$$\frac{33.64 - 1}{2 \times 2.35}$$

The AE sequence is 33.64 $\boxed{-}$ 1 $\boxed{=}$ $\boxed{\text{STO}}$ 2 $\boxed{\times}$ 2.35 $\boxed{=}$ $\boxed{\text{1/x}}$ $\boxed{\times}$ $\boxed{\text{RCL}}$ $\boxed{=}$ The RPN sequence is 33.64 $\boxed{\uparrow}$ 1 $\boxed{-}$ 2 $\boxed{\uparrow}$ 2.35 $\boxed{\times}$ $\boxed{\div}$.
The result indicates that her RP is 6.94 minutes per mile. But we usually don't talk about hundredths of a minute; we use seconds instead. Thus to complete her RP we want to know how many seconds are in 0.94 minute. We can evaluate 60×0.94 independently or use the display of 6.94 and then the sequence:

$\boxed{\text{AE}}$ $\boxed{-}$ 6 $\boxed{=}$ $\boxed{\times}$ 60 $\boxed{=}$

$\boxed{\text{RPN}}$ 6 $\boxed{-}$ 60 $\boxed{\times}$

This gives the number of seconds, which is 56.68. Thus her RP for the 2-lap run was 6 minutes, 56.68 seconds per mile.

The next example demonstrates how a calculator can handle very large numbers quite easily. You can use either the $\boxed{y^x}$ or the $\boxed{x^2}$ key.

Example 3

Problem Compute the Mersenne number $M_{23} = 2^{23} - 1$.

Solution This problem was an exercise in Section 1-6, but it was expected that only a few hardy individuals would attempt it by hand. However, by calculator it is easy, just press

$\boxed{\text{AE}}$ 2 $\boxed{y^x}$ 23 $\boxed{=}$ $\boxed{-}$ 1 $\boxed{=}$

$\boxed{\text{RPN}}$ 2 $\boxed{\uparrow}$ 23 $\boxed{y^x}$ 1 $\boxed{-}$

The display will read 8,388,606.98. This provides us with an instance of the effects of round-off error, for we know that M_{23} must be a natural number. So we use our common sense and write our answer as 8,388,607.

If you do not have a $\boxed{y^x}$ key, you can evaluate 2^8 by pressing 2 $\boxed{x^2}$ $\boxed{x^2}$ $\boxed{x^2}$, then inserting it in memory; then compute 2^{16} by again pressing $\boxed{x^2}$; now multiply 2^{16} by 2^8 in memory by pressing $\boxed{RCL}$. The display gives 2^{24}; so divide by 2 by pressing $\boxed{\div}$ 2 $\boxed{=}$ and you have 2^{23} on display. Then subtract 1.

Example 4
Problem Evaluate $2^{67} - 1$.
Solution This is M_{67}, which we discussed in Section 1-4, about which Cole gave his famous silent "lecture." Notice that after pressing

$$\boxed{AE} \quad 2 \;\boxed{y^x}\; 67 \;\boxed{=}$$

$$\boxed{RPN} \quad 2 \;\boxed{\uparrow}\; 67 \;\boxed{y^x}$$

the display reads

$$1.4757395 \quad 20$$

This means that the number is approximately 1.4757395 times 10^{20}, which in turn means the result contains 21 digits but the calculator cannot display all of them. This demonstrates one of the limitations of the use of hand calculators.

A Rational Connection

There is a connection between rational numbers and their decimal representations. For instance, $\frac{1}{2}$ is also equal to 0.5. In fact every rational number has a decimal representation which can be obtained by dividing the denominator into the numerator. Some examples are:

$$\frac{1}{4} = 0.25 \qquad \frac{1}{10} = 0.1$$

$$\frac{1}{5} = 0.2 \qquad \frac{235}{1000} = 0.235$$

$$\frac{1}{8} = 0.125 \qquad \frac{613}{100} = 6.13$$

Each of these numbers has a finite number of decimal places, and so we say that their decimal representation is *terminating*. Other rational numbers have decimal representations that are infinite. For example

$$\frac{1}{3} = 0.3333\ldots \qquad \frac{1}{9} = 0.1111\ldots$$

$$\frac{1}{6} = 0.16666\ldots \qquad \frac{4}{33} = 0.121212\ldots$$

Notice that each of these decimal expansions contains a digit, or a group of digits, that is repeated endlessly. That is why we call them *repeating* decimals. The usual way of denoting a repeating decimal is to place a bar above the repeating part. Thus

$$\frac{1}{3} = 0.\overline{3} \qquad \frac{1}{6} = 0.1\overline{6} \qquad \frac{1}{9} = 0.\overline{1} \qquad \frac{4}{33} = 0.\overline{12}$$

The process for converting a rational number to its decimal representation involves only long division. It is easy to determine whether a decimal representation of a rational number is terminating or repeating. If any remainder in the long division is ever zero, then it is terminating. If one of the digits in the quotient is repeated, then the sequence of digits starting from that digit is the repeating part of the decimal. Let us look at an example.

Example 5

Problem Determine a decimal representation of each of the following rational numbers.

$$(a)\ \frac{1}{8} \qquad (b)\ \frac{3}{7} \qquad (c)\ \frac{611}{4950}$$

Solution The procedure is to carry out the long division and to list the remainders beside it. We have circled the remainders in the long division for clarity. For $\frac{1}{8}$ the computations are

```
      0.125      Remainder
   8 ⟌1.000        1
      8
     ⟍20⟋          2
      16
     ⟍40⟋          4
      40
     ⟍0⟋           0
```

Since one of the remainders is zero, the decimal expansion is terminating.
We have $\frac{1}{8} = 0.125$.

Part (b) becomes

```
         0.428571      Remainder
    7 | 3.000000          3
        2 8
         2⃝0               2
         14
         6⃝0               6
         56
         4⃝0               4
         35
         5⃝0               5
         49
         1⃝0               1
          7
         3⃝                3
```

The remainder 3 is repeated, so the group of numbers 428571 in the quotient
is repeated. Hence $\frac{3}{7} = 0.\overline{428571}$.

Part (c) becomes

```
          0.1234        Remainder
  4950 | 611.0000         611
         495 0
        ⃝116 0⃝0          1160
         99 00
        ⃝17 0⃝00          1700
         14 850
        ⃝2 15⃝00          2150
         1 9800
         ⃝17000⃝          1700
```

The remainder 1700 is repeated, so the group of numbers 34 in the quotient
is repeated. Hence $611/4950 = 0.12\overline{34}$.

Example 5 illustrates the following general result.

Every rational number has a decimal representation which is either
terminating or repeating.

Let us see how to proceed in the opposite direction. That is, given a decimal that is terminating or repeating, find its representation as a fraction. The case when the decimal is terminating is easy, as the next example shows.

Example 6
Problem What rational number is 0.6875 equal to?
Solution By the definition of a decimal expansion we have $0.6875 = 6875/10,000$. This is one answer to the question. However, the preferred answer is to reduce the fraction to lowest terms by using the fundamental theorem of arithmetic. This becomes

$$0.6875 = \frac{6875}{10,000} = \frac{5^4 \cdot 11}{2^4 \cdot 5^4} = \frac{11}{16}$$

When a rational number has a decimal expansion that is repeating, the process used to determine its fractional form is a bit tricky. Let us look at a detailed example.

Example 7
We will determine a rational number which is equal to $0.8\overline{3}$. Set $x = 0.8\overline{3}$. The bar rests over one digit; so we multiply x by 10. Notice that

$$10x = 10(0.8\overline{3}) = 8.\overline{3} = 8.3\overline{3}$$

The reason why we multiplied by 10 is because it allows us to perform the subtraction below.

$$
\begin{aligned}
10x &= \quad 8.3\overline{3} \\
-x &= -0.8\overline{3} \\
\hline
9x &= \quad 7.50
\end{aligned}
$$

Then

$$x = \frac{7.5}{9} = \frac{75}{90}$$

The one possible answer is $0.8\overline{3} = \frac{75}{90}$. However, the preferred answer is to reduce the fraction; so

$$0.8\overline{3} = \frac{5 \cdot 15}{6 \cdot 15} = \frac{5}{6}$$

The process for converting from a repeating decimal number to a rational number is the following:

1 Set d equal to the number.

2 Multiply d by 10 to a power, where the power is the number of repeated digits.

3 Write the equation in step 2 above the original equation in step 1, and subtract.

4 Solve for d as a fraction.

5 If the numerator of the fraction is a decimal, make it a rational number by multiplying it by an appropriate power of 10.

6 Reduce the rational number if possible.

Example 8

Problem What rational number is $0.\overline{45}$ equal to?

Solution We will follow the steps above.

1 $d = 0.\overline{45}$

2 $100d = 45.\overline{45}$

3 $\begin{aligned} 100d &= 45.\overline{45} \\ -d &= -0.\overline{45} \\ \hline 99d &= 45.0 \end{aligned}$

4 $d = \dfrac{45}{99}$

6 $d = \dfrac{5}{11}$

Of course, you can check that $\frac{5}{11} = 0.\overline{45}$ by long division.

Summary

In this section we discussed Simon Stevin's introduction of decimal numbers, showed how to solve some everyday problems involving them, and related them to rational numbers. Decimal numbers are convenient for handling most computations, especially with the aid of a hand-held calculator or a computer.

Sometimes, however, decimal numbers lend themselves to human error. Who among us has never misplaced a decimal point? In fact, the Popeye legend about spinach is the result of one such error. The theory that eating spinach will make you strong came about when a scientist at the turn of the twentieth century was measuring that vegetable's iron content. He put the decimal point in the wrong place, an arithmetical error that has had extraordinary consequences. It turns out that spinach contains the same amount of iron as many other vegetables.

EXERCISES

In Problems 1 to 4, convert the given number from Stevin's notation to decimal notation.

1. 16 ⓪ 1 ① 0 ② 4 ③

2. 5 ⓪ 0 ① 3 ②

3. 2 ② 8 ③ 1 ④

4. 1 ① 2 ③

In Problems 5 to 8, express the given number, using Stevin's notation.

5. 843.017

6. 170.32

7. 0.109

8. 0.019

In Problems 9 to 14 convert the given rational number to its decimal representation.

9. $\dfrac{1}{11}$

10. $\dfrac{1}{6}$

11. $\dfrac{1}{7}$

12. $\dfrac{2}{13}$

13. $\dfrac{3}{4}$

14. $\dfrac{5}{8}$

In Problems 15 to 20 find a rational number that is equal to the given decimal number. Reduce all rational numbers to lowest terms.

15. 0.1023

16. 1.023

17. 0.7777...

18. 0.5555...

19. 0.363636...

20. 0.393939...

Problems 21 to 32 are concerned with everyday applications of decimal numbers. They are taken from a wide variety of interests.

21. How many miles per gallon did a car average on a trip that used 10.9 gallons of gas, if the odometer went from 56,782.4 to 57,023.9?

22. How many miles per gallon did a car average on a trip that used 15.9 gallons of gas, if the odometer went from 69,872.1 to 70,288.1?

23. *(Running)* Don Ritchie and 30 other "ultramarathoners" began running in the New York Invitational 100-mile run at 7 P.M. He finished the next morning at 6:51:11.
    (*a*) How long did it take him to complete this incredible distance?
    (*b*) What was the average time it took him to run each mile?
    (*c*) The course consisted of loops around a 2.27-mile path. How many laps were run?

24 *(Running)* Don Ritchie's world record of 11:30:51 for the 100-mile run was set on a quarter-mile track at London's Crystal Palace in 1977.
(*a*) What was his average time per mile in that run?
(*b*) How many laps did he run?

25 The 1977 EPA rating of the gas mileage for a Grand Prix was based on 17 mpg in the city and 25 mpg on the highway. The final rating is not obtained by averaging these two. Instead, the figure of 0.55 of the time for city driving is used to compute a car's rating. What is the Grand Prix's rating?

26 The cost of a telephone call was listed as $1 for the first 3 minutes and 30 cents for each additional minute. What is the total cost of an 8-minute call?

27 *(Running)* A marathon is 26 miles and 385 yards. Alberto Salazar won the 1982 Boston Marathon in 2:8:51. Runners measure their pace in minutes per mile. What was Salazar's pace?

28 *(Running)* If you run a pace of 8 minutes per mile, how long will it take you to complete a marathon (26 miles and 385 yards)?

29 *(Racecar driving)* Racecar driver Tom Sneva once owned the one-lap record of 203.62 miles per hour on the Indianapolis-500 track. If he maintained this speed for the entire race, how long would it take him to complete it?

30 *(Baseball)* In one span of 31 games during the 1977 baseball season Mike Schmidt slugged 18 home runs and had 38 runs batted in (rbi's). If he maintained this torrid pace over an entire season of 162 games, how many homers and rbi's would he have?

31 *(Baseball)* One newspaper listed the following standings for the National League East division during midseason, where "Pct." is the percentage of games won.

Team	W	L	Pct.	GB
Montreal	55	42	.567	—
Pittsburgh	56	44	.567	$\frac{1}{2}$
Chicago	54	44	.551	$1\frac{1}{2}$
St. Louis	50	47	.567	5
Philadelphia	52	49	.567	5
New York	42	55	.433	13

According to the standings there were two ties. But the respective records are different. Determine whether the standings are correct.

32 *(Golf)* The all-time Professional Golf Association score records are:

72 holes, 257, Mike Souchak—1955

54 holes, 189, Chandler Harper—1954

54 holes, 189, Mike Souchak—1955

36 holes, 125, Ron Streck—1976

18 holes, 59, Al Geiberger—1977

Compute the average number of strokes per hole for each one.

2-4 IRRATIONAL NUMBERS

As legend has it, Hippasus was a loyal member of the Pythagorean brother-hood. He loved the Pythagorean way of life, and he especially believed in the motto "All is number." Thus he was particularly excited when his turn came to present to the entire assemblage the most recent result he had discovered. A nagging voice within him urged him to keep the discovery to himself, but youthful exuberance won out. He presented it ever so slowly, allowing plenty of time for the other Pythagoreans to follow each step in his chain of reasoning. When he arrived at his final conclusion he wheeled around, expecting to see smiling, appreciative faces. Instead he encountered blank stares and abso-lute silence. After what seemed like an eternity, one of the brothers rose and said, "Hippasus, your result seems correct, but we will have to examine it closely. You understand, of course, that this poses a danger to our very exis-tence, so we remind you of your vow not to mention it outside the brother-hood."

Hippasus waited and waited, but no response was forthcoming. Instead of generating an intellectual exchange of ideas, he was ostracized. Frustration finally set in, and he broke the vow of silence by publicizing his discovery.

The Pythagoreans were outraged. Instead of reacting in their usual calm, rational manner they resorted to violence. One day a band of them kidnapped the unsuspecting youth, took him for a boat ride, and pushed him overboard.

The Flaw in the Pythagorean Philosophy

You will most likely be surprised to learn how seemingly harmless Hip-pasus' discovery was. He proved that there exists a number that is not a rational number! So what was the big deal, right? It was important to the Pythagoreans because they had based their whole philosophy on "all is number," and they had always *assumed* that *every* number is expressible as a fraction. Hence they assumed every number is a rational number. Hippasus found a fundamental flaw in their philosophy.

What led to his startling discovery? He most likely studied the sim-ple figure on page 120 and asked the question: What is the length of the diagonal $\overline{AB}$ of the square whose sides are 1 unit? [See Figure 1(a).] From the pythagorean theorem he knew that $\overline{AB}$ was that number whose square is 2, since the diagonal and two sides form a right triangle [see Figure 1(b)], and the sum of the squares of the legs is $1^2 + 1^2 = 2$,

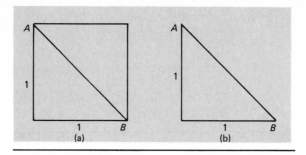

FIGURE 1

which is equal to the square of the hypotenuse, that is, $\overline{AB}^2$. In modern terminology, since $\overline{AB}^2 = 2$ we say that $\overline{AB}$ is equal to the *square root* of 2, written $\overline{AB} = \sqrt{2}$.

Because the Pythagoreans believed that every number is a rational number, Hippasus must have asked: What rational number—or what fraction—is equal to $\sqrt{2}$? His answer to this question sealed his fate. He proved that *there is no such fraction*. Thus he proved that $\sqrt{2}$ is not a rational number. If a number is not a rational number we say it is an *irrational* number. Hence Hippasus showed that

$\sqrt{2}$ is an irrational number

Let us study a proof of this result. The steps themselves are rather straightforward but you might have some difficulty seeing the whole proof all at once. The reasoning that we give is due to Aristotle, who wrote the proof some 50 years after Hippasus' final boat ride. We do not know the chain of reasoning that Hippasus used because the Pythagoreans destroyed all evidence of their knowledge of the result.

The Proof

Theorem $\sqrt{2}$ is an irrational number, that is, $\sqrt{2}$ cannot be expressed as a ratio of two natural numbers.

Proof Suppose $\sqrt{2}$ is a rational number. We are going to show that making this assumption will lead to a logical contradiction. Since logical contradictions cannot exist, this means that we cannot make the assumption that $\sqrt{2}$ is rational.

Under our assumption that $\sqrt{2}$ is rational, we can write

$$\sqrt{2} = \frac{c}{d} \tag{1}$$

for some natural numbers c, d. First reduce c/d so no common terms remain, and call the reduced fraction x/y. Then

$$\sqrt{2} = \frac{x}{y} \tag{2}$$

where *x and y have no divisor in common*. Keep this property of x and y in mind because it will be crucial to our conclusion; this is where the contradiction will lie.

Write Equation (2) twice:

$$\sqrt{2} = \frac{x}{y}$$

$$\sqrt{2} = \frac{x}{y}$$

Since you can multiply equals and still get equals, we multiply both left-hand sides and both right-hand sides to get

$$\sqrt{2}\,\sqrt{2} = \frac{x}{y} \cdot \frac{x}{y}$$

$$2 = \frac{x^2}{y^2} \tag{3}$$

Then multiply both sides by y^2 and reduce:

$$2y^2 = \left(\frac{x^2}{y^2}\right) y^2$$

$$2y^2 = x^2 \tag{4}$$

Notice that the left-hand side of (4) is an even number since 2 is one of its factors. Thus the right-hand side must be even too, so

$$x^2 \text{ is an even number.} \tag{5}$$

What about x? Can x be odd? No, since any odd number multiplied by itself yields another odd number. In other words, if x were odd then x^2 would be odd too. But (5) tells us otherwise. So we conclude that

x is an even number

Since x is even, x is equal to 2 times another number; so we can write

$$x = 2k \tag{6}$$

where k is some appropriate number.

Now substitute (6) into (4): $2y^2 = (2k)^2 = 4k^2$; so $\frac{1}{2}(2y^2) = \frac{1}{2}(4k^2)$ and so

$$y^2 = 2k^2 \tag{7}$$

The form of Equation (7) is identical to that of Equation (4). By a reasoning process just like the one we used in going from (4) to (6) we can conclude that

> y is an even number

and so

$$y = 2m \tag{8}$$

for some appropriate number m.

Look at Equations (6) and (8). They show that *x and y have the number 2 as a common divisor*. But recall in the earlier part of this proof the fact that x and y cannot have any common factors. So we have arrived at a contradiction since it is impossible for two numbers to have a common factor and yet to not have a common factor.

This means that the assumption that $\sqrt{2}$ could be written in the form c/d cannot be made. And this means that $\sqrt{2}$ is not a rational number.

QED

Many legends have grown from this episode, so catastrophic for the Pythagoreans. Perhaps the only concrete fact we have is that they did discover an irrational number. Nevertheless, the reason that the Hippasus legend is popular is that we do know that the Pythagoreans suppressed this knowledge. According to later writers, Hippasus was the "goat" in the story. Whether he was actually drowned as a result of his discovery has never been substantiated. What is important is that the Pythagoreans discovered the existence of a new type of number and they suppressed that knowledge because it was damning to their philosophy. Thus the ascent of irrational numbers had a very stormy inception.

A point that we would like to emphasize is that humanity's knowledge and understanding of irrationals has not increased so swiftly or so thoroughly as other aspects of mathematics. Irrational numbers were one of the first types of numbers to be discovered, but it was not until around 1875 that mathematicians felt they had a firm understanding of them.

In fact, even today most people are much less at ease with irrational numbers than other types of numbers, including, say, negative numbers. That is, you are probably more familiar with numbers such as -1, -2, and $-\frac{3}{4}$ than with $\sqrt{2}$, π, and e. Compare this with the fact that negative numbers were not recognized as legitimate numbers even 150 years ago, when the well-known mathematician De Morgan was ridiculed in a spoof play for using them freely in his work.

Square Roots

Now that we know that $\sqrt{2}$ is an irrational number, it seems natural to ask if there are other irrational numbers. The operation of taking the square root of a number often produces an irrational number, but other times it does not. Let us investigate this type of number in some detail.

TABLE 1

The First 20 Square Numbers

$1^2 = 1$	$6^2 = 36$	$11^2 = 121$	$16^2 = 256$
$2^2 = 4$	$7^2 = 49$	$12^2 = 144$	$17^2 = 289$
$3^2 = 9$	$8^2 = 64$	$13^2 = 169$	$18^2 = 324$
$4^2 = 16$	$9^2 = 81$	$14^2 = 196$	$19^2 = 361$
$5^2 = 25$	$10^2 = 100$	$15^2 = 225$	$20^2 = 400$

A knowledge of square numbers makes it easier to evaluate some square roots and to approximate others. Recall that the Pythagoreans studied this type of number a great deal. Table 1 lists some square numbers that will be helpful later.

Example 1
The square root of 225 is 15 because $15^2 = 225$. We write $\sqrt{225} = 15$. In terms of the definition of square root, we have $\sqrt{225} \times \sqrt{225} = 225$.

In order to evaluate $\sqrt{625}$ we have to go beyond the list of square numbers in Table 1. We know that $30^2 = 900$. Since 625 lies between 400 and 900 we know that $\sqrt{625}$ lies between $\sqrt{400} = 20$ and $\sqrt{900} = 30$. Moreover, the divisibility tests reveal that 625 is a multiple of 5, and the only number between 20 and 30 having this property is 25. Checking, we find that $625 = 25^2$. Thus $\sqrt{625} = 25$.

Sometimes $\sqrt{n}$ is a rational number, for example, when $n = 4$ or $n = 625$. For other values of n the number $\sqrt{n}$ is irrational. We proved that $\sqrt{2}$ is irrational. Very similar reasoning shows that $\sqrt{p}$ for p a prime number is an irrational number. In the exercises we will outline the proof.

In fact, whenever n is not a square number, one can prove that $\sqrt{n}$ is an irrational number. We outline this in the exercises.

Therefore you can generate many irrational numbers by using the square root operation. You can also get further irrationals by considering higher roots, such as cube roots and fourth roots. We will pursue one more way to generate irrationals.

Enter Stevin

It is very natural to be ill at ease with irrational numbers. Part of the problem stems from the very definition, which relies on the fact that we can *never* find a ratio of natural numbers that is equal to the given irrational. We can take some solace from the fact that almost 2000 years passed after the discovery of irrational numbers before mathematicians were convinced that they thoroughly understood them.

Stevin's decimal representation of numbers proved to be an important step in the evolution of irrational numbers. Recall that every rational number has a decimal equivalent that is either terminating or repeating, such as $\frac{1}{2} = 0.5$ or $\frac{1}{3} = 0.\overline{3}$. What about a decimal that is neither terminating nor repeating? Here is an example of one.

$$0.246810121416\ldots$$

The sequence of digits follows a definite pattern in that the digits can be viewed as the sequence of even natural numbers. But note that no part of them ever continuously repeats itself. Thus the number must represent an irrational number because it is a decimal that is not terminating or repeating.

This provides us with a description of irrational numbers.

> The decimal representation of an irrational number is neither terminating nor repeating.

This criterion allows us to construct a vast array of irrational numbers. The next example shows a few ways.

Example 2

Problem Which of the following numbers is irrational?

$$w = 0.010110111011110\ldots$$

$$x = 0.12345678910111121314\ldots$$

$$y = 0.123232323\ldots$$

$$z = 1 + \sqrt{2}$$

Solution The number w is composed of sequences of 1s, each sequence separated by a 0. The length of each sequence increases by 1 each time. Thus no set of digits repeats itself throughout w; so w is irrational.

Similarly, the decimal expansion of x follows a pattern, but the pattern does not contain any set of repeating digits. Thus x is irrational.

The number y is rational because it contains a set of repeating digits. Notice that $y = 0.1\overline{23}$. According to the procedure established in Section 2-3, you can write $y = \frac{61}{495}$.

The number z is irrational for the following reason. If z were rational then we could write $z = a/b$ where a and b are natural numbers. This would mean that $a/b = 1 + \sqrt{2}$. Then

$$\sqrt{2} = \frac{a}{b} - 1 = \frac{a}{b} - \frac{b}{b} = \frac{a-b}{b}$$

which would imply that $\sqrt{2}$ is rational. But $\sqrt{2}$ is not rational. Hence the assumption that z was rational leads to a contradiction. For this reason, z must be irrational.

We would like to emphasize the important distinction between a pattern and a repeating set that was drawn in Example 2. Every number containing a repeating set of digits forms a pattern, but the numbers w and x show that a number can follow a pattern without containing a repeating set of digits.

Decimal Approximations

Not all decimal expansions of irrationals follow such nice patterns as those we have looked at thus far. In fact, if we consider the decimal expansion of $\sqrt{2}$, we find that the sequence of digits follows no apparent pattern.

To explain this, use your calculator to evaluate $\sqrt{2}$. Use the key sequence 2 $\boxed{\sqrt{}}$. Correct to one decimal place the display will read 1.4. But $\sqrt{2}$ is not precisely equal to 1.4 because $1.4^2 = 1.96$. Since $1.5^2 = 2.25$ we know that $\sqrt{2}$ lies between 1.4 and 1.5. Similarly

$$1.41^2 = 1.9881 \qquad 1.42^2 = 2.0164$$

So $\sqrt{2}$ lies between 1.41 and 1.42. Table 2 lists further approximations of $\sqrt{2}$. Notice how the use of the decimal expansion improves our feeling for the relative size of $\sqrt{2}$, and makes us feel more comfortable with it as a number.

The decimal approximations of $\sqrt{2}$ in Table 2 reveal a number of important facts about decimals and irrationals:

1 Each approximation is a terminating decimal; hence each is a rational number and cannot be precisely equal to $\sqrt{2}$. This point is emphasized by noting that the square of each decimal is not equal to 2.

2 The approximations get closer to $\sqrt{2}$ as we include more decimal places. This is evident because x^2 gets closer to 2 as more decimal places are considered.

TABLE 2

Approximations of $\sqrt{2}$

Decimal Places	Approximation, x, of $\sqrt{2}$	The Square, x^2, of the Approximation
1	1.4	1.96
2	1.41	1.9881
3	1.414	1.999396
4	1.4142	1.9999616
5	1.41421	1.9999899
6	1.414214	2.0000012
7	1.4142136	2.0000001

3   You can tell whether the approximation is greater than or less than $\sqrt{2}$ by considering whether its square is greater than or less than 2. Hence the first five approximations in the table are less than $\sqrt{2}$ while the last two approximations are greater.

4   What does the latter point have to do with round-off? Each decimal is rounded off from the entire decimal representation of $\sqrt{2}$. Sometimes when you round off you get a number greater than the original number and sometimes less. The last significant digit is the one affected. For example, look at the sixth entry in Table 2. Since the square of 1.414214 is greater than 2, its last digit must have been rounded off one number higher so that the actual decimal must be of the form 1.414213 . . . , and the next digit must be at least 5. We see it is 1.4142136. Can you predict what the next approximation would be? It would be 1.4142135__, where the last digit is at least 5.

The skill of approximating numbers by decimals is a very important one. Before 1971, when the first hand-held calculator was marketed, a number like $\sqrt{300}$ was difficult to deal with. A decimal approximation was not easily obtained. Today we simply use a calculator with the key sequence 300 $\boxed{\sqrt{}}$ and get the approximation 17.320508 almost instantaneously. So as long as we recognize that this decimal is an approximation to the irrational number $\sqrt{300}$—that is, they are not equal but very close—the calculator enables us to feel much more comfortable with this type of irrational number.

It is possible to approximate square roots without a calculator, but, of course, the accuracy suffers. The method is to use Table 1 to locate the two square numbers between which the number lies and then approximate the decimal part of the number. Let us look at two examples.

Example 3
Use Table 1 to find a rough approximation of $\sqrt{60}$. Since 60 lies between 49 and 64 we know that $\sqrt{60}$ lies between 7 and 8. The difference between 64 and 49 is 15 ($= 64 - 49$) and the difference between 60 and 49 is 11; so 60 is about $\frac{11}{15}$ of the distance from 49 to 64. Since $\frac{11}{15}$ is about 0.7, we see that 60 is about 0.7 of the way from 49 to 64. Hence a fairly good approximation of $\sqrt{60}$ is 7.7. To check we see that $(7.7)^2 = 59.29$.

Example 4
Problem   Use Table 1 to approximate $\sqrt{210}$.
Solution   Since 210 lies between the two square numbers 196 and 225, $\sqrt{210}$ lies between 14 and 15. Computing the differences $210 - 196 = 14$ and $225 - 196 = 29$ enables us to say that $\sqrt{210}$ lies about $\frac{14}{29}$ of the distance from 196 to 225. Since $\frac{14}{29}$ is about 0.5 our approximation is 14.5. Checking yields $(14.5)^2 = 210.25$.

Summary

The ascent of irrationals began with a sudden fury. The discovery of an irrational number was followed immediately by the suppression of its existence. Throughout the development of our number system, irrational numbers have always played a strange role. Their existence was discovered well before other types of numbers such as negatives and decimals. Yet it was not until very recently that mathematicians were able to give a concise definition of irrational numbers.

For example, the number π is known to be irrational, but the proof of this fact is very difficult. Because it is irrational we know that the decimal expansion of π is nonterminating and nonrepeating.

The facts that you should be aware of from this section are:

1   $\sqrt{2}$ is an irrational number and it is possible to give a fairly straightforward argument to prove it.

2   A number is irrational if and only if its decimal equivalent is nonterminating and nonrepeating.

3   It is possible to generate irrationals by forming numbers of the form $\sqrt{p}$, where p is a prime.

4   Further examples of irrationals can be generated by forming decimals that follow a pattern that ensures that no set of digits is continuously repeated.

5   Very accurate approximations of square roots can be found by using a calculator with a square root key.

6   Less accurate approximations of square roots can be made by locating the two perfect squares that the number lies between and then approximating the decimal.

EXERCISES

In Problems 1 to 4 evaluate the square root.

1   $\sqrt{289}$ 3 $\sqrt{1225}$

2   $\sqrt{196}$ 4 $\sqrt{2025}$

In Problems 5 to 18 determine whether the number is rational or irrational.

5   $\sqrt{4}$ 8 $\sqrt{6}$

6   $\sqrt{625}$ 9 $0.151151115\ldots$

7   $\sqrt{3}$ 10 $0.123123312333\ldots$

11 0.1234567891011 . . .   15 $\sqrt{2} - 1$

12 0.3142731427 . . .   16 $3\sqrt{2}$

13 0.28652865 . . .   17 $4\sqrt{5}$

14 0.369121518 . . .   18 $\sqrt{5} + 1$

In Problems 19 to 22 use Table 1 to approximate the square root to one decimal place.

19 $\sqrt{72}$ 21 $\sqrt{200}$

20 $\sqrt{300}$ 22 $\sqrt{90}$

Problems 23 and 24 refer to a phrase in a *National Geographic* article by the eminent primate specialist Jane Goodall. She said the chimpanzee community at Gombe had a home range that "fluctuated between five and eight square miles." Assume the home range is square.

23 What is the length of the side of a home range with 8 square miles?

24 What is the length of the side of a home range with 5 square miles?

25 Use the table below to make a conjecture about which numbers have a square root that is rational, where Q stands for rational and I irrational.

Number	$\sqrt{1}$	$\sqrt{2}$	$\sqrt{3}$	$\sqrt{4}$	$\sqrt{5}$	$\sqrt{6}$	$\sqrt{7}$	$\sqrt{8}$	$\sqrt{9}$	$\sqrt{10}$
Type	Q	I	I	Q	I	I	I	I	Q	I

26 Use the table below to make a conjecture about which numbers have a cube root that is rational. (See Problem 25.)

Number	$\sqrt[3]{1}$	$\sqrt[3]{2}$	$\sqrt[3]{3}$	$\sqrt[3]{4}$	$\sqrt[3]{5}$	$\sqrt[3]{6}$	$\sqrt[3]{7}$	$\sqrt[3]{8}$	$\sqrt[3]{9}$	$\sqrt[3]{10}$
Type	Q	I	I	I	I	I	I	Q	I	I

Problems 27 and 28 refer to this table of approximations of $\sqrt{5}$.

Decimal Places	x	x^2
1	2.2	4.84
2	2.24	5.0176
3	2.236	4.999696

27 Is the approximation 2.24 greater or less than $\sqrt{5}$?

28 Is the approximation 2.236 greater or less than $\sqrt{5}$?

Problems 29 and 30 refer to $\sqrt{5}$, which, when approximated to five decimal places, is 2.23606.

29 What is $\sqrt{5}$ when approximated to four decimal places?

30 What is $\sqrt{5}$ when approximated to three decimal places?

31 (*a*) Evaluate $\frac{3}{2} \cdot \frac{3}{2}$.
  (*b*) Is $\frac{3}{2} \cdot \frac{3}{2}$ less than 2?
  (*c*) Is $\frac{3}{2}$ less than $\sqrt{2}$?

Use Problem 31 to answer Problems 32 to 34.

32 Which is larger, $\frac{9}{4}$ or $\sqrt{5}$?

33 Which is larger, $\frac{7}{5}$ or $\sqrt{2}$?

34 Which is larger, $\frac{7}{4}$ or $\sqrt{3}$?

35 Which is greater, $\dfrac{1}{\sqrt{2}}$ or $\dfrac{1}{\sqrt{3}}$?

36 Which is greater, $\dfrac{1}{\sqrt{5}}$ or $\dfrac{1}{\sqrt{6}}$?

37 Evaluate $\sqrt{\dfrac{1}{4}}$.

38 Evaluate $\dfrac{1}{\sqrt{9}}$.

Problems 39 to 42 are concerned with two "laws" of linearity.

39 Does $\sqrt{4} + \sqrt{9} = \sqrt{13}$?

40 Does $\sqrt{4} \cdot \sqrt{9} = \sqrt{36}$?

41 Does $\sqrt{a} + \sqrt{b} = \sqrt{a + b}$ for all numbers a, b?

42 Does $\sqrt{a} \sqrt{b} = \sqrt{a \cdot b}$ for all numbers a, b?

In Problems 43 and 44, use your calculator to make a table like Table 2 to approximate these numbers.

43 $\sqrt{8}$

44 $\sqrt{10}$

The Greek mathematician Heron of Alexandria (about A.D. 100) made use of unit fractions to approximate square roots. In Problems 45 to 48, use your calculator to approximate the square root and Heron's approximation. To how many decimal places is Heron's approximation accurate?

Root	Heron's Approximation
45 $\sqrt{32}$	$5 \ \frac{1}{2} \ \frac{1}{14}$
46 $\sqrt{63}$	$7 \ \frac{1}{2} \ \frac{1}{4} \ \frac{1}{8} \ \frac{1}{16}$
47 $\sqrt{593}$	$24 \ \frac{1}{4} \ \frac{1}{8}$
48 $\sqrt{126} \ \frac{1}{2} \ \frac{1}{4}$	$11 \ \frac{1}{4}$

49 Examine the proof that $\sqrt{2}$ is irrational. Alter the proof to show why $\sqrt{5}$ is irrational too.

50 Use Problem 53 as a guide to prove that $\sqrt{p}$ is irrational for every prime number p.

51 The proof that $\sqrt{2}$ is irrational cannot be used to prove that $\sqrt{12}$ is irrational. What is the first step in the proof that will fail if you try to place 12 wherever 2 occurs?

52 A horse called Clever Hans would reply to mathematical problems put to him with coded taps of his foreleg. When asked, "Hans, how much is twice the square root of 9, less 1?" He would dutifully raise his right foreleg and tap 5 times. Was he correct? (See C. Sagan, "Broca's Brain," Random House, New York, 1979.)

2-5 SETS

There are a number of uses of the concept of number. One is an identification: student numbers, social security numbers, car licenses, ZIP codes. Another is to put things in an order: customers in a bakery, places in a race, positions in the draft, etc. Others are to supply writers with descriptive phrases and to give them subjects on which to vent their frustration. As an example of the latter consider the following item which appeared in William Safire's column "On Language" in *The New York Times Magazine*:*

> A number of letters have crossed my desk (and when my desk is crossed, it retaliates with a furious memo) about a meaningless phrase, used by lazy writers, which is enjoying a boom. The phrase is "a number of."
>
> Complainants zero in on the number of times "a number of" is used in *The New York Times.* In a piece about editor Roy E. Larsen's retirement: "Mr. Larsen, who held a number of titles. . . ." In a financial story: "A growing number of other major bank holding companies. . . ." In a dispatch from overseas: "Pretoria had ordered the expulsion of a number of American embassy personnel. . . ." An irate editorial: "What is unseemly, however, is how a number of United Way chapters. . . ."

*"A Number of," June 9, 1979. © 1979 by The New York Times Company. Reprinted by permission.

What is unseemly is the way this locution fails to answer the simple question: How many? If the writer does not know, and has no time to find out, several more specific fudges are available: ''Several'' is one; ''a few'' is another; ''many'' and ''scores'' are available as well, and then we're off into ''a whole bunch,'' and on up to the Greek ''myriad,'' which means 10,000.

The best way to break ourselves of this habit is to think of phrases that might have wound up as ''the face that launched a number of ships,'' ''through the valley of death charged a number of soldiers,'' ''a number of years ago, our fathers brought forth on this continent. . . .'' Newsmen in biblical times might begin their stories: ''Moses descended from Mt. Sinai with a number of commandments,'' or ''It rained a number of days and a number of nights.''

How many? Give us readers a hint; ''a number of'' is too broad to mean anything. Therefore, change the opening of this entry to read ''Five letters have crossed my easily irritated desk,'' all from one man, Alex W. Burger of New Rochelle, N.Y., who gets all worked up at writers who do a number on him.

Georg Cantor, a mathematician in the late nineteenth century, had the same view of the state of mathematics, and in particular the number concept, as William Safire has about the common phrase "a number of things." Safire was worried that there was little agreement in the meaning of this simple phrase and that the word "number" was being misused. Cantor was worried that mathematicians did not have a firm understanding of the number concept. Safire called for more specific information about the term number. Cantor also wanted to find a way to unify and organize the many branches of mathematics, and to provide a common language for them. Mathematicians had been groping for a way to handle numbers, and Cantor accomplished this by creating set theory and using it to form a firm foundation for all of mathematics.

In the previous section we mentioned that the current concept of an irrational number was not fully understood until the late 1800s. It was Cantor and his set theory, along with his friend Dedekind, who paved the way for a synthesis of our concept of number.

We will use the rudiments of set theory to organize and synthesize the material in our first two chapters, in a similar way that Cantor did. As a result, we will also be able to answer one of the first questions posed in this text, namely, has your mathematical education paralleled the historical development of the subject?

Sets

We have studied a number of different sets of numbers in the first two chapters. The concept of a set can easily be extended to include sets of objects other than numbers, such as the set of all birds, the set of students at your school, or the set of all words in "Webster's New Collegiate Dictionary." For the most part we will deal with sets of numbers.

There are various methods to describe sets. The most direct way is the method used in the previous paragraph, the *descriptive method,* in which the phrase "the set of all . . ." precedes a description of the set. You must be certain that the set is *well defined* in that it is theoretically possible to tell whether every single object is in the set or not. For instance, the "set of all large numbers" is not a well-defined set because the word "large" is ambiguous. Usually we let a capital letter stand for the set. In Chapter 1 we wrote, "Let N stand for the set of all natural numbers." Likewise we will let Q represent the set of rational numbers.

Another way of specifying a set is by using braces, $\{\ \}$, to denote objects in the set or to describe the elements in the set. (We use the terms "objects" and "elements" synonymously.) For example, we can write

$$A = \{4, 5, 6\}$$

which is read "A is equal to the set containing 4, 5, and 6." This means that A is the name of the set whose elements are 4, 5, and 6. We can also use this notation to define the set by implying the characteristic property or properties of the set by using an ellipsis. For example, we now define the set of all *integers* to be the set I defined by

$$I = \{\ldots, -2, -1, 0, 1, 2, 3, \ldots\}$$

Thus I consists of the set of natural numbers N, where $N = \{1, 2, 3, \ldots\}$, together with their negatives and 0. We can also explicitly state the property using set-builder notation, whose form is $\{\underline{\ \ } \mid \underline{\ \ }\ \underline{\ \ }\ \underline{\ \ }\}$ and which is read "the set of all $\underline{\ \ }$ such that $\underline{\ \ }\ \underline{\ \ }\ \underline{\ \ }$." For example, the set of all rational numbers Q can be defined by

$$Q = \left\{\frac{a}{b} \mid a, b \text{ are integers, } b \neq 0\right\}$$

which is read "Q is equal to the set of all a/b such that a, b are integers, b not equal to 0." This definition extends the definition of rational numbers, given in Section 2-1, to include negative numbers.

The symbol $\in$ is used to represent "is an element of." Thus for the set $A = \{4, 5, 6\}$ we have $4 \in A$, $5 \in A$, and $6 \in A$. No other number is an element of A. In order to explicitly state that a number, say 7, is not an element of A we write $7 \notin A$.

A set with a finite number of objects is called a *finite* set. Otherwise it is an *infinite* set. The set A above is a finite set since it has three elements, whereas N, I, and Q are infinite sets.

Subsets

Notice that every natural number is an integer. We say that N is a subset of I and write $N \subset I$. In general, for two sets S and T we say S is a *subset* of T, written $S \subset T$, if and only if whenever $x \in S$ then $x \in T$. If $S \subset T$, we sometimes say that T *contains* S. If S is not a subset of T, we write $S \not\subset T$. Let us look at some examples.

Example 1
Consider the sets $A = \{4, 5, 6\}$, $B = \{-1, 1\}$, and $C = \{5, 6\}$. Of these sets only one is a subset of another: $C \subset A$. We also have $A \subset N$ and $C \subset N$, but because $-1 \in B$ and $-1 \notin N$ we have $B \not\subset N$. However, each set is a subset of I. We write $A \subset I$, $B \subset I$, and $C \subset I$.

Example 2
Problem   Consider the set $D = \{0, \frac{1}{2}, 1\}$. Is D a subset of N, I, or Q?
Solution   Note that $\frac{1}{2}$ is neither a natural number nor an integer. If we first consider N, we see that $\frac{1}{2} \in D$ but $\frac{1}{2} \notin N$ so $D \not\subset N$. Also $\frac{1}{2} \notin I$ so $D \not\subset I$. Each element of D is a rational number so $D \subset Q$

The most important subsets for our purposes involve the three sets N, I, Q. We stated that $N \subset I$. The next example shows that $I \subset Q$. Hence $N \subset Q$. Thus Q contains all natural numbers, their negatives, and 0.

Example 3
Problem   Show that $I \subset Q$.
Solution   We must show that every element $n \in I$ is expressible as a fraction. We have $n = n/1 \in Q$. Hence every element in I is in Q, so $I \subset Q$.

We can visually express the fact that $N \subset I \subset Q$ by a *Venn diagram*, which is a diagram using circles to represent sets. In Figure 1 the innermost circle represents N. It is contained in the circle representing I. Each is contained in the circle representing Q.

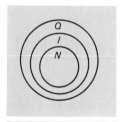

FIGURE 1

Operations on Sets

Are there any numbers outside of Q? The Pythagoreans thought not—until Hippasus demonstrated that $\sqrt{2}$ is such a number. The numbers that are outside of Q are the irrationals. We say that the set of irrational numbers is the complement of Q. In general, if S is a set, then the *complement* of S, written S' (read "S prime" or "S complement"), is the set of objects not in S. This definition raises another issue. If we start looking for objects that are not in Q, how far should we look? For instance, is the kitchen sink "not in Q"? Of course, we need to limit our search to

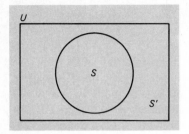

FIGURE 2

some set that we call the *universal* set, most often denoted by U. The Venn diagram in Figure 2 gives a visual depiction of a set and its complement S', where the universal set U consists of the area in the rectangle. Thus S is represented by the area inside the circle, and S' is the area inside U but outside S.

Usually the universal set is understood from context, but sometimes it must be specified. Let us look at an example.

Example 4

Problem For the following choices of the set A and the universal set U, find A'.

(a) $A = \{1, 2, 3\}$, $U = \{1, 2, 3, 4, 5, 6\}$

(b) $A = \{1, 2, 3\}$, $U = N$

(c) $A = \{2, 4, 6, 8, \ldots\}$, $U = N$

Solution

(a) $A' = \{4, 5, 6\}$

(b) $A' = \{4, 5, 6, 7, \ldots\}$

(c) $A' = \{1, 3, 5, 7, \ldots\}$

There are two more important operations on sets. We remarked before that I consists of the elements in N, their negatives, and 0. If we let

$$M = \{0, -1, -2, -3, -4, \ldots\}$$

then we say that I is the union of N and M and we write $I = N \cup M$. In general, if S and T are sets, then the *union* of S and T, which is written $S \cup T$, is the set of all elements that are either in S or in T (or in both). Let us consider an example.

Example 5

Problem Let $A = \{4, 5, 6\}$, $B = \{6, 7, 8\}$, and $C = \{2, 4, 6, 8\}$. Find $A \cup B$, $B \cup C$, and $A \cup C$.

Solution

$$A \cup B = \{4, 5, 6, 7, 8\}$$

$$B \cup C = \{2, 4, 6, 7, 8\}$$

$$A \cup C = \{2, 4, 5, 6, 8\}$$

Notice that for any set S and universal set U we have $S \cup S' = U$.

If Q is the set of rational numbers, then Q' is the set of irrational numbers. Forming the union of these two sets, $Q \cup Q'$, gives us the set of all numbers that we will consider. This set is called the set of *real numbers,* denoted by R. Thus

> The set of the real numbers R is the union of the set of rational numbers and the set of irrational numbers. Hence $R = Q \cup Q'$.

Sometimes it is important to refer to the set of elements that are common to two sets. If S and T are sets, we say that the *intersection* of S and T, written $S \cap T$, is the set of all elements that are in S and T. In other words, the elements of $S \cap T$ are the elements that are in both S and T. Let us consider an example.

Example 6
Problem Consider the sets $A = \{1, 2, 3, 4\}$, $B = \{2, 4, 6, 8\}$, $C = \{0, 1, 2, 9\}$. Find $A \cap B$, $B \cap C$, and $A \cap C$.
Solution

$$A \cap B = \{2, 4\}$$
$$B \cap C = \{2\}$$
$$A \cap C = \{1, 2\}$$

What is the set $Q \cap Q'$? There is no number that is a rational number and an irrational by definition of Q'. The set that has no elements is called the *empty* set and is denoted by $\varnothing$. Hence $Q \cap Q' = \varnothing$. If two sets S and T have no elements in common, so $S \cap T = \varnothing$, then they are called *disjoint*. For any set S we have that S and S' are disjoint.

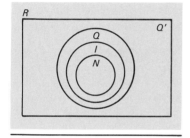

FIGURE 3

The Set of Real Numbers

The most important idea in this section is the organization of our number system according to its various subsets. Our number system is called the *real number system* and its important subsets are N, I, and Q. Thus

$$N \subset I \subset Q \subset R$$

In terms of a Venn diagram the three circles depicting these sets in Figure 3 give us a visual representation of the containment properties.

Does Ontogeny Recapitulate Phylogeny?

One of the first questions posed in Section 1-1 was whether your mathematical education paralleled the historical development of the subject. In this section we covered the rudiments of elementary set theory, and you should be able to give a definite answer to our question.

You most likely studied many aspects of set theory even in grade school. In fact, when you first encountered multiplication, you probably separated objects into various sets to understand the concept. For example, to see that $3 \cdot 4 = 12$ you most likely considered that 3 sets of 4 objects yields 12 objects. Yet mathematicians did not formalize the concept of set until Cantor's time. This provides us with one answer to the question (stated so philosophically in the subsection title). The two are not parallel since set theory is used in early mathematical education but was only a recent historical development.

There are many other instances where the development of mathematics did not follow what we might call the "natural course" in the sense of how we learn it. Let us consider some.

The Evolution of Numbers

We have seen how the four sets N, I, Q, and R are related logically, that is, in terms of containment. This aspect of the various number sets conveys the underlying structure which both describes and connects them. We shall refer to it as the *logical aspect*. But historically this approach is unwarranted. Let us review how and when these sets developed.

The natural numbers came first, naturally. Some artifacts from prehistoric times indicate that tally procedures were known a long time ago. Moreover, anthropological studies have revealed that every known culture possessed some kind of number system. Of course, numbers can be written in many ways (numerals), and we examined the systems used by the ancient Egyptians and Mayans. Moreover, we saw that the number concept can exist in a language without being reflected in a written system.

In particular it is significant to examine when the integers emerged. The oldest existing mathematical document, the Rhind papyrus, dates back to about 1800 B.C., and it presupposes a knowledge of the natural numbers and the positive rational numbers. The question is, then, when did negative numbers emerge? The answer to this question it is a counterexample to the prevailing notion that mathematics progresses continuously and linearly. Let us look at the checkerboard history of negative numbers.

The Chinese used rod numerals on counting boards about 300 B.C. They calculated with two sets of rods—a red set for positive numbers and a black set for negatives—and the idea of negative numbers did not seem to cause them difficulty. They were thus the first people to use negative numbers in computations. An Indian mathematician, Brahmagupta (about A.D. 628), presented a systematized arithmetic of negative numbers and zero, but rejected negative roots of equations. By 1500 a negative number appeared in an equation for the first time when Nicolas Chuquet wrote $4x = -2$. However, in the same work he rejected zero as being the solution of an equation. Two first-rate Renaissance mathematicians rejected negative numbers completely. Cardano, who presented the first general method of solving third- and fourth-degree equations in 1545, referred to them as "fictitious" numbers. Viète, who introduced the modern notation of letting letters denote unknowns, would neither admit negative coefficients into his equations nor consider negative numbers as roots of any equation. Even the famous French philosopher René Descartes called them "false" roots. Most eighteenth-century textbooks on algebra dwelled on the multiplication rules of negative numbers, but some rejected categorically the multiplication of two negative numbers.

The important point here is that the concept of a rational number, known to the ancient Egyptians, preceded negative numbers.

Recall also that the first definitive use of zero as a number occurred in India in 876. Hence the integers did not develop until many years after rational numbers were invented.

The evolution of the numeration system exhibits another, particularly fascinating anomaly in contrast with the logical development. Recall that the Pythagoreans knew of the existence of some irrational numbers during the fifth century B.C. (More precisely, they were forced to acknowledge their existence.) Thus irrational numbers preceded integers.

This brief historical presentation has shown that the concepts which had the most intuitive meaning were accepted and utilized first, while the less intuitive ones required many centuries either to be created or to be accepted. Henri Poincaré, one of the greatest twentieth-century mathematicians, wrote that "in a brief period the development of the embryo recapitulates the history of its ancestors of all geological epochs." Do you agree with this statement? Do you think that all people experience what their predecessors did? Our predecessors traversed a path similar to

$$N \to Q \to Q' \to I \to R$$

The logical path is

$$N \to I \to Q \to Q' \to R$$

What path did you follow in school? What path do you think is the best to follow?

EXERCISES

In Problems 1 to 6 determine if the set is well defined.

1   The set of all large dogs.

2   The set of all tall people.

3   The set of all positive integers.

4   The set of rational numbers between 1 and 2 inclusive.

5   The set of all nice numbers.

6   The set of all cities in the United States with a population of over 100,000.

In Problems 7 to 12 list the elements in the set.

7   The set of divisors of 12.

8   The set of prime numbers less than 20.

9   The set of perfect numbers less than 25.

10   The set of integers between 0 and 5 inclusive.

11   The set of integers between -3 and 0 inclusive.

12   The set of even integers between -10 and -4 inclusive.

In Problems 13 to 18 specify the sets by giving the defining property of the elements.

13 $\{1, 2, 3, 4, 5, 6, 7\}$

14 $\{2, 4, 6, 8, 10\}$

15 $\{3, 6, 9, 12, 15\}$

16 $\{2, 3, 5, 7, 11, 13, 17, 19\}$

17 $\{10, 11, 12, 13, 14\}$

18 $\{1, 4, 9, 16, 25\}$

In Problems 19 to 24 determine whether A and/or B is a subset of S.

19 $S = \{1, 2, 3, 4\}, A = \{0, 1, 2\}, B = \{2, 3, 4\}$

20 $S = \{-1, 0, 1, 2, 3, 4\}, A = \{0, 1, 4\}, B = \{1, 3, 4\}$

21 $S = \{1, 2, 3, 4\}, A = \{3, 4, 5\}, B = \{0, 1, 2\}$

22 $S = \{-2, -3, -4\}, A = \{-5\}, B = \{-2, -3, -4, -5\}$

23 $S = I, A = \{-2, 0, 2, 4\}, B = \{2, 4, 6, 8, \ldots\}$

24 $S = Q, A = \{x \mid x^2 = 4\}, B = \{x \mid x^2 = 2\}$

In Problems 25 to 30 find S' for the given choice of U.

25 $S = \{1, 2, 3\}, U = \{1, 2, 3, 4, 5, 6\}$

26 $S = \{1, 2, 3\}, U = \{0, 1, 2, 3, 4, 5, 6, 7\}$

27 $S = \{1, 2, 3\}, U = N$

28 $S = \{1, 2, 3\}, U = I$

29 $S = \{2, 4, 6, 8, 10, \ldots\}, U = N$

30 $S = N, U = I$

In Problems 31 to 38 find $A \cup B$.

31 $A = \{1, 2, 3\}, B = \{2, 3, 4\}$

32 $A = \{-1, 0, 1\}, B = \{-1, 1\}$

33 $A = \{2, 4, 8, 16, 32\}, B = \{64\}$

34 $A = \{1, 2, 3, 4\}, B = \{1, 2, 3, 4\}$

35 $A = \{-2, -1, 0, 1\}, B = N$

36 $A = \{\sqrt{2}, 2, \sqrt{3}, 3\}, B = Q$

37 $A = Q, B = \emptyset$

38 $A = Q, B = Q'$

In Problems 39 to 48 find $S \cap T$.

39 $S = \{1, 2, 3, 4\}, T = \{3, 4, 5\}$

40 $S = \{1, 2, 3, 4\}, T = \{4, 5\}$

41 $S = \{2, 4, 8, 16\}, T = \{32, 64\}$

42 $S = \{2\}$, $T = \{4\}$

43 $S = \{2, 4, 6, 8\}$, $T = \{0, 2\}$

44 $S = \{2, 4, 6, 8\}$, $T = \{0, 2, 4, 6, 8\}$

45 $S = \{-2, -1, 0, 1, 2\}$, $T = N$

46 $S = \{\frac{2}{2}, \frac{3}{2}, \frac{4}{2}, \frac{5}{2}\}$, $T = I$

47 $S = Q$, $T = \{\sqrt{2}, \sqrt{3}, \sqrt{5}, \sqrt{7}, \sqrt{11}, \ldots\}$

48 $S = Q'$, $T = \{1, \frac{1}{2}, \frac{1}{4}, \frac{1}{8}, \frac{1}{16}, \ldots\}$

In Problems 49 to 54 draw a Venn diagram of the sets S and T, with nonintersecting circles if $S \cap T = \varnothing$, with one circle inside the other if $S \subset T$ or $T \subset S$, or with intersecting circles if $S \cap T \neq \varnothing$ and neither is a subset of the other.

49 $S = \{1, 2, 3\}$, $T = \{4, 5, 6\}$

50 $S = N$, $T = Q$

51 $S = N$, $T = I$

52 $S = \{-2, -1, 0, 1, 2\}$, $T = N$

53 $S = Q$, $T = Q'$

54 $S = I$, $T = Q'$

55 Explain the difference between the terms *element* of a set and *subset* of a set.

56 Give an example of a set that has an element that is a set.

57 Here is a set that has an element that is also a subset of the set: $\{1, 2, \{1, 2\}\}$. Explain.

58 One of the reasons why Cantor's work was not accepted by the majority of the mathematical hierarchy is that some inconsistencies arose directly from his results. Here is one such paradox:

> Every set is either an element of itself or it is not. For example, the set of all infinite sets has an infinite number of elements so it is an infinite set, thus it is an element of itself. Every other set that we have mentioned is not an element of itself. Let S be the set of all sets that are not elements of themselves. Either $S \in S$ or $S \notin S$. However, if $S \in S$ then S is not an element of itself by definition of S so $S \notin S$. This is a contradiction. Hence it cannot be true that $S \in S$. Therefore $S \notin S$. But in this case S is not an element of itself so it is in S. Therefore $S \in S$, another contradiction.

This is one of the most interesting paradoxes in mathematics. It touches at the very heart of mathematics. The difficult part of this problem is to understand the paradox.

2-6 A DIGRESSION ON PROOF

We have described in some detail how the concept of number ascended from its earliest, primitive stages to the highly structured, modern approach using set theory. Along the way we encountered several proofs, including a proof of the fact that $\sqrt{2}$ is an irrational number. Thus it is appropriate for us to pause here to discuss what is meant by a mathematical proof, both because it differs from the general notion of proof and because the ascent of mathematical proof will be examined in detail in Chapter 3.

We will begin with a quote from a timely article by Robert Reinhold that appeared in *The New York Times** (January 8, 1980). It will serve as a thread that binds our discussion of mathematical proof to other commonly accepted types of proof.

Probably nowhere does the world of science intersect more directly with politics and the making of public policy than when it comes to deciding how convincing scientific evidence has to be to ban, label or otherwise restrict a substance thought to be hazardous to public health. And probably nowhere else are the foibles and limitations of both worlds more apparent.

Consider the case of caffeine. For years indirect evidence has been accumulating that this substance—by far the most widely consumed stimulant as well as a food ingredient of wide economic importance—may cause cleft palates and other birth defects when used by pregnant women. The Food and Drug Administration is expected to release soon the results of a lengthy caffeine study of its own, done on rats, and it is reliably understood that the findings will strongly confirm the suspected hazard.

All else being equal, prudence might suggest requiring a warning label on the coffees, teas, cola drinks and many other foods and drugs containing caffeine. But all else is seldom equal in such matters, and the agonizing in Washington these days over what to do about caffeine illustrates well the vast political, economic and technical complexities that underline the translation of scientific fact into public policy.

Is the evidence solid enough? How will Congress react? Should a move be made during an election year? What will be the economic consequences? Is the public numb to new reports of hazards in one food after another? Are more deformed babies being born for every day of delay?

The emphasis in the last paragraph of this quote is ours because the entire issue revolves about the question of determining whether the evidence is indeed convincing. After all, if it were found *beyond a shadow of doubt* that a certain level of caffeine intake among pregnant women *definitely* caused birth defects, then some kind of warning label (at the least) would be required, regardless of whether it were an election year or certain segments of the economy would suffer. True, the public might be numb to new reports of hazards in foods, but no human being can remain unmoved by definite adverse effects to unborn children.

*"Caffeine Quandary Illustrates FDA's Plight," January 8, 1980. © 1980 by The New York Times Company. Reprinted by permission.

Once again the important words have been emphasized. In fact, there are really two separate questions which lie at the heart of the basic issue:

1 How is knowledge obtained?

2 What is the degree of certainty that can be attached to that knowledge?

Let us take a general look at both these questions, because both relate to the essential spirit of mathematics.

Ways to Gain Knowledge

There are several ways in which we humans gain knowledge. The first undoubtedly is *innate.* Then there are *authority figures,* ranging from parents and teachers to religious leaders and the head of the Federal Reserve Board. Another way is by *experience,* which your automobile mechanic and personal physician use when making diagnoses, prescribing remedies, and charging for their services. Other ways involve *intuition* and *reason.* Let us look at two anecdotes from the physical sciences to illustrate some ways in which mathematical scientists use these and other ways of obtaining knowledge.

Anecdote 1 Mendeleev (Chemistry)

Up until about 1860 it was known that each chemical element could be defined by its characteristic atomic weight. But what was not known was how the properties of families of elements depended upon this number. What is the role of number in science? The modern physicist N. D. Mermin, of Cornell, wrote: "A number unaccounted for cries out for explanation, and the act of explaining is an act of conquest."

The act of conquering the underlying problem of the characteristic atomic weight of an element awaited the Russian chemist Dmitri Ivanovich Mendeleev. He was the youngest of at least 14 children, and his widowed mother drove her darling, brilliant son to science. He loved science, and he especially loved to investigate the properties of the elements. He wrote each element on a card and arranged the cards on a table, with the first column of 6 elements in order of increasing atomic weights. The next element in order of atomic weights is sodium, and since its properties closely resemble those of lithium, he began a second column with sodium next to lithium. Some of the elements are listed with their atomic numbers in Table 1.

Mendeleev discovered that each horizontal row consisted of a set of elements, each having common properties. He then discovered a very natural pattern, one which provided evidence for a mathematical key to the elements. He published his periodic table of elements in 1869, when he was 35 years old.

Even the briefest glance at Table 1 reveals a problem: Something is missing. And this was Mendeleev's stroke of genius, for he solved it by *interpreting* it as a gap. He wrote:

TABLE 1

Atomic Weights of Some Elements

Li	7	Na	23	K	39
Be	9	Mg	24	Ca	40
B	11	Al	27		
C	12	Si	28	Ti	48
N	14	P	31		
O	16	S	32		

> There is a missing element there, and when it is found its atomic weight will put it before titanium. Opening the gap will put the later elements of the column into the right horizontal rows; titanium belongs with carbon and silicon.

Mendeleev continued to make very practical forecasts; he predicted not only the existence of the missing element but also several of its properties. There were in fact three such gaps altogether, and within 15 years all three were discovered.

Notice how Mendeleev obtained his knowledge. First he observed a *pattern* based upon the *experimental data* he had accumulated, and then he explained the pattern. It is important to note that his theory preceded the discovery of the missing element. Indeed, Mendeleev's theory predicted its very existence and several of its properties. In the next anecdote we will see another example of theory preceding application. It concerns a discovery which led to many of the amenities we enjoy in our western technological society, including stereo music systems.

Anecdote 2 Maxwell (Physics)

James Clerk Maxwell was a contemporary of Mendeleev and, like Mendeleev, was a born scientist. As a child he was persistently inquisitive and constructed his own scientific toys at the age of 8. Ten years later he published his first research papers while still an undergraduate at the University of Edinburgh. After graduation from Cambridge he held a number of prestigious chairs at British universities until suddenly, at the age of 34, he gave up university life for the seclusion of his family estate in Scotland. It was here, 2 years later, that Maxwell arrived at the epochal conclusion that light is an electronic wave.

Maxwell came to this conclusion in a curious way. First he derived four mathematical equations relating electrical charges and currents to electric and magnetic fields. But he felt that the equations were not *aesthetic* because they lacked a certain *symmetry*. Based upon this criterion alone, and not upon experimental evidence, he theorized that one of the equations was missing an additional term. Thus it was Maxwell's *intuition* which led to the discovery of the existence of electromagnetic radiation, and this in turn caused a technical revolution that has produced electric lights, telephones, radio, television, cardiac pacemakers, computers, and high-fidelity electronic music systems.

Anecdote 2 adds two more ways of gaining knowledge, aesthetics and symmetry, both of which play vital roles in mathematics. The quest for symmetry often motivates research, and a certain kind of aesthetics is sometimes applied to certain investigations.

These two anecdotes illustrate the scientific method of obtaining knowledge. Roughly, the scientific method involves experimentation, the search for patterns, the statements of laws based upon the patterns, and the proofs of the statements. This is exactly how a mathematician proceeds too; a mathematician obtains knowledge as does everyone else. The significant difference, however, lies in the proof. Let us turn to this issue.

The Meaning of Proof

We have seen several ways in which knowledge is gained. Once knowledge is gained, the question arises:

> What degree of certainty can be attached to that knowledge?

Let us answer this from three perspectives. The general public becomes convinced when there is enough data given. The public would conclude that Fermat's conjecture about $x^n + y^n = z^n$ is true because it is true for the first 125,000 cases.

Experimental scientists employ a different set of criteria. Because they possess a healthy skepticism about their work, they insist that all results must occur within a prescribed margin of error, say 5 percent, and that other independent researchers must be able to duplicate the experiment within the same range of error.

Theoretical scientists use a much more stringent criterion. They insist upon 100 percent certainty. No amount of data is sufficient for them unless it is total, and no margin of error is acceptable unless it is zero. Thus Mendeleev's theory was not accepted until the missing element was discovered and its properties verified. Since that time his periodic table has formed the foundation for the study of chemistry. Similarly, Maxwell's theory had to await the discovery of electromagnetic radiation before it could be accepted. Today's exciting ventures into space continue to be the basis for confirming or rejecting numerous theories about the behavior of our universe; so in this way every space-shuttle voyage serves as an arbiter of whether certain theories will be accepted.

Mathematicians are theoretical scientists, of course, so they too rely on a standard of 100 percent certainty. But the nature of mathematics is different from that of such physical sciences as chemistry, physics, and astronomy in that mathematical results are based upon chains of reasoning that are called *proofs*. Let us define this concept, because it significantly distinguishes mathematical proofs from all other types of proof.

> DEFINITION: A *mathematical proof* consists of a sequence of statements such that the validity of each statement follows from an overwhelmingly compelling logic applied to statements already accepted.

The next example will illustrate this definition.

Example 1
Recall the result:

$\sqrt{2}$ is an irrational number

The proof of this result centers around three statements.

$$2y^2 = x^2 \text{ implies } x^2 \text{ is even} \tag{1}$$

$$x^2 \text{ is even implies } x \text{ is even} \tag{2}$$

$$2y^2 = x^2 \text{ implies } x \text{ is even} \tag{3}$$

Statement (1) is true because $2y^2$ is even and is equal to x^2. Statement (2) is a little more subtle, and involves what we call the contrapositive in Chapter 3, but the important point is that logic alone will convince any reader who is willing to exert the energy needed to follow the line of reasoning that x must be even. Statement (3) follows from (1) and (2) by what is called transitivity, which also will be described in Chapter 3.

The entire proof of the result consists of sequences of statements like these three. Note that the validity of (1) and (2) is based upon compelling logic, and that (3) follows from a combination of (1) and (2), which had already been established. Thus statements (1), (2), and (3) form one of the chains of reasoning that form the proof of the result.

Example 1 is typical of the proofs that mathematicians use to establish their theories. No result can be regarded as true until it is accompanied by a proof that any independent, knowledgeable reader can follow. The next anecdote relates a recent incident which resulted from this insistence upon verifiable proofs.

Anecdote 3   Khachian
A paper was published by L. G. Khachian in 1979 in a research journal of the Soviet Academy of Science. Because it was written in Russian and the author completely unknown outside the U.S.S.R., it took several months for word of his findings to reach mathematicians here. Translations of Khachian's paper began to circulate later that year, but since the paper contained only an outline of the proof, several interested mathematicians began attempts to reconstruct the entire proof. It was not until a pair of them, one in Rochester, New York, and the other in Szeged, Yugoslavia, supplied a proof that Khachian's result was accepted. In the course of supplying a proof the pair clarified the meaning of the result and even managed to simplify several of the original steps.

The area that Khachian was working in is called *linear programming*. It is of particular interest because it can be applied to problems in economic modeling and business planning involving the allocation of work force and resources at minimum costs and maximum profit. The simplex method of solving such problems was devised by George B. Dantzig of Stanford in 1947. It worked extremely well in practice; yet some mathematicians sought a new procedure which would be as efficient not only in practice but in principle as well. No such procedure was known until Khachian's paper appeared. We will discuss linear programming in Chapter 5.

The Ascent of Proof

As we stated previously, the standard that mathematicians apply today in order to determine whether a result is acceptable is the construction of a mathematical proof that can be understood by independent readers. This represents the present stage in the ascent of the concept of a mathematical proof. Earlier stages in the evolution will be discussed in Chapters 3 and 4. The next example will point the way toward a future stage in the ascent.

Example 2   The Four-Color Problem

If you examine a map of any region of the world in any atlas, you will see that only four different colors are needed to distinguish the borders of different countries in such a way that no two neighboring countries share the same color (bodies of water count as a country, too).

It is just this kind of evidence which whets a mathematician's appetite. The question arises, "Will this *always* be the case?" Note the importance of the italicized word. The degree of certainty must be 100 percent. The existence of thousands, even millions, of maps that require only four colors is not enough.

The four-color problem arose in the late nineteenth century. All of the evidence said that it was true. But overwhelming evidence does not convince a mathematician; only a proof does. Finally three American mathematicians published a proof in 1977. They showed that there were no exceptions, that the four-color problem was always true; *every map can be colored with four colors.* Figure 1 shows a postage meter indicia that reflects their proof.

FIGURE 1 Postage-meter indicia commemorating the four-color problem.

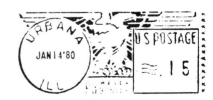

The proof of the four-color problem raised an issue that will have to be resolved in the future. The proof depends in a crucial way on the use of a computer. Consequently, some mathematicians have refused to accept it because it violates the standard criterion that independent readers be able to confirm every step. Proponents of accepting computer-based proofs counter that even some customary proofs are so long as to be practically unverifiable. They cite as an example one particular result in group theory whose proof required one whole issue of a research journal.

Where does this leave the status of the four-color problem? Those. who accept computer proofs call it the four-color *theorem;* those who oppose still regard it as an outstanding unsolved *conjecture.*

And where does this leave the state of mathematical proof? In a flux. It shows that the ascent of proof has not been completed, that there remain future stages yet to evolve as such controversial issues are resolved.

Caffeine

Let us end this section by returning to the problem that began it, namely, how should the Food and Drug Administration proceed with regard to the harmful effects of caffiene? In the human arena 100 percent certainty can almost never be achieved, so the FDA must take a calculated gamble. But just what margin of error is acceptable? The president of the National Academy of Sciences has complained that regulatory decisions have been made on "observations so preliminary that they could not find acceptance for publication in an edited scientific journal." But the margin of error for a scientist usually lies below 5 percent. Should this rigor be required to make public policy? Or should the regulatory cutoff be, say, 6 out of 100, or even 10 out of 100? What do you think?

EXERCISES

1   What do you think? (See the last line of the section.)

2   What evidence did the Surgeon General of the United States use to force cigarette companies to print the warning "may be dangerous to your health" on its labels and advertisements?

3   The FDA obtained its results by pumping caffeine directly into the stomachs of rats. Two central issues arise: (1) humans sip it over a period of time, and (2) rats are not humans. How do these factors affect the degree of certainty you would attach to the FDA's conclusion?

4   An advertisement for Remy Martin cognac reads, "Symbols of the World's most precious elements: Au, Ag, Pt, Rm." The last is Remy Martin. What are the first three?

5 Mendeleev named one of the three elements whose existence he had predicted: eka-silicon. What is it called today? Why? What are the other two elements? (See Bronowski's "Ascent of Man.")

6 A review of the book "The Brethren" (by Bob Woodward and Scott Armstrong) contains the following quote: "Judicial writing rests very largely on citing precedents, on saying: What we say today is more or less what we have always said; it's what our citizens have tacitly agreed; it's what the constitution meant." Do you see any differences between the mathematical method and the judicial method?

7 The city of Paris asserts that the second-platform elevators of the Eiffel Tower are unsafe. They run from 300 feet above ground to the top, 900 feet high. The operating company claims otherwise. To prove their point the company runs them regularly for their employees. Do you find the company's logic compelling?

8 Psychologists have long believed that humans are the only animals to seek out foods that cause pain (alcohol, coffee, kumquats, chili peppers). To prove this Professor Paul Rozin, of the University of Pennsylvania, fed nausea-inducing food to rats and found that the rats would never eat such foods again. Do you find Rozin's proof compelling? If not, how else would you test the hypothesis?

9 The philosopher Thomas Tymoczko, of Smith College, objects to the "proof" of the four-color problem. He points out that nobody, including the authors, can review every step of the proof because so many of the steps were done by a computer. Would you accept a proof of a statement if you were able to verify all the details (including checking that the programs were correct) but in which a computer performed the computations?

10 Modern physicists believe that in some subnuclear events "virtual particles" play an intermediary role. They are born and die in less time than it would take light to cross the diameter of an atomic nucleus. Therefore they can never be perceived, even theoretically. How can one go about proving the existence of things which cannot even be perceived?

11 The seventeenth-century Dutch physicist Christian Huygens presented a very curious argument that hemp can be found on Jupiter. He knew that the main function of the Earth's moon was to provide a navigational aid to mariners (aside from raising the tides and giving a little light at night). But Jupiter had four moons. (How many does it have by today's calculations?) Thus there must be many mariners there. Alas, mariners imply boats, boats imply sails, sails imply ropes, and ropes imply hemp. With what degree of certainty do you expect hemp to be found on Jupiter?

12 Black holes have been called the new superstars. What are they? If they cannot be seen then how do we "know" they exist? (Use any available source.)

13 This problem will enable you to apply scientific reasoning to answer an intriguing question. The Greek mathematician Archimedes discovered the following four properties of solids placed in a fluid:

A A solid lighter than a fluid will, if immersed in it, not be completely submerged, but part of it will project above the surface.

B Any solid lighter than a fluid will, if placed in the fluid, be so far immersed that the weight of the solid will be equal to the weight of the fluid displaced.

C If a solid lighter than a fluid be forcibly immersed in it, the solid will be driven upwards by a force equal to the difference between its weight and the weight of the fluid displaced.

D A solid heavier than a fluid will, if placed in it, descend to the bottom of the fluid, and the solid will, when weighed in the fluid, be lighter than its true weight by the weight of the fluid displaced.

Imagine yourself perched on a raft in your bathtub. Get off the raft and sit in the tub. The raft remains floating. Does the initial water level rise, fall, or remain the same?

14 Norman Bloom regards numerical relationships found in scripture and everyday life to be proof of the existence of God. Which of his arguments below do you find convincing? (For a fuller account of Bloom and his theories see Carl Sagan, "Broca's Brain," Random House, New York, 1979, pp. 128–136.)

(*a*) The angular size of the moon or the sun as seen from the Earth is half a degree. This is just $\frac{1}{720}$ of the circle of the sky. But $720 = 6!$ $= 6 \times 5 \times 4 \times 3 \times 2 \times 1$. Therefore God exists.

(*b*) Abraham was born 1948 years after Adam, at a time when Abraham's father was 70 years old. But the Second Temple was destroyed by the Romans in A.D. 70 and the state of Israel was created in 1948.   QED

(*c*) "Look, mankind, I say to you all, in essence you are living in a clock. The clock keeps perfect time, to an accuracy of one second/day.... How could such a clock in the heavens come to be without there being some being, who with perception and understanding, who, with a plan and the power, could form the clock?"

15 For each equation below find one counterexample which shows that the equation does not always hold.

(*a*) $(a + b)^2 = a^2 + b^2$

(*b*) $(a - b)^2 = a^2 - b^2$

(*c*) $\dfrac{1}{a} + \dfrac{1}{b} = \dfrac{1}{a + b}$

(*d*) $\sqrt{a + b} = \sqrt{a} + \sqrt{b}$

Geometry and Logic

3

Frequently geometry and logic are taught entirely independently of each other, with geometry presented in high school and symbolic logic in college. We will see, however, that they evolved together; our presentation will be true to the historical development. It will show the natural context in which symbolic logic originated, in much the same way that set theory appeared naturally in the nineteenth century.

Some believe that geometry originated in ancient Greece. In Section 3-1 we will see that the roots run deeper than that, to classical Egypt. The study of Egyptian geometry shows the empirical origins of the subject, thus linking it to the physical sciences. Various kinds of geometrical errors cropped up in Egypt, and in Section 3-2 we see how they influenced the brand of geometry that emerged in Greece. Section 3-2 casts Euclid's book, the "Elements," as a model for deduction by distinguishing rigorously between what is given and what has to be proved.

Section 3-3 discusses how symbolic logic evolved during the period from 600 to 300 B.C. It uses Euclid's "Elements" as a vehicle for introducing conditional statements, their logical variants, and equivalent statements. Section 3-4 continues this theme, investigating mathematical arguments. By the end of Section 3-4 you should have a good conception of what constitutes deduction. The deductive method is used in Section 3-5 to introduce parallel lines; here a painstaking analysis is applied to some results in the "Elements." Further development of this theme is carried out in Chapter 4, where Euclid's results are shown to yield graphic applications for Renaissance artists. Section 3-6, which is optional, discusses some other aspects of geometry which also originated in Greece, including Platonic solids and conic sections.

3-1 THE INCEPTION OF GEOMETRY

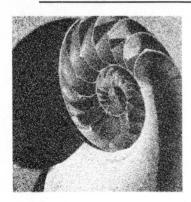

This section opens with a fictional account that is based upon fact. To understand the significance of the mathematics of Sections 3-1 and 3-2 you must have a feeling for the philosophy of the ancient Egyptians—a graceful and tranquil people who, by the majestic Nile River, took the first steps in the ascent of geometry. They were one of the first civilizations to use formulas and functions implicitly to describe geometric shapes. It is important at this juncture for you to understand how they might have derived these first formulas.

Let us briefly sketch a portrait of a day in the life of one of these people. We will choose a scribe, much like Ahmes, and call him Amos. (See Illustration 1.)

The setting is in the town of Beni-Hasan, located along the Nile some 2000 miles from the Mediterranean Sea. The time is 1985 B.C. As Amos walks to work early in the morning he is disturbed by the noise and hubbub of thousands of laborers. His daily routine used to be serene and peaceful, but now the laborers are toiling every day building a tomb for Amenemhet, the ruler who caused the arts to flourish to such an unprecedented degree. Amos rues the fact that it will be a rock-cut tomb instead of a pyramid because he loves his ruler and feels that his path to the afterlife should be adorned with as much pomp as earlier pharoahs. On the other hand, he realizes that the recent spate of labor strikes makes the building of an immense pyramid a veritable impossibility.

By now Amos reaches the school where he spends his early morning hours instructing the sons and daughters of nobility in the methods of unit fractions. He finds this part of his job rewarding, even if he does complain frequently that his students prefer partying to studying. It is the other part of the job he dreads, the monotony of counting corn and computing the tax on it. This task causes him to daydream about the historic role played by earlier scribes whose computations were essential to the erection of vast pyramids, the construction of elaborate sculptures, and the derivation of the precise laws for painting.

Life has been good to Amos and his compatriots. They have lived peacefully for some 2000 years, and only a small standing army is necessary for defending the border. He has heard that conditions within the distant land of Sumer are so bad that police are necessary for maintaining order, and this makes him feel fortunate that the respect for his Pharaoh alone mitigates against such a state of affairs.

The ever-present sun beats down on Amos as he reaches home at the end of his day. He glances at his modest dwelling and reflects on the irony of its impermanence in contrast to the Pharaoh's permanent tomb under construction. Why, he wants to know, are the preparations for the afterlife more important for some than others? His private writings might express such thoughts, but his public demeanor would never reveal them.

The ascent of geometry begins with the Rhind papyrus in ancient Egypt. As we saw with number, however, mathematics had already reached an advanced stage by that time. We can only guess what were the earlier influences on the development of mathematics. Illustration 2, for example, shows an Egyptian wall painting from 3500 B.C. containing strictly Sumerian elements, and recalls the reasons for Thor Heyerdahl's expedition from Sumer to India to the Red Sea.

Our portrait of the life of a scribe in Egypt about 2000 B.C. contains many elements from the culture of that period. The two most important characteristics are the continuity of tradition and the practical bent of the people. We will see these characteristics reflected below in a brief discussion of the arts in ancient Egypt. The mathematics of a civilization will always reflect its culture. Can you guess now what kind of geometry problems were contained in the Rhind papyrus that reflect the practical nature of Egyptian culture?

Egyptian Art

The dominant cultural element in the Egyptian civilization was its art. Amos daydreamed about the glories of earlier scribes in three of the major areas of art: architecture, sculpture, and painting. Let us discuss these areas.

Architecture is the first area that comes to mind when we think of Egypt. View the Great Pyramids of Gizeh (Illustration 3), which date to 2600 B.C. Their mammoth size is overwhelming. They were not built for sheer beauty; they served a very practical purpose. The Egyptians believed that each human was made of flesh and spirit, that the spirit lived after the flesh died, and that the better the preservation of the flesh the longer the spirit would live. For this reason the pyramids served as tombs which guaranteed immortality by being built of huge blocks of stone and by being filled with food, clothing, and other amenities to enhance the life of the spirit.

Egyptian sculpture usually took the form of visages of the pharaoh and his court. View the temple of Rameses II (Illustration 5), which dates to 1257 B.C. The figures are erect, stiff, unblemished, and symmetric. They clearly do not give an accurate depiction of the individuals. Their purpose was not to present an actual image, but to adorn the pharaoh's burial plot by presenting him in a deified form. Once again, art serves a practical purpose.

The third area, painting, frequently was centered around the hunt. The Nile provided Egypt with bountiful game and crops, and the painting reflected this. View Illustration 7, which is a painted limestone relief from about 2500 B.C. Notice how unnatural the setting is. The hunter is presented as a superior being, his helpers are much smaller than he, and the prey beckons his call. The purpose of the painting is to ensure a bountiful catch.

The Egyptians tried very hard to gain a realistic view of the human form in their art and sculpture. Perhaps they tried too hard! They invented a code of rules that every artist had to follow when reproducing certain forms. They called this set of laws the *canon*. Through the canon they hoped to depict as much information as possible about the three-dimensional world either onto a two-dimensional entity, such as a vase or wall, or in a stone sculpture.

A very specific, rigid grid was adopted, requiring explicit dimensions. Only on a very rare occasion would the artist deviate from the established code. To produce the human figure, a grid as in Illustration 8 was used. The torso was always triangular in shape and occupied the same amount of space in each figure. The size of the squares in the grid could be enlarged or made smaller depending upon the size of the artwork, but the relative dimensions in the work remained the same.

The code was a very pragmatic means of depicting reality. It afforded the artist a formula to follow, even if it squelched individuality. But the result, to a modern observer, is quite unnatural. In Illustration 7, for instance, the hunter's torso is facing us but his head is facing right, and he appears to have two left feet. Moreover, the relative sizes of hunter and helpers are out of proportion.

However, Egyptian art served its purpose. It endured for centuries, experiencing only minor variations. It was intended for very pragmatic ends, and it met them satisfactorily.

Egyptian mathematics also served its purpose. It too endured for centuries and was intended for practical means. Let us turn to it now.

Cylinders and Circles

The Rhind papyrus is the oldest existing mathematical document. By examining the geometry contained in it we see the first stage in the evolution of the subject. Altogether the papyrus contains 84 problems, 20 of which deal with geometry. Here is the first geometry problem.

> **PROBLEM 41:** Find the volume of a cylindrical granary of diameter 9 and height 10.

Note that the oldest geometry problem deals with a cylinder (see Figure 1). Example 1 presents Ahmes' solution to the problem.

Example 1

The solution in the Rhind papyrus reads as follows. "Take away $\frac{1}{9}$ of 9, namely, 1; the remainder is 8. Multiply 8 times 8; it makes 64. Multiply 64 times 10; it makes 640 cubed cubits."

This method involves three steps. The first is to multiply the diameter by $\frac{8}{9}$ and the second to square this product. The third is to multiply the square by the height. This produces

$$V = \left(\frac{8}{9} \cdot 9\right)^2 \cdot 10 = 8^2 \cdot 10 = 64 \cdot 10 = 640$$

Notice that the literal translation introduces cubed cubits, which means that the scribe intended the granary's measurements to be in terms of units. This indicates that he did not have merely a mental picture of an abstract cylinder in mind, but a specific granary in terms of particular units.

The solution to Problem 41 shows that the volume of a cylinder depends on two linear measurements, the diameter and the height. To

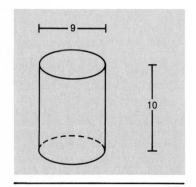

FIGURE 1 A cylinder in Problem 41.

illustrate his procedure, Ahmes added two similar problems (see Problems 1 and 2 in the exercises). He most likely assumed that any reader who could solve the three problems could find the volume of any cylinder once the diameter and height were determined.

Today we use functions and equalities to solve such problems. Since the volume V depends on the two variable quantities, the diameter d and height h, we say that V is a *function of two variables*. We express this by writing $V = V(d, h)$, which is read "V equals V of d and h"; it simply means that V depends on d and h. By writing the procedure from Example 1 as a formula, we obtain

$$V = V(d, h) = \left(\frac{8}{9} \cdot d\right)^2 \cdot h$$

The next example will illustrate how to find volumes by using this formula.

Example 2
Problem Using Ahmes' formula find the volumes of the cylinders having (*a*) diameter 9 and height 20, (*b*) diameter 27 and height 50.
Solution (*a*) We put $d = 9$ and $h = 20$ into the formula above. This yields the volume

$$V = V(9, 20) = \left(\frac{8}{9} \cdot 9\right)^2 \cdot 20 = 8^2 \cdot 20 = 1280$$

(*b*) Put $d = 27$ and $h = 50$ into the formula. We get

$$V = V(27, 50) = \left(\frac{8}{9} \cdot 27\right)^2 \cdot 50 = 24^2 \cdot 50 = 28{,}800$$

The letter f is customarily used to denote a function. If f is a function of two variables x and y, we write

$$f = f(x, y)$$

When we evaluate $f(x, y)$ at specific values of x and y, say $x = 1$ and $y = 2$, we express this by writing "find $f(1, 2)$," which is read "find f of 1, 2." The next two examples illustrate how to evaluate a function of two variables when values for x and y are given.

Example 3
Problem Evaluate $f(x, y) = (x + y)/xy$ when

(*a*) $x = 1, y = 2$ and   (*b*) $x = 2, y = \frac{1}{2}$

Solution (*a*) To evaluate $f(x, y)$ when $x = 1$, $y = 2$ means to find $f(1, 2)$. You substitute $x = 1$, $y = 2$ into the right-hand side of the equation. Thus

$$f(1, 2) = \frac{1 + 2}{1 \cdot 2} = \frac{3}{2}$$

(*b*) When $x = 2$, $y = \frac{1}{2}$ we obtain

$$f\left(2, \tfrac{1}{2}\right) = \frac{2 + \frac{1}{2}}{2 \cdot \frac{1}{2}} = \frac{\frac{4}{2} + \frac{1}{2}}{1} = \frac{5}{2}$$

Example 4

Problem Evaluate the function $f(x, y) = (x^2 + y^2)/(x + y)$ at these values.

(*a*) $x = 1$, $y = 2$ (*b*) $x = 2$, $y = \frac{1}{2}$

Solution   (*a*) $f(1, 2) = \dfrac{1^2 + 2^2}{1 + 2} = \dfrac{1 + 4}{3} = \dfrac{5}{3}$

(*b*) $f\left(2, \tfrac{1}{2}\right) = \dfrac{2^2 + \left(\frac{1}{2}\right)^2}{2 + \frac{1}{2}} = \dfrac{4\frac{1}{4}}{2\frac{1}{2}} = \dfrac{\frac{17}{4}}{\frac{5}{2}} = \dfrac{17}{10}$

From problems involving volume, the scribe moved to problems involving area. Example 5 shows how the following problem was solved.

PROBLEM 50:   What is the area of a round field of diameter 9 khet?

Example 5

The literal solution of Problem 50 reads almost exactly like that of Problem 41. "Take away $\frac{1}{9}$ of the diameter, namely 1; the remainder is 8. Multiply 8 times 8; it makes 64. Therefore it contains 64 setat of land."

The way in which the area of a circle is found using this procedure is

$$A = \left(\frac{8}{9} \cdot 9\right)^2$$

Once again, notice that the problem is practical. The circle is not abstract; it is a "round field." The diameter too has specific units, not just abstract numbers.

The formula that emerges from Example 5 is

$$A = \left(\frac{8}{9} \cdot d \right)^2$$

It shows that the area of a circle depends only on its diameter d; so it is a function of one variable. Thus we write $A = A(d)$, read "A equals A of d." Thus

$$A = A(d) = \left(\frac{8}{9} \cdot d \right)^2$$

Example 6 illustrates how functions of one variable are evaluated.

Example 6
Problem Evaluate the function $f(x) = x^2 - x + 2$ at these values of x:
    (*a*) 1   (*b*) 0   (*c*) -1
Solution   (*a*) $f(1) = 1^2 - 1 + 2 = 2$

    (*b*) $f(0) = 0^2 - 0 + 2 = 2$

    (*c*) $f(-1) = (-1)^2 - (-1) + 2 = 1 + 1 + 2 = 4$

Scientific Method

Let us return to the procedure that Ahmes used to find the volume of a cylinder. It can be summarized by the formula

$$V = \left(\frac{8}{9} \cdot d \right)^2 \cdot h$$

It is natural to ask where the number $\frac{8}{9}$ came from. Rephrasing this question we get:

> QUESTION 1: How could Ahmes have discovered such a formula?

There is another question which seems rather natural:

> QUESTION 2: Why did volume problems precede area problems, when the opposite order seems so natural?

The answer to Question 2 will explain why we introduced functions of two variables before functions of one variable.

Questions 1 and 2 seem entirely unrelated, but we will argue that they have the same answer—one that can be found by examining the civilization in which Ahmes lived. We noted that the Egyptians had a very practical bent. Notice that neither Problem 41 nor Problem 50 is concerned with an abstract figure but rather a specific figure with particular units. In fact, almost all the problems in the Rhind papyrus are taken from everyday life.

The scribe Amos had the task of counting corn. His task was probably similar to Ahmes'. The scribes did not actually count kernels. Instead they probably encountered many granaries of corn and recorded as many relevant quantities as they could. Somehow they had to relate the volume to these quantities.

Let us consider some specific numbers. Try to picture the scribes sitting in their office in the Pharaoh's palace, after measuring the corn stored in a specific granary. Suppose its volume was 640 cubed cubits. They knew that the granary had diameter 9 and height 10. What, they wonder, is the connection between 640 on the one hand and 9 and 10 on the other? It dawns on them that $640/10 = 64$, but the connection between 64 and 9 is certainly not obvious. Somehow one of them notices that $(\frac{8}{9} \cdot 9)^2 = 64$. The other is skeptical. He challenges, "Sure it works for that granary, but how do I know that your method will work for every granary?" The first accepts the challenge. "Let's try it on some of the other granaries." They try it, it works, and the result is a collection of problems exactly like Problem 41.

You might ask how the scribe knew his method always worked. That was easy for him, however, because he was very practical. He would reply, "Because it works for every granary that presently exists." Would this be enough evidence to convince you?

This little incident is our way of showing that, in early times, mathematics was no different from what we call the physical (or natural) sciences today. We have described three steps that the scribes probably followed to derive their method. These steps are:

1 Record data by experimentation.

2 Derive a relationship (or formula) between the known and unknown quantities.

3 Verify the formula by further experimentation.

The exercises contain some problems that will help you understand the scientific method by seeking a formula by experimentation.

Question 1 has now been answered: Formulas were discovered by experimenting and then drawing conclusions based upon the experiments. But what about Question 2? Why did volume precede area? This answer too lies with the practical nature of the Egyptians. It is much easier and more natural to experiment with a cylindrical granary than a circular piece of cloth because the instruments needed for measuring them are much easier and more natural to construct in three dimensions. To illustrate, draw a circle of radius 3 inches and try to cut it into square inches to find its area.

Approximately Right, Exactly Wrong

The formulas that the scribes discovered for the volume of a cylinder and the area of a circle are ingenious. There is only one thing wrong with them. *They are wrong.*

Consider the circle. It is well known that its area is $A = \pi r^2$, where r is the radius. The formula from the Rhind papyrus is

$$A = \left(\frac{8}{9} \cdot d \right)^2$$

Since $d = 2r$, this reduces to

$$A = \left(\frac{8}{9} \cdot 2r \right)^2 = \left(\frac{16}{9} \cdot r \right)^2 = \frac{256}{81} r^2$$

The value of π, correct to four decimal places, is $\pi \approx 3.1416$. However $256/81 \approx 3.1605$. Thus the Egyptian formula is not exact, but is a very good approximation. The same is true for the volume formula.

The Egyptian civilization demanded an accurate means of measuring the amount of grain in a granary. The scribes met this societal need by developing a high level of accuracy in their methods. It would require an extremely sophisticated measurement to show that the volume formula is incorrect.

This discussion might lead you to ask how *we* know that the formula $A = \pi r^2$ is correct, and not $A = (\frac{8}{9} \cdot d)^2$. The Egyptians would never have asked such a question. Practicality was all they sought, and their formula worked well in practice. We will have to await further stages in the ascent of geometry before such a natural question can be answered.

The next example provides another instance of an Egyptian formula that was wrong. This one, however, is not even approximately accurate. We will then discuss how such a practical people could accept such an impractical formula.

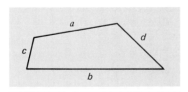

FIGURE 2

Example 7
A quadrilateral is a four-sided figure. Consider a quadrilateral whose sides are of length a, b, c, d (see Figure 2). The Egyptian formula for the area of it is

$$A = \frac{(a + b)(c + d)}{4}$$

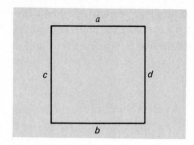

FIGURE 3

Is this formula correct? One way to test whether any formula is correct is to test it on some special cases. First try a square: $b = a$, $c = a$, and $d = a$ (see Figure 3). Then

$$\frac{(a + a)(a + a)}{4} = \frac{(2a)(2a)}{4} = a^2$$

which is correct. Next try a rectangle with $b = a$ and $d = c$ (see Figure 4). Then

$$\frac{(a + a)(c + c)}{4} = \frac{(2a)(2c)}{4} = ac$$

which is also correct. So far, so good.

Next try a trapezoid. Rather than proceed with the general case, consider the trapezoid in Figure 5 with base 10, top 4, and slanted height 5. The area is obtained by dropping perpendiculars, as in Figure 6, thus dividing the trapezoid into two triangles and a rectangle. The height of each triangle is 4, from the pythagorean theorem. Now move the left triangle atop the right, as in Figure 7, obtaining a 4×7 rectangle. Its area is 28, so the area of the trapezoid is 28.

Now substitute $a = 4$, $b = 10$, $c = 5$, $d = 5$ into the Egyptian formula. It yields area

$$\frac{(4 + 10)(5 + 5)}{4} = \frac{14 \cdot 10}{4} = 35$$

Thus the Egyptian formula yields a significant error. This means that the formula for the area of the quadrilateral is incorrect.

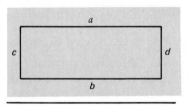

FIGURE 4

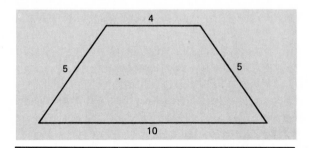

FIGURE 5

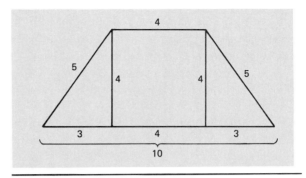

FIGURE 6

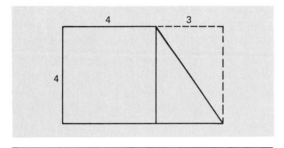

FIGURE 7

How could the Egyptians have used formulas that yielded such glaring errors? We do not know the answer, but we suspect that the quadrilaterals they encountered in practice deviated only slightly from rectangles. Unfortunately the Egyptians never indicated the derivations of their methods, so we cannot be sure how such a formula arose. It seems reasonable, however, that they tested it on rectangles and then averaged the lengths of the opposite sides for those figures which deviated from a rectangle. This would produce

$$\frac{a + b}{2} \cdot \frac{c + d}{2} = \frac{(a + b)(c + d)}{4}$$

which agrees with the formula they adopted.

The idea of a clear-cut proof never occurred to the Egyptians. But the errors of their ways set the stage for a civilization that would place much emphasis on the issue of proof. We will turn to this civilization in the next section.

EXERCISES

Problems 1 and 3 are taken directly from the Rhind papyrus.

1 Find the volume of a cylindrical granary of diameter 10 and height 10.

2 The solution that Ahmes gave to Problem 1 was 790 $\frac{1}{18}$ $\frac{1}{27}$ $\frac{1}{54}$ $\frac{1}{81}$. Is this solution correct?

3 Find the volume of a cylindrical granary of diameter 9 and height 6.

4 The solution that Ahmes gave to Problem 3 was 455 $\frac{1}{9}$ klar. A *klar* is equal to $\frac{2}{3}$ cubed cubit. Is this solution correct?

In Problems 5 and 6 find the volumes of the cylinders of the given dimensions.

5 Diameter 18 and height 5   **6** Diameter 36 and height 2

In Problems 7 to 10 evaluate the function at the stated value.

7 $f(x, y) = x^2 + 2xy + y^2$ when $x = 1$, $y = 2$

8 $f(x, y) = x^2 - y^2$ when $x = 2$, $y = 1$

9 $f(x, y) = (x + y)^2$ when $x = 1$, $y = 2$

10 $f(x, y) = (x + y)(x - y)$ when $x = 2$, $y = 1$

In Problems 11 to 14 let $f(x, y) = (x^2 - y^2)/(x + y)$ and find the stated value.

11 $f(2, 1)$ **13** $f(2, -1)$

12 $f(1, 2)$ **14** $f(-2, 1)$

In Problems 15 to 18 let $f(x) = x^2 + x - 1$ and find the stated value.

15 $f(3)$ **17** $f(0)$

16 $f(2)$ **18** $f(-2)$

19 Calculate the area of a circle of diameter 2, using each of the four formulas:
(*a*) Ahmes' formula (see Example 5)
(*b*) $A = 3r^2$
(*c*) $A = 3.1416r^2$
(*d*) $A = \pi r^2$

20 Which of the formulas in Problem 19 is easiest to use?

21 Which formula in Problem 19 yields the best answer for a mathematician?

22 Which formula in Problem 19 yields the best answer for an engineer?

The area of a right triangle depends on its base b and height h:

$$A = A(b, h) = \tfrac{1}{2}bh$$

In Problems 23 and 24 find the area of the stated triangle.

23   Base 5 and height 2

24   Base 2 and height 5

In Problems 25 and 26 find the volume of a cube having the given side s. The volume is a function of s:

$$V = V(s) = s^3$$

25 $s = 4$

26 $s = 6$

27   The Egyptian formula for finding the area of a triangle is

$$A = A(a, b, c) = \frac{(a + c)b}{4}$$

where a, b, and c denote the lengths of the three sides. Is this formula correct?

28   Is there enough information given on the trapezoid below to determine its area? If so, find the area; if not, explain why.

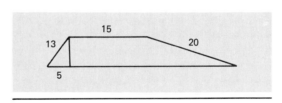

29   Each of the Great Pyramids of Gizeh (Illustration 3) has a square base. Here is the way Ahmes explains how to find the slope, called the *seked*, of such pyramids.

Problem   If a pyramid is 250 cubits high and the side of its base 360 cubits long, what is its seked?

Solution   Take $\frac{1}{2}$ of 360; it makes 180. Multiply 250 so as to get 180; it makes $\frac{1}{2}$ $\frac{1}{5}$ $\frac{1}{50}$ of a cubit. A cubit is 7 palms. Multiply 7 by $\frac{1}{2}$ $\frac{1}{5}$ $\frac{1}{50}$. The seked is 5 $\frac{1}{25}$ palms.

(*a*)   Problem 58 in the Rhind papyrus asks for "the seked of a pyramid 93 $\frac{1}{3}$ cubits high and with the side of its base 140 cubits." Verify Ahmes' solution that the seked is 5 palms, 1 finger.

(*b*)   Problem 59 in the Rhind papyrus asks, "If a pyramid is 8 cubits high and the side of its base 12 cubits long, what is its seked?" Solve it.

(*c*)   What is the seked of the Great Pyramid at Cheops? It is 280 cubits high and 440 cubits long.

(*d*)   Can you find Ahmes' formula for the seked of a pyramid h cubits high and base b cubits long?

30 Solve Problem 44 of the Rhind papyrus, which indicates that rectangular-base granaries were used as well as cylindrical ones. "What is the amount of grain that goes into a rectangular granary, its length being 10, its width 10, its height 10?"

31 (*a*) If the height of a cylindrical granary is doubled, what happens to its volume?
    (*b*) Suppose the original height is tripled?
    (*c*) Suppose the original height is increased by a multiple of *n*?

32 (*a*) If the diameter of a cylindrical granary is doubled, what happens to its volume?
    (*b*) Suppose the original diameter is tripled?
    (*c*) Suppose the original diameter is increased by a multiple of *n*?

33 If the radius of a cylindrical granary is doubled, what happens to the (*a*) diameter, (*b*) volume?

34 If both the diameter and height of a cylindrical granary are doubled, what happens to the volume?

35 A formula is a shortcut. It is a way of solving a problem without going through the lengthy rigmarole that the problem calls for. For example, consider the problem: Sum the first 10 positive integers. Now sum the first 100 positive integers. The first is much easier to do than the second. However, if you know the formula, the shortcut makes them equal in difficulty. The formula is: $1 + 2 + \cdots + n = n(n + 1)/2$. Use the formula to sum the first 10 and then the first 100 positive integers.

3-2 EUCLIDEAN GEOMETRY

Let us meet a scribe who resided in the Egyptian port of Alexandria about 300 B.C. In light of the long continuity of tradition in Egypt it is tempting to conclude that the new scribe lived under conditions similar to Ahmes. The opposite is true, however. The new scribe, named Euclid, is Greek, not Egyptian, and is representative of an entirely different civilization. He came to Egypt when he was recruited from Athens to the newly established library in the Museum of Alexandria. Thus his work reflects the culture of the Golden Age of Greece, the so-called Age of Pericles.

We will begin this section by contrasting the cultures of Egypt and Greece. Then we will investigate the revolutionary geometry that emerged from Euclid.

**A Tale of
Two Cultures**

Following the great naval victory in 485 B.C., the statesman Pericles led Greece through a period that is usually regarded as the most cultured in history. The emphasis of the arts was on the ideal aspects of humans, nature, and reason. Let us examine the architecture, sculpture, and painting from the Age of Pericles.

One of the most outstanding examples of Greek architecture is the Parthenon, which was built at Athens between 448 and 432 B.C. Its main purpose was to house the statue of Athena Parthenos. In this sense it is similar to the Egyptian pyramids, but the similarity ends there. View the Parthenon in Illustration 4. Its lines appear perfectly straight; its columns seem to be perpendicular to the ground, uniformly spaced, and of the same diameter. They are not—the lines are actually curves, and the columns tilt inward, are closer to each other at the corners, and have greater girth at the corners. These deviations, however, were intended to offset marginal distortions in human vision. The intrusion of reason into architectural design would never have occurred in Egypt, nor would the emphasis on the ideal relations between the length, width, and height.

The Parthenon was intended to be more than an engineering marvel; it was to be viewed as a great work of sculpture. Let us compare some of its statuary (Illustration 6) with the Temple of Rameses (Illustration 5). Recall that the Egyptian figures are erect, stiff, symmetric portrayals of a Pharaoh and his family. The Greek subjects are men and women in natural clothing and in a natural setting. But their faces express ideal characteristics; they are devoid of expression and imperfection. The purpose of this condition is to impart timelessness.

Also contrast the practical bent of the Egyptians with the ideal form of the Greeks. Compare the discus thrower (Illustration 9) with the limestone relief of the hippopotamus hunt (Illustration 7). The Egyptian hunt is unrealistic, the Greek athlete quite realistic. Yet the athlete's proportions are ideal; he is no mere mortal. If the artist had been capturing a natural pose he surely would have drawn straining muscles and a contorted face. His right arm, his throwing arm, would be much larger and more muscular than his left arm. Instead we see a serene, beautiful form in the act of glorifying athletic achievement.

The "Elements"

In Section 3-1 we studied the Rhind papyrus for two reasons. First, it is representative of the state of mathematics that existed in Egypt during that period. Second, it reflects the culture of that great civilization.

In this section we will study Euclid's book, titled the "Elements," for similar reasons. It is representative of the state of mathematics that existed in Greece about 300 B.C., and it reflects the emphasis on reason and ideal forms, which are the two characteristics that dominated Greek culture.

In contrasting the cultures of Egypt and Greece we find that they are quite different. Thus it is not surprising that the geometry in the "Elements" is vastly different from the geometry in the Rhind papyrus. However, it is not just the difference in their contents that is important but also their approach. The "Elements" is based upon deduction, which means that every statement must be proved from certain assumptions. It does not contain any measurements. This approach is totally unlike the intuitive, inductive, empirical approach in the Rhind papyrus.

You might wonder why there is so much emphasis on the "Elements." There are several reasons. It serves as a paragon for deductive thinking, which is used today in almost all areas of human endeavor. Besides, it is the world's oldest textbook—it is still being used today in various forms, some 23 centuries after it was written—and it has sold more copies than any other book in history except the Bible. Let us now examine the contents of this classic textbook.

The Definitions and Axioms

Immediately following the title, the "Elements," Euclid lists 23 definitions. There is neither a dedication nor an introduction. We will discuss a few of the definitions here; the complete list is given in an appendix at the end of this chapter. The first definition reads:

A *point* is that which has no part.

There is an obvious problem here because the word "part" has not been defined. To avoid such circularity, the modern approach is to begin with a list of undefined terms which are known intuitively, and then to define the remaining objects in terms of them.

Euclid's second definition raises the same issue:

A *line* is a breadthless length.

However, his definitions should be viewed as ideal forms. Thus a point is not a small dot nor is a line the result of a finely sharpened pencil. Instead both terms are mental concepts having no place in the physical world. The same is true of Definition 4:

A *straight line* is a line which lies evenly with the points on itself.

The way in which Euclid uses these terms shows that a straight line refers to a line segment, whereas a line is endless in two directions.

Following the definitions comes a list of 10 properties of the defined terms. We call such properties *axioms* today, but Euclid divided them into *common notions* (which refer to all of mathematics) and *postulates* (which refer to the specific subject of geometry). The common notions are concerned with equalities and inequalities, whereas the postulates refer to geometric objects. The complete list is given in the appendix.

The first two postulates read:

1   A line segment can be drawn from any point to any other point.

2   A line segment can be extended indefinitely in a straight line.

They indicate that line segments can be constructed and extended in either direction. The next postulate reads:

3 A circle with any center and radius can be drawn.

The instruments used to construct these figures are called a *straightedge* and a *compass,* or sometimes collectively "Platonic instruments." Lines and circles were regarded as ideal figures, and they are the only figures whose existence was postulated.

There is a marked difference between the first four postulates and the fifth. The first four are succinct, clear, and easy to understand; the fifth is verbose, initially unclear, and difficult to understand. We will study it in detail in Section 3-5.

The First Two Results

After stating the definitions and axioms, Euclid proved a list of further properties, called *propositions*. The "Elements" contains some 500 propositions altogether.

Proposition 1 illustrates the distinction between defining an object and knowing that it exists. An equilateral triangle is defined in Definition 20; Proposition 1 proves that such triangles exist. The only figures that are assumed to exist are the line and circle. After we prove Proposition 1 we will discuss the form of the proof.

Proposition 1
An equilateral triangle can be constructed on a given line segment.
Proof Let AB be the given line segment. It is required to construct an equilateral triangle on AB. Refer to Figure 1.

1   Draw the circle BCD with center A and radius AB [Post. 3].

2   Draw circle ACE with center B and radius BA [Post.].

3   Draw line segments CA and CB, where C is a point at which the two circles meet [Post.].

4   $AC = AB$ since the point A is the center of circle CDB [Def. 15].

5   $BC = BA$ since B is the center of circle CAE [Def.].

6   $AC = BC$ since things which are equal to the same thing are also equal to one another [CN.].

Therefore the three line segments CA, AB, and BC are equal to one another. Therefore the triangle ABC is equilateral, and it has been constructed on the given line segment AB.

The form of the proof of Proposition 1 will serve as a model for all subsequent proofs. Note first that it resembles the kind of formal essay that is written for composition courses; it consists of an introduction, a

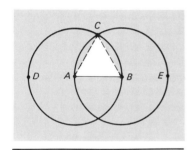

FIGURE 1

body of supporting evidence, and a conclusion. The introduction restates the proposition in terms of symbols, so it adds further clarification to the original wording, which might not have been fully understood upon first reading. The body of supporting evidence is the logic that is aimed at convincing the reader that the reasoning is sound. We shall examine this logic in detail in the next two sections. The logic in a proof is presented as a chain of statements, each of whose justifications is enclosed in square brackets [ ]. We have supplied some of the specific justifications and will leave the remaining ones for the exercises.

Table 1 introduces the notation used. The conclusion of the proof restates what was to have been proved in terms of the symbols. The notation for "the end" is QED, which is Latin for "that which was to be proved."

TABLE 1

Notation Used in Proofs

Notation	Example	Meaning
Def.	Def. 15	Definition 15 (circle)
CN	CN 1	Common Notion 1
Post.	Post. 2	Postulate 2
Prop.	Prop.10	Proposition 10

Many proofs are accompanied by figures. The purpose of the figures is to guide your reasoning through the logic in the proof. It is important to keep in mind that the figure does not constitute a proof; only the logic does.

The second proposition states that it is possible to draw a line segment of any given length anywhere in the plane. Its proof is long but important. Refer to Figure 2 as you read through it.

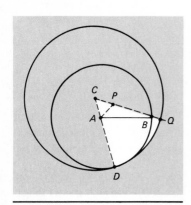

FIGURE 2

Proposition 2
A line segment can be constructed from a given point so that it is equal to a given line segment.
Proof Let *AB* be the given line segment and *P* the given point. It is required to construct a line segment *PQ* whose length is equal to the length of *AB*.

1 Draw line segment *AP* [Post. 1].

2 Construct an equilateral triangle on *AP* [Prop. 1]. Label the remaining point *C*.

3 Draw the circle with center *A* and radius *AB* [   ].

4 Extend the line *CA* until it meets this circle, and label the point of intersection *D* [   ].

5   Draw the circle with center C and radius CD [   ].

6   Extend CP to the point where it intersects the circle constructed in step 5 [   ]. Label Q the point of intersection.

7   $CQ = CD$ since they are radii of the same circle [Def. 15].

8   $CP = CA$ since they are sides of an equilateral triangle [   ].

9   $CQ - CP = CD - CA$ [   ].

10   $PQ = AD$ [   ].

11   $AB = AD$ since they are radii of the same circle [   ].

12   $AB = PQ$ [   ].

QED

The proof of Proposition 2 is important for two reasons. First, it illustrates the fact that Euclid's proofs consist of two parts, the construction and the deduction. Steps 1 to 6 show how to construct the point Q that is sought. Steps 7 to 12 supply the deductive proof that Q has the desired property.

The second reason why the proof is important is because it reflects the differences between Egyptian and Greek mathematics. The Egyptians would have been content with steps 1 to 6 if they worked for various lines AB and points P. But the Greeks would not accept such empirical evidence. They demanded the deductive steps contained in steps 7 to 12.

Proving the Obvious

Do you believe that the following statement is true?

If you add the lengths of any two sides of a triangle, the sum is greater than the length of the third side.

The Epicureans did. So did Euclid. But Euclid demanded a proof of it. The Epicureans ridiculed him for bothering to prove such an obvious statement. They said that it was known even to an ass, for "if fodder is placed at one point and the ass at another, he does not, in order to get to his food, traverse the two sides of the triangle but only the side separating them." See Figure 3.

Euclid viewed his task as keeping the number of assumptions to a bare minimum. Thus he endeavored to prove any statement that could be shown to follow from his 10 axioms. It is important to keep in mind that all statements are either assumed or proved; there is no middle

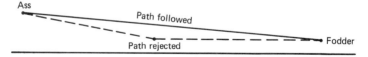

FIGURE 3

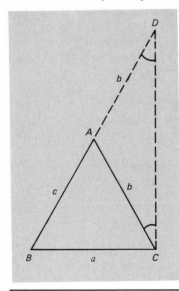

FIGURE 4

ground. This insistence on pure deduction yielded the purest form of logical reasoning.

Before proceeding to Euclid's proof of this obvious statement, let us emphasize two aspects of the deductive method. First, no numerical measurements can be made; the only instruments permitted in the construction are the straightedge and compass. Second, each statement in the proof must be accompanied by one of four possible justifications:

1 Definition

2 Axiom (common notions and postulates)

3 Previously proved proposition

4 A hypothesis of the proposition

The obvious statement above turns out to be Proposition 20 in the "Elements." We will supply the justifications for the steps that refer to Propositions 3 to 19. Recall that according to Rule 3 any previously proved proposition can serve as a justification, so it is possible to use Propositions 1 to 19 in the proof of Proposition 20.

Proposition 20

In any triangle the sum of any two sides is greater than the remaining side.
Proof Let A, B, C be the vertices of a triangle, and let a, b, c be the lengths of the opposite sides, respectively. We will show that $b + c$ is greater than a. Similar proofs would show that $a + c$ is greater than b and $a + b$ greater than c. Refer to Figure 4.

1 Extend BA through A [Post.].

2 Choose a point D on the extended line such that $AD = b$ [Prop.].

3 Let DC be joined [Post.].

4 Since $DA = b$, angle ADC = angle ACD [Prop. 5].

5 Angle BCD is greater than angle ACD [CN].

6 Angle BCD is greater than angle ADC [CN].

7 DB is greater than length a [Prop. 19].

8 Hence $b + c$ is greater than a [CN].

QED

Summary

Euclid's "Elements" reflects the arts that dominated Greek culture during the classical period. It also represents the mathematics of that time period.

The most important characteristic of the "Elements" is its total reliance upon deduction. The method of deduction begins with a set of definitions of the basic terms and a list of assumptions about them, called axioms. Additional properties about the terms are proved rigorously and are called propositions or theorems.

The proof of Proposition 20 illustrates the absolute reliance on the deductive method. The statement itself is obvious, yet Euclid proved it anyway. The Epicureans ridiculed him harshly for his folly, but history has vindicated Euclid. What appears to be obvious can sometimes turn out to be subtle, and sometimes wrong.

There is one other important issue about the deductive method—if the axioms are changed, then the results will be changed. In the next chapter we will discuss spherical geometry, which arose when one mathematician altered the meaning of "indefinitely" in the axioms. The resulting geometry later formed the basis for Einstein's theory of relativity.

EXERCISES

In Problems 1 to 4 supply justifications for the steps listed in the proposition.

1 Proposition 1, steps 2, 3, 5, 6.

2 Proposition 2, steps 3 to 6 and 8 to 12.

3 Proposition 20, steps 1, 2, 5.

4 Proposition 20, steps 3, 6, 8.

5 How do you know that the two circles in step 3 of the proof of Proposition 1 actually intersect?

6 Draw a line segment AB 1 inch long and mark the point $P \frac{1}{4}$ inch from B and $\frac{1}{4}$ inch above AB. Follow the steps in the proof of Proposition 2. Where does the desired point Q lie?

7 (*a*) Draw a figure to illustrate Proposition 3. (It is stated in the appendix.)
   (*b*) Write a proof of the proposition by following this outline:
   1 Let A be an endpoint of the longer line. Construct on A a line equal to the shorter one.
   2 Draw a circle with center A and radius the length of the shorter line.

8 In step 3 of the proof of Proposition 1, the point C is one of the two points where the circles meet. Would the remainder of the proof be changed if the other point had been chosen?

9 If you are willing to describe all the steps, can each proposition be proved from the definitions and axioms without referring to other propositions?

10 The quotation below is taken from a dissenting opinion of former U.S. Supreme Court Justice William O. Douglas, who served in that capacity for $36\frac{1}{2}$ years. Into which of the four categories does he place the amendments to the constitution: undefined term, defined term, axiom, or proposition?

> The First Amendment makes confidence in the common sense of our people and in their maturity of judgment the great postulate of our democracy.

11 Can Proposition 456 be used in the proof of Proposition 345? Why?

12 Can Axiom 4 be used in the proof of Proposition 3? Why?

Problems 13 to 16 refer to Proposition 20.

13 The perimeter of a triangle is the total length of its three legs. Is it possible for a triangle to have a perimeter of 6 inches if one leg has a length of 1 inch and the lengths of the other legs are natural numbers?

14 Is it possible for a triangle to have legs of length 3, 4, and 8? Why?

15 How many different triangles can have perimeter 12 and legs whose lengths are natural numbers?

16 How many different triangles can have perimeter 6 and legs whose lengths are natural numbers?

17 Definitions 1 and 2 use the words "part" and "breadth," neither of which had been defined beforehand. Find other undefined words in the 23 definitions listed in the appendix.

18 Make up a "better" definition than Euclid for the term "circle."

19 Read Proposition 13 in the appendix. It is certainly obvious. Why do you suppose that Euclid bothered to prove it instead of just assuming it as one of the postulates or common notions?

20 Read Isaiah 44:13 to find out what tools carpenters were permitted to use.

Problems 21 to 24 make use of Definition 20 and Propositions 5 and 6, which are stated in the appendix. In each problem, determine the justification for the statement.

21 An isosceles triangle has two equal legs.

22 A triangle that has two equal legs is isosceles.

23 A triangle that has two equal angles is isosceles.

24 An isosceles triangle has two equal angles.

25 Construct an equilateral triangle on a line 2 inches long. Can you construct more than one triangle on the given line?

26 In the book "Finnegan's Wake," by James Joyce, there is a line which is a favorite with Euclid lovers: "Ferst construct am aquilittoral dryankle!" Which proposition assures the construction can be carried out?

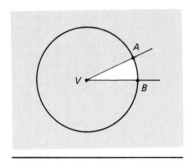

FIGURE 5

27 Which of the first 20 propositions proves that right angles exist?

28 What are vertical angles? Which of the first 20 propositions proves that they are equal?

29 Draw a figure to illustrate Postulate 5.

30 The word "isosceles" comes from two Greek words: *iso,* meaning equal, and *skeles,* meaning leg (note: "skeleton" has the same source). There are many other modern words with Greek origins. What mathematical or semi-mathematical words of this type can you find? Try, for instance, "abacus," "arithmetic," "decagon," "decathlon," "kilometer," "kilogram," "logarithm," "myriad," "pentagon," "pentathlon." Use a dictionary that gives the origins of words, such as "The Oxford English Dictionary."

31 Carry out the construction in parts (*a*) to (*d*) on Figure 5. Then answer the question in part (*e*).

  (*a*) Draw a circle with center *V*. Label *A* and *B* the two points where the circle intersects the angle.

  (*b*) Draw a circle with center *A* and radius *AB*.

  (*c*) Draw a circle with center *B* and radius *AB*.

  (*d*) Draw line *VC*, where *C* is the intersection of the two circles in (*b*) and (*c*).

  (*e*) How are angles *AVC* and *BVC* related?

32 Repeat Problem 31 for Figure 6. Does a problem arise in part (*d*)?

Mark each part in Problems 33 and 34 true or false.

33 __(*a*) The "Elements" was written about 600 B.C.

  __(*b*) Definitions do not require precise wordings.

  __(*c*) Axioms must be proved.

  __(*d*) Undefined terms should precede the defined ones.

  __(*e*) Postulates are the same as propositions.

  __(*f*) The definition of a term assures its existence.

  __(*g*) There is no difference between a postulate and an axiom.

34 __(*a*) The "Elements" was written about A.D. 300.

  __(*b*) Definitions must be proved.

  __(*c*) Propositions must be proved.

  __(*d*) Undefined terms should follow defined terms.

  __(*e*) Postulates are the same as axioms.

  __(*f*) A ruler is different than a straightedge.

  __(*g*) Axioms are essential to deduction.

  __(*h*) There is no difference between a postulate and a proposition.

35 The Greeks had a type of "canon" similar in many respects to that of the Egyptians. It differed, however, in one important aspect. The purpose of the Egyptian canon was to define exactly how a figure was to be created. Once the size was determined, the grid was made to fit that size and then the parts of the body were made in a specific way. On the other hand, the "canon of Polyclitus" was a rule of thumb for the Greek artist; it provided

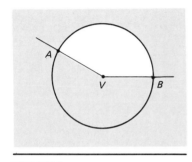

FIGURE 6

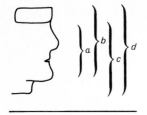

the artist with objective proportions of a typical human figure, and the artist was expected to adjust these guidelines as needed. In fact, if the rules were followed rigidly they produced a contradiction. Here are some measurements in the "canon of Polyclitus" as recorded by Vitruvius. Can you locate contradictions in them?

(*a*)   Face (hairline to chin): $\frac{1}{10}$ of the total body length

(*b*)   Head (crown to chin): $\frac{1}{8}$

(*c*)   Pit of throat to hairline: $\frac{1}{6}$

(*d*)   Pit of the throat to crown of the head: $\frac{1}{4}$

3-3   CONDITIONAL STATEMENTS

In the preceding section we viewed the "Elements" as a reflection of the classical Greek civilization and as a representative of its mathematics. The deductive approach that it is based upon was not the invention of Euclid. The roots of deduction go back to Egypt, but it developed primarily in Greece between 600 and 300 B.C. The result was a revolution that transformed the earlier empirical approach into the strict deductive method. The revolution occurred in three stages, which we will trace now.

Sometime during the sixth century B.C. two Greek citizens traveled extensively to Egypt and the Near East. One was Thales (pronounced thay´-lees), the first of the Seven Wise Men. He is reputed to have proved the following result:

An angle inscribed in a semicircle is right.

Exercise 43 will help you understand the meaning of this statement. The other was Pythagoras, whose merry band we met in Chapter 1. The pythagorean theorem is named after him because he was the first person to prove it.

What do we mean by saying that Thales and Pythagoras *proved* the results named after them? How might the Greeks have reasoned in the sixth century B.C.? We are not sure how Thales proceeded, but Jacob Bronowski has suggested that Pythagoras based his result on the mere shuffling of tiles that exist in abundance where he lived.

Bronowski's reconstruction of Pythagoras' proof offers one instance of *how* the early Greeks reasoned. But *why* did they resort to this method?

This is where the Egyptian connection comes in. By the time Thales and Pythagoras traveled there, it was known that some significant errors had crept into mathematics. When each returned to Greece, he brought along the desire to sift the correct results from the erroneous ones. This explains why the Greeks sought proofs in the first place. It marked the first stage in the revolution from empiricism to deduction.

The second stage is due to Zeno, who presented an argument contradicting the assumption that a line is composed of points. It is called Zeno's paradox. Zeno considered a 1000-meter race between the fleet-footed Achilles and the slow-moving tortoise (see Figure 1). Because Achilles was a sporty sort, and because he was 10 times faster than the tortoise, he spotted the tortoise a

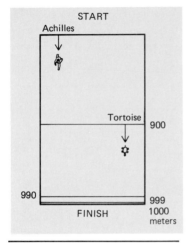

START
Achilles

Tortoise 900

990

990

999
FINISH 1000
meters

FIGURE 1

900-meter lead. The race began and Achilles sprinted out of the blocks. By the time he reached the tortoise's initial position at 900 meters, the tortoise had moved forward by 90 meters. By the time Achilles reached this new point the tortoise had again advanced slightly, 9 meters. By the time Achilles reached 999 meters the tortoise had progressed again. The process continued over and over. Achilles became so exhausted trying to catch the tortoise that he pulled a tendon (guess which one!) and had to drop out of the race.

The point that Zeno's argument raised is whether there is a smallest unit of space. It attacked the pythagorean notion that the whole line is greater than its part; in fact, it showed that the whole can be equal to its part. The Pythagoreans responded in a curious way. They dropped their assumption that a line is composed of points and idealized the terms "point" and "line." However, they retained the common notion that the whole is greater than the part. Thus the wheels were set in motion which were to have an everlasting effect on the nature of mathematics—it became idealized, or, as is sometimes said, *abstract*. This development in geometry reflects the ideal nature of the arts we examined in Section 3-2.

The third, and final, stage in the revolution that transformed Egyptian empiricism into Greek deduction was caused by Hippasus' discovery of irrational numbers. The fact that $\sqrt{2}$ is an irrational number means that there is no line segment that can divide line segments of lengths 1 and $\sqrt{2}$. This caused a grave problem for geometry because many results had been proved using ratio arguments, i.e., using similar triangles.

What this so-called "crisis in mathematics" did was to send the geometers back to their drawing boards. They had to examine all the assumptions upon which their science was based and then erect a new structure which would avoid the pitfalls. This was the final stage in the birth of deduction. It was delivered some time between 500 and 300 B.C.

Euclid's "Elements," written about 300 B.C., is a paragon for deductive thinking. Its contents reflect all three stages in the revolution: it is based entirely upon proof (the first stage), its basic concepts are idealized (the second stage), and it relies on a set of assumptions that must be explicitly stated beforehand (the third stage). Once mathematics evolved from empiricism to deduction it became an entirely new subject. Today the deductive approach dominates not only mathematics but almost all areas of human endeavor.

Statements

We introduced deduction in Section 3-2, where we saw that its major component is the logic that forms the core of all proofs. We will discuss logic in this section and the next. The type of logic we will introduce is sometimes called *symbolic logic* because it uses symbols in order to remove bias from our ordinary language.

The term "statement" in mathematics means a sentence that is either true or false. It does not include sentences that are vague or nonsensical, or that express a question or a feeling. Some examples of sentences that are statements are:

1 The figure is a circle.

2 Seven is a prime number.

3 Woodrow Wilson was assassinated.

Some examples of sentences that are not (mathematical) statements are:

1 Mathematics is fun.

2 The Empire State Building is today.

3 Please stop!

The first sentence is vague, the second makes no sense, the third expresses a feeling.

All proofs are composed of strings of statements, one following logically from the other, whose purpose is to demonstrate that a complex statement is true. Let us first look at how we combine simple statements, that is, statements that convey one essential idea, into more complex statements.

Conditional Statements

The building blocks of deduction are *conditional statements*. Often they are called *if-then statements* because that is their standard form.

Table 1 contains a list of statements that appear in the first part of the "Elements." All five statements have one thing in common: they are conditional statements. We will examine several aspects of conditional statements in this section in order to understand the "Elements" better and to introduce you to symbolic logic.

TABLE 1

Def. 20	An isosceles triangle has two of its sides alone equal.
CN 2	If equals are added to equals, the sums are equal.
CN 4	Things which coincide are congruent.
Prop. 6	If two angles of a triangle are equal, then the triangle is isosceles.
Prop. 18	In any triangle, the greater angle is opposite the greater side.

Let us look at some examples before considering the statements in Table 1.

Example 1
Examples of conditional statements are:

1 If a figure is a square, then it is a rectangle.

2 If you are in Los Angeles, then you are in California.

3 If you add salt to water, then its boiling point increases.

4 If the Dodgers win the fifth game, then they become the World Series champions.

5 If the figure is a rectangle, then it is a square.

Look at the statement of Proposition 6 in Table 1. It is a conditional statement because it is in the standard if-then form. Let us introduce some symbolism to help study the statement. Let

p be "two angles of a triangle are equal"

q be "the triangle is isosceles"

Then Proposition 6 can be shortened to

if p then q

To achieve complete symbolism we use an arrow to stand for if-then. Now Proposition 6 can be symbolized.

$p \rightarrow q$

It is read "if p, then q."

The notation $p \rightarrow q$ is the symbolic form for all conditional statements. We call p the *antecedent* and q the *consequent*.

Example 2

Consider Common Notion 2 in Table 1.

Problem Determine its antecedent and consequent, and write it in symbolic form.

Solution In general, the part of the statement preceding the word "then" is the antecedent, and the part after it the consequent. However, the word "then" is implicit in Common Notion 2, so you might want to rewrite it first making "then" explicit:

If equals are added to equals, then the sums are equal.

This is the standard form. The antecedent is p

"equals are added to equals"

and the consequent is q

"the sums are equal"

The symbolic form is $p \rightarrow q$.

There are various English forms in which a conditional statement may appear. For instance, the statement "If it is a fire engine, then it is red" can take on any of the following forms:

Given that it is a fire engine, it is red.

It is a fire engine only if it is red.

It is red if it is a fire engine.

It will be red whenever it is a fire engine.

Suppose it is a fire engine, then it is red.

In general, these alternate forms for the conditional $p \rightarrow q$ are:

Given p, then q

p only if q

q if p

q whenever p

Suppose p, then q

You must be careful when one of these forms is used. In particular, when the conclusion is mentioned first, as in "q if p" and "q whenever p," you must not confuse $p \rightarrow q$ with $q \rightarrow p$ because they have different meanings. We will see why after examining two more examples of conditional statements.

Example 3
Problem   Write Common Notion 4 in if-then form. (See Table 1.)
Solution   If things coincide, then they are congruent.

Example 4
Problem   Write Proposition 18 in if-then form. (See Table 1.) Draw a figure to illustrate it, and symbolize it by using the figure.
Solution   The meaning of Proposition 18 is rarely clear upon first reading. To understand it requires translating it into if-then form, even if this translation is carried out subconsciously. Notice that the logical order is the reverse of the written order. The meaning flows from the fact that the sides come first and the angles second. One possible way of expressing Proposition 18 in if-then form is as follows:

In any triangle, if one side is greater than another, then the angle opposite the first side is greater than the angle opposite the second.

Illustrations

Illustration 1

Illustration 2

Illustration 3

Illustration 5

PROCESSION DES PANATHÉNÉES
FRAGMENT DE LA FRISE DU PARTHÉNON
442 - 438 AVANT J.C.

Illustration 6

Illustration 9

Illustration 8

Illustration 10

Illustration   11

Illustration   12

Illustration 13

Illustration 14

Illustration 15

INRI

Illustration 17

Credits

ILLUSTRATION 1   *The Seated Scribe.* Egyptian, 2500 B.C. Musée du Louvre, Paris. *(Scala/Art Resource.)*

ILLUSTRATION 2   Tomb of Menna (detail). Luxor, Egypt. *(Borromeo/Art Resource.)*

ILLUSTRATION 3   Pyramid of Khufu. Dynasty IV. Gizeh, Egypt. *(Art Resource.)*

ILLUSTRATION 4   The Parthenon. Athens, Greece. *(Scala/Art Resource.)*

ILLUSTRATION 5   The Temple of Rameses II. Abu-Simbel, Egypt. *(Art Resource).*

ILLUSTRATION 6   Head of the Procession. Musée du Louvre, Paris. *(Universal Color Slide Company.)*

ILLUSTRATION 7   Hippopotamus Hunt. *(Universal Color Slide Company.)*

ILLUSTRATION 8   The Canon of Egyptian art. Temple of Queen Hatshepsut. Deir el Bahri, Egypt. *(Borromeo/Art Resource.)*

ILLUSTRATION 9   Discobolo Lancillotti. Musco Nazionale, Rome. *(Scala/Art Resource.)*

ILLUSTRATION 10   Filippo Brunelleschi, Santa Spirito. Florence, Italy. *(Scala/Art Resource.)*

ILLUSTRATION 11   Byzantine Icon, Monastery of Mt. Athos, Greece. *(Ricos/Voutsas/Art Resource.)*

ILLUSTRATION 12   Giotto, *The Death of St. Francis,* 1320. Church of Santa Croce, Florence, Italy. *(Scala/Art Resource.)*

ILLUSTRATION 13   Masaccio, *St. Peter Resuscitating the Son of the King of Antioch.* Brancacci Chapel of Santa Maria del Carmine, Florence, Italy. *(Scala/Art Resource.)*

ILLUSTRATION 14   Piero della Francesca, *Resurrection,* 1463. Palazzio Comunale, Borgo, San Sepolcro, Italy. *(Scala/Art Resource.)*

ILLUSTRATION 15   Leonardo da Vinci, *Last Supper,* 1498. Santa Maria delle Grazie, Milan, Italy. *(Scala/Art Resource.)*

ILLUSTRATION 16   Raphael, *School of Athens,* 1511, Vatican. *(Scala/Art Resource.)*

ILLUSTRATION 17   Montegna, Crucifixion. Museé du Louvre, Paris. *(Scala/Art Resource.)*

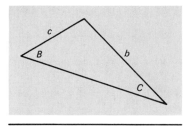

FIGURE 2

Logical Variants: The Converse

Figure 2 illustrates this proposition. The uppercase letters stand for angles and the lowercase for sides. Let

s be "side b is greater than side c"

t be "angle B is greater than angle C"

The symbolic form for Proposition 18 is $s \rightarrow t$.

By a *logical variant* of a conditional statement we mean one of three possible variations of the statement. We will examine the converse and contrapositive here, and reserve the inverse for the exercises. First we will cover the converse.

> The *converse* of the conditional statement $p \rightarrow q$ is the conditional statement $q \rightarrow p$.

Thus the converse is obtained by reversing the roles of the antecedent and consequent.

What is the logical connection between a statement and its converse? Do they have the same meaning? Let us look at an example.

Example 5
Consider the statement

(S) If you are eligible to vote, then you are at least 16 years of age.

Its converse is

(C) If you are at least 16 years of age, then you are eligible to vote.

The original statement (S) is true, but its converse (C) is false because there may be circumstances that prevent a person from being allowed to vote, such as not being a U.S. citizen. Also, you must be 18 years of age in order to vote. Notice that (S) is still true.

The purpose of the next example is to show that the truth of a conditional statement is independent of its converse, and vice versa. In other words, even though one is true the other can be either true or false.

Example 6
Problem Determine whether each of these four statements is true. Write the converse of each and determine whether it is true.
    (a) If $x + 3 = 7$, then $x = 4$.
    (b) If $x = -4$, then $x^2 = 16$.

(*c*) If someone is a member of Congress, then that person is a U.S. Senator.

(*d*) If x is even, then x^2 is odd.

Solution   (*a*) The conditional statement is true because $x = 7 - 3 = 4$. Its converse is:

If $x = 4$, then $x + 3 = 7$

It is also true.

(*b*) The conditional statement is true because $(-4)^2 = 16$. Its converse is:

If $x^2 = 16$, then $x = -4$

It is false, since it is possible that $x = 4$.

(*c*) The conditional statement is false because Congress consists of senators and representatives, so the legislator might be a representative. The converse is:

If someone is a U.S. senator, then that person is a member of Congress.

It is true.

(*d*) The conditional statement is false; let $x = 2$, for example. The converse is:

If x^2 is odd, then x is even.

It is false, since $3^2 = 9$ is odd but 3 is not even.

Table 2 summarizes the results of Example 6. It emphasizes the independence of a conditional statement and its converse by showing all four logical possibilities.

TABLE   2

Statement	Conditional	Converse
(*a*)	True	True
(*b*)	True	False
(*c*)	False	True
(*d*)	False	False

Logical Variants: The Contrapositive

The next logical variant of a conditional that we will study, the contrapositive, is defined in terms of negation. The negation of a statement p means "not p." We denote it by $\sim p$. It is true precisely when p is false, and it is false when p is true. The contrapositive involves both $\sim p$ and $\sim q$.

> The *contrapositive* of the conditional statement $p \rightarrow q$ is the conditional statement $\sim q \rightarrow \sim p$, read "if not q, then not p."

The next example demonstrates how to form the contrapositive. You must negate both p and q and then reverse their order. We will use it to explain the logical relationship between a conditional statement and its contrapositive.

Example 7
Problem   Write the contrapositive of each of the two statements below. Determine the truth of each statement and its contrapositive.
   (*a*) If two lines are parallel, then they do not intersect.
   (*b*) If two lines intersect, then they are perpendicular.
Solution   (*a*) The contrapositive is:

If two lines intersect, then they are not parallel.

Both the conditional statement and its contrapositive are true.
   (*b*) The contrapositive is:

If two lines are not perpendicular, then they do not intersect.

Both the conditional statement and its contrapositive are false.

Example 7 indicates the logical connection between a conditional statement and its contrapositive: if the statement is true, then its contrapositive is also true; and if it is false, so is its contrapositive. Thus the relationship between a conditional statement and its contrapositive is quite different from its relationship with the converse. Essentially, a conditional statement and its contrapositive convey the same idea, but its converse might indicate a different idea. For example, the statement

If $x = -4$, then $x^2 = 16$

is true (see Example 6), and its contrapositive,

If $x^2 \neq 16$, then $x \neq -4$

says essentially the same thing and is also true. However, the statement's converse,

If $x^2 = 16$, then $x = -4$

is false; so it certainly does not convey the same meaning.

Equivalent Statements

Suppose a conditional statement and its converse are both true. That is, suppose $p \rightarrow q$ and $q \rightarrow p$ are both true. Then we say that p and q are *equivalent statements*. This is an important definition so we will highlight it.

Two statements r and s are *equivalent* if both $r \rightarrow s$ and $s \rightarrow r$ are true. We write $r \leftrightarrow s$, read "r if and only if s."

Let us look at an example.

Example 8

Problem Which of the pairs of statements are equivalent?

(a) "$x + 3 = 7$," "$x = 4$"

(b) "You are in Chicago," "you are in Illinois"

Solution (a) The statement

If $x + 3 = 7$, then $x = 4$

is true, and its converse

If $x = 4$, then $x + 3 = 7$

is also true, so the statements are equivalent.

(b) These statements are not equivalent because one conditional statement

If you are in Chicago, then you are in Illinois.

is true but its converse is false.

Usually, two statements that are equivalent say the same thing. For instance, every definition involves two equivalent statements, even though the if-then form is generally used. Recall Euclid's Definition 20 in Table 1.

An isosceles triangle is one which has two of its sides alone equal.

The definition consists of two parts, each of which is a conditional statement:

(1) If a triangle is isosceles, then two of its sides are equal.

and

(2) If two sides of a triangle are equal, then it is isosceles.

By expressing the definition this way we see that the first statement gives a property of isosceles triangles and the second indicates that this property describes them uniquely; that is, they are the only triangles with this property.

The last example in this section asks you to single out the two conditional statements that comprise a definition. In this case the definition of the basic concept of congruence is listed as a common notion.

Example 9

The definition of congruence is given as follows:

Things which coincide are congruent.

Problem   Determine the two if-then statements that comprise this definition.

Solution   We have already determined one of the if-then statements in Example 3. It is:

If things coincide, then they are congruent.

The other statement is the converse of this one:

If things are congruent, then they coincide.

The precise way of writing definitions is to use the phrase "if and only if." Thus Common Notion 4 could read:

Things are congruent if and only if they coincide.

However it is a convention to be a little lax in the wording of definitions and use "if" instead of "if and only if."

Summary

Logic, the core of deduction, rests upon conditional statements. Their standard form is "If p, then q," denoted by $p \rightarrow q$. Symbols are used to avoid ambiguity and bias.

Two variants of the conditional statement $p \rightarrow q$ are:

The converse: $q \rightarrow p$

The contrapositive: $\sim q \rightarrow \sim p$

The truth of a conditional statement is independent of the truth of its converse. However, a conditional statement and its contrapositive are equivalent, meaning that they are either both true or both false. Hence they are two different ways of expressing the same thought.

EXERCISES

Determine whether each sentence in Problems 1 to 6 is a statement.

1 Life is great.

2 The Rolling Stones are great!

3 Whales are mammals.

4 George Balanchine was a choreographer.

5 Whales are fish.

6 George Washington was the second president.

In Problems 7 to 16 express the statement in if-then form.

7 Each time it rains I drive to school.

8 He studies only if it is the night before the exam.

9 All humans are mortal.

10 Every snowflake is unique.

11 Two lines are parallel if each is perpendicular to a line.

12 You are in Washington only if you are in the District of Columbia.

13 A Rorschach test can be considered a valid test of personality only if its results show positive correlations with the results of other personality tests.

14 Each Moslem desires to make a yearly trip to Mecca.

15 Things which are equal to the same thing are equal to one another.

16 A given line segment can be bisected.

In Problems 17 and 18, write the proposition in if-then form, draw a figure to illustrate it, and symbolize it by using the figure. The propositions are stated in the appendix.

17 Proposition 19

18 Proposition 15

In Problems 19 to 24 state the converse of the given conditional statement.

19 If it is red, then you must stop.

20 If it is a mammal, then it suckles its young.

21 If it is raining, then I drive to work.

22 If she is running for governor, then she must resign her post as mayor.

23 If all three angles are equal, then the triangle is equilateral.

24 If he plays on the basketball team, then he can dunk.

In Problems 25 to 32 determine whether the statement and its converse are true.

25 If $x - 3 = 7$, then $x = 10$.

26 If $x - 3 = 4$, then $x = 7$.

27 If $x^2 = 9$, then $x = -3$.

28 If $x = 3$, then $x^2 = 9$.

29 If the planet is closer to the Sun than to Earth, then it is Mars.

30 If the city is Houston, then the population exceeds 1 million.

31 If x^2 is odd, then x is even.

32 If $x = 3y$, then x is even.

In Problems 33 to 36 state the contrapositive and determine if the statement and its contrapositive are true.

33 If you are in New York City, then you are in New York State.

34 If it is a rectangle, then it is a quadrilateral.

35 If he was a member of the Red Guard, then he proclaimed strict adherence to capitalist doctrines.

36 If she is in the District of Columbia, then she is not in Maryland.

In Problems 37 to 42 a definition is given. What are the two conditional statements that comprise it?

37 A circle is a set of points equidistant from a given point.

38 Two lines that never intersect are parallel.

39 Abnormal psychology is the study of all forms of abnormal behavior in humans.

40 Photographic film is a thin sheet of plastic with a thin coating of a gelatin of silver salts.

41 Rubella is called German measles.

42 Niobium is the chemical element whose atomic number is 41.

43 (*Thales' theorem*) The figure below is a semicircle with diameter *AB*. The points *C, D, E* lie on the circumference. Draw triangle *ACB*. What special type of triangle is it? Do the same for triangles *ADB* and *AEB*. What general property does this evidence suggest?

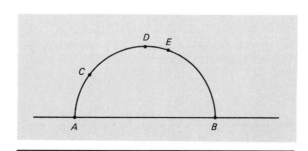

3-4 ARGUMENTS

The classical Greeks were reacting to a lack of rigor and precision of thought when they devised the deductive method of reasoning. Their mathematics and philosophy were based on strict logical reasoning. Aristotle was the first person to thoroughly study the subject as he, along with other Greek philosophers, searched for irrefutable truths. In the previous section we studied the building blocks of logic—conditional statements. Now we will see how to string conditionals together to create logical arguments.

Today we use almost the same techniques of logical arguments that Aristotle created over 2000 years ago. The rudiments of logic have changed very little since then. We will augment his system with more modern terminology and symbolism that simplify complicated arguments. This was first done by the great German universalist Gottfried Leibniz (1646–1716), who, at the age of 14, tried to revamp Aristotelian logic so that errors of thought would surface as computational mistakes rather than bias. This was his main reason for introducing symbols into logic.

No one recognized Liebniz's contributions in this realm until George Boole (1815–1864) wrote an outstanding treatise on logic entitled "An Investigation of the Laws of Thought." Boole improved Leibniz's symbolic form of logic by introducing symbols for mathematical operations. This idea culminated in the classic book, "Principia Mathematica," by Bertrand Russell (1872–1970) and Alfred North Whitehead (1861–1947). They used just a few assumptions and three undefined terms to create a system of symbolic logic that developed the formal systems of arithmetic. It is ironic that even though geometry was built upon an axiomatic basis as early as 300 B.C., it was not until the late nineteenth century that the number system received such a firm foundation.

The book that motivated all these later developments and served as a model for them is Euclid's "Elements." In essence, it consists of a series of if-then statements. Since the conclusions reached in it came to be regarded as truth, it is reasonable to ask:

QUESTION: When can we conclude that a statement is true according to the rules of logic?

We will answer this question by the end of the section. First we must introduce a further element of precision into our logic.

Definitions

Logic can be defined as the science of drawing valid inferences. It is the study of the methods and principles used in distinguishing valid from invalid arguments. Let us define the key terms formally. An *argument* is the assertion that a statement follows from other statements. It is called *valid* when it follows from certain rules of logic, three of which we will describe in this section. Otherwise it is called *invalid*. We will examine some invalid arguments that are sometimes misconstrued as being valid.

Modi Operandi

The easiest type of argument combines a conditional statement with a simple statement. We will consider two of them here, *modus ponens* and *modus tollens*. To convince you that they are intuitive we will precede their definitions with everyday examples.

> **Example 1**
> Consider this statement
>
> If Bill gets one more traffic ticket, then he will lose his driver's license.
>
> First we symbolize it:
>
> Let p be "Bill gets one more traffic ticket."
> Let q be "Bill will lose his driver's license."
>
> The conditional statement is then symbolized $p \rightarrow q$. Assume that the statement $p \rightarrow q$ is true. Suppose that we learn that Bill got another ticket; this means that statement p is true.
> What can we conclude? We conclude that Bill will lose his driver's license, meaning that q is true. Thus we see that when $p \rightarrow q$ and p are both true, it follows that q is true too.

The type of reasoning used in Example 1 is called *modus ponens*. It is often written in the following form.

$$p \rightarrow q \qquad \text{(If } p, \text{ then } q)$$

$$p \qquad (p \text{ is true})$$

$$\therefore q \qquad \text{(Therefore } q \text{ is true)}$$

where $\therefore$ stands for "therefore" and the line separates the *hypotheses* from the *conclusion*. This is the most elementary rule of logic. It means that if we combine a true conditional statement ($p \rightarrow q$) with the fact that its antecedent (p) is true, we can conclude that the consequent (q) is true.

Let us consider another example.

Example 2

Cook County, Illinois, consists of the city of Chicago and its suburbs. Thus a true conditional statement is, "If you live in Chicago, then you live in Cook County." Let p and q have the following representations:

Let p be "you live in Chicago."

Let q be "you live in Cook County."

An argument utilizing *modus ponens* is

$p \to q$ (If you live in Chicago, then you live in Cook County.)

p (You live in Chicago.)

$\therefore q$ (Therefore you live in Cook County.)

Let us look at another rule of logic. This one uses the fact that a conditional statement $p \to q$ is equivalent to its contrapositive, $\sim q \to \sim p$.

Example 3

The culmination of the baseball season is the World Series, which is played between the winners of the American League and the National League. The first team to win four games wins the series. Suppose the headlines in the morning paper read, "If the Dodgers win today then they win the Series." That evening you hear on the radio that the Dodgers did not win the Series today. What conclusion can you draw about the game played earlier? Certainly, the Dodgers did not win the game, because if they had won it they would have won the whole series.

The type of reasoning used in Example 3 is called *modus tollens*. Its symbolic form is

$p \to q$ (If p, then q)

$\sim q$ (Not q)

$\therefore \sim p$ (Therefore not p)

Modus tollens is often referred to as indirect reasoning, while *modus ponens* is often referred to as direct reasoning.

Transitivity

There are three logical forms which underlie symbolic logic. We have just met two of them. Like them, the third is simple and intuitive. Unlike them, it is formed by combining two conditional statements. Example 4 will motivate the definition.

Example 4

Consider the argument that consists of these two conditional statements:

If you are a political science major, then you must take a statistics course.

If you take a statistics course, then you need a good mathematical background.

What conclusion can be drawn? We begin by symbolizing each statement:

Let p be "You are a political science major."

Let q be "You must take a statistics course."

Let r be "You need a good mathematical background."

The two conditional statements become

$p \rightarrow q$

$q \rightarrow r$

The conclusion we reach is the conditional statement

If you are a political science major, then you need a good mathematical background.

In symbols, the conclusion is $p \rightarrow r$.

Example 4 shows that two conditional statements can be combined to form an argument whose conclusion is yet another conditional statement. The symbolic form of this rule of logic is

$$p \rightarrow q$$

$$q \rightarrow r$$

$$\overline{}$$

$$\therefore p \rightarrow r$$

It is read, "If p then q, and if q then r, therefore if p then r." This kind of argument is called *transitivity* and it forms the foundation of a mathematical proof, as the next example shows.

Example 5

In Section 2-4, we proved that $\sqrt{2}$ is an irrational number. A key step in that proof was to establish that the following statement is true:

If $x^2 = 2y$ then x is an even number.

To show that this conditional statement is true requires three simple statements

Let p be "$x^2 = 2y$."

Let q be "x^2 is even."

Let r be "x is even."

The aim is to conclude that $p \rightarrow r$ is a true statement. It is achieved by transitivity. First, the statement $p \rightarrow q$ is true because if $x^2 = 2y$ then x^2 is a multiple of 2, meaning that x^2 is even. Next, $q \rightarrow r$ is true because if x is an odd number then so is x^2, but x^2 is even and so x cannot be odd. Thus we have two true conditional statements. We combine them into an argument and draw the desired conclusion by transitivity.

$$p \rightarrow q \qquad \text{(If } x^2 = 2y, \text{ then } x^2 \text{ is even.)}$$

$$q \rightarrow r \qquad \text{(If } x^2 \text{ is even, then } x \text{ is even.)}$$

$$\therefore p \rightarrow r \qquad \text{(Therefore, if } x^2 = 2y, \text{ then } x \text{ is even.)}$$

Thus the conditional statement $p \rightarrow r$ is true.

Combining Argument Forms

A mathematical proof often uses a combination of the three argument forms that we have described: *modus ponens, modus tollens,* and transitivity. Many such proofs resemble the one in Example 5, except that more than three simple statements are involved. For instance, if the aim is to prove that $p \rightarrow q$ is true and it is known that $p \rightarrow r$, $r \rightarrow s$, and $s \rightarrow q$ are true, then we proceed as follows:

$$p \rightarrow r$$

$$r \rightarrow s$$

$$\therefore p \rightarrow s \qquad \text{(by transitivity)}$$

and

$$p \rightarrow s$$

$$s \rightarrow q$$

$$\therefore p \rightarrow q \qquad \text{(by transitivity)}$$

Generally such an argument is shortened to

$$p \rightarrow r$$

$$r \rightarrow s$$

$$s \rightarrow q$$

_____

$$\therefore p \rightarrow q$$

Let us look at two examples to illustrate this type of argument. The first will be a result in arithmetic whose proof is straightforward. The second is from geometry and is similar to the types of proofs you will encounter later in this chapter.

Our first theorem is concerned with odd numbers. Try to make a conjecture about the sum of two odd numbers. Is the sum always odd, always even, or sometimes odd and sometimes even? Try some examples: $3 + 5 = 8, 5 + 13 = 18, 21 + 33 = 54$. This might lead you to guess that the sum of two odd numbers is always even. Can you prove that this is always true?

The next example will present a proof of this result. We will then translate the proof into symbolic notation in order to demonstrate that the argument relies heavily on transitivity. The crux of the argument is that an odd number can always be expressed as an even number, say $2n$, plus 1. So the general form of an odd number is $2n + 1$ for some integer n.

Example 6

Prove the following theorem and express the proof in symbolic notation.

Theorem The sum of two odd numbers is an even number.

Proof The proof consists of five conditional statements. Let x and y be two odd numbers and let $x + y$ be their sum.

If x and y are odd numbers, then $x = 2n + 1$ and $y = 2m + 1$.

If $x = 2n + 1$ and $y = 2m + 1$, then $x + y = 2n + 1 + 2m + 1$.

If $x + y = 2n + 1 + 2m + 1$, then $x + y = 2n + 2m + 2$.

If $x + y = 2n + 2m + 2$, then $x + y = 2(n + m + 1)$.

If $x + y = 2(n + m + 1)$, then $x + y$ is an even number.

Therefore, if x and y are odd numbers, then $x + y$ is an even number.

This completes the proof. In order to translate the proof into symbolic notation, we will let

p be "x and y are odd numbers"

r be "$x = 2n + 1$ and $y = 2m + 1$"

s be "$x + y = 2n + 1 + 2m + 1$"

t be "$x + y = 2n + 2m + 2$"

u be "$x + y = 2(n + m + 1)$"

q be "$x + y$ is an even number"

Then the theorem reads in symbolic form $p \rightarrow q$ and the proof becomes

$$p \rightarrow r$$
$$r \rightarrow s$$
$$s \rightarrow t$$
$$t \rightarrow u$$
$$u \rightarrow q$$
$$\underline{\hspace{3cm}}$$
$$\therefore p \rightarrow q$$

Here we used transitivity four times to reach the final conclusion.

In actual practice this proof would be a bit shorter because you do not have to repeat the consequent as the antecedent in the next conditional statement as long as it is clearly understood. For example, the proof would probably read like this:

If x and y are odd numbers, then $x = 2n + 1$ and $y = 2m + 1$ for some integers n and m. Then $x + y = 2n + 1 + 2m + 1 = 2n + 2m + 2 = 2(n + m + 1)$, which is an even number.

We put in all the steps to emphasize that transitivity is the rule of logic that makes the proof valid.

The next example examines a result discussed in Section 3-2.

Example 7

Consider Euclid's proof of Proposition 1. We restate it as follows:

If AB is a given line segment, then an equilateral triangle can be constructed with AB as its base.

The construction part of the proof entails drawing two circles, each with radius AB, one with center A and the other with center B. The circles intersect at a point C and the desired triangle is ABC, as in Figure 1. The remainder of the proof demonstrates that the following statement is true:

If AB and AC are radii of one circle and AB and BC are radii of another circle, then triangle ABC is equilateral.

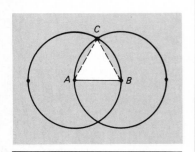

FIGURE 1

The proof utilizes the following statements: let

p be "AB and AC are radii of the same circle and AB and BC are radii of the same circle."

r be "$AB = AC$ and $AB = BC$."

s be "$AC = BC$."

q be "ABC is an equilateral triangle."

In its symbolic form the proof becomes

$$p \rightarrow r$$

$$r \rightarrow s$$

$$s \rightarrow q$$

$$\therefore p \rightarrow q$$

Examples 6 and 7 illustrate the general framework of a mathematical proof. We start with the intention of showing that a certain statement $p \rightarrow q$ is true. First we establish a string of true conditional statements, one of which begins with p, another ends with q, and the others connect them. By repeatedly using transitivity and the other rules of logic, we conclude that $p \rightarrow q$ is a true statement.

The next example illustrates how another combination of logical forms can be used to reach a conclusion. It shows how symbolic logic enters into everyday situations.

Example 8

Consider the following sequence of statements.

If Jane gets an A on the final exam, then she will get an A for the course.

If Jane gets an A for the course, then she will make the Dean's list.

Jane did not make the Dean's list.

Problem What conclusion can you draw?
Solution We will use the following symbols: let

p be "Jane gets an A on the final exam."

q be "Jane gets an A for the course."

r be "Jane made the Dean's list."

The statements then become

$$p \rightarrow q$$

$$q \rightarrow r$$

$$\sim r$$

By combining the second and third statements and using *modus tollens,* we get

$$q \rightarrow r$$

$$\sim r$$

$$\overline{}$$

$$\therefore \sim q$$

Then consider the first conditional statement with the above conclusion and apply *modus tollens* again:

$$p \rightarrow q$$

$$\sim q$$

$$\overline{}$$

$$\therefore \sim p$$

Hence the conclusion is $\sim r \rightarrow \sim p$. Thus if Jane did not make the Dean's list, then she did not get an A on the final exam.

Invalid Arguments

So far, all the arguments encountered in this section have been valid, meaning that the conclusion was obtained by the rules of logic. Sometimes, however, a conclusion is drawn that is not warranted. We will end this section with a few examples of these invalid arguments. We will first present an argument that is obviously invalid. Then we will show you two others that appear convincing but are invalid for the same reason as the first.

Example 9
Problem Express the argument in symbolic form and determine whether it is valid.

If you live in London, then you live in England.

You live in England.

$$\overline{}$$

$\therefore$ You live in London.

Solution If we let

p be "you live in London."

q be "you live in England."

then the argument is

$$p \rightarrow q$$

$$q$$

$$\therefore p$$

It is easy to understand why this argument is invalid. There are many people who live in England but do not live in London, so the argument is clearly not valid.

Example 10

Problem Express the argument in symbolic form and determine whether it is a valid argument.

If you are beautiful then you use Ivory soap.

You use Ivory soap.

Therefore you are beautiful.

Solution If we let

p be "you are beautiful."

q be "you use Ivory soap."

the argument is

$$p \rightarrow q$$

$$q$$

$$\therefore p$$

It is invalid, but it is exactly what many commercials would like us to conclude.

Here is a much more serious example.

Example 11

Mathematics has found its way into our legal system, as it has in many everyday situations. A trial that has caused much mathematical controversy is "People v. Collins"* because of its misuse of probability. The prosecution used many arguments similar to the following. After a mugging in a Los Angeles suburb, a couple was seen running from the scene. They were

---

*People v. Collins, 68 Ca. 2 319, 438 p. 2d 33, 66 Ca.

described as a black man with a beard and a white woman with a blond ponytail. Malcolm and Janet Collins, who fit the description, were later arrested. The district attorney then used some faulty probability theory to show that it was very unlikely that another couple in LA would fit the description. The jury convicted the Collinses on this scanty evidence.

Let us analyze the DA's argument. This is what the DA proved:

If they are the guilty parties, then they fit the description.

The Collinses fit the description.

Therefore the Collinses are the guilty parties.

The argument can be expressed symbolically: let

p be "they are guilty."

q be "they fit the description."

Then the argument reads

$p \rightarrow q$

q

_____

$\therefore p$

It is invalid.

This type of argument, unfortunately, is used a great deal to persuade us into making erroneous conclusions. The court case cited above is a prime example. It is called circumstantial evidence, and if it is used many times in a trial it can be very convincing. Malcolm Collins was eventually freed after an appeal to the California Supreme Court, but only after he had spent a good deal of time in prison.

EXERCISES

In Problems 1 to 4 a conditional statement is given. Use *modus ponens* to draw a valid conclusion.

1   If you live in the United States you are an American citizen.

2   You like beef if you are a meat lover.

3   You live in Miami only if you live in Florida.

4   You will fail the course if you receive an F.

In Problems 5 and 6 use *modus tollens* to draw a valid conclusion.

5 A circle can be drawn if three noncollinear points are given.
A circle cannot be drawn.

6 If an angle is a right angle then it has 90°.
Angle A has 45°.

In Problems 7 to 10 determine which arguments are valid.

7 If he gets the orange pits, then he is the next to be murdered.
Jones got the orange pits.
Therefore Jones is next to be murdered.

8 If the dough rises too much, then the bread will be ruined.
The bread was ruined.
Therefore the dough rose too much.

9 If a parallelogram has two right angles, then it is a rectangle.
It is not a rectangle.
Therefore it is not a parallelogram.

10 If she works for the government, then she is a civil servant.
She is not a civil servant.
Therefore she does not work for the government.

In Problems 11 to 16 use transitivity to draw a valid conclusion.

11 If Mary is in Montreal, then she is in Quebec.
If Mary is in Quebec, then she is in Canada.

12 If John gets one more ticket, then his insurance will be canceled.
If John's insurance is canceled, then he will not be able to drive to work.
If John cannot drive to work, then he will lose his job.

13 If a four-sided figure has three right angles, then all four angles are right angles.
If all four angles of a four-sided figure are right angles, then it is a rectangle.

14 If x and y are consecutive integers, then one is even and one is odd.
If one integer is even and the other is odd, then their sum is an odd integer.

15 If x and y are consecutive integers, then one is even and the other is odd.
If one integer is even and the other is odd, then their product is an even integer.

16 If x, y, and z are consecutive integers, then at least one is even and one is divisible by 3.
If one number is divisible by 2 and another number is divisible by 3, then their product is divisible by 6.

In Problems 17 to 20 some true conditional statements are given. If they are rearranged in proper order, they constitute a proof of a fact about either even numbers or odd numbers. Rearrange them into a proper order and determine what the fact is.

17 (1) If a and b are even numbers, then $a = 2n$ and $b = 2m$ for some numbers n and m.
    (2) If $a + b = 2(n + m)$, then $a + b$ is an even number.
    (3) If $a = 2n$ and $b = 2m$, then $a + b = 2n + 2m$.
    (4) If $a + b = 2n + 2m$, then $a + b = 2(n + m)$.

18 (1) If $a + b = 2(n + m + 1)$, then $a + b$ is an even number.
    (2) If a and b are odd numbers, then $a = 2n + 1$ and $b = 2m + 1$.
    (3) If $a + b = 2n + 1 + 2m + 1$, then $a + b = 2n + 2m + 2 = 2(n + m + 1)$.
    (4) If $a = 2n + 1$ and $b = 2m + 1$, then $a + b = 2n + 1 + 2m + 1$.

19 (1) If $ab = (2n + 1)(2m + 1)$, then $ab = 4nm + 2n + 2m + 1$.
    (2) If a and b are odd numbers, then $a = 2n + 1$ and $b = 2m + 1$.
    (3) If $ab = 4nm + 2n + 2m + 1$, then ab is an odd number.
    (4) If $a = 2n + 1$ and $b = 2m + 1$, then $ab = (2n + 1)(2m + 1)$.

20 (1) If $a = 2n$, then $ab = 2nb$.
    (2) If $ab = 2nb$, then ab is an even number.
    (3) If a is an even number and b is any number, then $a = 2n$.

21 This problem is twofold. There is given below a "proof" consisting of six conditional statements. Five of them are true and one is false. They are arranged in the proper order to constitute a proof, but since one is not true, the whole proof is invalid, even though it appears to be sound when it is first encountered. Can you determine what "theorem" is "proved"? Take a guess.

Proof

If $a = b = 1$, then $a = ab$ and $a - b = (a - b)(a + b)$[   ].

If $a = ab$, then $a - b = ab - b$ [   ].

If $a - b = ab - b$, then $(a - b)(a + b) = ab - b$ [   ].

If $(a - b)(a + b) = ab - b$, then $(a - b)(a + b) = (a - b)b$ [   ].

If $(a - b)(a + b) = (a - b)b$, then $a + b = b$ [   ].

If $a + b = b$, then $1 + 1 = 1$ [   ].

The theorem that is proved is "if $a = b = 1$ then $1 + 1 = 1$." Obviously the conclusion is false, but that is not why the proof is incorrect. First determine which of the six reasons below goes with each of the conditional statements in the proof. Then determine which conditional statement is false.
(1) Substitute $a = 1$ and $b = 1$.
(2) Equals multiplied by equals are equal.
(3) $a - b = (a - b)(a + b)$.
(4) Equals subtracted from equals are equal.
(5) $ab - b = (a - b)b$.
(6) Equals divided by equals are equal.

**An Interlude
from "Wonderland"**

Most people have never heard of Reverend Charles Lutwidge Dodgson, but hardly anyone is unfamiliar with his pen name and his most famous book. Dodgson is the author of "Alice in Wonderland" under the pseudonym of Lewis Carroll. Logic and deduction fascinated Carroll, and many of the passages in "Alice in Wonderland" have their comedic foundation in the use and abuse of logic, with Alice eternally championing right and correct reason while the strange inhabitants of her new topsy-turvy world befuddle her with their careless use of logic. While he is best remembered for "Alice in Wonderland," Dodgson wrote many works pertaining to the deductive process of reasoning. Most notable is his book "Symbolic Logic," containing many cleverly designed puzzles that afford the reader an opportunity to solve reasoning problems while being entertained at the same time. The objective of each problem is to use all the information given in the statements to determine the "theorem" inferred. To solve the puzzles, put each statement in if-then form, assign an appropriate letter to each statement, and then string the conditional statements together, perhaps using the contrapositive where applicable to find the intended conclusion. Carroll phrased the problems not only to be entertaining but to border on the absurd, so that emotion and intuition would not interfere with the understanding of strict logical reasoning.

We first solve two of the puzzles as examples and then present several more for your reasoning entertainment.

Example 1

Consider the following three statements:

1 No ducks waltz.

2 Every military officer dances an excellent waltz.

3 All my poultry are ducks.

Assign the following letters to the statements:

d: it is a duck

w: it dances the waltz

m: it is a military officer

p: my poultry

Putting each statement in if-then form together with its corresponding symbolic notation yields:

1 If it is a duck then it does not waltz. ($d \rightarrow \sim w$)

2 If it is a military officer then it does the waltz. ($m \rightarrow w$)

3 If it is one of my poultry then it is a duck. ($p \rightarrow d$)

In order to string the statements together we note that p is mentioned once; so we can start with the third statement $p \rightarrow d$. Next list the first conditional statement, $d \rightarrow \sim w$. By transitivity we get

$$p \rightarrow d \qquad (3)$$

$$d \rightarrow \sim w \qquad (1)$$

$$\overline{\qquad\qquad}$$

$$\therefore p \rightarrow \sim w$$

In order to use the second conditional statement in our string we form the contrapositive of $m \rightarrow w$, which is $\sim w \rightarrow \sim m$. By using transitivity twice:

$$p \rightarrow d \qquad (3)$$

$$d \rightarrow \sim w \qquad (1)$$

$$\sim w \rightarrow \sim m \qquad \text{(contrapositive of 2)}$$

$$\overline{\qquad\qquad}$$

$$\therefore p \rightarrow \sim m$$

Hence Carroll's intended theorem is, "If it is one of my poultry, then it is not a military officer."

As an alternate approach we could have noted that m is mentioned only once and started with statement 2 followed by the contrapositive of 1, which is $w \rightarrow \sim d$, followed by the contrapositive of 3, which is $\sim d \rightarrow \sim p$, to get the string

$$m \rightarrow w$$

$$w \rightarrow \sim d$$

$$\sim d \rightarrow \sim p$$

In this case the answer would be $m \rightarrow \sim p$, which is the contrapositive of our original answer and hence logically equivalent to it.

Example 2
This puzzle has five sentences, but its solution is achieved by the same process. Carroll was a lover of animals. Perhaps this problem whimsically describes his pets.

1 No kitten that loves fish is unteachable.

2 No kitten without a tail will play with a gorilla.

3 Kittens with whiskers always love fish.

4 No teachable kitten has green eyes.

5 Kittens that have tails have whiskers.

Assign the following letters to the properties of the kittens:

f: loves fish

t: teachable

T: has a tail

g: will play with a gorilla

w: has whiskers

e: has green eyes

Putting the statements in if-then form and expressing them symbolically yields

1   If it loves fish, then it is teachable. ($f \rightarrow t$)

2   If it does not have a tail, then it will not play with a gorilla. ($\sim T \rightarrow \sim g$)

3   If it has whiskers, then it loves fish. ($w \rightarrow f$)

4   If it is teachable, then it does not have green eyes. ($t \rightarrow \sim e$)

5   If it has a tail, then it has whiskers. ($T \rightarrow w$)

The statements *g* and *e* are mentioned only once. We can start the string of statements with either one. We arbitrarily choose *g*. To begin with *g* we must use the contrapositive of statement 2, which is $g \rightarrow T$. Then use statement 5, then 3, then 1, and then 4, which yields

$$g \rightarrow T$$
$$T \rightarrow w$$
$$w \rightarrow f$$
$$f \rightarrow t$$
$$t \rightarrow \sim e$$

Therefore Carroll's intended theorem is, "If a kitten will play with a gorilla, then it does not have green eyes."

Now you are ready to get a glimpse of the world through Lewis Carroll's "logic-colored" glasses. Put each puzzle in if-then form, symbolize the statements, and combine them to determine Carroll's intended theorem.

22   Anyone who is sane can do logic.
     No lunatics are fit to serve on a jury.
     None of your relatives can do logic.

23   Promise-breakers are untrustworthy.
     Wine-drinkers are very communicative.
     Teetotalers are trustworthy.

24   All unripe fruit is unwholesome.
     All these apples are wholesome.
     No fruit that is grown in the shade is ripe.

25 Animals with pouches carry extra weight.
Animals that hop well do not carry extra weight.
Every kangaroo has a pouch.

26 All my sons are slim.
No child of mine who takes no exercise is healthy.
All gluttons who are children of mine are fat.
No daughter of mine takes any exercise.

27 Babies are illogical.
Anyone who can manage a crocodile is not despised.
The person who cannot manage a crocodile is logical.
Anyone who is not despised is a saint.

28 All sharks believe they are brave.
Any fish that cannot dance the jig is despicable.
A fish is brave only if it has three rows of teeth.
No shark is kind.
A fish with three rows of teeth is not despicable.

29 Animals that do not kick are always unexcitable.
Donkeys have no horns.
A buffalo can always toss one over a gate.
No animals that kick are easy to swallow.
No hornless animal can toss one over a gate.
All animals are excitable except buffaloes.

30 Here is a puzzle that has what seems to be a simple solution, but most people find it more difficult than it first appears. In other words they get it wrong on the first guess. Suppose you are the director of a tennis tournament consisting of 64 players. It is a single elimination tournament, which means that when you lose a match you are out of the tournament and the winners keep playing until there is only one person left, and that person is the champion. As director you must schedule court time and order tennis balls for each match; so you must determine how many matches there are. First take an intuitive guess as to how many matches there are in the tournament. Then try a more systematic approach by counting the number of matches in each round. In the first round there are 32 matches with 32 winners, then in the second round they square off in 16 matches, etc. Next solve the problem for 32 players, 16 players, and 128 players. If you see a pattern, you can make an educated guess (via induction) to the solution of the general problem when there are n players. The solution to the general problem with n players is not easily proved by using the counting technique described above. You should recognize that your guess, although probably correct (especially if you guessed $n - 1$) does not constitute a proof of your assertion. Use the following conditional statements to formulate a proof of your conjecture.

(1) Every match has a unique loser.

(2) Everyone in the tournament, except one (the champion) loses exactly one match.

3-5 THE THEORY OF PARALLELS

The phrase "the fifth" evokes different images to people from various walks of life. Music lovers think of Beethoven's famous symphony. Lawyers envision courtrooms where witnesses plead it in defense. It reminds party-goers of a bottle of whiskey. Social misfits often think of themselves as the fifth wheel.

And what about *mathematicians?* They think of Euclid's fifth postulate. In fact, Euclid's is the only "fifth" that has had a life of its own: it was created reluctantly, it faced early criticism and controversy, it was labeled extraneous, and throughout history it has been loved, rejected, and expanded upon.

How could such a seemingly insignificant statement have such a turbulent history? This is what we will explain now and in Chapter 4. Perhaps you are worried that we are making the proverbial mountain out of a molehill, but we guarantee that if you stick with us you will be amazed at the ramifications that Euclid's fifth has caused—not just in mathematics but in many areas that touch our daily lives.

Euclid's Fifth Postulate

Let us take a closer look at the "renegade" postulate. It is:

> EUCLID'S FIFTH POSTULATE: If a line intersects two other lines in such a manner that the sum of the interior angles on the same side is less than 180°, then the two lines, if extended indefinitely, meet on that side where the angle sum is less than 180°.

In Section 3-2 we contrasted this postulate with the five common notions and the other four postulates. We said that it was neither particularly clear nor easy to understand. If you agreed, then you are in heady company because for over 2300 years mathematicians have been saying the same thing. In fact, one of the most vociferous criticisms of Euclid by his contemporaries was that the fifth postulate seemed bizarre compared with the other axioms. It was much longer, much too wordy, and its meaning lacked the succinct character of the others. It simply did not conform. There is ample evidence to suggest that Euclid himself was uncomfortable with it. If you think of the 10 axioms as the building blocks of his geometry, as the 10 fundamental rules that form the foundation of geometry, and you realize that he uses each of the other 9 axioms over and over throughout the manuscript, even in the early postulates, it is surprising to realize that he never uses the fifth postulate until Proposition 29. He could easily have called upon it earlier and used it efficiently, so it appears to have been a conscious and calculated decision to delay its use until it was absolutely necessary. (See Problem 13 in the exercises.)

Yet one of the recurring questions of Euclidean geometry is just that—is the fifth postulate necessary? Even Euclid's peers questioned whether he could have done away with it. In other words, Euclid proved 28 propositions without it—could the remaining ones also be proved without it? Notice also that the fifth postulate reads more like a proposition than an axiom, both in its length and its content. This led Euclid's critics and devotees alike to wonder whether, rather than *assume* it was a true statement, one could *prove* that it was true by showing that it followed as a logical conclusion from the other nine axioms and the first 28 propositions. Doubtlessly many tried. In fact, for more than 2000 years people have been searching for a proof that Euclid's fifth can be regarded as a proposition instead of an axiom. This quest has been the springboard for many fruitful revolutions in the evolution of geometry. We will explain a number of them in detail in Chapter 4. It has made some men famous, cost others their careers, and is a root of one of our major societal problems today.

Here is the key question:

Is Euclid's fifth postulate necessary?

By "necessary" we mean that if you reject it as an axiom, what are the consequences? One answer that we just touched upon is to try to show that it is a logical consequence of the other nine axioms. Many a sleepless night has been devoted to this task and to date no one has been successful.

Another answer is to dispense with it altogether. Euclid proved 28 results without it, and many more could be proved similarly. But the resulting geometry is not very rich nor interesting without parallel lines.

Before we tell you about its storied past, can you guess how others have attempted to answer the question?

If you were to draw a figure in order to get a better understanding of the fifth postulate, it would probably look like the one in Figure 1. We have labeled the lines as L_1 and L_2 and we let l represent the line that intersects them. We call l the *transversal*. The angles referred to in the postulate are labeled y and w, respectively. According to the figure, Postulate 5 states that if the angle sum of w and y is less than $180°$, then L_1 will intersect L_2. If we use the symbolic notation → for "if-then" and if

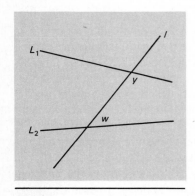

FIGURE 1

we let $L_1 \cap L_2$ represent "L_1 *intersects* L_2," then we can state Postulate 5 in a succinct symbolic form as follows:

> POSTULATE 5: $w + y < 180° \rightarrow L_1 \cap L_2$

We mentioned earlier that Postulate 5 refers to parallel lines. In order to explain this we consider its logical equivalent—the contrapositive. It involves the negation of the statement $L_1 \cap L_2$, which is "L_1 does not intersect L_2." This means that L_1 is parallel to L_2, which we express as $L_1 \parallel L_2$. To determine the negation of $w + y < 180°$, which, strictly speaking, is $w + y \geq 180°$, we need to interpret Postulate 5 more thoroughly. If the angle sum on one side of the transversal is greater than $180°$, the angle sum on the other side is less than $180°$. Therefore the negation of $w + y < 180°$ cannot include $w + y > 180°$ because Postulate 5 refers to the angle sum on *either* side. Thus the negation requires that *neither* side have an angle sum less than $180°$. This means that $w + y = 180°$ is the negation. Hence we can state the contrapositive of Postulate 5 in symbolic form as

> CONTRAPOSITIVE OF POSTULATE 5:
> $L_1 \parallel L_2 \rightarrow w + y = 180°$

Therefore Postulate 5 includes a property of parallel lines. If we assume that Postulate 5 is true, then it is also true that if a transversal intersects parallel lines the sum of the interior angles on the same side of the transversal is $180°$.

Parallel Lines

FIGURE 2 How would you determine whether these lines are parallel?

Consider the two lines in Figure 2. How would you determine whether or not they were parallel? One way, and perhaps the most immediate one, would be to measure the distance between the lines at several points to see if the distances were the same. But you would have to be careful because if you intended to use a ruler and if you were abiding by the tenets of Euclidean geometry, this would be an improper procedure. Rulers are *verboten*. The Greeks established their brand of geometry on the premise that measurements involve human error. In fact such measurements are confounded in this case because they depend upon perpendiculars, which in turn rely on an angular measurement. It is possible to compare the distances at various points between the lines, but it would be a somewhat complicated procedure. You would first have to select two points on one line, drop a perpendicular from each point to the other line

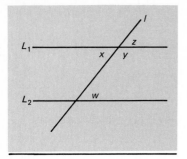

FIGURE 3

by using the procedure in Proposition 12, and then compare the distances, using the procedure in Proposition 3.

Euclid found more direct means for determining when two lines are parallel. They are contained in Propositions 27 and 28. Consider Figure 3. Proposition 27 states that if $x = w$ then L_1 and L_2 are parallel. Using symbolic notation we can express this idea as

---

PROPOSITION 27: $x = w \rightarrow L_1 \parallel L_2$

Proposition 28 is similar. It gives two more conditions for determining when L_1 and L_2 are parallel.

PROPOSITION 28: $z = w \rightarrow L_1 \parallel L_2$ and

$$w + y = 180° \rightarrow L_1 \parallel L_2$$

Altogether Propositions 27 and 28 contain three conditional statements—three ways to tell whether two lines are parallel. What about the converse of each statement? In Proposition 29 Euclid shows that each converse is true. In other words, if L_1 and L_2 are parallel, then each condition concerning the angles is true; that is, $x = w$, $z = w$, and $w + y = 180°$. In symbolic form we have:

PROPOSITION 29: $L_1 \parallel L_2 \rightarrow \begin{cases} x = w \\ z = w \\ w + y = 180° \end{cases}$

Proposition 29 is the cornerstone of our ascent theme in geometry. It appears to be an innocent result at first glance. After Euclid proved Proposition 27, the natural question to ask was whether its converse was also true. As we saw, sometimes the converse is true and sometimes it is false. In this case Euclid found a clear and succinct proof of the converse. But he discovered more. With only a few more steps he could prove the other two converses.

For this reason alone Proposition 29 is significant; its simple and direct proof yields a bounty of results—three properties of parallel lines. But it is not the statement of the proposition itself that is of most interest for us. It is the proof, and in particular one line in the proof, for it is in Proposition 29 that Euclid finally resorts to the fifth postulate.

**The Proofs of
Propositions 27, 28,
and 29**

We will now take a close look at the proofs of these three important propositions. At first glance the proofs seem similar. Each is about the same length and uses the same figure. Propositions 27 and 29 employ a proof by contradiction. While Proposition 28 has a direct proof, it depends heavily on Proposition 27. As you read the proofs, and especially while you supply the reasons for each step, try to determine where the pivotal step is. We will point it out after the proof of Proposition 29.

Each proof is based upon Figure 4, the so-called pencil diagram. The figure consists of two lines L_1 and L_2 that sometimes represent parallel lines and sometimes not. You will have to stretch your imagination in the proof of Proposition 27 because we will assume that L_1 and L_2 are not parallel; they meet in point C. Then we think of ABC as a triangle even though the "lines" AC and BC look more like curves. You should picture it this way: L_1 and L_2 are "almost" parallel in that they appear parallel to the naked eye, but if we extend the lines out indefinitely they eventually meet at point C. The problem with our intuitive grasp of the situation is that our page is not large enough to have two lines that are "almost" parallel and yet still meet.

Here is Proposition 27 and its proof. We insert the justifications that have not been previously mentioned and leave the others for the exercises.

Proposition 27
If $x = w$ then $L_1 \parallel L_2$.
Proof Refer to Figure 4. The method of proof is reductio ad absurdum. You should supply the missing justifications.

1 If L_1 and L_2 are not parallel, then they meet at some point, say C [Def.   ].

2 ABC forms a triangle in which w is interior and opposite to the exterior angle x [Def.   ].

3 x is greater than w [Prop. 16].

4 This contradicts the hypothesis that $x = w$. Therefore the original assumption in the proof, that L_1 and L_2 are not parallel, is false. Hence $L_1 \parallel L_2$.

QED

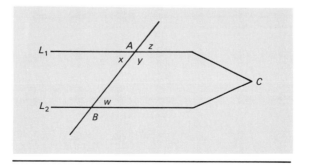

FIGURE 4

The statement and proof of Proposition 28 also use Figure 4. The proof uses Proposition 27. In fact, armed with this knowledge, perhaps you can prove it yourself.

Proposition 28
    (*a*) If $z = w$, then $L_1 \parallel L_2$.
    (*b*) If $w + y = 180°$, then $L_1 \parallel L_2$.
Proof (*a*)

1   $x = z$ [Prop. 15]

2   $z = w$ [Hypothesis]

3   $x = w$ [CN]

4   $L_1 \parallel L_2$ [Prop.]

   (*b*)

1   $z + y = 180°$ [Prop.]

2   $w + y = 180°$ [Hypothesis]

3   $w + y = z + y$ [CN]

4   $w = z$ [CN]

5   $L_1 \parallel L_2$ [by part (*a*)]

Proposition 29 is the converse of Propositions 27 and 28. It proves three results. As you go through the proof, note where each of the three properties of parallel lines is proved. Note also how each result is used to prove the next one.

Proposition 29
If $L_1 \parallel L_2$, then $x = w$, $z = w$, and $w + y = 180°$.
Proof We show that $x = w$ by the method of reductio ad absurdum. The next two results follow directly from the first.

1   Assume $x \neq w$. Then one is greater, so suppose $w < x$.

2   $w + y < x + y$ [ ].

3   $x + y = 180°$ [ ].

4   $w + y < 180°$ [ ].

5   L_1 meets L_2 if both are extended indefinitely [ ].

6   This contradicts the hypothesis that L_1 and L_2 are parallel. Hence x cannot be greater than w. Similar reasoning shows that w cannot be greater than x. Hence $x = w$.

7   $x = z$ [ ].

8   $z = w$ [ ].

9 $z + y = w + y$ [ ].

10 $z + y = 180°$ [ ].

11 $w + y = 180°$ [ ].

We did not give you many hints for the justifications in the proof of Proposition 29 because we wanted you to do the thinking yourself. We now want to center our attention on step 5 because it is the pivotal point in this section. Here Euclid used the fifth postulate for the first time. Notice that it seems rather inconsequential in this setting.

The key is to realize that Propositions 27, 28, and 29 do not form as compact a group as it appears. Propositions 27 and 28 are significantly different because their proofs do not rely on Postulate 5. In fact, the first 28 propositions required only the other nine axioms in their proofs. If we could somehow find a proof of Postulate 5 that depended only on three types of statements—Postulates 1 to 4, the five common notions, and Propositions 1 to 28—then the proof of Proposition 29 would follow from these statements also. In fact all of Euclid's subsequent propositions would then follow from them and we would have decreased the number of axioms from 10 to 9. But no such proof exists even though many attempts have been made over the last 2300 years.

Let us look at how Propositions 27, 28, and 29 fit together. Proposition 27 states that $x = w \rightarrow L_1 \parallel L_2$. The first part of Proposition 29 states that $L_1 \parallel L_2 \rightarrow x = w$. Recall that two statements p and q are equivalent if $p \rightarrow q$ and $q \rightarrow p$ are true. The equivalence is denoted by $p \leftrightarrow q$. By setting p equal to $L_1 \parallel L_2$ and q equal to $x = w$, we have

$$L_1 \parallel L_2 \leftrightarrow x = w$$

This equivalence provides an alternate way of defining parallel lines, though it is certainly not preferable. Propositions 28 and 29 provide two more equivalents. We can write all four in one string:

$$L_1 \parallel L_2 \leftrightarrow x = w \leftrightarrow z = w \leftrightarrow y + w = 180°$$

This form indicates a "one for all and all for one" type of behavior, in that if one property holds then all the rest do too, while if one does not hold then neither do the others.

Prologue

Let us now pull together the ideas in this section and thereby lay the groundwork for Chapter 4.

Euclid proved almost 500 propositions in the "Elements"—all from just 10 assumptions. Each axiom was essential, but it seems as though nine were more essential than the other. Propositions 1 to 28 were proved without once referring to Postulate 5. Certainly Euclid's geometry would have been flat and drab with so little about parallel lines. The question that undoubtedly confounded him was how to include them. Must that tenth axiom be included or might it be possible to prove a key result on parallel lines from the first nine axioms? Most likely it was against his better judgment that Euclid included the parallel postulate in his list. It seems as though he tried to make amends by avoiding it until it became necessary, which was in the proof of Proposition 29.

Perhaps you are getting the impression that we are making much ado about nothing. After all, Euclid provided us with a beautifully rich geometry. He even provided us with a means to assess its veracity via a rigid logical foundation. It has withstood the test of time; even today we study his geometry in school, and many professionals, including architects, surveyors, and builders, use his results daily. It is aesthetically pleasing and useful. So what is all the fuss about? Does it really matter whether Postulate 5 is an axiom or a theorem? It is obviously true . . . or is it?

Remember that the Greeks rebelled against such reasoning. They saw nothing wrong in proving the "obvious." Their adherence to strict logical reasoning prompted Euclid to adopt that format in the "Elements." The essence of the work was proof. The Greeks would never assume a result is true on the basis of intuition or empirical evidence alone. Of course one has to start somewhere. Some things must be assumed true, but the number of them must be kept to a minimum. Nine axioms would have been preferable, but Euclid found it necessary to include the tenth.

This then is the first step in the evolution of the concept of parallel lines. Euclid probably recognized that his parallel postulate was an albatross, but he had to include it in his list of assumptions. The story unfolds from here as others pick up the gauntlet and try to either support him or discredit him.

One extremely important point that might be overlooked is the importance of logic. The ascent of geometry has taken many diverse paths since Euclid's time. The reasons why we chose to center attention on the evolution of this one aspect, parallel lines, are twofold. First, the history of the subject is colorful and fascinating. Second, to fully understand and appreciate the evolution you must have a firm grasp on the rudiments of logical reasoning. Of essential import is the recognition of the fundamental difference between an axiom, something assumed to be true, and a theorem or a proposition, whose truth depends upon a rigorous chain of conditional statements that follow logically from the axioms.

The classical Greeks devised this seemingly foolproof method of reasoning. Euclid took geometry from empiricists like the ancient Egyptians and put it in this logical setting. A slight problem revolving around Postulate 5 remained, but it seemed that it too would be cleared up soon. The problem was virtually ignored for 2000 years. We will pick it up again in Chapter 4.

EXERCISES

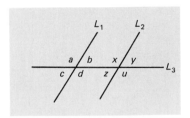

1 Fill in the missing justifications in the proof of Proposition 28.

2 Fill in the missing justifications in the proof of Proposition 27.

3 Fill in the missing justifications in the proof of Proposition 29.

Problems 4 and 5 refer to the accompanying figure.

4 Which line is a transversal?

5 Determine the pairs of angles that are alternate interior angles.

Problems 6 to 8 refer to the accompanying figure.

6 If angle a = angle b, which lines are parallel? Why?

7 If angle b = angle c, which lines are parallel? Why?

8 If angle b = angle d, can you conclude that any lines are parallel?

9 Give the justifications to the following proof.

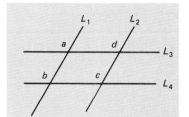

Theorem

If two lines form supplementary (their sum is 180°) interior angles on the same side of a transversal, then the lines are parallel.

Proof In the accompanying figure we are given that $x + y = 180°$. We are asked to show that $L_2 \parallel L_3$.

1 $x + y = 180°$ [   ]

2 $y + z = 180°$ [   ]

3 $x = z$ [   ]

4 $L_1 \parallel L_2$ [   ]

10 Give the justifications to the following proof.

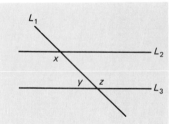

Theorem

Two lines perpendicular to a third are parallel to each other.

Proof In the accompanying figure we are given that L_1 and L_2 are perpendicular to L_3. We are asked to prove that $L_1 \parallel L_2$.

1 x and y are right angles [   ]

2 $x = y$ [   ]

3 $L_1 \parallel L_2$ [   ]

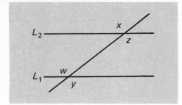

11 Prove the following theorem. In the accompanying figure, if $x + y = 180°$ then $L_1 \parallel L_2$.

12 Throughout history people have tried to prove the following theorem, without resorting to the parallel postulate. (We will explain why in Chapter 4.)

Theorem
Through a point not on a line there is no more than one line parallel to the line.

No one yet has been able to construct a valid proof. Here is one such attempt. Can you find the flaw? It is a proof by contradiction. See the accompanying figure.

Proof Suppose L_1 and L_2 pass through point P and are parallel to L_3. Choose any point Q on L_3.

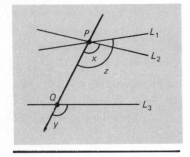

1 Construct PQ.

2 PQ is a transversal for L_1 and L_3 as well as for L_2 and L_3.

3 $x = y$

4 $z = y$

5 $x = z$

6 This is a contradiction because $x < z$.

13 (a) We mentioned that Euclid could have effectively used Postulate 5 in an earlier proof rather than wait until Proposition 29. For example, consider Proposition 16. Here is the proof that Euclid gave. Fill in the missing justifications.

Proposition 16
In any triangle an exterior angle is greater than either interior and opposite angle.

Proof Refer to the figure, where the given triangle is ABC and δ is the exterior angle. We will show that $\delta > \alpha$.

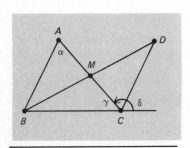

1 Bisect the line AC at M [Prop. 10].

2 Join the line BM [   ].

3 Extend BM [   ].

4 Let D be the point on the extended line such that $BM = MD$ [   ].

5 Draw the line DC [   ].

6 $AM = MC$ [   ].

7 Angle $AMB = $ angle CMD [Prop. 15].

8 Triangle *AMB* is congruent to triangle *CMD* [   ].

9 Angle *MCD* = angle *MAB* [   ].

10 Angle *MCD* = α [   ].

11 But δ > angle *MCD* [   ].

12 $\delta > \alpha$ [   ].

(*b*) Here is an alternate proof using the parallel postulate. Fill in the justifications.

Alternate proof

1

2    $\delta + \gamma = 180°$ [   ].

3   *AB* and *BC* are cut by the transversal *AC*; so $\alpha + \gamma < 180°$ [   ].

    $\delta > \alpha$ [   ].

3-6   AFTERMATH

We have been discussing the details of the "Elements" for several sections in order for you to understand formal logic and the method of deductive reasoning. We chose this book for three reasons: (1) it reflects many aspects of Greek culture during the classical age, (2) it represents the pinnacle of Greek mathematics during that time period, and (3) it illustrates logic and deduction without requiring extensive background. Let us emphasize, however, that the rational approach as the primary method for acquiring truth predated Euclid by almost 300 years.

The goal of this section is to place Euclid's work in a historical perspective. We will describe some other Greek mathematics, not only to see how the "Elements" fits into the whole scheme of things but also to describe the remaining contents of this classic textbook. We will also describe some more recent developments of the material from the nineteenth and twentieth centuries. This section is informative rather than demonstrative: The material is for appreciation instead of mastery.

The "Elements" (Continued)

The first advances in geometry were made by the Egyptian and Babylonian civilizations, especially their carpenters, stonemasons, and surveyors, from about 4000 B.C. The material remained static until about 600 B.C., when a revolution occurred. Initially it caused a shift from empiricism to proof. The likely cause was the discovery of errors in results thought to be true in general. But then some additional errors in the first proofs were revealed too. In an attempt to erect a new structure which would rid the system of these flaws and of others caused by Zeno's arguments, the classical Greek philosophers of 500 to 300 B.C. went back to their foundations and, as a result, devised an abstract, deductive system that typifies modern mathematics. The three revolutions produced a vast structure of results, culminating in Euclid's "Elements," which were dedicated to understanding the universe. This structure was to reign supreme for over 2000 years until a nineteenth-century revolution would topple it too. But we are getting ahead of ourselves; let us return to classical Greece.

The "Elements" was used as a paragon for deductive reasoning. Like the earlier Rhind papyrus it was selected because it typifies the mathematics of a certain civilization at a certain point in time. But it should be remembered that it is neither an elementary textbook, for there is, say, no arithmetical computation in it whatsoever, nor is it a compendium of the most advanced mathematics then known. The three problems of antiquity and the curves invented to solve them are not even mentioned. The treatment of the curves awaited Apollonius' "Conics" in 200 B.C. Moreover, the foundations for the calculus were laid by Eudoxus over a century before Euclid; yet this too is missing. So the "Elements" remains on the middle road of difficulty but on the high road of deduction.

We have discussed several of the first 29 propositions in the "Elements" in considerable detail. A continued examination of the first 48 propositions, which Euclid labeled Book 1, will reveal that these results constitute a general outline of the proof of the pythagorean theorem, using the concepts of congruence and parallelism. Similarity was not used because it relies on ratios of numbers and hence reduces to incommensurables. The last two propositions in Book 1 are given below. Euclid's proof is accompanied by a figure that has been described as a bride's chair. (See Figure 1.)

PROPOSITION 47: In each right triangle the square of the hypotenuse is equal to the sum of the squares of the other two sides.

PROPOSITION 48: If the square of one side of a triangle is equal to the sum of the squares of the other two sides, then the triangle is a right triangle.

The remainder of the "Elements" consists of over 400 propositions drawn from several areas of mathematics. It is a common fallacy that it deals only with geometry. There are 13 books altogether. (A book is roughly a chapter.) The first six books deal with geometry. Book 5, on ratios, is due to Eudoxus, who resolved the "crisis" caused by Zeno's paradoxes. It is applied in Book 6 to similar triangles. Books 7 to 9 concern number theory and include the fundamental theorem of arithmetic, the formula for producing perfect numbers, and the proof that the primes

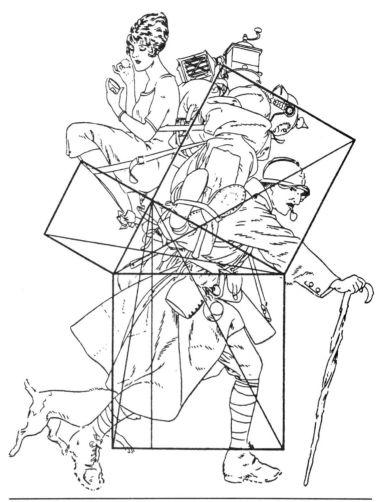

FIGURE 1 The "Bride's Chair" diagram of Euclid's Proposition 47 in a World War I setting.

are infinite, as well as the formula for a geometric progression. (We covered all these topics in Chapters 1 and 2.) Book 10 is on a topic in the area we call *mathematical analysis* today, and it contains a discussion of irrational numbers. The last three books deal with solid geometry. We will discuss the principal result below.

Before getting to solid figures we would like to discuss some of the figures that Euclid constructed in the plane. This topic illustrates the distinction between constructible and nonconstructible figures, and it contrasts constructible figures in two and three dimensions.

Recall that Proposition 1 showed that an equilateral triangle can be constructed using a straightedge and compass. Euclid also proved that a square can be constructed. These two figures are examples of what are

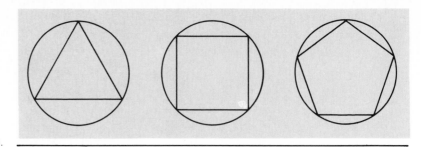

FIGURE 2   Three regular polygons.

called regular polygons, where a *polygon* is a figure whose sides are lines (examples are rectangles, trapezoids, quadrilaterals). A polygon is called *regular* if all the lines are of the same length and all interior angles are equal. A regular five-sided polygon is called a *pentagon,* and Euclid proved that it too can be constructed. (See Figure 2.) In fact, altogether Euclid showed how regular polygons with 3, 4, 5, 6, and 15 sides can be inscribed in a circle.

It is easy to see how to construct the regular six-sided polygon (hexagon). First inscribe an equilateral triangle in a circle, then bisect its sides, next erect perpendiculars at these midpoints until they meet the circle, and lastly join the points. (See Figure 3.) This same method can be applied to construct regular polygons with 12, 24, 48, . . . sides. Beginning with the square and pentagon you can construct regular polygons with

4, 8, 16, . . .

5, 10, 20, . . .

sides

It is possible to define a regular *n*-sided polygon for each $n > 2$. For instance, a *septagon* is a regular seven-sided figure. But bestowing a name upon an object does not mean it exists, and according to Euclid an object exists only if it can be constructed with straightedge and compass. The question of which regular polygons can be constructed was not answered until the nineteenth century, and even then it took a combination of algebra and geometry to answer it. The solution showed that a septagon cannot be constructed with straightedge and compass. Table 1 summarizes some regular polygons which have specific names.

The contrast between regular figures in two dimensions (polygons) and three dimensions (polyhedra) is a bit surprising. We have seen how to construct regular polygons with 3, 6, 12, 24, . . . sides; so there is an infinite number of regular polygons. Not so with regular polyhedra, those three-dimensional figures made up of regular polygons. Euclid showed how to construct five of them. The "Elements" ends with a proof of this proposition: No other solid, besides the said five solids, can be constructed which is made up of regular many-sided polygons.

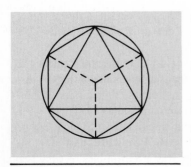

FIGURE 3

TABLE 1

Regular Polygons

Number of Sides	Name	Constructible?
3	Equilateral triangle	Yes
4	Square	Yes
5	Pentagon	Yes
6	Hexagon	Yes
7	Septagon	No
8	Octagon	Yes
9	Nonagon	No
10	Decagon	Yes

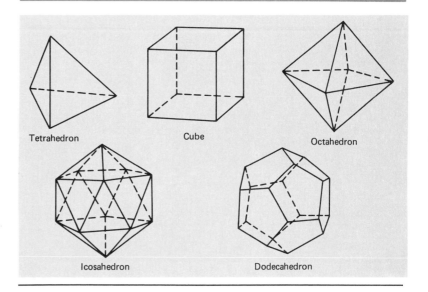

Tetrahedron   Cube   Octahedron

Icosahedron   Dodecahedron

FIGURE 4 The platonic solids.

Plato admired these figures greatly. He associated each of the first four of them with one of the four basic elements because he felt that God would make them central to the overall scheme of things. He associated the fifth figure with the universe itself. As a result the five regular polyhedra are sometimes called the *platonic solids;* they are also called the *regular solids.* Table 2 below summarizes this correspondence.

TABLE 2

Polyhedron	Shape of Faces	Number of Faces
Tetrahedron	Triangle	4
Cube	Square	6
Octahedron	Triangle	8
Dodecahedron	Pentagon	12
Icosahedron	Triangle	20

The renowned astronomer Johann Kepler found in these regular solids evidence for the mathematical design of the universe. In the preface to his book "The Mystery of the Cosmos" he wrote,

I undertake to prove that God, in creating the universe and regulating the order of the cosmos, had in view the five regular bodies of geometry known since the days of Pythagoras and Plato.

His scheme was that the planets move along spheres and that the regular polyhedra separate the spheres. (See Figure 5.) Thus Saturn moved along the outer sphere. Inside this sphere was a cube and inside the cube

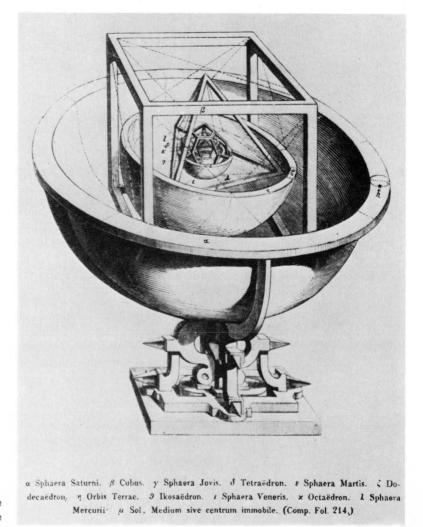

α Sphaera Saturni. β Cubus. γ Sphaera Jovis. δ Tetraëdron. ε Sphaera Martis. ζ Dodecaëdron. η Orbis Terrae. ϑ Ikosaëdron. ι Sphaera Veneris. ϰ Octaëdron. λ Sphaera Mercurii· μ Sol, Medium sive centrum immobile. (Comp. Fol. 214.)

FIGURE 5 The orbits of the planets as determined by the five regular solids.

was another sphere, along which Jupiter moved, and so on through the five regular solids. The beauty of this mathematical relation had such a spellbinding effect on him that he insisted for some time on the existence of just six planets because there were only five regular figures to fit between them.

Conics

So far we have seen that the geometry contained in the "Elements" was typical of the main current of Greek geometry, but there is one major area that Euclid did not consider at all, the conics. This topic has a rich history, reflecting the interface between geometry and algebra, and it has found numerous applications.

To place the story in historical perspective we will have to take you to jail. Not a modern jail, but a fifth-century-B.C. prison in Athens. There the philosopher Anaxagoras was serving a sentence for having the audacity to suggest that the sun is a red-hot stone and the moon a body like earth which receives its light from the sun. (Recall that Homer and Hesiod, who lived some three centuries earlier, described the heavenly bodies as deities.)

What is the best way to spend your time in jail? Some inmates become experts in law. A few pursue mathematics. Anaxagoras was one of the latter. He occupied himself by trying to square the circle. Here we have the first occurrence of one of the so-called "three problems of antiquity" which emerged around 450 B.C. and were to fascinate mathematicians for 2000 years. Unfortunately we have no further information regarding Anaxagoras' progress on the problem, but we do know that—fortunately for him—an influential friend arranged for his release. That friend was Pericles.

The three problems of antiquity are listed below. The first apparently started in jail. The second, sometimes called the Delian problem, is rooted in Greek mythology. Nothing is known about the origin of the third.

1 Given a circle, construct a square having the same area ("square the circle").

2 Given a cube, construct another cube whose volume is twice the first one ("duplicate the cube").

3 Given an angle, construct another angle which is one-third the first angle ("trisect an angle").

Angle	Curve
Right	Parabola
Acute	Ellipse
Obtuse	Hyperbola

A Greek named Menaechmus (mē-nah-eck′-muss) was the first person to make any progress on these problems. In attempting to solve the Delian problem he introduced three curves obtained by slicing a cone with a plane which meets the cone at various angles. (See Figure 6 on page 220.) The table on the left relates the curve to the angle at which the cutting plane meets the cone.

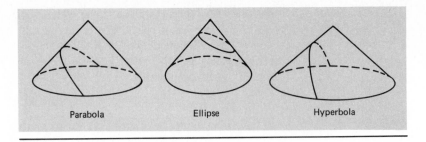

FIGURE 6   The conics.

Menaechmus had a student who was destined to become one of the greatest rulers in history, Alexander the Great. Later in life the famed ruler asked his former teacher for a shortcut to geometry, and the response is classic: "O King, for traveling over the country there are royal roads and roads for common citizens; but in geometry there is one road for all."

The three curves, the parabola, ellipse, and hyperbola, have come to be called collectively the *conic sections* for obvious reasons. Their names, however, are not due to Menaechmus but to Apollonius, whose book the "Conics" (*circa* 225 B.C.), defined them by means of a double cone. (See Figure 7.) Appolonius was an outstanding third-century-B.C. mathematician, but he had the misfortune of being a contemporary of the great Archimedes and his lack of fame reflects the results of the rivalry. However, his principal book, the "Conics," is an extraordinarily broad and deep examination of numerous properties of these curves. So detailed and extensive was his treatment that it played a role similar to the "Elements" in that it superseded all its rivals and made copying earlier works unnecessary. Moreover, his results were not improved upon until the seventeenth century, when the intrusion of algebra into geometry enabled mathematicians to obtain additional results.

Appollonius' viewpoint on his work is interesting. He wrote "[My results] are worthy of acceptance for the sake of the proofs themselves," which is a mathematical version of "art for art's sake." Ironically his results have led to fruitful applications in diverse areas, including parabolic paths of projectiles and cross sections of reflecting beams, elliptical orbits of planets and designs of the stylus on your turntable, and hyperbolic problems in engineering and economics.

We mentioned above that the three problems of antiquity emerged during the fifth century B.C. The conics were introduced in order to solve them. But Plato added another restriction about a century later. He insisted that the constructions involve only a straightedge and compass, so the geometers were sent back to their drawing boards. In searching for solutions they discovered numerous other curves which worked, but none of these new curves could be constructed with the platonic instru-

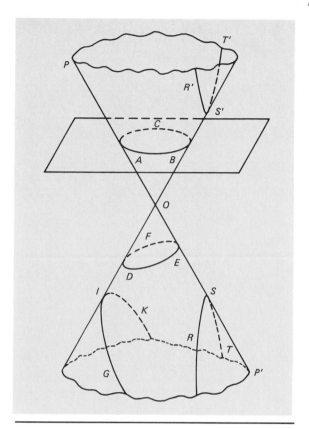

FIGURE 7 A conical surface and the sections made by intersecting planes.

ments. The solutions eluded all of the Greek geometers. In fact, they eluded all mathematicians until the nineteenth century, when, once again, the close relationship between geometry and algebra produced the result: It is impossible to construct solutions to any of the three problems of antiquity by using only a straightedge and compass. And so a 2300-year search came to an end with a negative conclusion. Along the way, however, some positive, beautiful, and very useful mathematics was discovered.

Logic Revisited

One of the reasons that the study of Euclidean geometry has remained one of the staples in the educational smorgasbord is because it teaches a person how to reason properly. Underlying the reasoning process is logic, whose formal aspect we introduced as part of our development of Euclidean geometry. Take a look at the test below. Has the study of geometry had the salutary side effect of improving your reasoning to the lofty heights of being able to solve the problem? Try it!

A Quick Test of How Your Brain Functions

"Most people simply do not listen enough and make decisions too quickly," says Dr. Evelyn Golding, a psychologist at the National Hospital for Nervous Diseases in London. Having administered a simple test of logic to a group of normal individuals, Dr. Golding is also saying that man's mind "does not follow the straightforward pattern of reasoning one would expect."

In fact, Dr. Golding believes that in most people, the right hemisphere of the brain, which is skilled in visual, spatial tasks, overpowers the less well-developed left side of the brain, which is good at abstract, logical and verbal tasks. And she has the data to prove it.

You can prove it, too, by trying to solve the four-card problem, presented here, which Dr. Golding used in controlled tests on 130 individuals. These included undergraduates, mathematicians and ordinary citizens.

If you get the answer wrong, do not despair. Only 4 percent of Dr. Golding's subjects got it right, a failure rate known to psychologists for many years. But Dr. Golding believes she can explain it.

She tried the test on people a part of whose brains were put temporarily out-of-action by electric shock treatment. Six out of 10 answered the problem correctly; the

remaining four got it partially correct. Dr. Golding notes that these people had lesions in that part of the brain involved in perceptual classi-

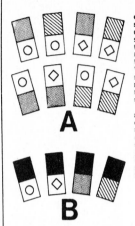

The Test: Look at the top eight cards [A]. Note that each has a shape (a circle or a diamond) on one half and a pattern (grey or striped) on the other half. Cover the cards so you cannot see them. Look at the four cards below [B], each half-masked. What could you find under the mask of each in turn? The alternatives are grey or striped for the first two; a circle or a diamond for the other two. Agreed?

Now read the following test sentence: *"Whenever there is a circle on one half of the card, there is grey on the other half of the card."* Your task is to name those cards and only those cards you would need to unmask in order to find whether the test sentence is true or false for the four cards.

fication — the right side. This, she argues, freed their logical left brains to deal with what is a problem in logic — the four-card test.

EXERCISES

1   How are Propositions 47 and 48 related logically? Can they be combined into one statement?

2   The construction of a hexagon involves bisecting a line, erecting a perpendicular, and joining two points. What justifications allow these constructions to be performed?

3   The accompanying figure shows a square inscribed in a circle. Show how to inscribe an octagon in the circle.

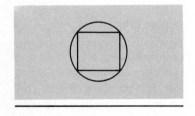

4   What octagonal directional sign do you see at the end of some streets which do not have a traffic light in the city?

5   What is a septet? What are the corresponding terms for groups with 3, 4, 5, and 6 members?

6   Each corner of a polygon is called a *vertex;* the plural is *vertices.* Label the vertices of a hexagon by the numbers 1 to 6 consecutively. Join vertices

1 and 3, 1 and 5, 2 and 4, 2 and 6, 3 and 5, 4 and 6. The resulting figure is a *hexagram* (also called Magen David, the Star of David). What figure lies inside the hexagram?

7   An ad for the new Canon A-1 camera ran under the caption, "hexa-pho-tocybernetic." It described the exposure control as being _____-mode, where a number fills in the blank. What number is it?

8   Which platonic solid does a soccer ball most closely resemble?

9   Kepler found evidence in the regular solids for the existence of six planets. How many planets are known to exist today?

10   A new and startling discovery was made recently in a marine lab in Wales: a square bacterium. Ordinarily the shapes of bacteria are limited to spheres, ellipsoids, and cylinders. Can you sketch what these three usual shapes of bacteria look like?

Problems 11 and 12 present simple properties of the triangle which, surprisingly, eluded Greek geometers.

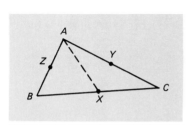

FIGURE 8

11   The line joining the point A in the triangle in Figure 8 to the point X, the midpoint of BC, is called a median of the triangle. Draw the other two medians. (The midpoints Y and Z are given.) What do you conclude? Draw another triangle and follow this same process. Do you reach the same conclusion? (See Figure 8.)

12   (*a*)   What is the area of a 3-4-5 triangle? Evaluate the formula

$$\sqrt{s(s - a)(s - b)(s - c)}$$

where $a = 3$, $b = 4$, $c = 5$, and $s = \frac{1}{2}(3 + 4 + 5)$.

(*b*)   What is the area of a 5-12-13 triangle? Evaluate the above formula with $a = 5$, $b = 12$, $c = 13$, and $s = \frac{1}{2}(5 + 12 + 13)$.

13   Which of the following are acceptable spellings?
(*a*)   hypotonose
(*b*)   hypothenuse
(*c*)   hypoteneuse
(*d*)   hypotheneuse
(*e*)   hypotinuse
(*f*)   hupoteneus
(*g*)   hypotenuse

14   Which of the following is the correct spelling?
(*a*)   isoseles
(*b*)   isosseles
(*c*)   issoseles
(*d*)   isosceles
(*e*)   isosoles
(*f*)   isoscoles

APPENDIX: EUCLID, THE "ELEMENTS"

Definitions

1 A *point* is that which has no part.

2 A *line* is a breadthless length.

3 The extremities of a line are points.

4 A *straight line* is a line which lies evenly with the points on itself.

5 A *surface* is that which has length and breadth only.

6 The *extremities* of a surface are lines.

7 A *plane surface* is a surface which lies evenly with the straight lines on itself.

8 A *plane angle* is the inclination to one another of two lines in a plane which meet one another and do not lie in a straight line.

9 And when the lines containing the angle are straight, the angle is called *rectilineal*.

10 When a straight line set up on a straight line makes the adjacent angles equal to one another, each of the equal angles is right, and the straight line standing on the other is called a *perpendicular* to that on which it stands.

11 An *obtuse angle* is an angle greater than a right angle.

12 An *acute angle* is an angle less than a right angle.

13 A *boundary* is that which is an extremity of anything.

14 A *figure* is that which is contained by a boundary or boundaries.

15 A *circle* is a plane figure contained by one line such that all the straight lines falling upon it from one point among those lying within the figure are equal to one another.

16 And the point is called the *center* of the circle.

17 A *diameter* of the circle is any straight line drawn through the center and terminated in both directions by the circumference of the circle, and such a straight line also bisects the circle.

18 A *semicircle* is the figure contained by the diameter and the circumference cut off by it. And the center of the semicircle is the same as that of the circle.

19 *Rectilinear* figures are those which are contained by straight lines, *trilateral* figures being those contained by three, *quadrilateral* those

contained by four, and *multilateral* those contained by more than four straight lines.

20 Of trilateral figures, an *equilateral triangle* is that which has its three sides equal and an *isosceles triangle* that which has two of its sides alone equal.

21 Further, of trilateral figures a *right-angled triangle* is that which has a right angle, an *obtuse-angled triangle* that which has an obtuse angle, and an *acute-angled triangle* that which has its three angles acute.

22 Of quadrilateral figures, a *square* is that which is both equilateral and right-angled; an *oblong* that which is right-angled but not equilateral; a *rhombus* that which is equilateral but not right-angled; and a *rhomboid* that which has its opposite sides and angles equal to one another but is neither equilateral nor right-angled. And let quadrilaterals other than these be called *trapezia*.

23 *Parallel straight lines* are straight lines which, being in the same plane and being produced indefinitely in both directions, do not meet one another in either direction.

Postulates

1 A line segment can be drawn from any point to any other point.

2 A line segment can be extended indefinitely in a straight line.

3 A circle with any center and radius can be drawn.

4 All right angles are equal.

5 If a straight line falling on two straight lines makes the interior angles on the same side less than 180°, the two straight lines, if extended indefinitely, meet on that side where the angles are less than 180°.

Common Notions

1 Things which are equal to the same thing are also equal to one another.

2 If equals are added to equals, the sums are equal.

3 If equals are subtracted from equals, the remainders are equal.

4 Things which coincide with one another are congruent to one another.

5 The whole is greater than the part.

Propositions

1 An equilateral triangle can be constructed on a given line segment.

2 A line segment can be constructed from a given point so that it is equal to a given line segment.

3 Given two unequal line segments, the shorter one can be superimposed upon the longer.

4 If two triangles have two sides equal to two sides, respectively, and have the angles contained by those sides equal too, then the triangles are congruent and the remaining angles are equal, respectively.

5 If a triangle is isosceles, then the angles which are opposite the two equal sides are also equal.

6 If two angles of a triangle are equal, then the triangle is isosceles.

8 If the three sides of one triangle are equal, respectively, to the three sides of another triangle, then the triangles are congruent.

9 A given angle can be bisected.

10 A given line segment can be bisected.

11 A perpendicular can be constructed from a given point on a given line.

12 A perpendicular can be constructed to a given line from a point not on it.

13 If one line meets another, either the two lines are perpendicular or the sum of the angles formed is 180°.

15 Vertical angles are equal.

16 In any triangle, if one of the sides is extended, then the exterior angle is greater than either of the interior and opposite angles.

17 In any triangle, the sum of any two angles is less than 180°.

18 In any triangle, the greater angle is opposite the greater side.

19 In any triangle, the greater side is opposite the greater angle.

20 In any triangle, the sum of any two sides is greater than the remaining side.

23 An angle equal to a given angle can be constructed at a given point on a given line.

Refer to the accompanying figure for 27 to 29.

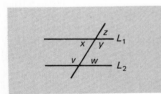

27 If $x = w$ then L_1 is parallel to L_2.

28 If $w + y = 180°$ or $w = z$, then L_1 is parallel to L_2.

29 If L_1 is parallel to L_2, then $x = w$, $w + y = 180°$, and $w = z$.

30 Two lines which are parallel to a given line are also parallel.

31 Given a point not on a line, another line can be constructed to pass through the point and be parallel to the given line.

32 If one side of a triangle is extended, then the exterior angle is equal to the sum of the two interior and opposite angles, and the sum of the three interior angles is $180°$.

The Power of Logic

This chapter extends historically the material presented in Chapter 3. Its title was chosen because it shows how two tremendous advances in geometry arose as the result of the rigorous deductive examination of Euclid's "Elements."

The material in Section 4-1 shows how a small array of results, taken solely from the first part of the "Elements," opens up numerous techniques which enable artists today to place their work in proper perspective. The section begins with a survey of Renaissance art. Then it provides instructions for drawing figures in perspective. Several classical illustrations show how the Renaissance artists Giotto, Masaccio, Piero, Leonardo, and Raphael mastered the geometrical techniques.

Sections 4-2 and 4-3 are collectively called *non-Euclidean geometry,* and the conclusions reached in them may surprise you greatly. Secion 4-2 presents the dilemma faced by a lone priest who devoted his life to freeing Euclid of all flaws. The geometry he developed is quite strange; so strange that he momentarily lapsed from strict deduction to make it consistent with common sense. Three mathematicians working independently derived similar results 100 years later without making the same mistake. Their results were revolutionary, for they implied that mathematics need not be limited by physical space. They too are not without applications, as airline routes and the theory of relativity attest.

4-1 REBIRTH OF MATHEMATICS

In 1425 a diminutive Florentine artist, Filippo Brunelleschi, conducted a seemingly inconsequential experiment. It lasted for only a few summer mornings. His "laboratory" was a cobbled cathedral piazza. His "apparatus" included two small mirrors, a wooden panel, an easel, and a paintbrush. His theory consisted of ordinary Euclidean geometry in which straight lines were interpreted as lines of vision. The results were extraordinary: they revolutionized art and revitalized mathematics.

Brunelleschi was born in Florence in 1377. His initial vocation was as a goldsmith, but he soon graduated into sculpture. In 1401 he entered a sculpture competition, but as soon as he saw a rival's superior design, he withdrew his own and moved to Rome. While there he became fascinated with Roman architecture, and endeavored to record accurately all that he saw. In so doing he set the stage for the revolutionary system of geometric linear perspective that was realized in his 1425 experiment. The time was ripe for change, and other fifteenth-century artists eagerly adopted his advance and refashioned it to their own needs.

Brunelleschi returned to Florence to vie for a commission to put a dome on the city's unfinished cathedral. The sheer size of the cathedral was so awesome that a conference of architects and engineers was held to judge the competition. Brunelleschi's plan of attack was ingenious. He challenged his competitors to make an egg stand on end and, after they all failed, he succeeded by pressing the blunt, empty end down upon the table. They protested that they could have done the same thing. Brunelleschi responded that they would make similar claims *after* his dome was built atop the cathedral. They ended up with egg on their faces; he ended up with the commission.

Brunelleschi exemplifies the Renaissance spirit of a compelling curiosity in all branches of human endeavor. We mentioned his contributions to painting, sculpture, and architecture, but his interests were many and varied. Today, however, he is regarded as the most renowned architect of his time.

In Illustration 11 we see the interior of a basilican church that Brunelleschi built in Florence. This modern photograph of the church captures the main ingredients of linear perspective that reflect the revolution he helped launch. We will now outline the evolution of perspective in painting and see how it relates to the ascent of geometry.

The basic problem that confronted Brunelleschi is the same problem that artists continually must face: how to depict an accurate image, subject to severe space limitations. A painter is limited by the confined area of a canvas, wall, vase, or dome; yet the subject of the painting is usually much larger. Shape and color are used to convey the intended meaning. Since geometry studies shape and form, the heart of the problem is geometrical. The marriage of art and geometry is a natural outgrowth of their mutual spheres of interest.

We have seen that Egyptian artists adhered to a rigid set of rules, called the *canon,* which defined a strict code for the shape of all figures. Because their aim was to depict reality as we *know* it instead of as we *see* it, their paintings contained as much information as possible. This explains why Egyptian figures had a frontal view of the torso, a profile of the head, a frontal view of one eye, a side view of the legs, and two left feet. In addition, the sizes of figures were determined by their social status and not by their real-life setting. (Recall Illustration 7.)

The classical Greeks made use of their knowledge of geometry, especially proportion and symmetry, to depict life in its ideal form. They also made use of foreshortening to capture the beauty of the human body in a natural setting. *Foreshortening* loosely means that figures in the foreground are painted larger than those in the background.* (Recall the discus thrower in Illustration 9.)

The artists of the Holy Roman Empire embraced the Greek ideal of art and copied many Greek originals. Moreover, they attempted to add a three-dimensional quality to their paintings by using light and shade.

After the fall of Rome in the late fifth century, however, the arts ground to a near standstill. The emphasis in the Middle Ages was on religious themes, so no attempt at depicting real settings was made. The paintings were essentially flat, lacking natural depth and dimension. Illustration 12 demonstrates this character. Note especially the uses of "false perspective" in two places: the buildings in the background and the cradle of the infant. To emphasize the lack of optical reality, compare Illustration 12 with the cartoon in Figure 1.

While the decline in the arts was significant during this period, the conditions were even worse in mathematics, for even though the Romans embraced Greek art, they ignored almost all the brilliant successes made by Greek mathematicians. Consequently, the period from 200 B.C. until A.D. 1300, when the interest in mathematics was spurred by the revival in art, can best be described as a descent of mathematics.

The rebirth of geometry occurred with Brunelleschi's experiment, which was the crowning point of linear perspective. However, it did not arrive totally unannounced. For a century or so before Brunelleschi, early Renaissance artists were breaking away from the essentially flat medieval look. The narrow function and use of space in the Middle Ages was gradually widened as artists began to place figures in a stagelike setting, in essence releasing them from the canvas or wall by placing them in a more lifelike environment.

Giotto is recognized as one of the first to break the medieval bonds of space. In *The Death of St. Francis* we see how the composition is set up like a stage (see Illustration 13). The painting reminds one of a dramatic play complete with stage scenery. For the first time since antiquity, objects were foreshortened into a perspective volume of three-dimensional space; yet the background of the picture still looks flat. To remedy this, another artist, Masaccio, penetrated Giotto's confined space and went beyond the back wall of the stage. In *St. Peter Resuscitating the Son of the King of Antioch* (see Illustration 14), the trees and a partial building in the background demonstrate this new space structure by suggesting more depth. Masaccio enlarged the space by using a courtyard in his work to replace the boxed-room setting.

Giotto, Masaccio, and other early Renaissance painters used various means of bringing an element of depth to their paintings, but their achievements remained strictly intuitive until Brunelleschi's experiment. It was Brunelleschi who began a revolution in art by making a scientific study of per-

---

*The concept of foreshortening is a central theme in the history of art. It is defined formally as the "seeming visual contraction of an object viewed as extended in a plane not perpendicular to the line of sight."

FIGURE 1 Cartoon illustrating
false perspective.

"Look here, men, shouldn't that be the other way around?"

spective based upon mathematical principles. The revolution applied to
scuplture and architecture as well as painting.

Ironically, Brunelleschi did not publish the results of his invention. The
honor of writing the first treatise providing a theoretical basis for art went to
another artist, who is also known today mainly as an architect, Leon Battista
Alberti. His book "Della Pictura," which was written in 1435 but not printed
until 1511, found an immediate adherent in Piero della Francesca. Piero's
work itself reflects the history of perspective in art. His early paintings demonstrate a supreme mastery of the techniques of geometric perspective (see
Illustration 15), but it was not until the end of his long career that he wrote
about them. It is noteworthy that with Piero the theory followed the practice.
This is frequently the case with mathematics.

Theoretical studies multiplied after Alberti's treatise was written, and within a century artists from the high Renaissance demonstrated a complete mastery of the techniques of perspective. The high point occurred with Leonardo and Raphael. Leonardo's impressive *The Last Supper* (Illustration 16) presented the first great figure composition in which a central religious theme is enhanced by its geometrical setting. Raphael's *The School of Athens* (Illustration 17) represents the perfection of pictorial science. In it Raphael created a vast perspective space in which the principal mathematicians and philosophers from antiquity appear so natural and realistic that you might think it was a photograph instead of a painting. For an unusual viewpoint, see Mantegna's *St. James Led to Martyrdom* (Illustration 18). It illustrates the extent to which artists had mastered even the nontraditional art of perspective.

The Problem

The problem that Renaissance painters grappled with was how to capture a three-dimensional setting on a two-dimensional canvas so that all the figures were in proper proportion. Brunelleschi's discovery showed how to proceed intuitively. Alberti transformed this intuition into a strict geometric code based entirely upon the first 32 propositions in Euclid's "Elements." By following Alberti's rules, later artists were able to extend the principles to fit their own needs. This can be seen graphically in Raphael's fresco *The School of Athens,* in which linear perspective is used to draw the viewer's eyes directly to the central figures, Plato and Aristotle.

Today all aspiring artists must master the classical techniques of perspective during their formal education. Our aim here is to present the first stage in the process. We hope that you will find it as exciting and rewarding as we did when we first encountered it.

View Raphael's *School of Athens* (Illustration 17). Notice that the four squares in the foreground, broken only by the reclining figure of Diogenes, do not really form a square, but a trapezoid. The basic problem is to draw the squares so that they recede into the background in proper perspective. The squares are shown in Figure 2(*a*). The problem is to draw the trapezoid in Figure 2(*b*) so that it conveys a sense of the squares receding into the background realistically.

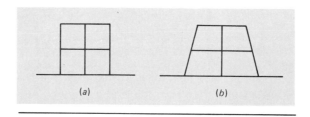

(*a*)   (*b*)

FIGURE 2

The Solution

We will describe Alberti's solution to this problem. Then we will show you how to draw four and eight squares in perspective. The exercises will ask you to draw other numbers of squares.

Some steps in Examples 1 and 2 are numbered for later reference. We have adopted the convention of labeling points by uppercase letters and lines by lowercase letters. The line joining points X and Y is denoted by XY; the notation $\overline{XY}$ stands for the distance between X and Y. Refer to Figure 3 as you read through the proof. In fact, it is instructive to draw your own figure.

Example 1   Four squares

Begin with two points A and B.

1   Draw the line through A and B. Label it g. Line g is called the *ground line*. Select a point V that does not lie on g. The point V is called the *vanishing point*.

2   Draw lines AV and BV.

3   Construct the line that passes through V and is parallel to g. This line is called the *horizon* because in practice it represents the line where the earth meets the sky. We will denote it by h. Now select any point C on h that lies to the left of V.

4   Construct the point D on h on the right side of V so that $\overline{CV} = \overline{DV}$.

5   Draw lines AD and BC. Label E the point of intersection of BC and AD. Write this as $E = BC \cap AD$. Then label $F = AV \cap BC$ and $G = AD \cap BV$.

6   Draw line FG.

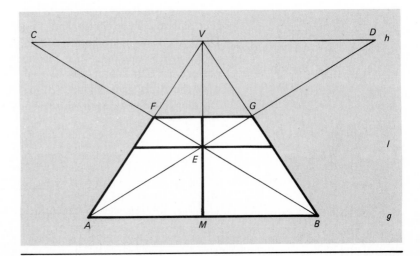

FIGURE 3

7 Construct the line that passes through E and is parallel to g. Label it l.

8 Line l is parallel to h.

9 Line l is parallel to FG.

10 Draw line VE and extend it through g. Let M be the point of intersection.

11 The point M is the midpoint of AB.

The trapezoid $AFGB$ contains the four squares in perspective.

Example 1 shows how to draw four squares in perspective. Figure 3 illustrates Alberti's procedure with the point V lying on the perpendicular bisector of line AB. You might want to experiment with other placements of V to see what other kinds of perspectives arise. We will pursue a few such possibilities in the exercises.

The justifications for the numbered steps in Example 1 have also been left for the exercises. All but two of them are immediate. However, steps 9 and 11 require proofs themselves.

The procedure for drawing eight squares in perspective involves a slight extension of the procedure for four squares. We will adhere to all of the notation developed in Example 1. Refer to Figure 4 as you read through Example 2.

Example 2   Eight squares

12 Bisect line segments AM and MB. Label the midpoints H and I, respectively.

13 Draw lines CI and DH. Let J be the point of intersection of CI and DH.

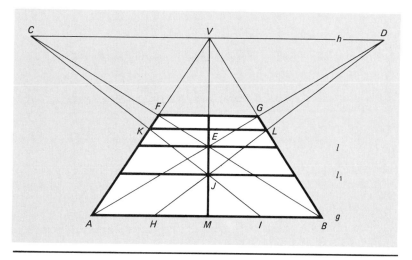

FIGURE 4

14 *J* lies on *VM*. Label $K = AV \cap CI$ and $L = BV \cap DH$.

15 Draw line *KL*.

16 *KL* is parallel to *FG*.

17 *KL* is parallel to the ground line *g*.

18 Construct the line that passes through *J* and is parallel to *g*. Label this line l_1.

19 l_1 is parallel to *KL*.

The trapezoid *AFGB* contains the eight squares in perspective.

The Viewer's Perspective

The choice of where to place the vanishing point *V* is completely arbitrary to a mathematician. For artists, however, there is a strict rule on where to mark it. Alberti wrote:

> We must now select a point on the horizon line a certain distance to the left of the center point [vanishing point]. The distance is determined by the following rule: the distance from the point to the center point is to be the same as the distance from the center point to the eye of the observer. To the right of the center mark another point on the horizon line that is this same distance from the center point.

To understand the effect Alberti was trying to reach with this rule consider the two diagrams in Figure 5. Figure 5(*a*) has the point *C* farther away from *V* than in Figure 5(*b*). The grid in Figure 5(*a*) appears as though the observer is viewing the scene from a point closer to ground level than in Figure 5(*b*). Thus the choice of where to select the point *C* is determined by whether the artist wants the painting to depict a scene as viewed by someone on the ground or in the air.

The grid that Alberti defined is reminiscent of the ancient Egyptian canon. The same type of rectangular array of squares used by the Egyptians to rigidly order the shape of all figures was used by Alberti to establish spacial order to the canvas.

A Glance Ahead

The area of mathematics that was spawned by the Renaissance revolution in art has come to be called *projective geometry*. Its present character is quite unlike the art of perspective, so it is difficult to recognize its roots.

One of the themes in Chapter 3 was the ascent of parallel lines, beginning with Euclid's treatment of them in his "Elements." The artists of the Renaissance viewed parallel lines somewhat differently. They drew them as if they would meet when extended indefinitely, because this is the way the eye sees them. Just visualize railroad tracks receding into the horizon.

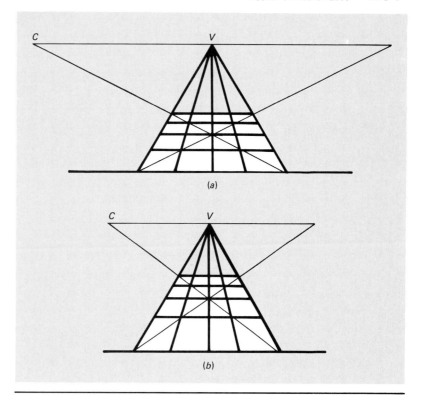

FIGURE 5

Mathematicians of the late eighteenth and early nineteenth centuries regarded parallel lines from an entirely different viewpoint, one that seems upon first impression to be totally divorced from reality. Yet their revolutionary creation provided the groundwork for Einstein's theory of relativity. The remaining sections in this chapter will explore this part of the ascent of mathematics.

EXERCISES

1   Supply the justifications for steps 1 to 8 and 10 in Example 1.

2   Supply the justifications for steps 12, 13, 15, 17, 18, and 19 in Example 2.

3   Supply a proof of step 9 in Example 1.

4   Supply a proof of step 11 in Example 1.

5   Supply a proof of step 14 in Example 2.

6   Supply a proof of step 16 in Example 2.

In Problems 7 to 12 follow Alberti's procedure for drawing four squares in perspective by choosing the points V and C as indicated.

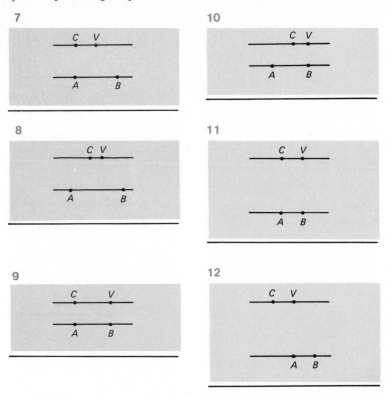

7

8

9

10

11

12

13 Describe how you would draw 16 squares in perspective.

14 Describe how you would draw 32 squares in perspective.

Construct several figures like Figure 4 on graph paper. Then answer Problems 15 to 18 empirically.

15 How does E divide line segment l?

16 How does J divide line segment l_i?

17 How does $\overline{VE}$ compare to $\overline{EM}$?

18 How does $\overline{EJ}$ compare to $\overline{JM}$?

19 How many squares are drawn in perspective if you proceed as follows: Extend Example 1 by bisecting CV and VD at points R and S, draw lines RB and SA, and continue as in Example 2?

20 Start with Figure 6 and use Alberti's method to construct the ceiling and the floor of a room, where each has the tiled effect of Example 1.

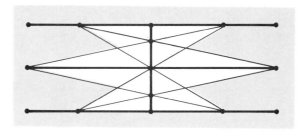

FIGURE 6

Problems 21 to 24 refer to Alberti's method for drawing a picture in "angular perspective." Problem 9 is an example of angular perspective. The letters refer to Figure 4.

21 Is VM perpendicular to the ground line g?

22 Is E the midpoint of line l_1?

23 Is M the midpoint of line AB?

24 Does J lie on VM?

25 What procedure would you follow to paint the ceiling of Leonardo's *Last Supper*, which consists of nine squares?

26 What procedure would you follow to paint six squares in the perspective in Figure 7?

In Problems 27 to 30 find the principal vanishing point. Mathematically, the *principal vanishing point* is the point to which all lines converge; artistically it is the point your eyes are attracted to subconsciously.

27 Leonardo's *Last Supper* (Illustration 16).

28 Piero's *Resurrection* (Illustration 15).

29 Raphael's *School of Athens* (Illustration 17).

30 Mantegna's *St. James Led to Martyrdom* (Illustration 18).

31 Here is another construction that yields the same grid as that of Alberti. Start with two parallel lines with points A and B on one; mark point V on the other so that the perpendicular bisector from V to AB is VM. (Thus $VM \perp AB$ and $\overline{AM} = \overline{BM}$.) Mark a point C on the line containing V and to the left of V. Let E be the point of intersection of BC and VM. Extend AE to VC and call the point of intersection D. Prove that $VC = VD$.

32 In his solution to the problem of linear perspective, Alberti did not explicitly state that the line in Figure 3 bisects AB; yet he implicitly used this fact. That is, he extended the line VE to meet AB and assumed it is the perpendicular from V to AB, and that $\overline{AM} = \overline{BM}$. These two facts are equivalent in the sense that if one is true the other follows. This is a theorem of Euclidean geometry. Prove part of the theorem: Given two parallel lines with points A and B on one line, and C, V, and D on the other such that $\overline{CV} = \overline{VD}$, with E the point of intersection of AD and BC, with VE intersecting AB in M; if $VM \perp AB$, then $\overline{AM} = \overline{BM}$.

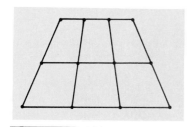

FIGURE 7

4-2 SACCHERI AND EUCLID'S FLAW

Our excursion through geometry has taken us on a tortuous path. We started in Egypt with some very impressive results. But the classical Greeks found that some of them were only approximations and not exact descriptions of the real world; so they set about constructing a deductive system which would describe the universe precisely without falling prey to such errors. The paragon of the deductive approach was Euclid's "Elements." During the Middle Ages, knowledge of geometry, like most other subjects, sank to abysmal levels. Thus when early Renaissance painters sought a geometric underpinning on which to base their works, they found that none existed; so they created their own. In this way the art of perspective, and what came to be known as projective geometry, emerged. In the early seventeenth century the study of geometry made a quantum leap forward when the methods and symbolism of algebra were combined with geometry to forge a new powerful tool, coordinate geometry. Like most alloys, the composition turned out to be stronger than its individual components, and it soon formed the basis for the calculus, that subject which burst onto the scene in the late seventeenth century with such fanfare that it dominated the intellectual milieu of the eighteenth and early nineteenth centuries.

While these advances were being made at the front, some lonely soldiers were engaging in a rear guard action. They were concerned about flaws in the foundations, fearing that the whole structure would crumble if its basement developed leaks. Most of the trouble stemmed from a lingering doubt about Euclid's fifth postulate, the so-called parallel postulate. Since the calculus rested upon coordinate geometry, and coordinate geometry upon Euclidean geometry, and Euclidean geometry upon its axioms, it was essential to show that the 10 postulates and common notions were self-evident. Only one axiom, the fifth postulate, remained unresolved. It certainly looked more like a proposition than an axiom. Could it be deduced from the other axioms or must it be listed as an assumption?

Many top-notch mathematicians tried proving that the parallel postulate was indeed true. One of the very best, the eighteenth-century geometer Joseph Louis Lagrange (la-gra͞nzh′, does *not* rhyme with range), once was convinced that he had discovered an airtight proof. He presented it formally to the French Academy of Science. In the midst of the very first paragraph of his lecture he stopped suddenly and exclaimed, "I shall have to think it over again." With that he put his papers in his pocket and abruptly left the hall, never to return to the subject again.

Another mathematician, Wolfgang Bolyai, who had spent much time trying to prove the same result, found out that his son, Janos, had inherited his passion. In typical paternal style ("do as I say, not as I do") he advised his son:

> You should detest it just as much as lewd intercourse; it can deprive you of all your leisure, your health, your rest, and the whole happiness of your life. This abysmal darkness might perhaps devour a thousand towering Newtons, it will never be light on earth.

Rarely do sons heed such advice. We will meet Janos again.

Saccheri

The most comprehensive attack on the problem took place a century before Bolyai's warning to his son. Girolamo Saccheri, a devotee of Euclid who yearned to prove that Euclid's choice of listing the parallel postulate as an axiom was not the correct course, wrote a book entitled "Euclid Freed of Every Flaw," published in 1733.

What was the "flaw" that Saccheri saw in the "Elements"? It was not a worry that the parallel postulate might be false; scientists had no doubt that it was a correct idealization of the behavior of actual lines. What gnawed at them was that it did not appear as obvious as the other nine axioms, and so it would be aesthetically preferable to be able to list it as a proposition.

Most scientists ignored the problem, subscribing to the Epicurean mode of reasoning: The statement is obvious and it works (that is, a beautiful and useful geometry is derived from it); so it doesn't matter whether it is an axiom or a proposition—it's no big deal!

But it was a big deal in the minds of a few. Some felt that it was enough to find an equivalent form of the parallel postulate that would appear to be much easier to understand. In fact you probably will recognize the most notable attempt, due to John Playfair (1748–1819), as it is the common substitute for Euclid's fifth postulate in modern plane geometry texts. It is known as Playfair's axiom, and a quick sketch will show you how easy it is to understand. (See Figure 1.)

PLAYFAIR'S AXIOM: Given a line and a point not on it, there is one and only one line that passes through the point and is parallel to the given line.

One can argue, however, whether Playfair's axiom is indeed preferable to Euclid's fifth. After all, it refers to the behavior of lines far out in space and is thus not part of our everyday experience.

Note, however, that merely rephrasing the parallel postulate does not prove that it is a necessary axiom. Saccheri's approach was radically different. He attempted to find a statement equivalent to the parallel postulate that would not only be clearer to understand but would also demonstrate that it was a proposition rather than an axiom.

Saccheri, a Jesuit priest and a professor of mathematics at the University of Pavia, purported to have solved the problem in his book. He felt that the "Elements" had just one blemish, namely, that it was unnecessary for Euclid to *assume* the parallel postulate—he set out to *prove* it from the other nine axioms and thus free Euclid of his sole flaw.

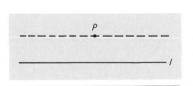

FIGURE 1   Only one line can be drawn through the point *P* and parallel to *l*.

Saccheri's Geometry

Let us consider in some detail Saccheri's brand of geometry. He first defined a specific type of quadrilateral, called an *S-quad* (where S can stand for specific, special, Saccheri, or strange, depending upon your inclination).

DEFINITION: An S-quad is a quadrilateral in which the base angles are right angles and the perpendicular heights are equal. In the quadrilateral *ABCD* in Figure 2 the angles at *A* and *B* are right angles and $\overline{AD} = \overline{BC}$.

Saccheri's argument centered on the *summit* angles, that is, the angles at *C* and *D*. Can you predict his first result concerning angles *C* and *D*?

Before answering that question we have to clarify Saccheri's rules of the game. Remember that he is not only restricting himself to the rules of Euclidean constructions but he is also utilizing only the elements of *absolute geometry,* which are:

1   The first nine axioms of Euclid (all except Euclid's fifth postulate)

2   Propositions 1 to 28 (those propositions that can be derived from the first nine axioms)

He could also use all of Euclid's definitions as well as the definitions of an S-quad. He then wanted to make certain substitutions for the parallel postulate and see where it would lead. Let us get started by labeling his first result Theorem S1. As before, we will fill in some of the justifications and leave the others for the exercises.

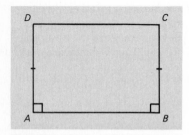

FIGURE 2   An S-quad. Angles *A* and *B* are right angles and $\overline{AD} = \overline{BC}$.

> **Theorem S1**
> An S-quad can be constructed.
> ***Proof***   Refer to Figure 2.
>
> 1   Let *AB* be a given line segment. Erect a perpendicular at *A* [Prop. 11].
>
> 2   Erect a perpendicular to *AB* at *B* [   ].
>
> 3   Choose any point *D* on the perpendicular at *A*. Mark the point *C* on the perpendicular at *B* such that $\overline{BC} = \overline{AD}$ [Prop.   ].
>
> 4   Connect the points *C* and *D* [Post.   ].
>
> 5   *ABCD* is an S-quad [Def. of S-quad].
>
> QED

Have you made a conjecture about angles *C* and *D*? It sure looks as though they must not only be equal but they must be right angles, right? Saccheri did prove that they are right angles, but his proof used the parallel postulate. At this juncture it appears that he is back where he started—assuming the fifth postulate. But this is where his approach takes a significant twist. He also proved the converse—if you assume that angles C and D are right angles, then the parallel postulate is true. Now he had the parallel postulate right where he wanted it—as a proposition that followed from the results of absolute geometry and a seemingly evident result. If he could now show that the summit angles had to be right angles, using only absolute geometry, he would have proved that the parallel postulate also follows from absolute geometry. Hence it would be a proposition instead of an axiom!

Showing that the angles at the summit were right angles appeared to Saccheri to be an easy, straightforward approach to the problem. It was simple to show that

> **Theorem S2**
> The summit angles are equal; i.e., angle C = angle D.

The proof is outlined in the exercises. It is clear that angles C and D are acute, obtuse, or right angles. That is, they are either less than 90°, greater than 90°, or equal to 90°. If the first two cases could be dispensed with—according to Euclid's rules—he had the problem licked.

In other words, Saccheri had three hypotheses. He hoped that two of them would lead to a contradiction, leaving the other as a necessary logical conclusion. He referred to the hypotheses as:

RIGHT ANGLE HYPOTHESIS (RAH):   The summit angles of an S-quad are right angles.

OBTUSE ANGLE HYPOTHESIS (OAH):   The summit angles of an S-quad are obtuse angles.

ACUTE ANGLE HYPOTHESIS (AAH):   The summit angles of an S-quad are acute angles.

Saccheri showed the the OAH produces a contradiction because it implies that the parallel postulate is true (this is not obvious, it took him many pages), which implies that RAH is also true. You cannot have both RAH and OAH true because no angle can be right and obtuse at the same time. Exit OAH!

A Strange Geometry

The AAH was more stubborn. This part of Saccheri's work is revealing; so we will present a few of his theorems, which will appear rather strange. Remember that all Saccheri was assuming was absolute geometry together with the acute angle hypothesis. We will refer to Saccheri's results that follow from AAH as a *Strange Geometry*. (If you draw a picture or two, you will see why we chose the name.)

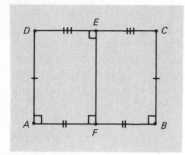

FIGURE 3

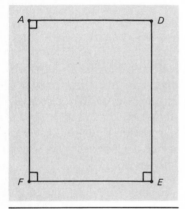

FIGURE 4

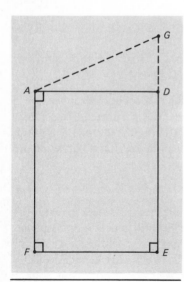

FIGURE 5

The Strange Geometry depends upon the S-quad given in Figure 3. Sides AB and CD have been bisected by F and E, respectively, so that $\overline{AF} = \overline{BF}$ and $\overline{DE} = \overline{CE}$. Saccheri showed that

Theorem S3
Angles AFE and DEF are right angles.

Theorem S3 is not itself entirely unexpected, but it is needed in the following theorems, which certainly appear to be.

Theorem S4
In an S-quad $ABCD$ the base AB is shorter than the summit CD.
Proof Consider the quadrilateral $ADEF$. We first want to show that $\overline{AF} \neq \overline{DE}$. We do this by showing that $ADEF$ is an S-quad with EF as its base and AD as its summit. Hence view $ADEF$ as in Figure 4. Suppose $\overline{AF} = \overline{DE}$.

1 $ADEF$ is an S-quad [Def. of S-quad].

2 Angles A and D are acute [AAH].

3 This contradicts the hypothesis that angle A is a right angle. Hence $\overline{AF} \neq \overline{DE}$. Suppose AF is longer then DE.

4 Extend ED beyond D [Post.   ]. See Figure 5.

5 Choose the point G on this extension such that $\overline{EG} = \overline{AF}$ [Prop.   ].

6 Angle GAF is obtuse [CN.   ].

7 $AFEG$ is an S-quad [Def. of S-quad].

8 Angle GAF is acute [AAH].

9 This is a contradiction because angle GAF cannot be both acute and obtuse. Hence it cannot be true that AF is longer than DE. Therefore AF is shorter than DE. Similarly, you can show that BF is shorter than CE. Hence AB is shorter than CD [CN.   ].

QED

Exercise 12 will ask you to show why the proof of Theorem S4 requires that we split in half the original S-quad rather than work with the whole thing. Can you see why? Here is a hint: to say that $ADEF$ is an S-quad required DEF to be a right angle. Could we say that if we started with $ABCD$?

In defense of our title, Strange Geometry, we ask you to draw a figure representing Theorem S4. Start with any line AB and draw a line CD above AB and longer than AB as in Figure 6. Now draw a quadrilateral. You must have drawn something like Figure 7(a) or (b). The latter is not an S-quad (since the base angles are not right angles), and the former is not a quadrilateral since the perpendiculars are curved.

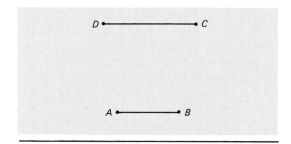

FIGURE 6

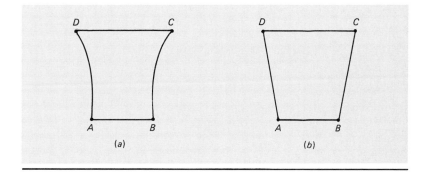

FIGURE 7

Saccheri's next result uses Euclid's Proposition 25.

Theorem S5

In any right triangle the sum of the other two angles is less than 90°.
Proof Let ABC be a triangle such that angle B is a right angle. See Figure 8.

1   Erect a perpendicular to AB at A [Prop.   ].

2   Choose the point D on this perpendicular so that $\overline{AD} = \overline{BC}$ [Prop.   ].

3   Draw CD [Post.   ].

4   $ABCD$ is an S-quad [Def. of S-quad].

5   $\overline{AB}$ is less than $\overline{CD}$ [Theorem S4].

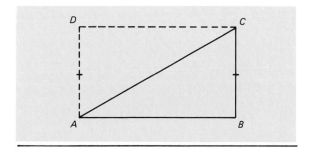

FIGURE 8

6   Angle ACB is less than CAD [Prop. 25].

7    $\angle ACB + \angle BAC$ is less than $\angle CAD + \angle BAC$ [CN.   ].

8    $\angle CAD + \angle BAC = 90°$ [Post.   ].

9    $\angle ACB + \angle BAC$ is less than $90°$ [CN.   ].

QED

Saccheri Not Freed of Every Flaw

With a little bit of work, Theorem S5 yields:

Theorem S6
In any triangle the sum of the three angles is less than $180°$.

We presented Theorems S4, S5, and S6 to impress upon you the rigor that Saccheri used in his proofs and also the repugnance of his results. And he did not stop there. He rattled off many more bizarre theorems. Yet—and this is the key—he never once found a proposition that contradicted an earlier one. He had generated a series of dubious results, and he probably felt that if he continued he eventually would find his contradiction. But the nature of his results was enough to convince him that his main thesis was valid; so he concluded:

> The acute angle hypothesis is absolutely false because it is repugnant to the nature of straight lines.

Notice his reliance on the redundant phrase "absolutely false." Do you see a flaw in his statement? He never reached a logical contradiction and thus he failed. In his proofs he used meticulous and impeccable logic, but when he came to his main objective, he threw deduction to the wind and relied wholly on his intuition. This was exactly the faulty reasoning that prompted the Greeks to invent deductive reasoning.

Saccheri's whole approach to the problem was to use pure logic to prove Euclid was correct. However, in the end, he fell into the Epicurean trap, resorting not to deductive reasoning but to his intuition about real-world facts.

What are the lessons of this episode? First, as a scientist, Saccheri should have either continued searching for his contradiction or else have given up the whole project and admitted defeat. He would be praised if someone else extended his results and found the contradiction. No one has, however. Why, then, do we even mention him? It is because his work *was* extended but not with the outcome that Saccheri visualized. In fact, no contradiction has ever been found—even 250 years after Saccheri's attempt.

This will help us explain Saccheri's niche in the history of mathematics. His book was ignored, both for its critical shortcoming and because few were interested in the problem at that time. It remained virtually unread for over 150 years. Meanwhile, three other mathematicians obtained similar results. Yet when credit is given, Saccheri's name is conspicuously absent.

So, was Saccheri a failure? The answer is qualified: yes and no. It is yes because his hazy, unconvincing final contradiction failed to show that the parallel postulate could be derived from the remaining axioms. Yet, if we ignore his ill-conceived contradiction and focus attention on the Strange Geometry, we find a logically sound mathematical discourse, even though it differs from Euclidean geometry at several significant places. He had created a logical alternative to Euclidean geometry, but he did not realize it because of the blinders he wore.

Saccheri's contribution to the ascent of geometry is that his seemingly insignificant, woefully unsound book was the springboard for the most dramatic revolution in the history of scientific thought. His work lay dormant for 150 years because mathematicians were "in the nine dots."* The next section describes the fertile environment that was needed for the revolution to reach fruition.

EXERCISES

1   Which of the figures below are S-quads?

(*a*)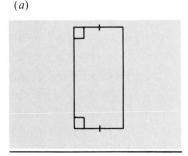

(*c*)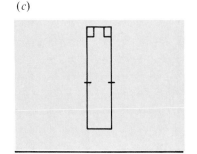

(*b*)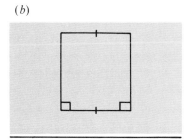

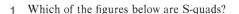

*This expression is synonymous with—but much more dramatic than—the phrase "had blinders on." Try Exercise 7 to see what we mean.

2   In the figure below the quadrilateral *QUAD* is an S-quad. Supply the missing justifications for the proof of the following theorem.

Theorem
The summit angles of an S-quad are equal.
Proof   We are given that *QUAD* is an S-quad.

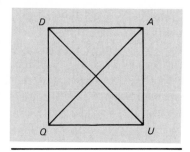

1   Angle *DQU* = angle *AUQ* [   ].

2   $\overline{DQ} = \overline{AU}$ [   ].

3   Triangle *DQU* is congruent to triangle *AQU* [Prop. 4].

4   $\overline{DU} = \overline{AQ}$ [   ].

5   Triangle *ADQ* is congruent to triangle *ADU* [Prop. 8].

6   Angle *ADQ* = angle *DAU* [   ].

3   In Euclidean geometry prove that an S-quad is a parallelogram.

4   In Euclidean geometry prove that an S-quad is a rectangle.

5   Assuming the acute angle hypothesis, is an S-quad a parallelogram? a rectangle?

6   Saccheri was only one in a long line of people who tried to prove that the parallel postulate followed from Euclid's other nine axioms. His major claim to fame (of course he never realized it) was that he had a long list of logically correct theorems from his hypothesis. Even his starting point does not differ a great deal from many others. Nasir Eddim was a thirteenth-century Persian mathematician—the court astronomer for the grandson of Genghis Khan—who tried his hand at the problem by starting with the following assumption: if two lines l_1 and l_2 are such that perpendiculars *m* and *n* from l_1 to l_2 make unequal angles with l_1, then the line segment on the base is shorter than the summit line segment. This can be proved by first showing that the legs are unequal, then constructing *E* in the figure so that $\overline{AD} = \overline{BE}$, and then using the theorem in Problem 2. What would then be the last step in the proof? What is the fundamental difference between Saccheri's substitute for the parallel postulate and Eddim's? (See the accompanying figure.)

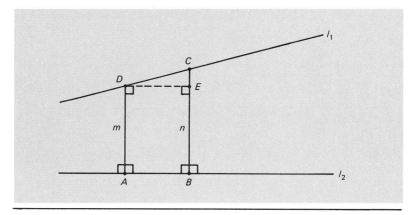

7 In the accompanying figure connect the nine dots with four lines without lifting your pencil from the paper.

• • •

• • •

• • •

In Problems 8 to 10, supply justifications for the given steps in the given theorem.

8 Theorem S-1, steps 2, 3, 4.

9 Theorem S-5, steps 1, 2, 3, 7, 8, 9.

10 Theorem S-4, steps 4, 5, 6, 9.

11 The proof of Theorem S-5 is not complete. Explain how the statement of the theorem follows from the proof.

12 Explain why the proof of Theorem S-4 requires that the original S-quad be split in half rather than working with the whole thing.

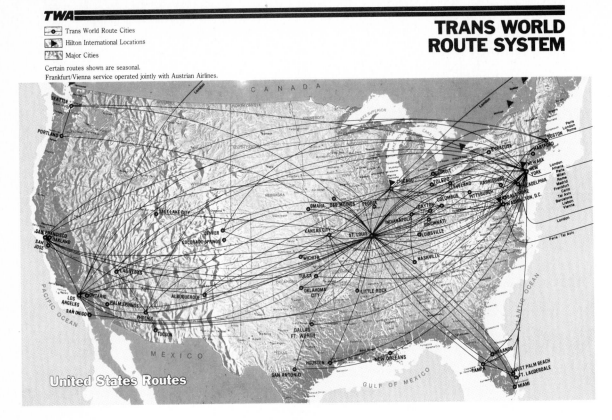

TWA

—●— Trans World Route Cities

▶ Hilton International Locations

Major Cities

Certain routes shown are seasonal.
Frankfurt/Vienna service operated jointly with Austrian Airlines.

**TRANS WORLD
ROUTE SYSTEM**

United States Routes

4-3 SPHERICAL GEOMETRY

Nicholas Lobachevsky (1793–1857) was a Russian mathematician whose fate was to be universally regarded as a little looney. Not just because he was a mathematician, of course. In 1829 Lobachevsky felt that he had made a significant breakthrough in the study of geometry. He published a book that purported to contain a new geometry, wholly different from Euclid's. His contemporaries considered him to be bordering on lunacy.

Why study a new geometry? Everyone accepted Euclid's text as the gospel truth. It was already 2000 years old. Its vitality had been tested over and over again. Yet if we consider some of Lobachevsky's theorems, we must sympathize with his detractors. His starting point was quite unassuming. Just as Saccheri had done 100 years before him, he began his geometry with the same first nine axioms of Euclid, but for the tenth, the parallel postulate, he substituted one of its two negations:

> LOBACHEVSKY'S TENTH AXIOM: Given a line *l* and a
> point *P* not on *l*, there is more than one line passing through *P* and
> parallel to *l*.

It is not much different from Euclid's starting point. However, just as Saccheri arrived at "repugnant" statements, so too Lobachevsky proved results that appeared to be in total conflict with our knowledge of the real world. Just as the Epicureans ridiculed Euclid for *proving* something that is *obviously* true, Lobachevsky's detractors ridiculed him for *proving* something that is *obviously* false.

Some Very Strange Theorems

Let us look at a few of Lobachevsky's results. He *assumed* that there were at least two lines parallel to *l* and passing through *P*. From that axiom he proved:

THEOREM L1:   **For any line *l* and point *P* not on *l*, there are an infinite number of lines parallel to *l* and passing through *P*.**

Here is another of Lobachevsky's results that rankles our intuitive notion of geometry.

THEOREM L2:   **The sum of the angles of any triangle is less than 180°.**

Lobachevsky's results were very similar to those of a contemporary of his, Janos Bolyai. Recall in Section 4-2 we mentioned that Bolyai's father had advised him to drop this "abysmal darkness." Instead he created a new geometry much like Lobachevsky's.

Enter Gauss

Karl Friedrich Gauss (1757–1855) played an important role in the creation of non-Euclidean geometry. Gauss was one of the greatest mathematicians and scientists that ever lived. Perhaps his name is not the household word that Newton's and Einstein's are, but his accomplishments nonetheless place him on as high a pedestal. Both Lobachevsky and Bolyai sought his blessing for their work, reasoning that if they could convince Gauss to publicly condone their work it would be accepted by the mathematical community simply because the great Gauss had accepted it.

Bolyai's father, a friend of Gauss, tried to intercede for his son. Gauss' response was very discouraging, however. He told Bolyai that he himself had dabbled with the notion of a new geometry just like Bolyai's. He even went further. After proving many of the same maverick results, he suspected that this new geometry might actually have some validity in the real world. Would it be possible to find a triangle somewhere whose angles really did add to less than 180°? It was clear that he could not look for the legs of such a triangle in the normal way that we think of straight lines, as resembling a stretched string. He knew that he needed another type of straight line. It has always been assumed that light travels in a straight line, in the sense that it takes the shortest path from one point to another. Perhaps, thought Gauss, if a triangle were constructed using light rays as straight lines forming the three sides, the sum of the angles might total less than 180°. He constructed his triangle using three mountain tops as the vertices, with assistants flashing light beams from each peak. The angle sum was so close to 180° that the difference was easily attributed to error of measurement. Gauss gave up his pursuit of the new geometry. He knew that without a clear demonstration that the apparently weird results actually were valid in the real world, he would be ridiculed for studying such "worthless" material.

Bolyai argued, as did Lobachevsky, that the new geometry did not have to bear any relationship to the real world. It was worthwhile in itself, as a purely mental exercise. Gauss disagreed, at least in public. They did not get their blessing from the master and their geometry was rejected. It is sad to note that the careers of both men never recovered from this blow. Lobachevsky gave up the study of mathematics and returned to teaching and administration. It was not until much later that his accomplishments were recognized. Bolyai gave up mathematics altogether and joined the Austrian army.

Enter Riemann

The concept of this new brand of geometry lay dormant for about 20 years. Then a student of Gauss, Bernhard Riemann, regarded later as one of the greatest mathematicians of all time, decided to study still another type of geometry. His whole thrust was radically different from Bolyai's and Lobachevsky's, and it appeared to have applications in many fields other than mathematics. Hence its study was justified.

Riemann was also young and energetic, and had already won the favor of the mathematicians of the day, including his mentor Gauss. He did not "fear the Boetians," that is, the ridicule of the commoners, as Gauss did.

Riemann altered Euclid's parallel postulate by choosing the negation that Lobachevsky perceived as fruitless and ridiculous. Recall that the parallel postulate states that exactly one line is parallel to a given line

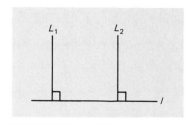

FIGURE 1

through a given point. In negating the statement, one gets that it is not true that there is exactly one line, so there is either more than one line (enter Saccheri, Lobachevsky, and Bolyai) or there is *no* line that is parallel to *l*. This is Reimann's contribution:

RIEMANN'S TENTH AXIOM: Given a line *l* and a point *P* not on *l*, there is no line passing through *P* that is parallel to *l*.

Riemann's 10 axioms are not much different from Euclid's or Lobachevsky's. The theorems, however, are very different.

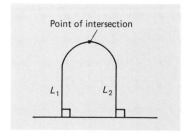

FIGURE 2

Theorem R1

If *l* is any line and if two other lines are perpendicular to *l*, then the two other lines will intersect.

If you draw a sketch of the hypothesis of the theorem, you get one like that in Figure 1. If you are still hung up on demanding that sketches of geometrical objects subscribe to your intuitive (Euclidean) notions, then you will necessarily demand that the two lines L_1 and L_2 be parallel. However, we know that they are not parallel in Riemann's geometry because of its tenth axiom. How would you complete the sketch in Figure 1 in order to satisfy the conclusion of Theorem R1? Is your sketch like the one in Figure 2? The one in Figure 2 is the one we will refer to later.

The proof of Theorem R1 is very easy. You might see immediately that it follows directly from Riemann's tenth axiom. In fact, Riemann's tenth axiom implies that *all* lines intersect. The theorem then merely accentuates this fact for a situation where you might not expect it.

Let us now look at three more theorems of Riemann. They will give us a good feeling for his geometry.

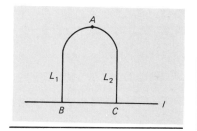

FIGURE 3

Theorem R2

If two lines are perpendicular to a third line, then the three line segments form an isosceles triangle.
Proof Let L_1 and L_2 be the two lines perpendicular to *l*.

1 L_1 and L_2 intersect [   ]. Call the point of intersection of L_1 and L_2 the point *A*. Call the points of intersection of *l* with L_1 and L_2 the points *B* and *C*, respectively. (See Figure 3.)

From here on the proof is indirect. We want to show that $\overline{AB} = \overline{AC}$. Suppose $\overline{AB} \neq \overline{AC}$. We now try to reach a contradiction. Suppose $\overline{AB} > \overline{AC}$.

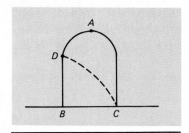

FIGURE 4

2 Mark the point D on AB such that $\overline{BD} = \overline{AC}$ [ ]. (See Figure 4.)

3 Construct CD [ ].

4 Angle DBC = angle ACB [ ].

5 We now have $\overline{BD} = \overline{AC}$ [ ], angle DBC = angle ACB [ ], and $\overline{BC} = \overline{BC}$; therefore triangle ABC is congruent to triangle BCD [Prop.].

6 This is a contradiction [ ].

Therefore, our last suppostion, which was $\overline{AB} > \overline{AC}$, must be false. Of course, we might have supposed that $\overline{AB} < \overline{AC}$, but then the same steps would have led to a contradiction. Hence we are led to conclude that $\overline{AB} = \overline{AC}$.

Theorem R3 states that all lines perpendicular to another line intersect in the same point. Try to sketch what this means by drawing four lines that are perpendicular to l and all meeting in the same point. Are all five lines (l and the four perpendicular lines) straight? Be careful how you answer this question. If you said to yourself "yes," we are winning you over. If you said, "Of course not; any fool can see that you can't do that—they'd have to be curved lines," you have a long way to go. Yes! They *are* straight. Remember that a figure is only a crutch.

In the proof of Theorem R3 we will be aiming at choosing *any* arbitrary point on l and demonstrating that the conclusion follows, but we will have to be content with first choosing specific points and showing how the process continues so that every point is attainable.

Theorem R3

If l is any line, then all lines perpendicular to l pass through the same point.
Proof Let L_1 and L_2 intersect l at points B and C, respectively, and intersect themselves at point A.

1 Construct D on l so that $\overline{BC} = \overline{CD}$ [ ].

2 Construct the line L_3 passing through D and perpendicular to l [ ].

We first show that the theorem is true for the point D; that is, we want to show that L_3 passes through A. The proof is indirect. Suppose L_3 does not pass through A. We will reach a contradiction. (See Figure 5.)

3 Let E be the point of intersection of L_2 and L_3 [ ].

4 Let F be the point of intersection of L_1 and L_3 [ ].

5 Then $\overline{BF} = \overline{DF}$ and $\overline{CE} = \overline{DE}$ [ ].

6 Triangle BCE is congruent to triangle CDE [ ].

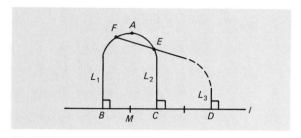

FIGURE 5

7   Therefore angle *EBC* = angle *EDC* [   ]. Then angle *EBC* is a right angle since angle *EDC* is a right angle.

8   Thus line segment *BE* lies on line segment *BF* [   ].

9   *E* is defined to be on L_2; by (8) *E* is on L_1; so *E* = *F* = *A* [   ]. Hence L_3 passes through *A*.

We will now consider the midpoint of *BC*—call it *M*—and show that *M* satisfies the theorem.

10   Construct the midpoint *M* of *BC* [   ].

11   Construct the line L_4 perpendicular to *l* and passing through *M* [   ].

We will now use an indirect argument to show that L_4 passes through *A*. Suppose it does not.

12   L_4 intersects L_1 and L_2 [   ]. Call the points of intersection *G* and *H*, respectively. (See Figure 6, which gives one possibility.)

13    $\overline{GB} = \overline{GM}$ and $\overline{HM} = \overline{HC}$ [   ].

14   Triangle *GBM* is congruent to triangle *GMC* [   ].

15   Angle *GBM* = angle *GCM* [   ].

16   *CG* lies on *CA* [   ].

17   *G* = *H* = *A* [   ].

The proof now requires us to bisect *BM* and *MC* and show that each of these points satisfies the theorem. In the exercises we ask you to fill in the details. Each of these four segments can be bisected, and the points can be shown to satisfy the theorem. In this manner all the points between *B* and *C* can be shown to satisfy the theorem.* (See Figure 7.) The points to the left of *B* and to the right of *C* are handled in the same way.

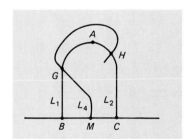

FIGURE 6

---

*The purists would shout loudly at this juncture—with good reason. The complete process needed to show that the theorem works for *every* point requires calculus. For expediency, we will not cover calculus now, but will still consider the proof to be complete.

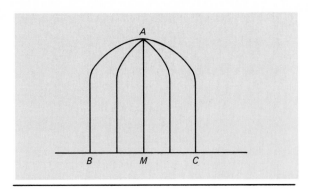

FIGURE 7

Consider again Figure 2. You see a triangle. How many degrees are in the triangle? We cannot say exactly, but we do know that two of the angles are right angles because *AB* and *AC* are assumed to be perpendicular to *l*. Therefore the sum of these two angles is 180° (= 90° + 90°) and so the triangle must have more than 180° because angle *BAC* will have some positive measure. This strange fact is true for *all* triangles in Riemann's geometry.

Theorem R4
The sum of the angles of any triangle is greater than 180°.

A Strange World: Earth

We are now ready to answer a question that we hope has been nagging you since the start of this section: Why did this section open with a map of the United States showing TWA routes? We will need the map to convince you that Figure 2 is not quite as strange as you might have thought.

Find the route TWA takes from Boston to San Francisco. Would you say that the route depicted in the map is the shortest? Draw a straight line on the map from Boston to San Francisco. Is your line closer to Chicago or Fargo? If you were the navigator of an airplane and you charted your course via the line you just drew, the one that *appears* straight on the map, and another plane flew exactly the same speed as yours but took the route indicated on the TWA map, you would take much longer to get from Boston to San Francisco. Your route is much farther. What appears to be a straight line between two points, in the sense of the shortest distance, is not a straight line.

Does this seem like double talk? It is not! The apparent double talk occurs because you are referring to two different entities, two different straight lines. When you drew a straight line on the map, it was a straight line *on the map,* which is situated on a two-dimensional flat plane. The earth, however, is not flat; it is more like a globe (it is, in fact, more egg-shaped than spherical, but for our purposes spherical will do). If you translate the straight lines on the map into a path on the globe, you are going from a flat surface to a curved surface. If you measure, *on the globe,* the distance from Boston to San Francisco along the path defined by your straight line on the plane (it goes just north of Chicago), you will get a longer distance than if you measure, *on the globe,* the distance along the path defined by the TWA route that is curved on the map.

The shortest distance between two points on a globe is along what is called a great circle. A *great circle* is a circle on a sphere whose diameter passes through the center of the sphere. On the earth the equator is an example of a great circle, as are all the lines of longitude, called *meridians,* that pass through the north and south poles.

Let us try one more example from the TWA map to see if you are getting the hang of it. What is the great circle route from Los Angeles to London? Of course, London is not on the map, but you probably have a good idea of where it lies in relation to this map. Now draw a straight line on the map that would take you from Los Angeles to London. Such a line would go through an eastern city such as Washington or Newport News. Do you think the straight line on the map is close to the great circle route? Not at all! The great circle route is depicted on the map. It is the line that proceeds north by northeast from Los Angeles, just passing by Butte, Montana, and then going into Canada.

What produces this distortion? It is caused by the way the map is made. Remember that a map is a flat representation of a spherical surface. It is impossible to get an accurate rendering of distances. No matter how the map is made (there are many ways of projecting a globe onto a flat surface), certain types of distances are distorted.

The best way to convince yourself of this fact is to take a globe, the larger the better, and stretch a string as tightly as you can between Los Angeles and London. You will see that the string passes through Canada. Mark off the length of string used. Now put the string on the path marked off by the straight line you drew on the map, the one that passes through an eastern city. (We cannot say exactly which city because different maps will yield different cities.) You will see that the string is not taut. It does not rest firmly on the surface. The length of string used is much greater than before. Hence the latter route is much farther than the great circle route.

We will now tie our ideas about great circles, the straight lines on the earth, together with Riemann's geometry. Picture yourself on the north pole. Start walking in a straight line (it might be better to take an airplane because you are in for a long journey). Walk south (that is the only direction to go) to Chicago; keep walking south in a straight line until you reach the equator. Then turn left and walk due east on the equator. You just made a right-angle turn. Go about 1000 miles and turn left again and walk north. You made another right-angle turn. As you head directly north you pass by New York and eventually get back to the north pole. You have traveled in three straight lines. Draw a picture of your journey. Which of the sketches, Figure 8(*a*) or (*b*), yields a better depiction? Since you are on a sphere, usually the trip is described as in Figure 8(*b*). Does the figure look familiar? It is the same as Figure 2, the triangle in Riemann's geometry whose base angles are both right angles. Recall that you made two right-angle turns, when you arrived at the equator and when you turned north. The triangle you traversed on the earth has more than 180°.

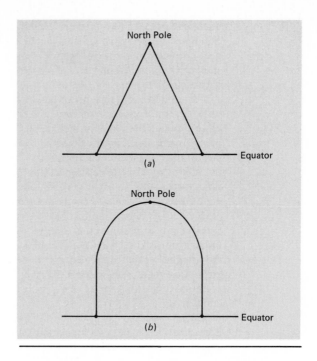

FIGURE 8

Note that any two great circles intersect. Hence there are no parallel lines on a sphere. All the meridians are perpendicular to the equator—and they all meet in the same point, the north pole. Every triangle constructed with great circles has more than 180°. Thus all four of the theorems we covered from Riemann's geometry are true theorems on a sphere. This is true for all theorems in Riemann's geometry. This is why it is often called *spherical geometry.*

Every theorem of Riemann's geometry has a physical reality when interpreted on a sphere with great circles considered as straight lines. This is not merely a mental exercise like the one Lobachevsky and Bolyai clung to in order to justify their geometries. Riemann's geometry is the geometry of the earth. Whenever large distances are traversed, such as long voyages or airplane routes, spherical geometry must be used rather than Euclidean geometry.

There Is Truth and Then There Is Truth

How can both geometries be true? If there is only one truth, then which geometry, Euclid's or Riemann's, is the geometry of our physical world? And what about Lobachevsky and Bolyai? Where do their geometries fit into the scheme of things?

The answers to these fundamental questions rely more on philosophy than on mathematics. In fact the answer that a pure mathematician might give you is that all three geometries are "true" from a mathematical sense in that each is an independent closed system that starts with 10 axioms, together with certain undefined and defined terms, and uses deduction to develop relationships, called theorems, among the entities of the system. Each system is "true" because it follows the rules of deduction.

But which gives us correct information about our physical world? The mathematician is tempted to say that that is a totally different problem, one that the physicist and the surveyor must wrestle with. Do you see how things have changed in little over 100 years? In 1830 Gauss would never publicly accept a mathematical theory such as Lobachevsky's and Bolyai's unless it could be demonstrated that it provided knowledge of our physical world. To Gauss a mathematical theory and its application were inseparable. Today scientists view them as totally separate entities. In many diverse fields of learning, including sociology, psychology, political science, biology, economics, and physics, mathematical theories are being constructed via deduction and based on assumptions, or axioms, that appear reasonable. Once the theory is developed, its theorems are compared with known data from the real world. If the theory produces major discrepancies from known results, the theory is altered; in other words the axioms must be reexamined and changed. A new set of theorems is produced and then compared with real data.

The nagging question remains: "What is the truth?" There is a simple answer: There is no *absolute* truth. When geometrical facts are sought in small regions of the globe or on the plane of a carpenter or architect, Euclidean geometry yields correct results. On larger regions of the earth when the curvature of the earth is significant and the definition of a straight line is taken to be a "stretched string" representing the shortest distance between two points, Riemann's geometry must be used.

How can it be that there are two different truths? When lines, triangles, and circles are drawn on a plane, then Euclid's plane geometry is the proper tool. An architect constructs a drawing on a flat drafting board. Engineers and carpenters make great use of a level and a plumb line because a structure whose foundation is not level soon succumbs to the force of gravity. The Leaning Tower of Pisa is testament. Hence in a world assumed to be flat, Euclid is truth.

However, navigators of the globe will make serious errors if they try to apply Euclidean theorems to their surface. The surface of the earth is curved, not flat, so Riemann's geometry yields proper results.

Even though geometry arose from measurements of the surface of the earth, a surface that is spherical, it was Euclid's brand of plane geometry, which is applied only to a flat surface, that was developed first. The answer is not that the ancients believed that the earth was flat. Indeed, the classical Greeks not only knew that the earth was round but also were able to use trigonometry to compute the circumference of the earth to an astounding degree of accuracy. The answer was that they lived in very limited regions of the globe. They did not venture very far and hence the earth was flat for all intents and purposes. In every physical triangle that they could measure or reasonably conceive, the sum of the angles was always $180°$. Remember that Euclid's "Elements" was heralded as a masterpiece of thought, the perfect mode to ensure that scientific results were correct. It was the paragon of deduction. In addition, whenever a geometric pattern was constructed on the

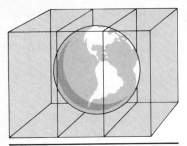

FIGURE 9

surface of the earth, its measurements agreed exactly with Euclid's theorems, because the earth is virtually flat in limited regions.

Let us consider one more explanation as to how these two different geometries can coexist, with each yielding true facts of our physical world.

One can view spherical geometry from two perspectives. The clearest view for most people is to picture our physical world as a Euclidean space, with straight lines represented as the edges of a ruler or as girders in a construction site. Step back and view the earth in such a setting. Picture a grid enclosing the sphere as in Figure 9. Then the great circles are curves in a three-dimensional geometry. This is a Euclidean view of space. In order to picture the great circles it is necessary to consider all three dimensions of our space. Euclidean geometry can be extended easily from the two-dimensional variety on the plane we have considered here to a three-dimensional brand.

The Acceptance of Non-Euclidean Geometry

There are a great many lessons to be learned from the discovery of non-Euclidean geometry. Perhaps the most important is the blindness of the human species. We are creatures of habit, and the more ingrained the habit becomes the less likely we are to accept new ideas. The "Elements" had been the epitome of truth for 2000 years. Its method of thought was copied in many diverse fields. But we know now that it gave us true theorems only in a specific area of human experience, only on a plane. When Bolyai and Lobachevsky proposed a new type of geometry, they not only were ignored but their work was ridiculed. The primary reason for this universal dismissal was that their theories were contrary to accepted thought. No one in the scientific community was willing to accept their work as simply a worthwhile logical demonstration and leave open the possibility that later investigations might show whether it had any application to the real world. Instead, their work was summarily rejected as being worthless.

It was not until 30 years later that the theory of non-Euclidean geometry was accepted, when Riemann showed that it can be interpreted as the geometry of our physical world. People were prepared to accept the idea of a straight line in a geometrical study being represented in the real world as a great circle on a sphere because, for the first time, it was common to chart long journeys across wide expanses on the globe. Thus Riemann's theory had a practical application, and so it was accepted.

But what about Lobachevsky and Bolyai? Remember that their geometry was different from Riemann's. Their geometry still had no physical representation, no real-world application to lend credence to its theorems. Even after Riemann's creation, Lobachevsky's geometry remained a purely theoretical exercise.

Here is the most important step in the revolution. The scientific community reversed itself and accepted Lobachevskian geometry. In fact, in one of the classic "return of the prodigal son" stories, Lobachevsky was heralded as the founder of non-Euclidean geometry. The torment and grief he suffered through his rejection were forgotten. His geometry became a central ingredient in the development of the subject. Scholars studied it and improved its representation; soon it became an approved segment of the graduate curriculum. This meant that a wide range of scientists learned of it, which improved the chances that someone might find a use for it.

The primary lesson that scientists learned from this episode was that theories should not be summarily dismissed simply on the grounds that they lack a physical application.

The Power of Logic

There is one more chapter to this story of the development of non-Euclidean geometry. It involves Lobachevsky's brand of geometry, often called hyperbolic geometry. We tried to give you a feeling for the skepticism confronting the pioneers of the subject. We then hoped you would be somewhat surprised to learn that the seemingly repugnant theorems of Riemann are actually valid results in our physical world. Recall that Lobachevsky's geometry has different theorems than Euclid's and Riemann's geometry. Around 1850, when Riemann presented his results, Lobachevsky's geometry was accepted on purely theoretical grounds. A theory that has no physical interpretation is often called pure mathematics. From 1850 until 1905 hyperbolic geometry was accepted as pure mathematics. This acceptance of a theory on a purely theoretical basis unlocked many doors to science. Today a great deal of science, in many diverse fields, is done first on a theoretical basis and then tested to see whether it has an empirical basis.

When Einstein created his theory of relativity, he used non-Euclidean geometry—hyperbolic geometry—to describe the world. In 1905 his first paper appeared. It was not until 1919 that his theory was proved to be a correct depiction of the universe. A British expedition traveled to Africa to view an eclipse, and they saw that the large mass of the sun appears to bend the light from a star that appears close to the sun in the heavens. If you used the light ray to form a triangle, it would have fewer than 180°. This is exactly the type of evidence that Gauss searched for 100 years earlier. He was on the right track, but the geometry of the universe becomes non-Euclidean only when very large or very small distances are considered. Einstein's theory of relativity had many notable ramifications—including nuclear energy.

Notice how it all started. An obscure Jesuit—Girolamo Saccheri—tried to free Euclid of an apparent flaw; the parallel postulate should be *proved* not assumed. Two others picked up the gauntlet—Lobachevsky and Bolyai—and produced a "weird" geometry. As they clung to their ideas they were ridiculed. Riemann showed that this new geometry did have an application. Einstein showed that this "weird" geometry was actually the geometry of the universe and used it to unlock fascinating mysteries. . . .

All because of logic.

EXERCISES

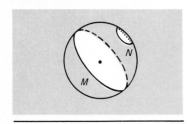

1 Explain what is meant by a *great circle* of a sphere.

2 Why is a great circle considered a straight line in spherical geometry?

3 We say that lines in Euclidean geometry are boundless. Are lines in spherical geometry boundless?

4 (*a*) Are all lines of latitude straight lines in spherical geometry?
   (*b*) Does either of the lettered curves in the accompanying figure seem like a straight line in spherical geometry?

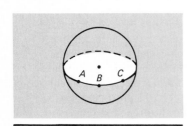

In Problems 5 and 6 refer to the accompanying figure, where the points *A*, *B*, and *C* all lie on the same great circle.

5 Are *A*, *B*, and *C* collinear in spherical geometry?

6 Is *B* between *A* and *C*? Is *A* between *B* and *C*?

7 There is a theorem in Euclidean geometry that states that if two distinct lines are not parallel then they intersect in exactly one point. Thus two distinct lines intersect in either no points or in one point. Draw two lines in spherical geometry (for example, draw the equator and a meridian). Do you think this theorem is true in spherical geometry?

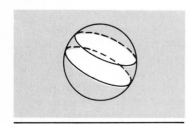

8 The accompanying figure seems to illustrate two straight lines in spherical geometry that are parallel. This must be incorrect because we assumed there are no parallel lines in spherical geometry. Can you explain what is wrong?

Consider the accompanying figure in Problems 9 to 11. Angles *ABC*, *CBD*, and *CAB* are right angles.

9 Compare the number of degrees of triangle *ABC* with 180°.

10 How many right angles can a triangle have in spherical geometry?

11 In Euclidean geometry Proposition 16, the so-called exterior angle theorem, states that an exterior angle is greater than either interior and opposite angle. Show that this theorem is not true in spherical geometry.

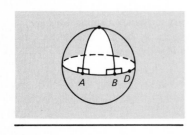

12 Define a parallelogram in Euclidean geometry. Do you think parallelograms exist in spherical geometry?

13 Define a rectangle in Euclidean geometry. Do you think rectangles exist in spherical geometry?

14 Fill in the following table by completing each statement according to the three geometries.

Statement	Euclid	Lobachevsky	Riemann
(a) Through a point not on a line there is	_____ parallel to the line	_____ parallel to the line	_____ parallel to the line
(b) In any triangle there are	_____ 180°	_____ 180°	_____ 180°
(c) The summit angles of an S-quad are	_____ angles	_____ angles	_____ angles

15 Supply the justifications for the steps in the proof of Theorem R2.

16 Supply the justifications for steps 1 to 9 in the proof of Theorem R3.

17 Supply the justifications for steps 10 to 17 in the proof of Theorem R3.

5 Linear Programming

Linear programming is one of the newest yet most potent of all mathematical tools. It is rare to be able to pinpoint the birth of a particular subject to a specific year, but it was precisely 1947 when linear programming burst on the scene. "Burst" is an appropriate word because World War II had produced complex problems both in military affairs and the business world, and since many of the problems were unsolvable in 1947, researchers were painstakingly seeking new methods of solution. The promised advent of electronic computers with their ability to ease the tedium of massive calculations prompted them to look in wholly new directions.

George Dantzig, then working for the U.S. Air Force comptroller as the mathematical adviser, was spurred by the revolutionary work of the Harvard economist Wassily Leontief to look for solutions of large-scale, dynamic problems. Dantzig first called his method "programming in a linear structure," but he later shortened it to *linear programming.*

The term "program" comes from the military reference to various plans and schedules of training, logical supply, and deployment of combat units as a *program.* Thus we see that the original application of the method was to military problems, but almost concurrently the economists in Dantzig's circle appreciated its value in the business sector. Today linear programming is a fundamental tool in such diverse industries as shipping, the gas, oil, and coal industries, and transportation.

In this chapter we start with the requisite skills needed to understand linear programming. These include graphing, straight lines, solving systems of equations, and solutions of linear inequalities. We then tackle some elementary applications of Dantzig's method. We have divided the latter material into two sections. Section 5-5 explains the main ingredients of a linear programming problem and how they are translated into mathematics. Section 5-6 then describes how to solve such problems geometrically.

5-1 DESCARTES AND HIS ANALYTIC GEOMETRY

The battlefields of Bohemia seem like a strange place for the inception of one of the greatest achievements of science. Even stranger is that an idea that would propel science in a new and fruitful direction would come to a young philosopher-turned-soldier in the form of a dream. It is not so unusual for a soldier to experience a disturbing dream. However, once awakened, René Descartes (1596–1650; pronounced day-cart′) claimed he was "filled with enthusiasm." One of his biographers, E. T. Bell, wrote that Descartes dreamed that he was reciting a poem of Ausonius which begins "which way of life shall I follow?"

Out of this night's sleep Descartes designed the foundation of his philosophy and his mathematics. His approach to philosophy had an immediate and profound influence on the sages of the day, but he is best known for his accomplishments in science. This can be seen in the quote of John Stuart Mill: "[Analytic geometry], far more than any of his metaphysical speculations, immortalized the name of Descartes, and constitutes the greatest single step ever made in the progress of the exact sciences."

The "greatest single step" that Mill referred to is the simple means that Descartes created to bridge the two age-old disciplines—algebra and geometry. He rendered each far more beautiful and powerful.

René Descartes lived in one of the greatest ages for intellectual development. The Age of Reason, as the period is sometimes called in Euopean history, boasted such greats as Shakespeare, Milton, Marlowe, Cervantes, Galileo, Newton, and Harvey. Descartes' mathematical contemporaries included Pascal and Fermat.

Descartes' Philosophical Bent

Descartes' major quest was for the proper method of obtaining truth. If it could be found, there would be no limit to the amount of knowledge that could be discovered. Since the classics were emphasized at that time, it is not surprising that he would turn to the ancient Greek civilization in search of his method. He lived in an age that questioned all accepted truths, even those over 1000 years old. His critical mind led him to conclude that there was no body of knowledge that was absolutely true and that the search for certainty must not depend upon any previously attained knowledge; it must start anew. It must rest upon a few simple irrefutable facts, to which the rigors of logical deductive thought would be applied.

Descartes based his new rational approach to philosophy on the simple assumption "I think therefore I am," which he considered to be obviously true. Armed with just a few such axioms, which he assumed to be irrefutable, he then developed new truths, using, as he stated, "long chains of simple and easy reasonings by means of which geometers have been accustomed to reach conclusions of their most difficult demonstrations."

Hence we see Descartes lifting a page from the Greeks and, more specifically, from Euclid. Euclid started with just 10 "irrefutable" assumptions and created his whole theory, with hundreds of propositions derived from these axioms by means of deductive logic. Descartes aimed to do the same for philosophy. He was the first to establish a rational approach to philosophical thought. In our context, however, we are more interested in his invention called analytical geometry, which set the scientific world on fire.

The results of Descartes' study were published in 1637 in a work titled simply "Geometry." It first appeared as an appendix to his major book, "Discourse on the Method of Reasoning Well and Seeking Truth in the Sciences," but later it was published separately.

The Marriage of Algebra and Geometry

We have studied the evolution of geometry since its inception, but we have only briefly mentioned the other segment of Descartes' great achievement—algebra. Algebra is generally regarded as that branch of mathematics that uses symbols, usually letters of the alphabet, to represent other quantities, often numbers. Thus the numbers themselves are not studied in algebra but rather the properties of the number system.

For example, a familiar statement in algebra says that for all real numbers a and b,

$$(a + b)^2 = a^2 + 2ab + b^2$$

Here the letters a and b represent any real number. The equation says that if we do what the left-hand side indicates, that is, if we first add the numbers and then square the result, we get the same result that we would if we do what the right-hand side indicates; that is, sum the squares of the two numbers together with twice their product.

The classical Greeks were aware of such algebraic properties of numbers too. However, because they lacked an appropriate symbolism, they viewed such properties geometrically. For instance, Figure 1 on page 268 shows immediately why $(a + b)^2 = a^2 + 2ab + b^2$.

Another property of real numbers that can be expressed in algebraic terms is as follows: For all real numbers a and b,

$$(a + b)(a - b) = a^2 - b^2 \qquad (1)$$

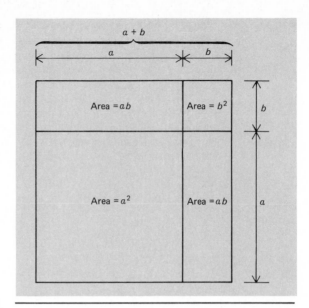

FIGURE 1 Total area $= (a + b)^2$.

What does it mean for this to be true for all real numbers a and b? Let us look at a specific example that demonstrates the meaning of the rule in Equation (1).

Example 1

Multiply 31 by 29. This product can be done in your head by using the formula in Equation (1). Let $a = 30$ and $b = 1$. Then the left-hand side of the equation is $(30 + 1)(30 - 1) = (31)(29)$. Few people can perform this multiplication in the usual way in their head. If, however, you use the formula you can note that

$$(31)(29) = (30 + 1)(30 - 1) = 30^2 - 1^2$$

You can probably perform this mentally, since $30^2 = 30 \cdot 30 = 900$; so $30^2 - 1^2 = 899$. Therefore $31 \cdot 29 = 899$.

Example 2

Problem Multiply 32 by 28.
Solution Using the formula in Equation (1) we get

$$32 \cdot 28 = (30 + 2)(30 - 2) = 30^2 - 2^2 = 900 - 4 = 896$$

Algebra is certainly a different subject than geometry. Algebra studies the relationships and properties of systems, usually number systems, while geometry is concerned with the relationships and properties of shapes such as lines, circles, and angles. The two subjects seem to be concerned with vastly different concepts. Descartes' great creation linked the two disciplines in such a way that each attained new heights never before dreamed of. As Joseph Lagrange, one of the great scientists of the eighteenth century, said,

> As long as algebra and geometry proceeded along separate paths, their advance was slow and their applications limited. But when these sciences joined company, they drew from each other fresh vitality and thenceforward marched on at a rapid pace toward perfection.

The Cartesian Coordinate System

Let us look at Descartes' invention. Consider the algebraic entity, the formula $A = \pi r^2$, where A is the area of a circle, π is a constant, and r is the radius of the circle. This formula gives a relationship between areas and radii of circles. Descartes saw that it could be viewed as a geometric object, a curve in a plane, as well as an algebraic one.

To accomplish this he defined a *coordinate system.* He started with what we now call a *real line,* which is a straight line [see Figure 2(*a*)] with one point marked as the *origin* corresponding to the number 0. A point, usually to the right of 0, is marked and it corresponds to the number 1. The distance from 0 to 1 is called a *unit* distance. If another point 1 unit to the right of 1 is marked, it corresponds to the number 2. The number 3 corresponds to the point 1 unit to the right of 2. The number −1 corresponds to the point 1 unit to the left of 0. Any real number *a* then corresponds to the point that is *a* units to the right of 0 if *a* is positive or to the left if *a* is negative. [See Figure 2(*b*).]

Then Descartes constructed a vertical line perpendicular to the horizontal real line. The point of intersection is called the *origin* and the same unit length is given in each real line. (Later we will see that the unit lengths do not have to be the same.) The positive direction is upward. He then noted that every point in the plane has a unique name given by an *ordered pair* of numbers. The first number is the perpendicular distance from the point to the vertical axis, and the second number is the perpendicular distance from the point to the horizontal axis. Thus the

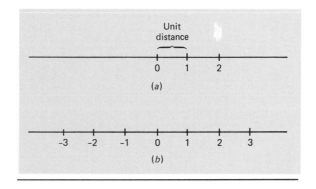

FIGURE 2

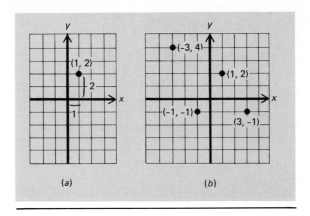

FIGURE 3

ordered pair (1, 2) corresponds to the point that is 1 unit to the right of the vertical axis and 2 units above the horizontal axis. [See Figure 3(*a*).]

In Figure 3(*b*) additional points are plotted in the coordinate system. The numbers in the ordered pairs are called the first and second *coordinates*. This type of coordinate system is referred to as a *cartesian coordinate system,* after Descartes.

It is customary to call the horizontal axis the *x axis* and the vertical axis the *y axis*. It should be understood that the choice of the letters *x* and *y* is arbitrary; in fact, we will sometimes change them as a particular problem dictates.

The Graph of an Equation

Let us return to our problem of the formula for the area of a circle, $A = \pi r^2$. To make the numbers more manageable, let us replace π by an approximation. We choose the approximation $\pi \approx 3$, even though it is not very accurate, because 3 is easier to work with than a more accurate approximation. Our new formula becomes $A = 3r^2$.

The geometric meaning that Descartes gave to the algebraic entity $A = 3r^2$ depends first upon listing some of the values of *r* and their corresponding values of *A,* as we do in the following table.

r	0	1	2	3	0.5	1.5
A	0	3	12	27	0.75	6.75

For example, when $r = 0$ then $A = 3 \cdot 0^2 = 3 \cdot 0 = 0$, and when $r = 2$ then $A = 3 \cdot 2^2 = 3 \cdot 4 = 12$. Now consider these numbers as ordered pairs: (0, 0), (1, 3), (2, 12), (3, 27), (0.5, 0.75), (1.5, 6.75). The first number is a value for *r* and the second is $3r^2$. The next step in Descartes' reasoning is to plot these ordered pairs as points in a coordinate system, as we do in Figure 4.

We plotted just six points, but note that we chose six values of *r* arbitrarily. Any value of *r* could have been selected; then its corresponding value of $3r^2$ could be found. If we plotted many points $(r, 3r^2)$ and

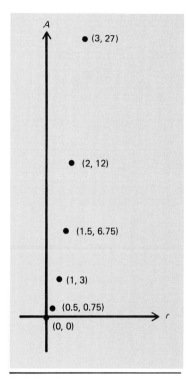

FIGURE 4

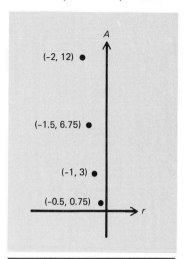

FIGURE 6

FIGURE 5

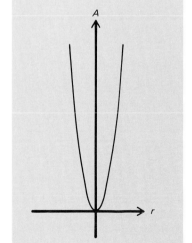

FIGURE 7   Graph of $A = 3r^2$.

then connected them with a smooth curve, the resulting figure would look like the curve in Figure 5.

We chose only nonnegative values for r because we started the discussion with a description of r as a radius of a circle. If we release that restriction and allow r to assume negative values, then we can plot points $(r, 3r^2)$ to the left of the vertical axis. (See Figure 6.) If we then plot all points, we get the curve in Figure 7, which represents the graph of $A = 3r^2$. In Figure 7 we call the horizontal axis the r axis and the vertical axis the A axis to emphasize that the r values are plotted on the horizontal axis (a specified distance away from the vertical axis) and the corresponding A values are plotted on the vertical axis. The type of curve in Figure 7 is called a *parabola*. Thus the graph of $A = 3r^2$ is a parabola.

The graph of the equation $A = 3r^2$ illustrates Descartes' scheme of molding algebra and geometry into one mathematical discipline called *analytic geometry* (or, sometimes, *algebraic geometry*). The essence of Descartes' method is this: To every equation with two variables there corresponds a curve in two dimensions, and to every such curve there corresponds an equation. In this section we will begin with an equation in two variables, x and y, and proceed to graph the curve. The procedure involves the following three steps:

1   Make a table of values for x and y.

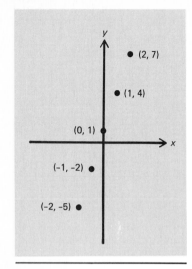

FIGURE 8

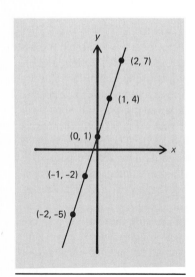

FIGURE 9 Graph of $y = 3x + 1$.

2 Treat the values as ordered pairs on a coordinate system.

3 Connect the points.

The curve that results from step 3 is called the *graph* of the equation. In the next section we will see how to proceed in the opposite direction, from the graph to the equation.

Why is this simple idea considered so revolutionary in the evolution of mathematics? The answer rests on the old adage that "a picture is worth a thousand words." Once algebraic quantities could be pictured, literally, as geometric entities, vast numbers of doors to understanding were opened. Let us consider a few more examples of how to apply Descartes' method of viewing an algebraic entity geometrically.

Example 3

Consider the relationship $y = 3x + 1$. In order to find the graph of this relationship we first make a table that gives the corresponding y values for selected x values. Thus if we let $x = -2$ in $y = 3x + 1$, we get $y = 3(-2) + 1 = -6 + 1 = -5$. Some other values are given in the table.

Table of values for $y = 3x + 1$:

x	-2	-1	0	1	2	3
y	-5	-2	1	4	7	10

We next consider the pairs of numbers in the table as points: $(-2, -5)$, $(-1, -2)$, $(0, 1)$, $(1, 4)$, $(2, 7)$, $(3, 10)$. Then we plot them in a cartesian coordinate system as in Figure 8. After we have plotted a few points, six in this case, we recognize that if we let x be *any* real number in the equation $y = 3x + 1$, we will get a value of y; so there are an infinite number of points $(x, 3x + 1)$ on the graph of the equation. The graph is then the totality of these points. In this example the graph is the straight line drawn in Figure 9.

When a point is on the graph of an equation we say that the ordered pair is a *solution* of the equation. Thus $(0, 1)$ is a solution of $y = 3x + 1$. The next example shows you how to use this terminology to help obtain the graph of an equation.

Example 4

Consider the equation $4x + 2y = 1$. To sketch the graph of this relationship we first construct a table of solutions. To get the entries in the table we choose a value for x, then substitute it into the equation to get the corresponding value of y. Thus if we let $x = 1$, we have $4 \cdot 1 + 2y = 1$; so $2y = 1 - 4 = -3$, $y = -\frac{3}{2}$. Therefore $(1, -\frac{3}{2})$ is a solution of $4x + 2y = 1$. The table provides five additional solutions.

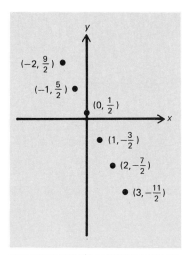

FIGURE 10   Six solutions to $4x + 2y = 1$.

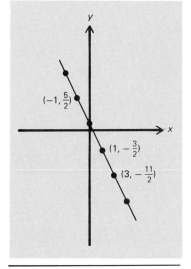

FIGURE 11   Graph of $4x + 2y = 1$.

FIGURE 12

Table for $4x + 2y = 1$:

x	-2	-1	0	1	2	3
y	$\dfrac{9}{2}$	$\dfrac{5}{2}$	$\dfrac{1}{2}$	$-\dfrac{3}{2}$	$-\dfrac{7}{2}$	$-\dfrac{11}{2}$
Solution	$\left(-2, \dfrac{9}{2}\right)$	$\left(-1, \dfrac{5}{2}\right)$	$\left(0, \dfrac{1}{2}\right)$	$\left(1, -\dfrac{3}{2}\right)$	$\left(2, -\dfrac{7}{2}\right)$	$\left(3, -\dfrac{11}{2}\right)$

If we plot these solutions as points in a coordinate system, we get Figure 10. If we then connect the points, we get the straight line in Figure 11, which is the graph of the equation $4x + 2y = 1$.

Let us take one more look at Descartes' method before we tackle the exercises. The equation in the next example, $y = 2x^2 - 1$, resembles the first equation we considered, $A = 3r^2$, in that one variable is squared (x^2 and r^2, respectively) and the other is not (y and A, respectively). This means that their graphs will be similar.

Example 5
Problem   Sketch the graph of the equation $y = 2x^2 - 1$.
Solution   In order to sketch the graph we construct the following table.

Table for $y = 2x^2 - 1$:

x	-2	-1	0	1	2	3
y	7	1	-1	1	7	17
Solution	$(-2, 7)$	$(-1, 1)$	$(0, -1)$	$(1, 1)$	$(2, 7)$	$(3, 17)$

If we plot the points that are given in the table as solutions to the equation, we get the graph by connecting these points with a smooth curve, as in Figure 12. The graph of $y = 2x^2 - 1$ is another example of a parabola.

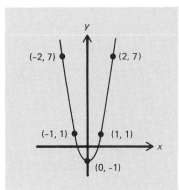

What's in a Name?

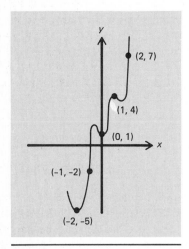

FIGURE 13 Could this be the graph of $y = 3x + 1$?

The examples of graphs given in this section have either been straight lines or parabolas. Of course, these are not the only curves that are generated by algebraic equations, according to Descartes' method. There is an endless variety of such curves, including circles, ellipses, and hyperbolas, whose geometric shapes were favored by the Greeks.

Once we are armed with algebraic names for these curves, our knowledge about them and our ability to use them to solve meaningful problems will be increased manyfold. For now, we hope that you are beginning to see the connection between the algebraic equation and its corresponding graph. When you see the equation $y = 3x + 1$, can you also picture its graph (Figure 9)? Perhaps you are not able to identify the two quantities immediately. You must at first memorize that the equation has a corresponding graph, but then, with a little practice and after reading the next few sections, you will also immediately be able to picture the graph when you see the equation and vice versa.

So far we have used sheer guesswork to arrive at the whole curve by connecting a few points from a table. We must base our conclusion that the graph of $y = 3x + 1$ is a straight line on more concrete information. It is conceivable that, given the meager information in the table for $y = 3x + 1$ of just those six points, the graph could have a wholly different shape entirely, as seen in Figure 13. In the next section we will investigate precisely what equations have graphs that are straight lines.

EXERCISES

In Problems 1 and 2 find the coordinates of each point.

1 $A(\ ,\ )$ $B(\ ,\ )$ $C(\ ,\ )$ $D(\ ,\ )$ $E(\ ,\ )$ $F(\ ,\ )$

2 $G(\ ,\ )$ $H(\ ,\ )$ $I(\ ,\ )$ $J(\ ,\ )$ $K(\ ,\ )$ $L(\ ,\ )$

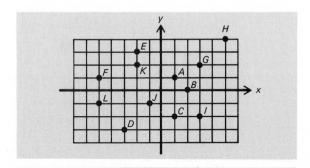

In Problems 3 and 4 plot each point on the grid.

3   A (2, 2) B (1, 4) C (3, 1) D (−1, 0) E (2, −1)
    F (−3, −5)

4   G (3, 2) H (4, 1) I (0, 5) J (2, −3) K (−5, 1)
    L (−1, −4)

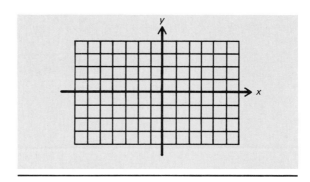

In Problems 5 to 8 plot the given points and then connect the points in the order given, joining the last point to the first. Then identify the polygon formed and find its perimeter and area. Here is an example.

Example
(0, 0)   (0, 4)   (2, 4)   (2, 0)

Polygon: It is a rectangle.

Perimeter: 4 + 2 + 4 + 2 = 12

Area: 4 × 2 = 8

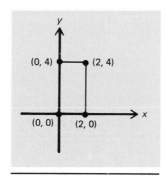

5   (0, 0), (0, 2), (2, 2), (2, 0)

6   (−2, 0), (5, 0), (5, 3), (−2, 3)

7   (−4, 0), (4, 0), (0, 3)

8   (1, 3), (5, 3), (1, 6)

In Problems 9 to 14 fill in the table for the given equation. In Problems 11 to 14 you must supply your own values of x.

9 $y = 2x - 3$

x	−2	−1	0	1	2	3
y						
Solution						

10 $y = x - 2$

x	−2	−1	0	1	2	3
y						
Solution						

11 $y = 4 - x$

x				
y				
Solution				

12 $3x - y = 2$

x				
y				
Solution				

13 $y = x^2 + 1$

x				
y				
Solution				

14 $y = 2x^2 - 3$

x				
y				
Solution				

In Problems 15 to 20 graph the equation using the points in Problems 9 to 14.

15 $y = 2x - 3$ 18 $3x - y = 2$

16 $y = x - 2$ 19 $y = x^2 + 1$

17 $y = 4 - x$ 20 $y = 2x^2 - 3$

In Problems 21 to 24 the graphs of the given equations are straight lines. Plot a few points and then sketch the graph.

21 $x - y = 3$

22 $2x + y = 3$

23 $3x - y = -1$

24 $3x + 4y = 1$

In Problems 25 to 28 the graphs of the given equations are parabolas. Plot a few points and then sketch the graph.

25 $y = x^2$

26 $y = 2x^2 - 4$

27 $y = \frac{1}{2}x^2 + 1$

28 $y = 4x^2 - 1$

In Problems 29 to 32 the graphs of the given equations are parabolas, but they are different from the graphs in the examples in the text and in Problems 13, 14, and 25 to 28. The parabolas in Problems 29 to 32 open downward, whereas the ones we have encountered previously open upward. Graph them.

29 $y = -x^2$

30 $y = 1 - x^2$

31 $y = 1 - 2x^2$

32 $y = -\frac{1}{2}x^2$

In Problems 33 to 36 the graphs of the given equations do not resemble straight lines or parabolas; so more points will be necessary to sketch them accurately. Plot the points for the given values of x and then sketch the curve.

33 $y = x^3, x = -2, -1, -\frac{1}{2}, -\frac{1}{4}, 0, \frac{1}{4}, \frac{1}{2}, 1, 2$

34 $y = x^3 + 2, x = -2, -1, -\frac{1}{2}, -\frac{1}{4}, 0, \frac{1}{4}, \frac{1}{2}, 1, 2$

35 $y = \frac{1}{x}, x = -2, -1, \frac{-1}{2}, -\frac{1}{4}, 0, \frac{1}{4}, \frac{1}{2}, 1, 2$

36 $y = 1 - \frac{1}{x}, x = -2, -1, -\frac{1}{2}, -\frac{1}{4}, 0, \frac{1}{4}, \frac{1}{4}, 1, 2$

In Problems 37 to 40 perform the computations mentally.

37 21×19

38 101×99

39 22×18

40 102×98

5-2   LINEAR EQUATIONS AND STRAIGHT LINES

The cartesian coordinate system is named in honor of its founder, René Descartes, even though Peirre de Fermat (1601–1665; pronounced fair-mah) deserves equal billing. There are striking similarities in the lives and works of these two great mathematicians. Both studied law, but Fermat, unlike Descartes, devoted most of his life to this vocation, serving as lawyer and counselor in the parliament at Toulouse. Both men also contributed to several branches of mathematics. Fermat did so only in his spare time, however, earning him the nickname "prince of the amateurs." Both were interested in the classical Greek civilization, and their attempts to solve two different problems from antiquity led them to the same method of solution: analytic geometry.

As we saw in Section 5-1, Descartes' discovery was published in the "Geometry" in 1637. But a year earlier Fermat had written a small pamphlet, entitled "Introduction to Plane and Solid Loci," which contained essentially the same material as Descartes' "Geometry." The two men made their discoveries completely independently of each other. In fact, there is no record of them ever having met.

Since Fermat's work was published a year before Descartes', why does Descartes receive the lion's share of the credit? For two reasons: first, Fermat's pamphlet was not published in book form until 1679, having circulated in manuscript form before then; second, Descartes' "Discourse on Reasoning" was such an overwelming success that it garnered a wide readership at once.

This explains why analytic geometry is generally regarded as the unique invention of Descartes. Ironically, Fermat's approach is much closer to the modern one. Another irony is that the original edition of "Discourse on Reasoning" was published without the name of the author, though the authorship was generally known.

There are two basic aspects to the historic invention of analytic geometry. One is the relationship between formulas and figures. We have seen one part of this relationship: how to draw the graph of an equation in two variables. This section will show you how to proceed in the opposite direction: given the graph of a curve, to construct its equation. This association of equations and curves has become so ingrained that one writer declared, "Say '$y = x^2$' to someone and odds are that the first image that pops into his head will be the graph of that equation."†

The other basic aspect of analytic geometry is a fixed frame of reference, called the cartesian coordinate system. Even though Fermat deserves half the credit, we will adhere to this standard terminology. In fact, we will continue to refer to the method as "Descartes' method."

The eminent historian of mathematics Carl Boyer described the mid-seventeenth century as "the time of Fermat and Descartes." Let us return to the analytic geometry that they discovered.

Linear Equations

In this section we will study equations whose graphs are straight lines. Lines are the simplest kind of curves in the plane, yet they have numerous important applications. We will apply Descartes' graphing techniques to them and show how they illustrate the correspondence between equations and curves.

At the end of the last section we plotted five points that were solutions to the equation $y = 3x + 1$. These points are shown in Figure 1.

---

†*The American Mathematical Monthly,* August–September 1982, p. 509.

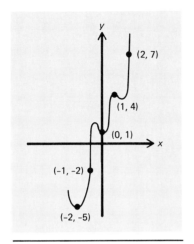

FIGURE 1 Could this be the graph of $y = 3x + 1$?

We asked whether the curve could look like the one in Figure 1. We will be able to answer this question once we know that the graphs of equations of this form are lines. In general, knowing the name of an equation enables you to graph it using a minimal number of points. For instance, if you know that the graph of an equation is a line, then you can draw the graph once two solutions of the equation are known.

The definition of a line is given below. Notice that this geometric object is defined by an algebraic equation.

The graph of an equation is a *line* if and only if it can be written in the form

$$y = mx + b \tag{1}$$

where m and b are real numbers.

The algebraic expression $y = mx + b$ is one form of a *linear equation*. Example 1 will illustrate how to graph this type of equation. Of course, they are called linear equations because their graphs are lines. Later examples will show how to graph lines which are defined by other forms of linear equations.

Example 1

The equation $y = 2x - 3$ is a linear equation with $m = 2$ and $b = -3$. Thus its graph is a line. Since a line is determined by any two points lying on it, we construct a table with two values of x and their corresponding values of y.

x	0	1
y	-3	-1

Thus the two points $(0, -3)$ and $(1, -1)$ lie on the graph. We plot these two points and draw the line through them to get the graph of $y = 2x - 3$. See Figure 2.

In the formula $y = mx + b$ the numbers m and b have special properties that describe the line. The first property, measured by the coefficient m of x, is intuitively described by the "steepness" of the line. In Figure 3 line l_1 is steeper than line l_2. To get an intuitive grasp of this idea, think of the lines as representing hills; then the hill corresponding to l_1 is certainly steeper than that of l_2. Steepness is an idea that is generally understood from an intuitive point of view. In fact, runners, bicyclists, and motorists in mountainous states have a concrete notion of this

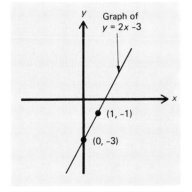

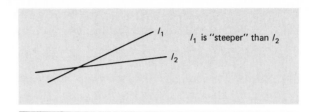

l_1 is "steeper" than l_2

FIGURE 3

fact. But mathematicians are fanatics about quantifying concepts. Thus we are led to the question: How do you measure this concept of steepness? We will illustrate it with the lines in Figure 3. First consider the point of intersection, then proceed in the horizontal direction some distance, and then proceed in the vertical direction until both lines are intersected, as we have done in Figure 4.

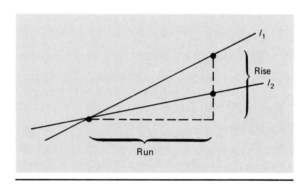

FIGURE 4

It is customary to call the horizontal distance the *run* and the vertical distance the *rise*. For the same run, line l_1 has a larger rise than l_2. This is why it is steeper. Hence the steepness of a line depends upon its rise compared to its run. Let us consider a concrete example of what we mean by rise and run.

Example 2

Consider the equation $y = 2x - 3$ that we graphed in Figure 2. To compare its rise and run we select two points on the curve, say $(1, -1)$ and $(4, 5)$. See Figure 5. The horizontal distance between these points, the run, is $4 - 1 = 3$; and the vertical distance, the rise, is $5 - (-1) = 5 + 1 = 6$. We now consider the ratio of the rise to the run and get

$$\frac{\text{Rise}}{\text{Run}} = \frac{6}{3} = 2$$

This quantity is called the *slope* of the line.

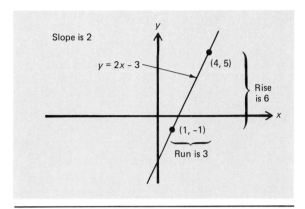

FIGURE 5

We are now ready to define formally the slope of a line. Notice that the rise is the difference in the y coordinates, and the run is the difference in the x coordinates. The rise and run depend upon which points are chosen, but the ratio does not.

DEFINITION: If (x_1, y_1) and (x_2, y_2) are two distinct points on a line, with $x_1 \neq x_2$, then the number m, where

$$m = \frac{y_2 - y_1}{x_2 - x_1}$$

is called the slope of the line.

The next example will demonstrate the meaning of the definition.

Example 3
Consider the linear equation $y = 3x + 1$. To find its slope we choose any two points on the graph, say $(0, 1)$ and $(3, 10)$. Then the slope is

$$m = \frac{10 - 1}{3 - 0} = \frac{9}{3} = 3$$

The value of the slope is independent of the two points chosen. In other words, if we choose any two points on the line $y = 3x + 1$, we will get the same value for the slope. For example, if we select $(-1, -2)$ and $(4, 13)$ on the line $y = 3x + 1$ and compute the slope, we get

$$m = \frac{13 - (-2)}{4 - (-1)} = \frac{13 + 2}{4 + 1} = \frac{15}{5} = 3$$

This agrees with the earlier result.

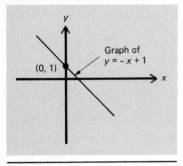

FIGURE 6

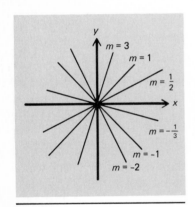

FIGURE 7

Another way of viewing the slope involves a table of solutions to the equation $y = 3x + 1$.

x	-2	-1	0	1	2
y	-5	-2	1	4	7

Notice that each time the value of x increases by 1, the value of y increases by 3.

Recall that we are looking for the interpretation of the number m in the formula for a linear equation, $y = mx + b$. Can you look back to Examples 2 and 3 and guess what it is? Exercise 41 outlines the proof of the following fact:

> The slope of the line $y = mx + b$ is the number m.

Thus far we have considered only positive numbers for the slope of a line. The next example will demonstrate the geometrical interpretation of a negative slope.

Example 4

The linear equation $y = -x + 1$ is a little different from the previous equations we have considered in that the coefficient of x, the slope, is negative. The graph of a linear equation with negative slope is inclined downward to the right. That is, starting at any point on the line, to get to another point on the line you move down 1 unit for every unit moved to the right. See Figure 6.

In Figure 7, a number of different lines are graphed. As you look from left to right, those with positive slope tend upward, or increase, whereas those with negative slope tend downward, or decrease.

The y Intercept

The number b in the linear equation $y = mx + b$ is a little easier to interpret than the slope m. If we let $x = 0$, we get $y = m \cdot 0 + b = b$; so the line passes through the point $(0, b)$. The point $(0, b)$ is on the y axis; so b is called the y *intercept* of the line.

> DEFINITION: For the linear equation $y = mx + b$ the number b is the second coordinate of the point where the line intersects the y axis. For this reason b is called the y intercept.

The next example shows how you can sketch the graph of a line if you know the slope and the y intercept.

Example 5
To graph the equation $y = -2x + 3$ we recognize that the slope is -2 and the y intercept is 3; so the line passes through (0, 3). We use these two pieces of information to sketch the graph in Figure 8. Since the slope is negative, the line slopes downward (from left to right).

Parallel Lines

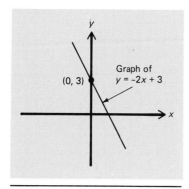

FIGURE 8

Recall that the geometric definition of parallel lines is that they do not meet in either direction when extended indefinitely. This definition is frought with difficulties, as Euclid and his successors discovered. The advent of analytic geometry removed such obstacles, as can be seen in Figure 9. It is obvious that parallel lines have the same steepness. This suggests the following result.

Two lines are parallel if and only if they have the same slope.

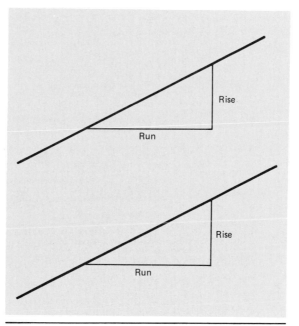

FIGURE 9 These two lines are parallel: they have the same slope.

Example 6
Problem What is the equation of the line that is parallel to the line $y = -2x + 2$ and passes through the origin?

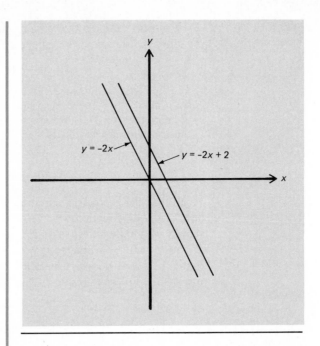

$y = -2x$

$y = -2x + 2$

FIGURE 10   Each line has slope
-2.

Solution   The given line $y = -2x + 2$ is a linear equation with slope $m = -2$ and y intercept $b = 2$. Its graph is sketched in Figure 10. The desired line will have the same slope $m = -2$. Since it passes through $(0, 0)$, its y intercept is $b = 0$. Substituting the values $m = -2$ and $b = 0$ into the linear equation $y = mx + b$ yields $y = -2x$. This is the equation of the line parallel to $y = -2x + 2$ and passing through the origin. The graph is shown in Figure 10.

Two special kinds of lines are those which are parallel to the x axis and those which are parallel to the y axis. The former are called *horizontal lines* and the latter *vertical lines*. Their equations have similar forms, but their slopes are entirely different.

Choose two points on the x axis, say $(x_1, 0)$ and $(x_2, 0)$ with $x_1 \neq x_2$. By definition the slope of the line through these points is

$$m = \frac{0 - 0}{x_2 - x_1} = 0$$

Thus the slope of the x axis is 0. Since parallel lines have the same slope, it follows that *the slope of any horizontal line is zero*.

The situation for vertical lines is quite different. Let $(0, y_1)$ and $(0, y_2)$ be any two points on the y axis. The value of the slope is

$$m = \frac{y_2 - y_1}{0 - 0} = \frac{y_2 - y_1}{0}$$

Since division by 0 is impossible, m is not a number. Thus the slope of the y axis is not defined. It follows that *the slope of a vertical line is not defined.*

Example 7

Problem What are the equations of the horizontal and vertical lines that pass through the point (2, 3)?

Solution Consider the horizontal line first. Its slope is 0. Since it passes through (2, 3) and is parallel to the x axis, its y intercept is 3. Substituting $m = 0$ and $b = 3$ into the equation $y = mx + b$ yields $y = 3$. This means that the horizontal line is located 3 units above the x axis.

The equations of vertical lines cannot be obtained from the linear equation $y = mx + b$ because m is not a number. To obtain the equation of the vertical line through (2, 3), you must realize that the line is located 2 units to the right of the y axis. Some points on it are (2, 2), (2, 1), (2, 0), and (2, −1). The value of x is always 2. Hence the equation of the line is $x = 2$.

Both lines are graphed in Figure 11.

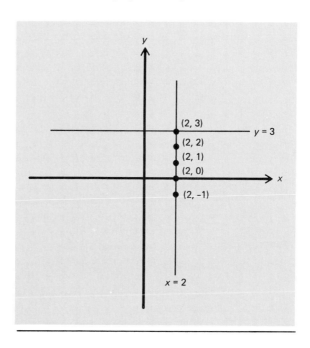

FIGURE 11

Example 7 suggests the following standard forms for all horizontal and vertical lines:

Horizontal lines: y = number

Vertical lines: x = number

Table 1 summarizes the results on the slopes of these lines.

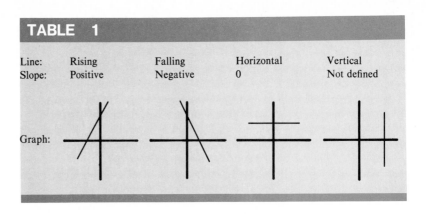

Line:	Rising	Falling	Horizontal	Vertical
Slope:	Positive	Negative	0	Not defined

Alternate Forms of a Line

The equation of a line that is neither vertical nor horizontal can be expressed in several forms. Thus far we have been relying solely upon the form $y = mx + b$. Because m denotes the slope and b the y intercept, this equation is said to be the *slope-intercept form* of a linear equation.

> The slope-intercept form of a linear equation is
>
> $$y = mx + b$$
>
> where m is the slope and b is the y intercept.

Example 8
Problem What is the equation of the line that passes through the point $(0, -1)$ with slope 2? Draw its graph.
Solution We are given $m = 2$. Because the line passes through $(0, -1)$ it follows that $b = -1$. Thus the equation in slope-intercept form is $y = 2x - 1$. Its graph is drawn in Figure 12.

Example 8 illustrates the fact that the equation of a line can be determined if its slope and y intercept are known. But what if its slope and another point are known? For instance, suppose a line has slope 4 and passes through the point $(3, 2)$. Certainly there is only one such line. However, there are various forms of equations that describe it. The most convenient is the aptly named *point-slope form*.

> The point-slope form of a linear equation is
>
> $$y - d = m(x - c)$$
>
> where m is the slope, and (c, d) is the given point on the line.

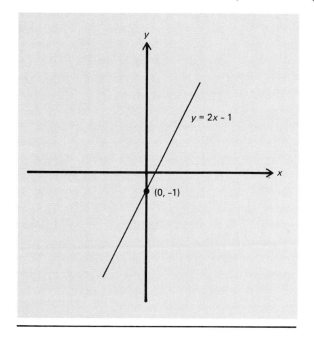

FIGURE 12

Let us look at an example that relates these two forms of a linear equation.

Example 9

Problem Find the point-slope form and the slope-intercept form of the equation of the line that passes through the point (3, 2) and has slope 4.

Solution We have $c = 3$, $d = 2$, and $m = 4$. The equation of the line in point-slope form is

$$y - 2 = 4(x - 3)$$

Now we make use of some properties of algebra, indicated on the right, to express the equation of the line in slope-intercept form. We arrange the terms as follows:

$y - 2 = 4(x - 3)$	point-slope form
$y = 4(x - 3) + 2$	add 2 to each side
$y = 4x - 12 + 2$	multiply to eliminate the parentheses (distributive law)
$y = 4x - 10$	add -12 and 2

Hence the y intercept is -10. The equation $y = 4x - 10$ is the slope-intercept form of the line. See Figure 13 on page 288.

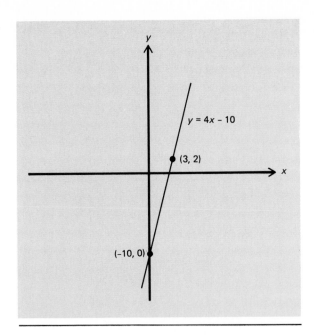

FIGURE 13

Another useful form of a linear equation is the *general form*. We will encounter it in numerous applications in later sections, especially in linear programming problems. It is important to know how to draw its graph and to derive pertinent information about the line.

The general form of a linear equation is

$$Ax + By = C$$

where A, B, and C are real numbers, with A and B not both 0.

The next example illustrates the most effective way to graph a line whose equation is in the general form. The following two examples show how to convert from one form to the others.

Example 10
Problem Draw the graph of the linear equation

$$3x + 2y = 4$$

Solution First find the y intercept by setting $x = 0$ and solving for y. The equation becomes:

$$3x + 2y = 4$$
$$3 \cdot 0 + 2y = 4$$
$$2y = 4$$
$$y = 2$$

Thus the line passes through $(0, 2)$.

Next find the point where the line crosses the x axis, called the x *intercept*, by setting $y = 0$. Then

$$3x + 2y = 4$$
$$3x + 2 \cdot 0 = 4$$
$$3x = 4$$
$$x = \frac{4}{3}$$

Thus the line passes through $(\frac{4}{3}, 0)$.

Then draw the line through the two intercepts. See Figure 14.

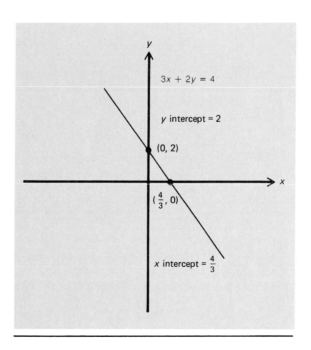

FIGURE 14

Example 11

The equation

$$3x + 2y = 4 \qquad \text{(general form)}$$

is expressed in the general form of a linear equation. To find the slope we can put it in slope-intercept form by solving for y. We get

$$2y = 4 - 3x$$

$$y = -\frac{3}{2}x + 2 \qquad \text{(slope-intercept form)}$$

Hence the slope is $-\frac{3}{2}$ and the y intercept is 2. Another point on the line is $(2, -1)$, and so the point-slope form of the line, using the point $(2, -1)$, is

$$y - (-1) = -\frac{3}{2}(x - 2)$$

$$y + 1 = -\frac{3}{2}(x - 2) \qquad \text{(point-slope form)}$$

This shows the three forms of the one line.

Notice that each line has a unique slope-intercept form of the equation, but since there is an infinite number of points on the line, there is a point-slope form corresponding to each point.

Example 12

Problem   Given the equation of the line in point-slope form,

$$y - 4 = 2(x - 1) \qquad \text{(point-slope form)}$$

express the equation in its (a) slope-intercept form and (b) general form.

Solution   (a) We isolate y on the left-hand side of the equation and get $y = 2(x - 1) + 4$. Then we multiply out the parentheses, which yields $y = 2x - 2 + 4$. Hence the slope-intercept form of the equation of the line is

$$y = 2x + 2 \qquad \text{(slope-intercept form)}$$

(b) The general form is obtained by moving the term $2x$ to the left-hand side of the equation, which yields

$$-2x + y = 2 \qquad \text{(general form)}$$

There are several ways that a line can be specified. That is, we can get the equation and the graph of a line if we are given any one of the following pieces of information:

1   One point and the slope

2   The slope and the y intercept

3   Two points on the line

Try to picture each case separately. Do you see that there can be only one line specified in each case? Example 9 demonstrated the first two. The next example illustrates the third.

> **Example 13**
>
> **Problem**   Find the equation of the line that passes through the points $(1, 2)$ and $(-1, 5)$.
>
> **Solution**   The slope of the line is
>
> $$m = \frac{5 - 2}{-1 - 1} = \frac{3}{-2} = -\frac{3}{2}$$
>
> By choosing $(1, 2)$ as our point (we could choose either point), the point-slope form of the equation is
>
> $$y - 2 = -\frac{3}{2}(x - 1)$$

Summary

Below is a list of the main results to remember from this section.

1   The slope of a line is $m = \dfrac{y_2 - y_1}{x_2 - x_1}$ (unless $x_1 = x_2$).

2   The slope measures steepness. It is the rise divided by the run. As x increases from left to right, y increases if the slope is positive and y decreases if the slope is negative.

3   Horizontal lines have form $y = b$, where b is a number. Their slope is 0. Vertical lines have form $x = a$, where a is a number. Their slope is not defined.

4   There are three principal forms for the equation of a line.

Name of Form	Equation	Variables
Slope-intercept	$y = mx + b$	m = slope b = y intercept
Point-slope	$y - d = m(x - c)$	m = slope (c, d) = point
General	$Ax + By = C$	A, B, C real numbers A, B not both 0

EXERCISES

In Problems 1 to 8 find the slope of the line passing through the two points.

1 $(1, 2), (2, 4)$

2 $(3, 1), (4, 3)$

3 $(5, 1), (2, 0)$

4 $(6, 4), (4, 1)$

5 $(-1, 2), (-3, 1)$

6 $(-3, 8), (3, -8)$

7 $(1.6, 4.1), (2.5, 5.9)$

8 $(\frac{3}{2}, 5), (-1, -\frac{3}{2})$

In Problems 9 to 14 sketch the graph of the line.

9 $y = x + 1$

10 $y = 2x + 1$

11 $y = 4x$

12 $y = -x + 2$

13 $y = -2x + 3$

14 $y = -3x - 2$

In Problems 15 to 20 determine the slope and the y intercept of the line.

15 $y = 2x - 1$

16 $y = 4x + 7$

17 $y = x$

18 $y = -5x$

19 $2y = 2x + 4$

20 $3y = -6x + 9$

In Problems 21 and 22 determine if the two lines l_1 and l_2 are parallel.

21 l_1 passes through $(0, 1)$ and $(1, 2)$; l_2 passes through $(-1, 6)$ and $(6, -1)$.

22 l_1 passes through $(1, 3)$ and $(5, -1)$; l_2 passes through $(1, 1)$ and $(2, 0)$.

In Problems 23 to 26 find the slope of the line.

23 $x = 2$

24 $y = 2$

25 $y = -7$

26 $x = 15$

In Problems 27 to 30 find the equation of the line passing through the given points in (*a*) point-slope form, and (*b*) slope-intercept form.

27 $(2, 1), (3, -6)$

28 $(2, -1), (-3, 6)$

29 $(2, 0.5), (3, 1.9)$

30 $(\frac{3}{2}, -2), (-1, -5)$

In Problems 31 to 34 find the equation of the line in general form.

31 $y = 5x - 1$

32 $y = \frac{3}{2}x + \frac{1}{5}$

33 $y - 2 = 3(x - 1)$

34 $y + 1 = 3(x + 4)$

In Problems 35 to 40 use the information to find the equation of the line in any convenient form. Then draw its graph.

35 $m = 2, b = 7$

36 $m = -6, b = 0$

37 The line has no slope; $b = 5$.

38 $m = -2$; the line passes through $(6, -1)$.

39 $b = -1$ and the line is parallel to $y = 2x + 1$.

40 The line passes through $(1, 1)$ and is parallel to $x - 3y = 4$.

41 Let (x_1, y_1) and (x_2, y_2) be any two points on the line $y = mx + b$. Then $y_1 = mx_1 + b$ and $y_2 = mx_2 + b$. Substitute these values of y_1 and y_2 into the formula $(y_2 - y_1)/(x_2 - x_1)$. Reduce the formula algebraically and determine what it is equal to.

42 Example 13 derived an equation of a line through the point $(1, 2)$.
(*a*) Determine the point-slope form, using the other point, $(-1, 5)$.
(*b*) Convert the point-slope form of both equations to slope-intercept form, and show that they represent the same line.

In Problems 43 and 44, determine which line is steeper, l_1 or l_2.

43

44

In Problems 45 to 50, determine if the slope is positive, negative, 0, or not defined.

45

48

46

49

47

50

In Problems 51 to 54 two intercepts of a line are given. Determine the equation of the line through these intercepts.

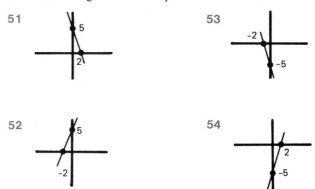

5-3 SYSTEMS OF LINEAR EQUATIONS

The history of systems of linear equations had a rather unusual beginning—it started on the back of a turtle. According to ancient Chinese tradition a turtle carried a special square from the river Lo to a man. The square contained these numbers.

4	9	2
3	5	7
8	1	6

Such a square is called a *magic square* because the three numbers in every row, column, and diagonal add up to 15.

The Chinese were especially fond of patterns, so it is not surprising that they would be intrigued by magic squares. About 250 B.C. a book called "Nine Chapters on the Mathematical Art" devoted one entire section to constructing them. It also asked for the solution to a set of three equations, each of which had three unknowns. This marked the first time in history that a system of linear equations was ever encountered, though Babylonian mathematicians had met related systems implicitly as early as 1800 B.C.

Chinese mathematicians continued to develop and refine techniques for solving systems of linear equations. The peak of this development occurred in A.D. 1303 with the publication of a mathematics book having the unlikely title "Precious Mirror of the Four Elements." It described a method for solving systems of four equations in four unknowns. The four unknowns were not given such nondescript names as *w, x, y,* and *z*; instead they were called heaven, earth, man, and matter.

The results of these Chinese advances remained unknown in the west; so they had to be discovered independently. During the early part of the nineteenth century the great German mathematician Karl Gauss introduced an effective method for reducing such systems by eliminating the variables, one at a time. Later in the century the French mathematician Camille Jordan modified Gauss' procedure to produce the method of solution employed today. It is called *Gauss-Jordan elimination* in honor of the two men who discovered and refined it.

The Carpenter's Problem

In this section we will show how to solve problems involving two unknowns and two pieces of information. The methods that we discuss apply to problems with more than two unknowns and pieces of information, but we will have no need for these extensions. Let us start with a practical problem.

> CARPENTER'S PROBLEM: A carpenter has a 2 × 4 which is 10 feet long and wants to cut it in such a way that one piece is 2 feet longer than the other. Where should it be cut?

Figure 1 shows the 2 × 4 schematically. This problem is typical of the problems we will encounter in this section, in that it involves two unknown quantities and two pieces of information. In solving it we will meet the terminology and procedure used in handling such problems.

The first thing to do when encountering a word problem is to reread it. You must identify the unknown quantities and the given pieces of information. The unknowns in the Carpenter's Problem are the lengths

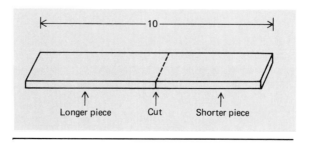

FIGURE 1

of the two pieces of the board. We will introduce variables to stand for each. Let

x = length of the longer board

y = length of the shorter board

The next step is to express the given pieces of information in terms of the variables. One given fact is that the original length of the board is 10 feet. This translates into the equation

$x + y = 10$

The second bit of information is that the longer piece is 2 feet longer than the shorter piece. There are several ways of writing this fact. We will use the form that corresponds to "the difference between the lengths of the two boards is 2," which becomes

$x - y = 2$

This procedure converts the Carpenter's Problem from a word problem to an algebra problem. Altogether there are two equations:

$x + y = 10$

$x - y = 2$

Taken together they comprise what is called a *system of two equations in two unknowns*. The object is to find values for x and y that satisfy both equations at the same time. We will discuss various methods of solution.

Solution by Inspection

The Carpenter's Problem is easy to solve by inspection. In other words, you can find ordered pairs of numbers (x, y) that satisfy the first equation and then check to see if they satisfy the second. For instance, $(1, 9)$ is a solution of the first equation because when we substitute $x = 1$ and $y = 9$ into it we get a true statement, namely, $1 + 9 = 10$. However $(1, 9)$ is not a solution of the second equation since $1 - 9 \neq 2$. Since $(1, 9)$ is not a solution to the second equation it is not a solution of the system.

Is $(8, 2)$ a solution of the system? It is a solution of the first equation but not of the second. Hence it is not a solution of the system. Can you find the solution? If you try a few ordered pairs it should be clear after a short time that $(6, 4)$ is a solution of the system because it is a solution of both equations, since $6 + 4 = 10$ and $6 - 4 = 2$.

The method of inspection worked only because this problem was easy. Generally this method does not yield the solution, but it should be tried first nonetheless. We introduced it here to indicate what a solution of the system looks like.

Solution by Graphing

Another method of solution is by the means of graphing the two equations in the same coordinate system. The ordered pairs that satisfy the equations are the points on the graphs of the equations. Since the graph of each equation is a straight line, the solution of the system is the point that is on both lines, that is, the point of intersection of the two lines.

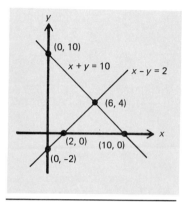

FIGURE 2

Example 1

Problem   Solve the Carpenter's Problem by graphing.
Solution   We graph the two lines $x + y = 10$ and $x - y = 2$ in Figure 2 and then locate the point of intersection. As we saw before, the solution is the point (6, 4).

Let us try our hand at another example to see if you understand these ideas.

Example 2

Problem   Given the system of two equations in two unknowns:

$$x + 2y = 5$$

$$x - y = 2$$

solve the system first by inspection by considering the ordered pairs (1, 2), (5, 0), and (3, 1), and then solve it by graphing.
Solution   The ordered pair (1, 2) is a solution of the first equation because $1 + 2 \cdot 2 = 5$ but not of the second because $1 - 2 \neq 2$. Similarly, (5, 0) is a solution of the first equation but not the second; neither is a solution of the system. The point (3, 1) is a solution of the system because $3 + 2 \cdot 1 = 5$ and $3 - 1 = 2$. The graphical solution is presented in Figure 3.

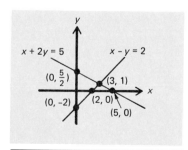

FIGURE 3

In Example 2 we said that (3, 1) was *a* solution rather than *the* solution. In other words, it may be the case that a system has more than one solution. Since the graph of each equation in a system is a straight line, we can interpret the types of solutions by considering the possibilities of the graphs. There are three cases for the straight lines: (1) they intersect in exactly one point, and so there is one unique solution (as was the case in Examples 1 and 2); (2) they are parallel, in which case they do not intersect, and so there is no solution; or (3) they coincide—in other words, the equations have the same graph—in which case all the points on the line are solutions, and so there is an infinite number of solutions. Let us look at an example of the latter two cases.

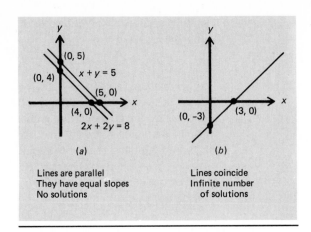

FIGURE 4

Example 3

Problem   Solve each of the following systems by graphing:

$$(a) \quad x + y = 5 \qquad\qquad (b) \quad x - y = 3$$
$$ \quad 2x + 2y = 8 \qquad\qquad \quad 3x - 3y = 9$$

Solution   The graphs of each system are shown in Figure 4. In system (a) the lines are parallel; so there is no point of intersection and hence no solution. In system (b) the equations have the same graph; so all of the points on the line are solutions. In particular, $(3, 0)$ and $(0, -3)$ are solutions.

The graphical method of solution yields a valuable picture of the problem, but it has a drawback that often produces a fatal flaw. Graphical methods only give rough approximations to the solutions of problems. It may be the case, as in the problems considered so far, that the solution can be guessed from the graph and then checked by substituting the ordered pair into the equations to determine if in fact it is a solution. But if the solution involves a decimal, say $(5.1, 3.3)$, the graphical method will only approximate the solution, and it may take a long time to sort through the points close to the one on the graph before the correct one is found. The next two methods, elimination and substitution, will produce correct solutions.

Solution by Elimination

The method of elimination is the procedure that is most often used to solve large systems of linear equations. Most computer programs that solve such systems follow this procedure. Besides, it is relatively easy to apply elimination by hand to small systems like the ones we will encounter here.

The object of elimination is to get rid of one of the variables from one of the two equations. The other variable is then determined from the reduced equation. Let us look at an example of the method.

Example 4

To solve the system

$$2x + 3y = 4$$

$$3x - 2y = 6$$

by elimination we choose one of the variables and eliminate it from one of the equations. After some practice it will become clear that sometimes it is easier to eliminate one variable than the other. In this case it does not matter; so we arbitrarily choose to eliminate y. We multiply the first equation by 2 and the second by 3 and obtain

$$4x + 6y = \ \ 8$$

$$9x - 6y = 18$$

Notice that the coefficients of y are negatives of each other. To eliminate y we add the two equations and get the reduced equation

$$13x = 26$$

We solve this for x and get

$$x = 2$$

Now substitute this value into either one of the original equations to solve for y. We will choose the first equation.

$$2x + 3y = 4$$

$$2 \cdot 2 + 3y = 4$$

$$4 + 3y = 4$$

$$3y = 0$$

$$y = 0$$

Thus (2, 0) is the solution to the system.

It is always advisable to check your answers. This is accomplished by substituting $x = 2$ and $y = 0$ into both of the given equations.

$$2 \cdot 2 + 3 \cdot 0 = 4$$

$$3 \cdot 2 - 2 \cdot 0 = 6$$

In Example 4, the variable y was eliminated by producing the numbers 6 and -6 as coefficients of y, and then adding the two equations. This two-step procedure is the essence of the method of elimination. The next display describes it in general.

Two steps in eliminating a variable v:

1   Multiply the equations by appropriate numbers to produce coefficients of v such that one is the negative of the other.

2   Add the two resulting equations.

The next example further illustrates this procedure.

Example 5
Problem   Use the method of elimination to solve the system

$$3x - 4y = 1$$
$$4x - 2y = 3$$

Solution   We choose to eliminate y because it can be done with only one multiplication. The object is to produce $+4$ as the coefficient of y in the second equation. This is achieved by multiplying the second equation by -2, which leads to this system.

$$3x - 4y = 1$$
$$-8x + 4y = -6$$

Adding these equations yields

$$-5x = -5$$
$$x = 1$$

Now substitute $x = 1$ into the first equation of the given system.

$$3x - 4y = 1$$
$$3 \cdot 1 - 4y = 1$$
$$-4y = -2$$
$$y = \frac{1}{2}$$

Hence the solution is $(1, \frac{1}{2})$.
   It checks since

$$3 \cdot 1 - 4 \cdot \frac{1}{2} = 3 - 2 = 1$$

$$4 \cdot 1 - 2 \cdot \frac{1}{2} = 4 - 1 = 3$$

Other Methods

There are two other methods that are useful for solving systems of two equations in two variables. If the coefficients of one variable are equal, then we solve both equations for that variable and set them equal. This produces one equation in terms of the other variable. The next example will illustrate this method.

Example 6

In order to solve the system

$$x + 2y = 2$$
$$x - y = 5$$

we solve both equations for x. This yields

$$x = -2y + 2$$
$$x = y + 5$$

Since we are searching for a value of x (and y) that satisfies both equations, we set the expressions equal to one another and then solve the resulting equation for y.

$$x = x$$
$$-2y + 2 = y + 5$$
$$-2y - y = 5 - 2$$
$$-3y = 3$$
$$y = -1$$

Part of the solution is $y = -1$. We can now substitute $y = -1$ into either of the original equations to find x. We choose the first equation.

$$x + 2(-1) = 2$$
$$x - 2 = 2$$
$$x = 2 + 2 = 4$$

The check is

$$4 + 2(-1) = 4 - 2 = 2$$
$$4 - (-1) = 4 + 1 = 5$$

You might have been able to solve the system in Example 6 by inspection or graphing. The method that we used is a variation of another useful method called *substitution*. This method involves solving one of the equations for one variable and then substituting this value into the other equation. Once again you obtain a reduced equation with only one variable. Example 7 will present a system that you probably cannot solve by inspection or graphing. If you think you can, try it.

Example 7

Problem   Use the method of substitution to solve the system

$$2x + 4y = 1$$
$$x - 5y = -2$$

Solution   It is easy to solve the second equation for x:

$$x = 5y - 2$$

Now substitute this expression into the first equation.

$$2x + 4y = 1$$
$$2(5y - 2) + 4y = 1$$
$$10y - 4 + 4y = 1$$
$$14y = 5$$
$$y = \frac{5}{14}$$

We determine x by substituting $y = \frac{5}{14}$ into the expression for x:

$$x = 5y - 2 = 5\left(\frac{5}{14}\right) - 2 = \frac{25}{14} - \frac{28}{14} = \frac{-3}{14}$$

Hence the solution is $(\frac{-3}{14}, \frac{5}{14})$.

Notice that it would have been virtually impossible to have obtained this solution by inspection or graphing. The check is

$$2\left(\frac{-3}{14}\right) + 4\left(\frac{5}{14}\right) = \frac{-6}{14} + \frac{20}{14} = \frac{14}{14} = 1$$

$$\frac{-3}{14} - 5\left(\frac{5}{14}\right) = \frac{-3}{14} - \frac{25}{14} = \frac{-28}{14} = -2$$

Systems without a Unique Solution

In all the systems that we have encountered so far there has been exactly one solution. This does not always have to be the case, as can be seen by examining the geometric meaning of a system of equations:

$$A_1x + B_1y = C_1$$
$$A_2x + B_2y = C_2$$

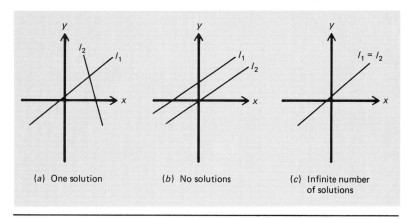

(a) One solution   (b) No solutions   (c) Infinite number of solutions

FIGURE 5

Each equation is the general form of a line, so the system consists of two lines, l_1 and l_2. Figure 5 shows that a system can have one solution, no solutions, or an infinite number of solutions.

A system with no solutions is called *inconsistent*. Figure 5(b) shows that a system of two linear equations in two unknowns that has no solution has two parallel lines for its graph. How do you recognize that a system has no solution in the other methods of solution? Let us consider an example of such a system.

> **Example 8**
> To solve the system
>
> $$x - y = 1$$
> $$2x - 2y = 3$$
>
> by elimination, we multiply the first equation by -2 and add it to the second equation, which yields
>
> $$\begin{aligned} -2x + 2y &= -2 \\ 2x - 2y &= 3 \\ \hline 0 &= 1 \end{aligned}$$
>
> This is a contradiction. In other words, there is no ordered pair (x, y) that satisfies $0 = 1$; so the system has no solution.

Therefore, if in the process of substitution or elimination a contradiction is produced, then the system is inconsistent.

If a system has an infinite number of solutions, then it is said to be *dependent*. Figure 5(c) shows that this occurs when the two equations represent the same line. To see how you can conclude that a system is dependent, consider an example.

Example 9

To solve the system

$$x - y = 1$$
$$2x - 2y = 2$$

by elimination we multiply the first equation by -2 and add it to the second equation, which yields

$$-2x + 2y = -2$$
$$\underline{2x - 2y = 2}$$
$$0 = 0$$

This is true for *all* ordered pairs (x, y) that satisfy the first equation. Hence the second equation has the same solution set as the first.

Therefore, whenever $0 = 0$ is obtained in the elimination process there is an infinite number of solutions.

Summary

In this section we presented various methods of solution of systems of linear equations in two unknowns, including inspection, graphing, substitution, and elimination. The first two methods are instructive, but they are not the best means to solve the problems. The method of inspection works only if the problem is very simple. The method of graphing can often yield only approximate answers; it can produce a good guess but it cannot distinguish between 5 and 5.1 or 4.9. Substitution is preferred to the first two methods, but the method of elimination is the one with the widest applicability. Although it may seem cumbersome, it is the method that is used to solve more complex problems.

EXERCISES

In Problems 1 to 4 solve the system by inspection.

1   $x + y = 2$
    $x - y = 2$

2   $x + y = 5$
    $x + 2y = 6$

3   $x - y = 4$
    $x + y = 4$

4   $x + 2y = 1$
    $x - y = -3$

In Problems 5 to 8 solve the system by graphing.

5   $x + y = 6$
    $x - y = 2$

6   $2x + y = 6$
    $x - 2y = 3$

7   $2x + y = 4$
    $x - y = 2$

8   $x + 3y = 1$
    $2x - y = 9$

In Problems 9 to 14 solve the system by elimination.

9   $2x + y = 3$
     $3x + 4y = 7$

12 $8x - 3y = 1$
     $3x + 4y = 9$

10 $2x + 5y = 2$
     $3x - 2y = 3$

13 $2x - 3y = 1$
     $3x - 2y = 3$

11 $5x + 2y = 1$
     $2x - 3y = 8$

14 $-4x + 2y = -3$
     $3x - 2y = 3$

In Problems 15 to 20 solve the system by substitution.

15 $x + y = -2$
     $x - y = 4$

18 $x + 3y = 4$
     $2x - 6y = 1$

16 $2x - 3y = 2$
     $3x - y = 3$

19 $2x - y = 3$
     $4x + y = 1$

17 $2x - 4y = 1$
     $x - 2y = 2$

20 $-x + 2y = 3$
     $2x + 6y = 1$

In Problems 21 to 24 use any method to solve the system.

21 $y = 8.5x$
     $x + 1.5y = 0$

23 $25x - 50y = 1500$
     $30x - 15y = 2700$

22 $15x - 110y = 4000$
     $10x + 55y = 1000$

24 $1.2x - 3.5y = 23$
     $8.5x + 7y = 1.5$

In Problems 25 to 30 determine whether the system is inconsistent or dependent.

25 $3x - y = 1$
     $6x - 2y = 5$

28 $12x - 6y = 3$
     $8x - 4y = 2$

26 $8y + y = 2$
     $8x + y = 3$

29 $15x - 5y = 20$
     $-12x + 4y = -16$

27 $2x - 4y = 2$
     $3x - 6y = 3$

30 $1.2x - 3.6y = 4.8$
     $1.8x - 5.4y = 6.2$

5-4 LINEAR INEQUALITIES

Thomas Harriot (1560–1621) is one of several mathematicians who are regarded as having bridged the gap between the Renaissance and the modern world. His most enduring contribution, however, came about as the result of a very nonmathematical journey. In 1585 he was sent as a surveyor on an expedition to North America with Sir Walter Raleigh, and upon his return the following year he published "A Briefe and True Report of the New-Found Land of Virginia" (it is now North Carolina). His "Report" described the condition of the land and the people, and still makes interesting reading today. It is accompanied by numerous drawings, one of which shows a design on the back

FIGURE 1

of an Indian male. Influenced by the natural beauty of the design, he adopted part of it in his mathematical writings, using the sign $<$ to denote "is less than" and $>$ to denote "is greater than." Figure 1 contains this drawing.

The symbols $<$ and $>$ were not accepted immediately by other writers. However, they gradually gained a wide acceptance, and today they form part of the list of standard mathematical symbols.

Brief Review of Elementary Properties

Expressions involving $<$ and $>$ are called inequalities. Thus far we have been concerned with *linear equalities*, whose general form is $Ax + By = C$. In this section we will consider *linear inequalities*, which are expressed in general form in one of the following four types of expressions:

$$Ax + By > C \qquad Ax + By < C$$
$$Ax + By \geq C \qquad Ax + By \leq C$$

These inequalities can be used to solve a wide variety of significant practical problems that we will study in Sections 5-5 and 5-6. We begin our discussion with the elementary properties of inequalities. You have probably studied them before, but a brief review is usually beneficial.

When we defined the cartesian coordinate system we began with the real number line (see Figure 2). Recall that we associated with each point on the line a unique real number, and with each real number a unique point. If a and b are any two real numbers with $a < b$, read "a is less than b," then a lies to the left of b on the real line, assuming that the positive direction of the line is to the right, as is usual. For example, $2 < 3$; so 2 is to the left of 3. Similarly we can write $b > a$, read "b is greater than a," to express the same idea. For example, $-1 < 3$ and $3 > -1$.*
Often we want to combine the inequality signs with an equality sign. Thus $a \leq b$ is read "a is less than or equal to b," while $b \geq a$ is read "b is greater than or equal to a." Some older books used the notation $\leqq$, which certainly corresponds more closely to the meaning, but this gradually evolved into $\leq$. For example, $6 < 8$, $6 \leq 6$, $-1 > -3$, and $-3 \geq -3$.

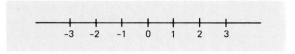

FIGURE 2

We will study inequality expressions involving variables such as x, y, A, and B. For example, an expression of the form $2x + 1 > 5$ involves the single variable x, while $3x + 2y \leq 1$ involves both x and y. We will first consider inequalities with one variable and review the permissible operations used to solve them.

PROPERTY 1:   If $a < b$ and c is any number, then

$$a + c < b + c$$

Let us look at an example that will demonstrate how to use Property 1 to solve an inequality.

Example 1
To solve the inequality

$$x + 3 > 5$$

*If the inequality signs are confusing, remember that the "arrow" points to the smaller number.

we must find all numbers x that satisfy the expression. Property 1 allows us to add -3 to each side of the inequality to get

$$x + 3 - 3 > 5 - 3$$
$$x > 2$$

Hence all the numbers greater than 2 are solutions to the inequality. The graph of this set of numbers on the real line is described in Figure 3 as a solid black line. The open circle at $x = 2$ indicates that 2 is not included in the solution set.

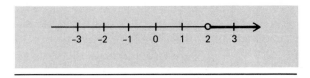

FIGURE 3   Graph of $x > 2$.

In handling inequalities much of the work is the same as dealing with equalities. The major difference is that when you multiply an inequality by a negative number the "sense of the sign" of the inequality reverses; that is, $<$ changes to $>$ and $>$ changes to $<$.

PROPERTY 2:   If $a < b$ and c is positive and d is negative, then

1 $ac < bc$ and $\dfrac{a}{c} < \dfrac{b}{c}$

2 $ad > bd$ and $\dfrac{a}{d} > \dfrac{b}{d}$

Thus you can multiply or divide an inequality by a positive number and keep the sense of the sign the same, but the sense of the sign must be reversed if you multiply or divide by a negative number.

Example 2
Problem   Solve the inequalities.

$$(a)\ 3x - 7 \le -3 \qquad (b)\ 8 - 4x > 2$$

Solution   We proceed, using Properties 1 and 2.

(a) $3x - 7\ \ \ \ \le -3$

$\qquad 3x - 7 + 7 \le -3 + 7$ \hspace{1cm} (Property 1)

$\qquad\qquad 3x \le 4$

$\qquad\qquad\quad x \le \dfrac{4}{3}$ \hspace{1.5cm} (Property 2)

(b) $8 - 4x > 2$

$$8 - 4x - 8 > 2 - 8 \qquad \text{(Property 1)}$$

$$-4x > -6$$

$$x < \frac{-6}{-4} \qquad \text{(Property 2)}$$

$$x < \frac{6}{4}$$

$$x < \frac{3}{2}$$

The graphs of these solutions are given in Figures 4 and 5. The solid circle at $x = \frac{4}{3}$ indicates that it is part of the solution.

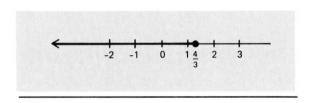

FIGURE 4   Graph of $x \leq \frac{4}{3}$.

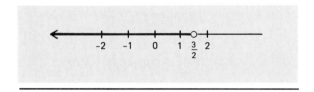

FIGURE 5   Graph of $x < \frac{3}{2}$.

Inequalities in Two Variables

Examples of inequalities in two variables are:

$$2x + 3y \leq 7 \qquad x - 4y > 6 \qquad y \leq 4 - 3x$$

We call such inequalities *linear inequalities*. Recall that a solution to a linear equality is an ordered pair. Let us look at an example that demonstrates that a solution to a linear inequality is an ordered pair.

Example 3

A solution to the inequality $2x + 3y \leq 7$ involves a number substituted for x and a number for y; thus it is an ordered pair. Let us test four such ordered pairs to see if they are solutions: (a) (0, 1), (b) (1, 3), (c) (−3, 2), and (d) (2, 1). We test them in order:

(a) Letting $x = 0$ and $y = 1$ yields $2 \cdot 0 + 3 \cdot 1 = 3 \leq 7$, which is true; so (0, 1) is a solution.

(b) Letting $x = 1$ and $y = 3$ yields $2 \cdot 1 + 3 \cdot 3 = 11 \leq 7$, which is not true; so $(1, 3)$ is not a solution.

(c) Letting $x = -3$ and $y = 2$ yields $2(-3) + 3 \cdot 2 = 0 \leq 7$, which is true; so $(-3, 2)$ is a solution.

(d) Letting $x = 2$ and $y = 1$ yields $2 \cdot 2 + 3 \cdot 1 = 7 \leq 7$, which is true; so $(2, 1)$ is a solution.

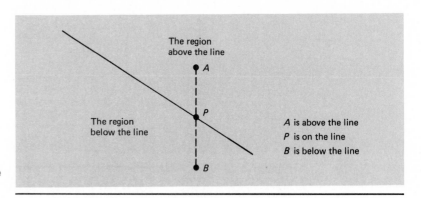

FIGURE 6 A line divides the plane into two regions.

The graphs of linear inequalities can be described as *half planes*. As in Figure 6, a nonvertical straight line divides the plane into two regions such that one region lies above the line and the other region below the line. If the line is vertical, the two regions are described as being to the right and left of the line.

In Figure 6 the point P is on the line. The coordinates of P satisfy the linear equation of the line. You can intuitively see that the point A is above the line and B is below the line. What can we say about the coordinates of A and B relative to the coordinates of P? When we say that A is above the line, we mean that A is above the point on the line that is on the same vertical line as A. In other words, A and P have the same x coordinate, while the y coordinate of A is larger than the y coordinate of P. Similarly, B has a smaller y coordinate than the point P that is on the line and having the same x coordinate. Let us consider an example that describes what we mean.

Example 4

The line $y = 2x - 1$ is graphed in Figure 7. Choose a point on the line, say $(1, 1)$. The coordinates of this point satisfy the equation. Any point with the same first coordinate and a larger second coordinate, say $(1, 3)$, is above the line. Any point with the same first coordinate and a smaller second coordinate, say $(1, -2)$, is below the line.

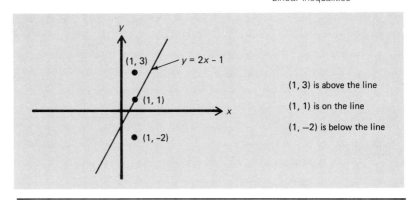

(1, 3) is above the line

(1, 1) is on the line

(1, –2) is below the line

FIGURE 7

Example 4 gives us a hint as to what the solution set of an inequality looks like. Consider the inequality

$$y > 2x - 1$$

What are the points that satisfy the inequality? In Example 4 we noted that (1, 1) satisfies the equation $y = 2x - 1$. All the points immediately above (i.e., vertically above) the point (1, 1) will have a y coordinate that is larger than the y coordinate of (1, 1), namely, 1. Hence they will satisfy $y > 2x - 1$. This is true for all the points above the line. In other words if (a, b) is any point on the line, so that $b = 2a - 1$, then any point with the same x coordinate, a, but with a larger y coordinate, say c, so $c > b$, will satisfy the inequality $y > 2x - 1$ because $c > b = 2a - 1$. Hence all points in the half plane above the line $y = 2x - 1$ satisfy the inequality $y > 2x - 1$. Similarly, all the points in the half plane below the line $y = 2x - 1$ satisfy the inequality $y < 2x - 1$. Altogether there are three sets of points:

1 The points on the line; they satisfy $y = 2x - 1$.

2 The points above the line; they satisfy $y > 2x - 1$.

3 The points below the line; they satisfy $y < 2x - 1$.

See Figure 8 on page 312.
In general, the solution to a linear inequality is determined by the corresponding equality formed by replacing the inequality sign by an equality sign. The solution to the inequality is one of the two half planes into which the line divides the plane. All that is necessary is to test one of the points in the plane not on the line to determine if it is a solution to the inequality. If it is a solution, then the half plane containing the point

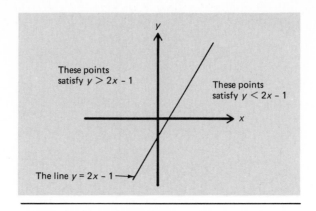

These points satisfy $y > 2x - 1$

These points satisfy $y < 2x - 1$

The line $y = 2x - 1$

FIGURE 8

is the graph of the inequality. If it is not a solution, then the other half plane is the graph.

Let us record this as an explicit procedure for graphing inequalities.

PROCEDURE FOR GRAPHING LINEAR INEQUALITIES

1 Replace the inequality sign by an equals sign and graph the straight line.

2 If the inequality sign is $\leq$ or $\geq$, then the line is also part of the graph. If the inequality sign is $<$ or $>$, then the line is not part of the graph.

3 Choose a test point that is not on the line and substitute the coordinates of the point into the inequality to determine if it satisfies the inequality. [Choose $(0, 0)$ if it does not lie on the line.]

4 If the test point satisfies the inequality, then the half plane containing the point is the graph of the inequality. If the test point does not satisfy the inequality, then the half plane not containing the point is the graph.

We will adopt the following two conventions when graphing inequalities:

1 We will indicate that the line is part of the graph by graphing it as a *solid line*. If the line is not part of the graph, we will graph it as a *broken line*.

2 The graph of the inequality will be the *unshaded* half plane. In other words, we will "throw away" the half plane that is not the graph by shading it.

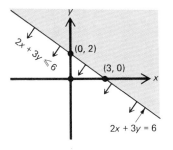

FIGURE 9

The second convention might appear strange to you at first because it may seem more natural to shade the graph itself. The reason we have adopted this particular method will become clear when we solve systems of linear inequalities later.

We will now consider some examples that demonstrate the technique of graphing linear inequalities. It is important to recall the procedure described in Section 5-2, Example 10, for graphing lines in the general form $Ax + By = C$.

Example 5

Problem Graph the linear inequalities.

(a) $2x + 3y \leq 6$ (b) $2x + 3y > 6$

Solution (a) We first replace the inequality sign $\leq$ by $=$ and obtain the linear equation $2x + 3y = 6$. We then graph the straight line, noting that the line passes through the points $(3, 0)$ and $(0, 2)$ as in Figure 9. Since the inequality sign includes the equals sign, we know that the line itself is part of the graph, and we indicate this by graphing it as a solid line. We can now choose any convenient point not on the line as a test point. As in this case, $(0, 0)$ is usually the easiest choice. We test $(0, 0)$ and get $2 \cdot 0 + 3 \cdot 0 \leq 6$, which is true, and thus $(0, 0)$ is a solution. Therefore the half plane containing $(0, 0)$, the region below the line, is the graph of the inequality $2x + 3y \leq 6$. It is shown in Figure 9.

(b) Notice that the only difference between part (a) and part (b) is that a different inequality sign is used. Hence the solution for (b) is very similar to that for (a). It differs in just two aspects. Can you tell what they are? We first graph the line $2x + 3y = 6$, as in (a), but note that the inequality sign for part (b) does not include the equals sign. Thus we graph the line as a broken line to indicate that the points on the line are not part of the graph. See Figure 10. Next we test $(0, 0)$ and get $2 \cdot 0 + 3 \cdot 0 = 0 > 6$, which is not true; so $(0, 0)$ is not a solution. Hence the graph is the half plane not containing $(0, 0)$, which is the region above the line as indicated in Figure 10. Remember that we are shading the region that is not the graph.

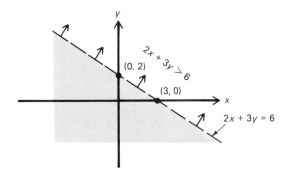

FIGURE 10

WARNING!

In Example 5 the first linear inequality was $\leq$ and the graph was the region below the line. The second inequality was $>$ and the graph was the region above the line. It might occur to you that this will always be the case. It sure would make life easier. Unfortunately it is not always true that the symbol $<$ yields a graph below the line. Sometimes it will be the region above the line. The same is true for $>$.

We will show you the meaning of the warning in the next example.

Example 6

Problem   Graph the inequality

$$2x - 5y \leq 10$$

Solution   We will follow the steps in the procedure.

1   We graph the line $2x - 5y = 10$ in Figure 11.

2   When we graph the line, we use a solid line because the line is part of the graph.

3   We choose (0, 0) as a test point and get $2 \cdot 0 - 5 \cdot 0 = 0 \leq 10$, which is true. Thus (0, 0) is part of the graph.

4   Since (0, 0) is in the half plane above the line, that region is the graph; so we shade the other half plane.

Notice that in Example 6 the inequality has a $\leq$ sign, but the graph is the half plane above the line.

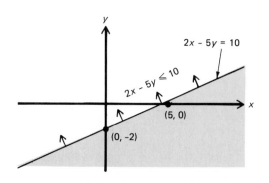

FIGURE 11

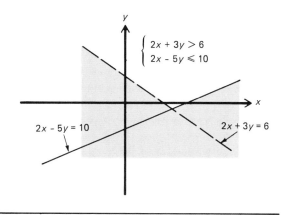

The system shown:
$$2x + 3y > 6$$
$$2x - 5y \leqslant 10$$

with lines labeled $2x - 5y = 10$ and $2x + 3y = 6$.

FIGURE 12

Systems of Linear Inequalities

What would happen if we were to graph both the linear equalities in Examples 5(*b*) and 6 in the same coordinate system? This would amount to solving the system of inequalities.

$$2x + 3y > 6$$
$$2x - 5y \leq 10$$

That is, the points in the resulting graph *simultaneously* satisfy the two linear inequalities. The graph of the system is simply the two individual graphs superimposed on the same coordinate system. The two lines separate the plane into four regions. The one region that remains unshaded is the graph of the system.* See Figure 12.

We will now consider two more examples of graphing systems of linear inequalities.

Example 7
Problem Graph the system of linear equalities.

$$x + y \leq 4$$
$$2x - y \geq -4$$
$$y \geq 0$$

---

*Now you can see why we chose to shade the unwanted region. When solving a system of linear inequalities, the graph is immediately seen from the graphs of the individual inequalities. This would not be the case if we had adopted the convention of shading the graph itself.

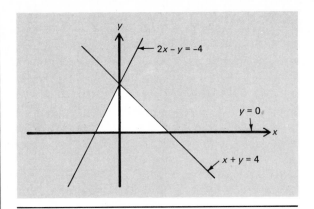

FIGURE 13

Solution We start by graphing each corresponding line: $x + y = 4$, $2x - y = -4$, and $y = 0$ in Figure 13. For the first two inequalities the test point $(0, 0)$ can be used. It is a solution to both: $0 + 0 \le 4$ and $2 \cdot 0 - 0 \ge -4$ are both true. Note that $y = 0$ passes through $(0, 0)$; so we must choose another point, say $(0, 4)$, which is a solution. The graph is given in Figure 13. It is the triangle whose vertices are $(4, 0)$, $(0, 4)$, and $(-2, 0)$.

In Example 7 we introduced the inequality $y \ge 0$. In the next two sections we will encounter this inequality and the inequality $x \ge 0$ a great deal. It will ease your work if you gain a familiarity with these ubiquitous but simple inequalities. Their corresponding lines are the x and y axes, respectively, so $y \ge 0$ is the half plane above the x axis and $x \ge 0$ is the half plane to the right of the y axis.

Consider one more example before you test your skill with the exercises. In the next example we will ask you to recall the technique for finding points of intersections of lines to help you understand the graph.

Example 8
Problem Graph the system of linear inequalities

$$2x + y \le 8$$
$$x + 2y \le 10$$
$$x \ge 0$$
$$y \ge 0$$

Solution We start by graphing the lines $2x + y = 8$, $x + 2y = 10$, $x = 0$, and $y = 0$. From the preceding two paragraphs we know that the last two inequalities give us the region above the x axis and to the right of the y axis. We can use the test point $(0, 0)$ on the other two inequalities and get: $2 \cdot 0 + 0 \le 8$ and $0 + 2 \cdot 0 \le 10$, which are both true; so $(0, 0)$ is in the graph of both inequalities. By shading those regions that are not the graphs of the individual inequalities, we get the region inside the quadrilateral depicted in Figure 14. Three of the vertices are easy to compute. They are

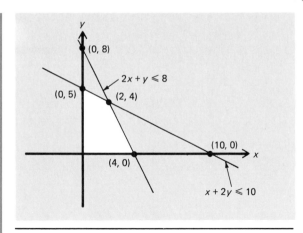

FIGURE 14

(0, 0), (4, 0), and (0, 5). The fourth vertex of the quadrilateral is the inter-section of the two lines $2x + y = 8$ and $x + 2y = 10$. We can determine the coordinates of the point of intersection by solving the system of equations simultaneously. We first multiply the equation $x + 2y = 10$ by -2 and then couple it with $2x + y = 8$, which yields

$$-2x - 4y = -20$$
$$2x + y = 8$$

Now adding the equations yields $-3y = -12$; so $y = 4$. Substituting $y = 4$ into the equation $2x + y = 8$ yields $2x + 4 = 8$. Then $2x = 4$; so $x = 2$. Hence the fourth vertex is (2, 4).

EXERCISES

In Problems 1 to 4 solve the inequality for x and sketch the graph of the solutions on the real line.

1 $2x < 3$ **3** $3 - x \geq 4$

2 $x + 5 \leq 1$ **4** $8 - 2x < 3$

In Problems 5 to 8 determine whether the given points satisfy the linear inequality.

5 $x + y > 4$ (1, 2), (2, 2)

6 $2x - y < 0$ (3, −1), (−2, 0)

7 $-x + 7y \geq 8$ (−1, 1), (0, 0)

8 $3x - y \leq 3$ (1, 0), (0, 1)

In Problems 9 to 12 determine whether the given points lie above, below, or on the line.

9 $x + y = 4$ $(0, 0), (1, 7)$

10 $2x - y = 0$ $(0, 0), (2, -1)$

11 $x - 3y = 1$ $(1, -1), (-1, -1)$

12 $2x + 3y = 6$ $(3, 0), (0, 0)$

In Problems 13 to 22 sketch the graph of the linear inequality by shading those points that do not satisfy the inequality.

13 $x + y > 2$ 18 $-x - 3y \le 6$

14 $x + y < -3$ 19 $4x + 3y \ge 12$

15 $2x + 3y \le 12$ 20 $4x - 3y \ge 12$

16 $2x + 3y > 12$ 21 $y \le 0$

17 $x + 3y > -6$ 22 $x \le 0$

In Problems 23 to 30 graph the system of linear inequalities.

23 $x + 3y \le 6$ 28 $2x + 7y \le 14$
    $x - 2y > 4$ $x \ge 0$
                                                           $y \ge 0$
24 $x + 2y < 4$
    $x - 3y \le 6$ 29 $x + y \le 8$
                                                   $2x + y \le 14$
25 $2x + y \ge 4$ $x \ge 0$
    $x - y < 5$ $y \ge 0$

26 $x + y < -2$ 30 $x + y \le 4$
    $2x - 5y \ge 10$ $x + 2y \le 6$
                                                           $x \ge 0$
27 $x + y \le 8$ $y \ge 0$
    $x \ge 0$
    $y \ge 0$

5-5 LINEAR PROGRAMMING PROBLEMS

The year 1947 was an important year for mathematics and many of its applied fields. Many complex problems that had cropped up during World War II had not yet been solved. While many of them were in the military sector, an increasing number pertained to the business and industrial world.

For example, one of the most important problems confronting the (British-American) Combined Shipping Adjustment Board involved the flow of cargo between various ports throughout the world. Of course, the efficient transportation of supplies and forces is critical during a war, but during peacetime similar problems affect many industries such as the iron and steel, petroleum, automotive, and food industries.

The variables that affected the solution to the transportation problem were many and varied: ship capacity, cargo security and perishability, ship availability, loading and discharging, sailing speed, availability for cargo on the return leg of the voyage, and prices, including rate of exchange. It was not uncommon to encounter a problem with 40 or more unknowns. Equations and inequalities alike were inherent in the problems.

One of the key steps in the solution of these problems came when both the military and the economic sectors saw a common thread running through their individual problems. Each problem contained an item that had to be either minimized or maximized, while other variables produced constraints or restrictions on how far the key variable could be optimized. For example, the Combined Shipping Adjustment Board wanted to minimize the amount of shipping used subject to a given transportation program. The restrictions on the program included ship capacity and the cost of escort ships for protection, as well as a host of other constraints.

This example is representative of the complexity of military and business affairs in the late 1940s. To understand how this complexity developed consider that before 1860 the commander of any armed forces could alone plan all military operations. This state of affairs ended in 1860 with the introduction of the general staff of the Prussian Army. During World War I military activities were subdivided among staff agencies. World War II saw a gigantic increase in the development of staff planning, as well as the introduction of a civilian component, the War Production Board. There was a clear need for an effective way to solve the complicated problems that arose. The Air Force, being the newest of the military branches and hence unfettered by tradition, was the first organization to seek scientific solutions by supporting the research of mathematical scientists. However, the war ended without solutions to most of the mathematical problems.

Many of the civilians on the War Production Board and similar agencies were economists. They recognized that their field also contained numerous problems in which a quantity was to be optimized subject to certain constraints. Although the New Deal sought to speed economic recovery during the 1930s, the classical economic models were not capable of producing viable solutions. The most important economic development at this time was the introduction of a quantitative model of the American economy by Harvard professor Wassily Leontief; yet the economists were no closer to solutions to these complex problems than the Air Force by the end of World War II.

The search ended in 1947, when George Dantzig of Stanford University invented the simplex method for solving such problems. Suddenly the study of linear inequalities took on more importance; a new field of geometry was born. It was initially called "programming in a linear structure," but by 1951 the term "linear programming" prevailed. It unified the seemingly different problems of the military and the economy under one mathematical framework that was capable of real solutions because of the advent of computers. So powerful was Dantzig's invention that within 2 years a conference was held to discuss linear programming solutions to problems in such areas as crop rotation, routing of ships between harbors, interindustry flow of commodities, and scheduling.

The first theoretical problem solved by Dantzig's method was the transportation problem. The applications of this solution were many and varied.

For example, the Department of Justice had been suspicious of the pricing policy of large steel companies, called the *basing-point* policy. It consisted of two separate parts—the manufacturing cost of the steel determined at the factory and a transportation cost. The latter was not the actual cost of loading, shipping, and unloading, but it was a 50 percent surcharge added to the manufacturing cost. The suspicion was that the surcharge was much greater than the actual shipping cost—so much so that it helped create a monopoly for the large companies. The company line was that the actual transportation cost was too complex to compute. Dantzig's simplex method silenced that alibi when his transportation model was applied to the problem. The result was an overhaul of the pricing policies in the industry.

The methods of solving linear programming problems were refined over the next 30 years, and the range of applications was greatly increased. Two economists, L. V. Kantorovich and T. C. Koopmans, were awarded the 1975 Nobel Prize in economics for their pioneering work in the field.

There is a double irony involved in Kantorovich's selection. In 1939 he wrote an extensive monograph on the subject, but it remained unread not only in the United States but in the Soviet Union as well. Dantzig's work was carried out independently of Kantorovich's. In an interview with *The New York Times* 20 years later Kantorovich said that Soviet economists lagged behind their American counterparts because of their fear of mathematics. Apparently that is no longer true, because a drastic improvement upon Dantzig's simplex method occurred with the Russian L. G. Khachian's ellipsoidal method in 1979 (see Section 2-6).

Two Specific Problems

In this section we will present four examples of typical linear programming problems. They are analogous to the ones which originated Dantzig's research in this area. The aim is to provide you with a method for translating a given problem into mathematical terms. In the next section we will show how to solve the problems once the translation has been carried out.

Let us consider two problems typical of the type of linear programming problems that we will study. After we pose these problems, we will show you how to formulate them mathematically. The first is a type of problem that is frequently encountered in the business world.

Example 1 A Manufacturing Problem
A clothing manufacturer makes two types of suits—a 100 percent worsted wool and a blend of 50 percent wool and 50 percent polyester. The worsted suit sells to the retail stores for $160 each and the blend for $90 each. It

costs $100 to make each worsted suit and $40 to make each blend suit. Thus a profit of $60 (= 160 − 100) is made on each worsted suit and $50 (= 90 − 40) on each blend suit. The manufacturer wants to determine how many suits of each type should be made per week in order to maximize profit. Because the company produces other types of clothing as well, the manufacturer is limited as to the number of suits that can be made. It has been determined that no more than 120 suits per week can be manufactured, and no more than $6000 per week can be budgeted to make them.

The second problem is taken from the health field. Similar problems confront all of us in our daily lives, though we usually either ignore them or simply guess their answers. Dantzig referred to such problems as *blend problems*.

It is not only health fanatics who contend that the state of our nutrition is poor and is swiftly declining. Leading nutritionists such as Jean Mayer of Harvard, among others, say that we are eating less and yet getting fatter. The apparent contradiction stems from our increased intake of sugar and fats, replacing more nutritious foods. Dr. Ernest L. Wynder, president of the American Health Foundation, has said that high-fat diets are linked to cancer deaths. A common problem for nutritionists (and, perhaps, for all of us) is to minimize fat content in our diets. Let us consider a concrete illustration.

Example 2   A Diet Problem

A hospital nutritionist is preparing a menu for patients on low-fat diets. The primary ingredients are chicken and corn. The meal must contain enough protein and iron to at least meet the U.S. recommended daily allowance (U.S. RDA) for each nutrient. The U.S. RDA for protein is 45 grams and, for iron, 20 milligrams. Each 3-ounce serving of chicken contains 45 grams of protein, 10 milligrams of iron, and 4 grams of fat. Each 1-cup serving of corn contains 9 grams of protein, 6 milligrams of iron, and 2 grams of fat. The problem is to determine how much chicken and corn should be included in the meal in order to minimize the amount of fat subject to the restriction that there is enough protein and iron to at least meet the U.S. RDA for each nutrient.

Each example is a good representative of a typical linear programming problem. Such problems emerge from a real-world setting, and thus the data are not always presented in an easily discernible way.* Once the problem has been articulated, as we have done above, we must express it in mathematical language. There are certain steps that we can follow that will lead us to the formulation.

---

*Students have been overheard referring to them as the dreaded

WORD PROBLEMS

1 We must identify the variable quantities, especially the quantity that is to be maximized or minimized. Assign a letter to each quantity, being sure to be careful about the appropriate units that are used.

2 Determine the equation that governs the quantity to be maximized or minimized. This equation is called the *objective function.*

3 The restrictions are translated into linear inequalities. You must be careful to ensure that each inequality deals with the same units and each term of the inequality has the same units. These inequalities are called the *constraints* of the problem.

We begin by reading the problem once or twice and organizing the data. Often it is most convenient to organize the data in a chart. Once the data are put in tabular form we can identify the various unknowns. Let us demonstrate the technique by reconsidering the manufacturing problem.

Example 1   (Revisited)

Mathematizing the manufacturing problem   There are two types of suits: worsted and blend. The objective function entails maximizing profit, so the first row of our chart consists of profit. The units in the first row are dollars. The two variables are seen to be:

x = number of worsted suits manufactured per week
y = number of blend suits manufactured per week

If we let P = profit, then the first row of Table 1 is interpreted as the equation

$$P = 60x + 50y$$

TABLE   1

	Worsted Suits	Blend Suits	Constraint
Profit, dollars	60	50	
Number manufactured	x	y	120
Cost, dollars	100	40	6000

Notice that each term represents dollars because the 60 represents dollars of profit per suit and x is the number of worsted suits; so $60 \cdot x$ is

$ profit of worsted suits = ($ profit per worsted suit)
$\times$ (number of worsted suits)

Similarly, $50 \cdot y$ is interpreted as

$ profit of blend suits = ($ profit per blend suit)
  × (number of blend suits)

Therefore, $P = 60x + 50y$ is the total profit from both types of suits. The second row of Table 1 represents the constraint that at most 120 suits can be made. This statement translates into the linear inequality

$$x + y \leq 120$$

since $x + y$ is the combined number of suits to be made. The bottom row of Table 1 represents another constraint, given by

$$100x + 40y \leq 6000$$

There are two additional constraints that are implicitly included in the problem. It makes no sense to allow x or y to take on negative values, and thus we must include the two inequalities $x \geq 0$ and $y \geq 0$ in the list of constraints. Most linear programming problems include constraints like these that are easy to overlook because they are not explicitly mentioned in the description of the problem.

The problem can be stated in mathematical terms by the following:

Maximize	$P = 60x + 50y$	Objective function
Subject to	$x + y \leq 120$	
	$100x + 40y \leq 6000$	Constraints
	$x \geq 0$	
	$y \geq 0$	

We have separated the objective function from the constraints because they play different roles in the solutions to linear programming problems.

Example 2 (Revisited)

Mathematization of the diet problem   We begin by arranging the data in tabular form in Table 2. Note that there are two types of food, chicken and corn. The data are given in units per serving, grams per serving for protein and fat, and milligrams per serving for iron. A serving of chicken is 3 ounces, and for corn it is 1 cup. (The U.S. RDA is not applicable to fat, of course.)

TABLE 2			
	Chicken per 3-Ounce Serving	**Corn per 1-Cup Serving**	**U.S. RDA**
Protein, grams	45	9	45
Iron, milligrams	10	6	20
Fat, grains	4	2	

The objective function entails minimizing the amount of fat in the meal, so the third row in the chart pertains to the objective function. The variables are:

x = number of servings of chicken

y = number of servings of corn

Note that for a solution of the problem to make sense each value of x and y must be close to the number 1, because only numbers close to 1 constitute reasonable amounts for a meal. In other words the value $x = 3$ requires the meal to contain 9 ounces of chicken, which is too much meat for most of us to eat. (Of course, if it is Thanksgiving, and we are eating turkey, common sense takes a hike!) The objective function is determined by letting F = amount of fat in the meal (in grams); so we want to minimize the equation

$$F = 4x + 2y$$

Note that each term has the same units, grams, since

$4x$ = (grams of fat per serving of chicken)
$\times$ (number of servings of chicken)

and

$2y$ = (grams of fat per serving of corn)
$\times$ (number of servings of corn)

The first constraint is represented by the first row in Table 2. It is required that the meal at least meet the U.S. RDA of 45 for protein. Hence

$$45x + 9y \geq 45$$

Similarly, the second row of Table 2 requires that the iron in the meal at least meet the U.S. RDA of 20. Hence

$$10x + 6y \geq 20$$

There are two additional constraints that are implicitly included in the problem. As in the previous problem it makes no sense to allow x and y to take on negative values; so we add $x \geq 0$ and $y \geq 0$ to the list of constraints. Therefore the problem can be stated in mathematical terms by the following:

Minimize $\quad F = 4x \quad + 2y \quad$ Objective function

Subject to $\quad 45x + 9y \geq 45$

$\qquad\qquad 10x + 6y \geq 20 \quad$ Constraints

$\qquad\qquad\quad x \geq 0$

$\qquad\qquad\quad y \geq 0$

**Key Steps
in Translating
the Problem**

Earlier we described in some detail three of the key steps in translating a linear programming problem into mathematical language. We will now list all the steps you should follow in order to translate the problem into mathematical terms.

STEPS IN TRANSLATING A LINEAR PROGRAMMING
PROBLEM INTO MATHEMATICAL LANGUAGE

1 Read and reread the problem to assess what is known and what is to be determined.

2 Organize the data, preferably into a chart.

3 Identify the unknown quantities and assign variables to them.

4 Determine the objective function.

5 Translate the constraints into linear inequalities.

We will now look at two examples in order to demonstrate how to implement these steps in translating a linear programming problem into mathematical terms. The one step that appears to us to be most overlooked by students is the first one: read *and reread* the problem. Most linear programming problems are somewhat complicated, and you are not expected to understand the statement of the problem on the first reading. When you are organizing the data in a chart, always check to make sure your description of the data agrees with the statement of the problem.

The next example shows how to apply this procedure to problems involving the allocations of personnel. The word "allocation" is important here because the term "programming" is derived from problems in the programming (or allocation) of personnel and supplies. The earliest applications of linear programming were to these problems. Dantzig referred to them as *scheduling activities*. The setting in Example 3 is somewhat elementary so that we can readily translate it into mathematical terms. You can easily see how the problem increases in complexity as the number of personnel becomes large and the number of constraints increases.

Example 3

Problem The computer director at a large university offers free consultant help for faculty and student users each day of the week. Most of the staff of the computer department participate in this program by taking shifts at the consultant stations. The director has determined that there are two types of service required. Faculty researchers and science majors need highly skilled consultant work, whereas most student jobs require less sophisticated help.

Thus the director has decided to staff the consultant stations with two different types of shifts or teams: Type 1 will consist of two senior programmers, who will primarily handle relatively complex problems, while type 2 will consist of one senior programmer and four student workers, who will primarily handle problems indigenous to novice programmers. Thus a type 1 shift will have a staff of two people and a type 2 shift will consists of five people. The director wants to use no more than 40 individuals to handle the consultant duties. The type 1 shift will be scheduled to work for 2-hour intervals, usually at the beginning and at the end of the day when demand is low. During the middle of the day, when demand is high, type 2 shifts will be scheduled to work at 1-hour intervals. Consultants will be on hand from 8:00 A.M. until at least 8:00 P.M., and there will always be two shifts working. Therefore there will be at least 24 hours to fill during each weekday. The director wants to determine the number of each type of shift to use in order to minimize the cost of the operation. In addition, the director has determined that the cost for a type 1 shift is $80 (a senior programmer makes about $20 per hour), while the cost of a type 2 shift is $36 ($20 per hour for the senior programmer and $4 per hour for each of the four student workers).

Solution We must organize the data into a chart, as is done in Table 3.

TABLE 3

	Type 1	Type 2	Constraint
Number of people	2	5	40
Number of hours	2	1	24
Cost	80	36	

The unknown quantities are the number of each type of shift and the cost. If we let

x = number of type 1 shifts

y = number of type 2 shifts

C = cost

then the rows of the chart can be translated into the constraints and the objective function. The first row becomes the constraint that at most 40 people are to be used. In mathematical terms it is

$$2x + 5y \leq 40$$

The second row becomes the constraint that at least 24 hours must be filled by the shifts. In mathematical terms it is

$$2x + y \geq 24$$

The third row yields the objective function

$$80x + 36y = C$$

You should check to make sure that each term of the respective relations has the same units. Hence the problem translates into mathematical terms according to the following:

Minimize	$C = 80x + 36y$	Objective function
Subject to	$2x + 5y \leq 40$	
	$2x + y \geq 24$	Constraints
	$x \geq 0$	
	$y \geq 0$	

The next example is a very common type of problem confronting many businesses and organizations. It is an example of a *transportation problem*. A transportation problem can have many different guises. We have chosen to illustrate this type of problem by maximizing profit subject to constraints of time and cost. It might be the case in actual practice that an entrepreneur would decide to minimize cost because of other limitations such as the number of vehicles or the amount of goods transported. We have tried to make the problems as realistic as possible while choosing the numbers so they "come out right," in the sense that the calculations do not interfere with your understanding of the techniques.

Example 4

Problem A bicycle manufacturer markets bicycles in two ways: at the company-owned retail outlet and at a large chain store. A truck transports the bicycles from the warehouse (1) to the retail outlet and (2) to the distribution center of the chain store. A round trip to the retail outlet, including time for loading and unloading, takes 2 hours, while it takes 4 hours for a round trip to the distribution center. It costs $50 per trip to the retail outlet and $200 for a trip to the distribution center, mainly because each item must be packaged when it is sent to the chain store. The profit of each truckload of bicycles sold through the outlet is $210. Because there is less overhead on the bicycles sold to the chain store, the profit for each truckload is $400. How many shipments should be made to each type of store per week in order to maximize profit if the manufacturer is limited to no more than 40 hours of travel time and a cost of $1600?

Solution The objective function and the constraints are found in the preceding sentence. Thus the objective function entails profit and the constraining conditions involve time and cost. We organize the data in the chart in Table 4. We label the variables, which are the number of trips per week to each type of store and profit, as follows:

x = number of trips per week to the retail outlet

y = number of trips per week to the chain store

P = profit

TABLE 4			
	Retail Outlet	Chain Store	Constraint
Time for round trip, hours	2	4	40
Cost of round trip	50	200	1600
Profit, dollars	210	400	

The last row in Table 4 yields the objective function

$$P = 210x + 400y$$

The first row becomes the constraint on the time that the truck is available during the week; that is, no more than 40 hours can be allotted for the transportation between the manufacturer's warehouse and the stores. Translating the first row into mathematical terms yields

$$2x + 4y \leq 40$$

The second row is the limitation on the amount of money that the manufacturer has decided to allot to the transportation of the product. The constraint becomes

$$50x + 200y \leq 1600$$

Hence the problem translates into mathematical terms according to the following:

Maximize $\quad \underline{P = 210x + 400y} \quad$ Objective function

Subject to $\quad 2x + 4y \leq 40$
$\qquad\qquad 50x + 200y \leq 1600$
$\qquad\qquad\qquad x \geq 0$
$\qquad\qquad\qquad y \geq 0$

Constraints

    The four examples considered in this section reflect the history of linear programming problems and show how to translate such problems into mathematical terms. We will show how to solve them in the next section. The solution to the transportation problem in Example 4 might surprise you. In the meantime we encourage you to try to solve them yourself, especially if your curiosity has been piqued by our presentation.

EXERCISES

In all the problems translate the linear programming problem into mathematical language. Take a guess as to what the solution might be. Make sure you save your work because in the next section the technique for solving these problems will be presented. Once you solve the problem you can compare the solution with your "guesstimate."

Problems 1 to 8 are very similar to the examples. Just the numbers have been changed. If you get stuck on one of the later problems, refer to the procedure you followed on these earlier ones.

1 A clothing manufacturer makes two types of outerwear—a 100 percent wool and a polyester all-weather coat. Each wool coat sells to the retail stores for $190 and each polyester coat for $130. The cost for manufacturing each wool coat is $150 and for each polyester coat is $100. The manufacturer can make no more than 100 coats per week and can budget no more than $12,000 per week. How many coats of each type should be made per week to maximize profit? Arrange the data in the following chart.

	Wool Coats	Polyester Coats	Constraints
Number			
Cost			
Profit			

2 A clothing manufacturer makes two types of jogging attire—regular and designer. The attire sells to the retail stores for $10 each for the regular suits and $40 each for the designer suits. The cost for manufacturing each regular suit is $5, and it is $10 for the designer suit. (The label costs $5.) If the manufacturer makes no more than 200 suits a week and budgets no more than $1500 per week, how many of each type should be made to maximize profit?

3 A hospital nutritionist prepares a menu for patients on low-fat diets. The primary ingredients are chicken and corn. The meal must contain enough iron and protein to meet at least three-fourths the U.S. RDA for each nutrient. Thus the meal should contain at least 45 grams of protein and 15 milligrams of iron. Each 3-ounce serving of chicken contains 45 grams of protein, 9 milligrams of iron, and 4 grams of fat. Each 1-cup serving of corn contains 9 grams of protein, 6 milligrams of iron, and 2 grams of fat. How much chicken and corn should be included in the meal in order to minimize the amount of fat? Arrange the data in the following chart.

	Chicken	Corn	RDA
Protein, grams			
Iron, milligrams			
Fat, grams			

4 Work Problem 3 with the constraint that the nutritionist wants the meal to contain at least the full U.S. RDA for protein but requires only that at least one-half the U.S. RDA for iron be met.

5 The computer center director at a large university wants to staff consulting stations with two types of shifts: type 1 will contain 2 senior programmers and 1 student worker and type 2 will consist of 1 senior programmer and 5 student workers. The director wants to use no more than 36 individuals. There will be at least 24 hours to be filled during the week, with a type 1 shift serving for 3 hours and a type 2 shift serving for 2 hours. The cost of a type 1 shift is $44 per hour and, for a type 2 shift, $40 per hour. Determine the number of shifts of each type in order to minimize cost. Arrange the data in the following table.

	Type 1	Type 2	Constraint
Number of people			
Number of hours			
Cost			

6 Work Problem 5 if a type 2 shift is to consist of 1 senior programmer and 7 student workers, so that the cost of a type 2 shift is $48, and all other figures remain the same.

7 A bicycle manufacturer has a retail outlet and, in addition, sells some bikes to a large chain store. One truck makes all deliveries, and it takes 3 hours to make a round trip to the retail store and 4 hours to make a round trip to the chain store. The cost of a round trip to the retail store is $100, and the cost of a round trip to the chain store is $200. The profit of each truckload sold through the retail store is $500; it is $800 for each truckload sold to the chain store. How many shipments per week should be scheduled to each store in order to maximize profit if the manufacturer is limited to no more than 48 hours per week and a cost of $2000? Arrange the data in the following chart.

	Retail Outlet	Chain Store	Constraint
Time for round trip			
Cost of round trip			
Profit			

8 Work Problem 7 if the profit on each truckload sold through the retail store is $600, and it is $800 for a truckload sold to the chain store.

9 A meat manufacturer mixes beef and pork in sausage links. Two types of links are made, regular and all-beef. Each pound of regular sausage meat contains 0.3 pound of beef and 0.2 pound of pork, while each pound of all-beef sausage meat contains 0.4 pound of beef and 0.1 pound of pork. There is 120 pounds of beef and 60 pounds of pork in stock. If the profit for regular links is 50 cents per pound and the profit for all-beef links is 70 cents per pound, how many pounds of each type of link should be made to maximize profit?

10 A consortium of travel agents has sold 1200 tickets to the Super Bowl. The weekend package includes air fare, and they have a choice of two types of airplanes for the charter flights. Type 1 can carry 100 passengers and type 2 can carry 150 passengers. Each flight of a type 1 aircraft will cost $9000 and each type 2 flight will cost $15,000. The consortium is allowed to lease no more than 10 planes. How many airplanes of each type should be leased in order to minimize the cost?

11 The administration of a money market fund wants to invest up to $15 million in two types of investments. The funds are to be divided between short-term bank notes and Treasury notes. The current yield for bank notes is 15 percent, and for Treasury notes it is 12 percent. Because the yield for bank notes fluctuates unpredictably, it has been decided that at least twice as much money is to be invested in Treasury notes as bank notes. At least $2 million must be allocated to each type of investment. How much money should be invested in each type investment to produce the largest return?

12 A window manufacturer produces windows in two styles, regular and thermopane. It costs $100 to make each regular window, which the manufacturer sells for $150. It costs $120 to make each thermopane, and each sells for $175. The daily production capacity is 110 windows and the daily cost cannot exceed $12,000. How many windows of each type should be made per day in order to maximize profit?

13 An automobile leasing company buys two types of cars, compacts and mid-sized. The company wants to purchase at most 1000 cars. The cost of each compact is $8000 and of each mid-sized car is $10,000. The company has decided to allocate no more than $9 million for the purchase of the cars. It must purchase at least 200 of each type of car. If the anticipated return from the sale of the cars at the end of the company's 3-year leasing period is $1000 for a compact and $1200 for a mid-sized car, how many of each should be purchased in order to maximize return?

14 Officials of the state high school basketball tournament must plan accommodations for at least 660 players. They have two types of rooms available, dormitory rooms that sleep three people and motel rooms that sleep two people. For meals, they have budgeted $10 daily per person for those in dorms and $20 daily per person for those in motels, and they must not spend more than $12,000 in meal money per day. They have available at most 200 dormitory rooms and 150 motel rooms. If the daily cost of a room is $20 for a dormitory room and $30 for a motel room, how many rooms of each kind must they schedule in order to minimize cost?

15 A heating oil delivery firm has two processing plants, plant A and plant B. Plant A processes daily 150 barrels of high-grade oil for commercial use and 50 barrels of low-grade oil for residential use. Plant B processes daily 100 barrels of high-grade oil and 50 barrels of low-grade oil. The daily cost of operation is $20,000 for plant A and $15,000 for plant B. An order is placed for 1000 barrels of high-grade oil and 450 barrels of low-grade oil. Find the number of days that each plant should be operated to fill the order and minimize the cost.

16 An automobile leasing company has warehouses in Los Angeles and San Francisco. A company in Las Vegas sends an order to lease 25 cars and a firm in Fresno orders 15 cars. The Los Angeles warehouse has 30 cars and the San Francisco warehouse has 20 cars. It costs $100 to ship a car from Los Angeles to Las Vegas, $90 from Los Angeles to Fresno, $150 from San Francisco to Las Vegas, and $80 from San Francisco to Fresno. Find the number of cars to be shipped from each warehouse to each city in order to minimize cost.

5-6 GRAPHICAL SOLUTIONS OF LINEAR PROGRAMMING PROBLEMS

Once a linear programming problem has been set up, that is, once its objective function and constraints have been determined, then the graphical solution of the problem can be computed. In this section we will demonstrate a method for solving this type of problem. Just as the technique described in the previous section for translating programming problems into mathematical language consisted of distinct stages, so too the method in this section will include various steps in the procedure. This method of solution is usually referred to as the *geometric approach* because is involves graphing. It is effective for solving problems that involve two variables like the ones we considered in Section 5-5. However, Dantzig's simplex method, which is much more involved and will not be discussed here, is preferred when more variables are involved.

The Manufacturing Problem

Let us illustrate the initial steps by recalling the first example in the previous section, the manufacturing problem. A clothing manufacturer made two types of suits. The problem was to maximize profit subject to certain constraints. We will start with the problem expressed in its mathematical terminology.

Example 1

Problem   Maximize $P = 60x + 50y$ subject to the constraints

$$x + y \leq 120$$

$$100x + 40y \leq 6000$$

$$x \geq 0$$

$$y \geq 0$$

That is, find the values of x and y that simultaneously satisfy the inequalities and yield the largest value of P.

Solution   We will separate the procedure into three steps. In the first two steps we will work only with the constraints. The first step is to determine the graph of the inequalities. Then we find the points of intersection of all the lines that form the boundary of the graph. Let us proceed with these two stages and then tackle the third.

Step 1. Graph the region defined by the inequalities. Use the method described in Section 5-4. Replace the inequality signs with equal signs and then graph the resulting straight lines. To graph the first two lines we plot the x and y intercepts and then draw a line through the two points. The other two lines, $x = 0$ and $y = 0$, are the two axes. For each line, shade the region that is not part of the solution of the inequality. The region that is unshaded is the graph of the system of inequalities. The graph of the constraints in this problem is given in Figure 1.

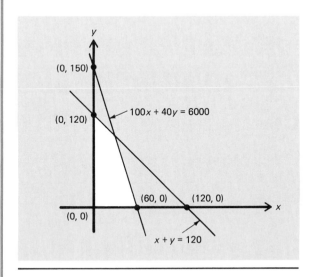

FIGURE 1   The graph of the system of inequalities

$$x + y \leq 120$$
$$100x + 40y \leq 6000$$
$$x \geq 0$$
$$y \geq 0$$

Step 2. Find the "corner points," that is, the points of intersection of the line segments that comprise the boundary of the graph. In this problem the boundary consists of line segments from the four lines $x + y = 120$, $100x + 40y = 6000$, $x = 0$, and $y = 0$. From Figure 1 you can see that the boundary of the region consists of four line segments, and thus there are four points of intersection. Three of them were determined when we graphed the lines. They are $(0, 0)$, $(60, 0)$, and $(0, 120)$. The remaining point is the point of intersection of $x + y = 120$ and $100x + 40y = 6000$. We can solve the two equations simultaneously by multiplying the first equation by -40 and adding it to the second, which yields

$$
\begin{array}{rl}
100x + 40y = & 6000 \\
\underline{-40x - 40y =} & \underline{-4800} \\
60x \quad\quad\;\; = & 1200
\end{array}
$$

and so $x = 20$. Substituting this value into the equation $x + y = 120$ yields $20 + y = 120$, and so $y = 100$. Thus the four corner points of the region that is the graph of the constraints are

$$(0, 0) \quad\quad (60, 0) \quad\quad (0, 120) \quad\quad (20, 100)$$

Step 3. Once the corner points of the graph have been determined we can use them to determine the solution of the linear programming problem (finally!). We simply evaulate the objective function at these points. The point that yields the largest (since we were asked to *maximize* the objective function) is the solution. We put the result of this computation in the following table.

Corner Point	Value of $60x$ $+ 50y$ $= P$
$(0, 0)$	$60 \cdot 0 + 50 \cdot 0 = 0$
$(60, 0)$	$60 \cdot 60 + 50 \cdot 0 = 3600$
$(0, 120)$	$60 \cdot 0 + 50 \cdot 120 = 6000$
$(20, 100)$	$60 \cdot 20 + 50 \cdot 100 = 6200$

The largest value of P is 6200. Therefore the solution is $(20, 100)$; it occurs when $x = 20$ and $y = 100$. Referring to the meaning of x, y, and P from the description of the manufacturing problem in the previous section, we see that the clothing manufacturer will make the largest profit, $6200, by making $x = 20$ worsted suits per week and $y = 100$ blend suits per week.

Justification of the Method

The natural question to ask at this point is *why* does this method work? How can we be sure that the maximum value of P is actually $6200? Could there be another point, somewhere else in the graph of the constraints, that yields a larger value for P?

We will answer these questions presently, but first let us introduce some terminology. Those points, and only those points, that satisfy the constraints have a chance to be a solution to the problem. For this reason we call the region defined by the constraints the region of *feasible solutions*. This region will have a boundary made up of line segments which are the lines determined by the constraints. The vertices of the boundary are called the *corner points* of the region of feasible solutions. The *optimum solution* is the point (or points) in the region of feasible solutions that, when substituted into the objective function, yields the largest value (if we were asked to maximize the objective function) or the smallest value (if we were asked to minimize the objective function).

Step 1 of our procedure in Example 1 requires us to determine the region of feasible solutions. Step 2 calls for us to determine the corner points of the region of feasible solutions. At this juncture it seems that we must substitute *all* the points in the region into the objective function and determine which yields the largest value. This appears to be a nearly impossible task, as there are an infinite number of points in the region. Step 3, however, entails substituting only the corner points into the objective function and locating the largest value in order to determine the optimum solution. This reduces our task to a manageable size. Let us now investigate why it works.

Thus far we have made use of the profit function $P = 60x + 50y$ by substituting values of x and y into it and determining the corresponding value of P. Suppose we change our tack a bit and arbitrarily assign P a value, say, for example, $P = 300$. We get the expression $60x + 50y = 300$. This is the equation of a straight line. In fact we can divide this equation by 10 and get an equivalent form of the equation, $6x + 5y = 30$. This line has slope-intercept form

$$y = -\frac{6}{5}x + 6$$

and thus the line has slope $-\frac{6}{5}$ and y intercept 6.

Now let P take on another value, say $P = 1200$. We get $60x + 50y = 1200$. Again we get the equation of a straight line. Its slope-intercept form is

$$y = -\frac{60}{50}x + \frac{1200}{50} = -\frac{6}{5}x + 24$$

Notice that this line has the same slope as the previous one.

In general, if we assign a specific value to P, say $P = P_0$, we get the equation of a straight line, $60x + 50y = P_0$, whose slope-intercept form is $y = -\frac{6}{5}x + P_0/5$. The important point is that no matter what value we assign to P we get the same slope, $-\frac{6}{5}$, but different y intercepts. This means that all expressions of the form $60x + 50y = P_0$ are the equations of parallel lines.

In Figure 2 we graph a number of these parallel lines. Some of them intersect the region of feasible solutions and some do not. How can we describe the optimal solution for our linear programming problem in this geometrical setting? First, the optimal solution must be a point in the region of feasible solutions. Second, when we substitute the optimal solution into the objective function we get a value for P which is the largest

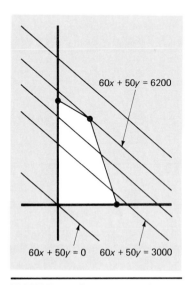

FIGURE 2 Some examples of the lines $60x + 50y = P_0$, the lines with slope $-\frac{6}{5}$.

possible. Thus we are considering expressions of the form

$$60x + 50y = P$$

only for points (x, y) in the region of feasible solutions. Geometrically this means that we are considering only those lines with slope $-\frac{6}{5}$ that intersect the region of feasible solutions.

What does it mean to say that we are searching for the largest value for P? Consider the expression $60x + 50y = P_0$ as representing a set of straight lines that have slope-intercept form $y = -\frac{6}{5}x + P_0/5$. The larger the value of P_0, the larger the y intercept is. Therefore we are searching for the line that intersects the region of feasible solutions that has the largest y intercept. This will always occur at a corner point of the region. Thus it suffices for us to inspect only the corner points and determine which one (or ones) yield the largest value of the objective function.

This outlines the geometric justification of the three-step procedure used to determine the optimal solution of the linear programming problem in Example 1. Even though this justification is valid only for this particular problem, the essence of the argument can be used to show that the method of solution works for any linear programming problem.

There are two additional points to keep in mind. First, if the problem calls for us to minimize the objective function, we still evaluate the objective function at each corner point, but the optimal solution is the point that yields the smallest value. Second, we have implied that there may be more than one corner point that yields the largest (or smallest) value of the objective function. This can happen in the following way. Consider Figure 3. The region of feasible solutions has a line segment of the boundary that is parallel to the objective function. Not only are both corner points A and B optimal solutions, but so are *all* the points on the line segment between A and B, since they all lie on the line defined by the objective function. Thus there is an infinite number of solutions.

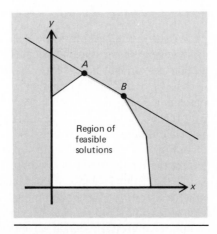

FIGURE 3

The Three-Step Procedure

We are now able to record the three steps in the method for solving linear programming problems. Once the linear programming problem has been translated into mathematical terms, proceed as follows to find the optimal solution.

> *Step 1.* Graph the constraints in order to find the region of feasible solutions. Use the techniques described in Section 5-4.
>
> *Step 2.* Find the corner points of the region of feasible solutions. They will consist of the points of intersection of the line segments comprising the boundary of the region of feasible solutions. Use the techniques described in Section 5-3.
>
> *Step 3.* Evaluate the objective function at each corner point. The optimum solution is that point that yields the
>
> (*a*) largest value if the problem requires that the objective function be maximized, or
>
> (*b*) smallest value if the problem requires that the objective function be minimized.

Applying the Procedure

We can now apply these three steps to solve the linear programming problems that were translated into mathematical terms in the previous section. Some of the exercises will ask you to solve the problems in the previous exercises that you have already translated.

Example 2

Problem Find the optimum solution of the diet problem in Example 2 of the previous section. That is, minimize the objective function

$$F = 4x + 2y$$

subject to the constraints

$$45x + 9y \geq 45$$
$$10x + 6y \geq 20$$
$$x \geq 0$$
$$y \geq 0$$

where x = the number of servings of chicken, y = the number of servings of corn, and F = grams of fat in the meal.

Solution

    Step 1. We first graph the constraints to find the region of feasible solutions. We replace the inequality signs by equality signs. Then we graph the resulting lines by locating the x and y intercepts. After testing a point on one side of each line and shading the region on the side of the line that is not the solution, we get the graph in Figure 4. Notice that this region is a bit different from the previous one in that it is unbounded. There are four line segments that comprise its boundary.

    Step 2. From Figure 4 you can see that two corner points are (2, 0) and (0, 5). The third is the point of intersection of $45x + 9y = 45$ and $10x + 6y = 20$. To find that point we divide the first equation by 9 and the second by 2, which yields

$$5x + y = 5$$
$$5x + 3y = 10$$

We can multiply the first equation by -1 and then add it to the second, which yields

$$-5x - y = -5$$
$$\underline{5x + 3y = 10}$$
$$\phantom{-5x + {}}2y = 5$$

and thus $y = 5/2 = 2.5$. Substituting this value into the equation $5x + y = 5$ yields

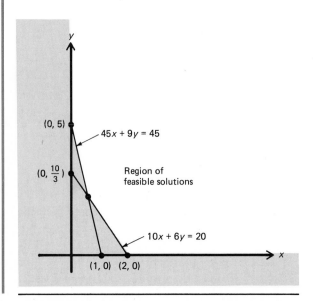

FIGURE 4

$$5x + 2.5 = 5$$

$$5x = 5 - 2.5 = 2.5$$

$$x = \frac{2.5}{5} = 0.5$$

Hence the point of intersection, the third corner point, is (0.5, 2.5). Thus the three corner points are

(2, 0)   (0, 5)   (0.5, 2.5)

Step 3. We now evaluate the objective function at each corner point and choose the point that yields the smallest value. Consider the following table.

Corner Point	Value of 4x + 2y = F
(2, 0)	$4 \cdot 2 + 2 \cdot 0 = 8$
(0, 5)	$4 \cdot 0 + 2 \cdot 5 = 10$
(0.5, 2.5)	$4 \cdot 0.5 + 2 \cdot 2.5 = 7$

Hence the optimum solution is (0.5, 2.5). This means that the meal should consist of 0.5 serving of chicken, that is $0.5 \cdot 3$ ounces $= 1.5$ ounces of chicken, and 2.5 servings of corn, that is, 2.5 cups of corn, to minimize the amount of fat in the diet.

The third example in the previous section required us to minimize the cost of a computer center consultant operation. The computer center director wanted to create two types of shifts that had constraints on the number of consultants available and the number of hours that the shifts would be available. We can now use our three-step technique to solve the problem.

Example 3

Problem Minimize the cost function $C = 80x + 36y$ subject to

$$2x + 5y \leq 40$$

$$2x + y \geq 24$$

$$x \geq 0$$

$$y \geq 0$$

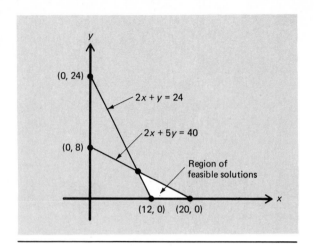

FIGURE 5

Solution

Step 1. The graph of the region of feasible solutions is given in Figure 5.

Step 2. Two of the corner points are (12, 0) and (20, 0) from Figure 5. The third is the point of intersection of the two lines $2x + 5y = 40$ and $2x + y = 24$. We can multiply the second equation by -1 and then add it to the first to get

$$2x + 5y = 40$$
$$\underline{-2x - y = -24}$$
$$4y = 16$$

Thus $y = 4$. Substituting this value into the equation $2x + y = 24$ yields

$$2x + 4 = 24$$
$$2x = 24 - 4 = 20$$

and so $x = 10$. Hence the three corner points are

$$(12, 0) \qquad (20, 0) \qquad (10, 4)$$

Step 3. We evaluate the objective function at each corner point and then choose the point that yields the smallest value. We record the computations in the following table:

Corner Point	Value of $80x + 36y = C$
(12, 0)	$80 \cdot 12 + 36 \cdot 0 = 960$
(20, 0)	$80 \cdot 20 + 36 \cdot 0 = 1600$
(10, 4)	$80 \cdot 10 + 36 \cdot 4 = 944$

Thus the optimum solution is (10, 4), and therefore the director should plan for 10 type 1 shifts and 4 type 2 shifts to minimize cost and still satisfy the constraints.

Example 4 of the previous section asked us to find the number of trips to a manufacturer's retail outlet and to the chain store that should be scheduled in order to maximize profit. The constraining conditions are limitations on time and cost.

Example 4

Problem   Maximize the profit function $P = 210x + 400y$ subject to

$$2x + 4y \leq 40$$

$$50x + 200y \leq 1600$$

$$x \geq 0$$

$$y \geq 0$$

Solution

Step 1. The graph of the region of feasible solutions is given in Figure 6.

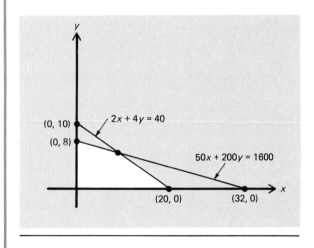

FIGURE 6

Step 2. Two of the corner points can be determined from the graph of the region of feasible solutions in Figure 6. They are (20, 0) and (0, 8). The third is the point of intersection of the lines $2x + 4y = 40$ and $50x + 200y = 1600$. We can divide the first equation by 2 and the second by 50, and we get

$$x + 2y = 20$$

$$x + 4y = 32$$

If we multiply the first equation by -1 and add it to the second, we get

$$2y = 12$$

and so $y = 6$. Substituting this value into $x + 2y = 20$ yields

$$x + 2 \cdot 6 = 20$$
$$x = 20 - 12 = 8$$

Hence the three corner points are

$$(20, 0) \qquad (0, 8) \qquad (8, 6)$$

Step 3. We now evaluate the objective function at each corner point and choose the point that yields the largest value. We record the computation in the following table:

Corner Point	Value of $210x + 400y = P$
(20, 0)	$210 \cdot 20 + 400 \cdot 0 = 4200$
(0, 8)	$210 \cdot 0 + 400 \cdot 8 = 3200$
(8, 6)	$210 \cdot 8 + 400 \cdot 6 = 4080$

Thus the optimal solution is (20, 0). Hence the manufacturer should schedule 20 trips to the retail outlet and no trips to the chain store each week to maximize profit.

In Examples 1, 2, and 3 the solution was the point of intersection of two lines other than the axes. Example 4 shows that the answer can lie along the x axis. We mention this fact to forewarn you against drawing an unjustified conclusion.

EXERCISES

In Problems 1 to 4 find the graph of the feasible set of solutions for the given set of constraints.

1
$$x + 4y \le 12$$
$$x + 3y \le 10$$
$$x \ge 0$$
$$y \ge 0$$

2
$$2x + y \le 10$$
$$3x + 2y \le 12$$
$$x \ge 1$$
$$y \ge 1$$

3 $x + 6y \leq 18$
  $2x + 3y \geq 12$
  $x \geq 0$
  $y \geq 0$

4 $x + 3y \leq 12$
  $2x + 5y \leq 10$
  $x \geq 2$
  $y \geq 1$

In Problems 5 and 6 find the corner points of the feasible set of solutions for the given set of constraints.

5 $2x + 3y \leq 10$
  $4x + 7y \geq 28$
  $x \geq 0$
  $y \geq 0$

6 $3x + 2y \geq 18$
  $x + 4y \leq 12$
  $x \geq 0$
  $y \geq 0$

In Problems 7 to 10 solve the linear programming problem as it is presented in its mathematical formulation.

7   Maximize $z = 4x + 10y$ subject to the constraints in Problem 1.

8   Minimize $z = 3x + 4y$ subject to the constraints in Problem 1.

9   Minimize $z = 20x + 18y$ subject to the constraints in Problem 3.

10   Maximize $z = 10x + 12y$ subject to the constraints in Problem 2.

In Problems 11 to 26 solve the indicated problem from the previous exercise set.

11   Problem 1

12   Problem 2

13   Problem 3

14   Problem 4

15   Problem 5

16   Problem 6

17   Problem 7

18   Problem 8

19   Problem 9

20   Problem 10

21   Problem 11

22   Problem 12

23   Problem 13

24   Problem 14

25   Problem 15

26   Problem 16

27   Solve Example 4 with the profit function $P = 150x + 400y$.

28   Solve Example 4 with the profit function $P = 200x + 400y$.

Probability

The curious evolution of probability had its origin in the seventeenth century, when a roguish friend of the great mathematician Blaise Pascal sent him a letter requesting a solution to a perplexing gambling problem. The ascent started when Pascal began a long and fruitful correspondence with Pierre Fermat, the lawyer and part-time mathematician we met in Chapters 1 and 5.

They laid the groundwork for probability theory, but it did not mature until late in the nineteenth century, when a host of biologists and astronomers realized that probabilistic reasoning provided the key to heredity and planetary motion. We start the first section with the story of Gregor Mendel, the father of genetics, and his application of probability to garden pea plants.

Except for the early work of Pascal, Fermat, and a few others, the ascent of probability and statistics is a fairly modern tale. Thus many of the examples used to explain the theory are the same as those used by its creators—especially games of chance. The primary application of probability is to describe natural phenomena that do not seem to obey any prescribed laws—the type that appear to occur haphazardly. So games of chance provide excellent models of seemingly random events. We will study the fundamentals of probability theory in this chapter, and then we will see how to apply these ideas to other disciplines in the following chapters.

6-1 A GENETIC APPROACH TO PROBABILITY

Heredity

The purpose of this section is to illustrate the basic principles of probability. Most of these principles were developed by two seventeenth-century French mathematicians, Pierre Fermat and Blaise Pascal, who dealt with problems pertaining to games of chance. Their theory was applied to a real-world problem for the first time in the revolutionary work of Gregor Mendel, who discovered the elementary principles of heredity and thus initiated the science of genetics. We will present the introductory methods of probability by discussing Mendel himself, by tracing the steps that led to his mastery of the process of heredity, and by comparing the two fundamental facets of probability: theory versus empiricism. The story of Mendel and the science of genetics elucidates the very basic type of reasoning that forms the foundation of probability theory.

The wonderment of heredity has dazzled humanity's curiosity for centuries. It has always been apparent that tall parents produce tall offspring and that red-haired parents often produce offspring with the same color hair. On the other hand, some parents bear children who are unlike each other entirely; one child might be scholarly but not athletic, while the other would be just the opposite. It was natural for people to wonder why. How are traits of the parents transmitted to the offspring? How is it possible to produce such wide variations among offspring?

Many theories were advanced that tried to explain the heredity process. Even the greatest philosophers and theologians grappled with the problem in vain. Aristotle, for instance, thought blood carried traits, flowing from parent to offspring. Vestiges of Aristotle's influence remain today, as is dramatized by the incorporation into our language of such phrases as "bad blood," "blue blood," and "blood relative." However, no basis of proof was ever given. The only suggested answers had a spiritual, almost holy, ring to them. You accepted the theory on faith, not fact.

Many scientific advances paved the way for a more detailed and sound account of the hereditary process. It was felt that a number of the important puzzles could be unlocked if the heredity maze could be mastered. There is little wonder that some of the greatest minds of the nineteenth century turned their attention to this problem (Pasteur, Leeuwenhoek, and others showed the close connection between heredity and disease). It is, however, curious that the one who ultimately provided the breakthrough came from a very disadvantaged background.

In the middle of the nineteenth century, a young farm boy, raised in poverty and isolation, became a monk in order to be educated and rise above his hapless economic state. He was trained to be a teacher, but he failed, so he was appointed caretaker of the monastery and died in obscurity. He is recognized today as the founder of modern natural science. By careful experiment, observation, and analysis of results, Gregor Mendel formulated the first laws of heredity and founded the science of genetics.

How could someone whose life was fraught with failure and ostracism come to be so highly regarded in the intellectual community? The key is the same one that will open our door to the mathematical disciplines of probability and statistics.

Mathematicians had created the tool of probability mainly to solve problems in gambling. In the sixteenth century Cardan solved many questions concerning gambling games with cards and dice. A century later Fermat and Pascal extended the fundamental notions of probability, but the problems solved always pertained to games of chance. While many a gambler and mathematician, especially some astronomers, took notice of the methods, probability theory had little application outside these restricted domains.

Mendel

Johann Gregor Mendel (men'-dull) was born into a poor farm family in Heinzendorf in 1822. There was little that could be done to break the economic cycle of the farm peasant. One ray of hope centered around the monasteries, which accepted and trained promising young men to enter the world of the intellectual. Mendel joined the Augustinian Order of St. Thomas at Brno (which is in present Czechoslovakia; it was Austria then). There he was given the name Gregor, by which he is known today. The Augustinians provided education to the rich and the poor alike. Mendel was to become a teacher, but during his training at the University of Vienna it became clear that he was unsuited for a classical education and he was sent back to the monastery. Those monks who did not teach were assigned various tasks to be performed for the upkeep of the monastery. The natural choice for Mendel, considering his childhood and his training, was maintenance of the garden. It was there that Mendel formulated his laws of heredity. These laws not only altered the course of biology but also demonstrated to the scientific community the usefulness and power of probability theory and statistics.

During his stay in Vienna, Mendel met a respected biologist by the name of Franz Unger. Unger's approach to the study of biology differed radically from that of his contemporaries, and his influence on young Mendel was profound. The education that Mendel disdained was based on a broadly philosophical, ethereal, head-in-the-clouds approach. Mendel's approach to his work, on the other hand, was a very practical, concrete, dig-in-the-dirt and grind-out-the-facts method. He was a naturalist, preferring practical problem-solving techniques. (This is the same kind of dichotomy that separated the approaches to mathematics taken by the ancient Egyptians and Greeks, which we discussed in Section 3-2.)

Unger emphasized the necessity for experiment; he felt that theory alone would not suffice. When Mendel found himself back in the garden of the monastery, he discovered a "laboratory" made to order for his intellectual pursuits. He was able to couple his early childhood farm life with his newly awakened intellectual curiosity. His practical experiments provided a simplistic yet natural view of the biological world.

The Experiment

Mendel decided to study pea plants, the kind that grow in numerous gardens. The reasons he focused his attention upon these plants are simple: the garden pea offers many varieties, it reproduces at a very fast rate, and each one produces definite traits. Mendel isolated seven characteristics: color of the flower, length of the stem, texture of the pod, color of the ripening pods, shape of the seed, position of the flowers on the stem, and color of the coat of the flower seed. He proceeded to study how these traits were transmitted from parent to offspring.

For our purpose it will suffice to discuss just one trait—the color of the flower. In the exercises we will consider several other traits. Here we will examine the ingenious and careful experiment that Mendel devised in order to study the way that the color of a flower is transmitted from parent to offspring.

He began by painstakingly taking the pollen from the red-flowered plants and—usurping the bee's work—fertilizing them for many generations until they repeatedly produced only red-flowered offspring. We will refer to these plants as "pure red." He repeated the process with the white-flowered plants.

The second step was to cross-fertilize the red-flowered plants with the white ones. This produced the first generation of hybrid plants. (The term "hybrid" means that the plant came from a cross-fertilization; that is, its parents had separate and distinct traits.) Undoubtedly Mendel knew exactly what to expect of this first generation. What do you think happened when pure red-flowered plants were crossed with pure white-flowered plants? We suggest that you mark the box that you think is appropriate before reading further.

☐ All will be pink-flowered.

☐ Each plant will have a mixture of red and white flowers.

☐ Exactly half the plants will have red flowers and half will have white flowers.

☐ It is not possible to determine the mixture ahead of time; i.e., each time the experiment is done a different percentage of red and white flowers will appear.

☐ None of the above.

In the section of the garden where the first generation matured, Mendel's anxious eyes observed only red flowers. No white flowers! Did you expect, as surely Mendel did, all red-flowered plants?

The next step was to fertilize this first generation of red-flowered plants among themselves. What do you think happened? Recall now that the parents (pure red and pure white) produced a first generation that had entirely red flowers, and that these plants were fertilized among themselves. What do you expect this second generation to look like? As before, mark the appropriate box.

☐ All will be red-flowered.

☐ All will be white-flowered.

☐ Exactly half will be red-flowered and half will be white-flowered.

☐ It is impossible to determine the proportion of red-flowered plants.

☐ None of the above.

The color of the plants in the first generation was surprising—a red parent and a white parent always produced red offspring. The second generation turned out to be surprising for just the opposite reason; namely, some of the offspring turned out to be red and the others white. (Did you mark the last box?) Most of us find the results obtained by this industrious monk surprising. Now let us investigate the two sides of his achievement, the biological and the mathematical, to see what kind of reasoning led him to uncover the underlying principles governing the experiment.

Mendel's Biological Achievement

The first part of Mendel's achievement can be seen in the outcome of the first generation. Recall that all the flowers in this generation were red, even though each one had one parent with red flowers and one with white. This led him to his major hypothesis that the red-flowered trait is *dominant* over the white-flowered trait. A similar result occurred in every one of the other six categories he considered. So the stage was set for his first law of genetics, that one trait is always dominant over another (or, equivalently, one trait is always *recessive*).

However, this first generation alone did not provide the key. It was the second generation that demonstrated the truth of Mendel's theory. In order to understand it we will have to review a few technical terms from biology.

In each cell there is an identical set of "instructions" that governs how the organism is to grow. These instructions are carried on tiny threadlike bodies called *chromosomes*, by even smaller objects called *genes*. (The Greek word *chroma* means color. The root of the word "gene" will be discussed in Problem 2.) Every cell contains two similar chromosomes, each containing all the genes necessary for growth. For example, each human cell has 46 chromosomes, two sets of 23. One set of 23 chromosomes comes from the sperm cell, one from the egg cell.

How much of this did Mendel know? Not much. He was the first to assume that the offspring receives a full set of instructions from *each* parent. He called the carrying agent a *factor;* we now call it a gene.

His great insight was to imagine what went on inside the reproductive apparatus of the pea without being able to see it. He did this before knowledge of the nucleus was even known. Even today with our highly sophisticated microscopes we can only barely see chromosomes, and we cannot see genes at all.

Mendel postulated that each of the two reproductive cells from the parents contributes one set of characteristics, or traits, to the offspring. Hence each offspring has two sets of traits, one from each parent. It might also help for you to picture a normal cell with two chromosomes splitting to create two reproductive cells, each with one chromosome, as in Figure 1. In this case a sperm cell fertilizes the egg cell to form the offspring.

Now let us return to Mendel's garden. When he inbred the pure red-flowered plants, each parent contributed the red-flowered trait to the offspring; so the offspring had the red-flowered trait on each chromosome. Hence each reproductive cell had the red-flowered trait on its single chromosome. Similarly, every white-flowered plant had the white-flowered trait on its chromosome.

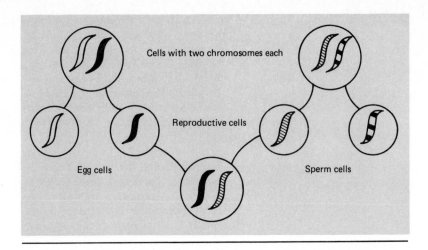

Cells with two chromosomes each

Reproductive cells

Egg cells

Sperm cells

FIGURE 1

Now we can understand why the first-generation hybrid plants were significantly different from their red-flowered parents even though they also had red flowers. When the reproductive cells from the parents united to form their first cell, one chromosome from the pure red-flowered strain had the red-flowered trait and the other had the white-flowered trait. Since the red-flowered trait is dominant over the white-flowered trait, only red flowers were produced.

This concept of one trait being dominant over another was Mendel's first breakthrough. Notice that there is no spiritual base for the theory and no necessity for resorting to faith. The second contribution, which, in fact, reinforced the first, occurred with the second generation.

Mendel's Use of Probability Reasoning

We are now in a position to explain what percentage of the second-generation plants had red flowers and what percentage had white flowers. We will use the reasoning that Mendel himself used in his historic research papers, a reasoning that both illustrates and motivates mathematical methods of probability.

It might be helpful to review briefly Mendel's experiment. He began with the parent generation, which consisted of a pure red-flowered strain and a pure white-flowered one. He cross-fertilized them, producing the so-called first generation of hybrid plants, all of which were red. Then he inbred these red-flowered plants, producing the second generation. We have already indicated that some of the second-generation plants were red and some were white. In fact, *three-fourths were red-flowered and one-fourth white-flowered.*

Mendel's brilliant discovery was the underlying principle of nature that governs the selection process (that is, the theory of genetics) and predicts the outcome beforehand. His contemporaries and his predecessors (including the great Darwin, who conducted similar experiments but was unable to explain their significance) would have said that the results in the garden were determined by some sort of averaging process left to chance or perhaps to nature. Mendel's explanation was much more convincing. His demonstration that his results were correct was systematic and conclusive.

Even though every first-generation plant visually resembled its red-flowered parent, his theory guaranteed that each plant actually contained two distinct traits, a red-flowered trait and a white-flowered trait. Biologists today speak of a "visible type" (*phenotype*) and a "hereditary type" (*genotype*) to describe this distinction. Mendel knew the difference. He imagined that the reproductive organs of the first-generation hybrids were different from those of their parents (genotype), even though they looked the same (phenotype). This is why the offspring of the first generation will not all have red flowers.

But the question is, what proportion of the second-generation plants will have red flowers? There are two ways of proceeding. Both have the same starting point, which can be seen in Figure 2.

One way of reasoning is as follows. There are three distinct possibilities for the traits:

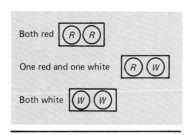

The two red traits in the cell would produce a red-flowered plant; the red and white would also produce a red-flowered plant because red is dominant over white; the two white traits would produce a white-flowered plant. This line of reasoning might lead you to predict that two-thirds of

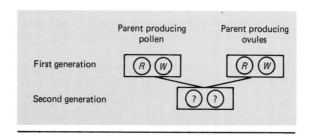

FIGURE 2

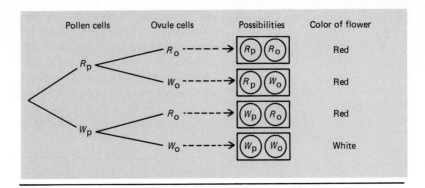

FIGURE 3

the plants in the second generation would be red-flowered and one-third would be white-flowered. But Mendel's experiment showed that three-fourths of the second-generation plants turn out to be red; so this line of reasoning does not hold.

Now let us analyze Mendel's reasoning, which led to a conclusion that coincided with the experimental facts. Let R and W denote red and white traits, and subscript them with p and o for pollen and ovules. Four possibilities result, as can be seen in Figure 3.

Can you see that the one possible gene configuration of the second generation, that of red-white, $\boxed{\textcircled{R}\ \textcircled{W}}$, can happen in two distinct ways: a "red" pollen cell can fertilize a "white" ovule, $\boxed{\textcircled{$R_p$}\ \textcircled{$W_o$}}$ or a "white" pollen cell can fertilize a "red" ovule cell. Hence there are four, not three, possibilities. Notice that the top three possibilities in the second generation correspond to red flowers, while only the bottom one corresponds to a white one. This agrees with the experimental evidence. Moreover, it confirms the theory of dominance.

The next two examples will illustrate this material.

Example 1
Begin with 400 pure red-flowered plants and 400 pure white-flowered plants. Cross-fertilize these 800 plants to produce 400 hybrid plants; then inbreed their offspring to produce 200 second-generation plants.
Problem (a) How many of the first-generation plants will be red? (b) How many of the second-generation plants will be red?
Solution (a) When the 400 pure red-flowered plants are cross-fertilized with the 400 pure white, they produce 400 hybrids, each of which contains a red trait and a white trait. By the theory of dominance, all 400 will be red.

(b) Now the 400 hybrids are divided into two sets of 200 each. When inbred they will produce 200 offspring. Of these, three-fourths will be red, so $\frac{3}{4}(200) = 150$ will be red.

Here is a simple experiment that you can try yourself.

Example 2

Begin with two decks of ordinary cards. Shuffle them and deal them out 2 at a time to get 52 pairs.

Problem   How many pairs will have 2 red cards? How many will have 2 black cards? How many will have 1 red and 1 black card?

Solution   Theoretically, the situation with cards is identical to the one with genes, since each card is either red (R) or black (B). Thus one-fourth of the pairs will have $2R$, one-fourth will have $2B$, and one-half will have $1R$ and $1B$. Since there are 52 pairs altogether, this translates into 13 pairs with $2R$, 13 pairs with $2B$, and 26 pairs with $1R$ and $1B$.

When you actually deal out the cards, however, you will not necessarily obtain these exact numbers. We will return to this consideration later.

The important point to be learned from Mendel's second-generation pea plants and from Example 2 is that in order to account for *all* possibilities, care must be taken to distinguish the order of the elements. Thus, with Mendel we had to differentiate $\boxed{R}\,\boxed{W}$ from $\boxed{W}\,\boxed{R}$, while in Example 2, red-black had to be differentiated from black-red. This principle will be elaborated on in the next section.

Theory versus Practice

We have described and illustrated Mendel's theoretical analysis of the problem of heredity. Now let us look at the results of the entire experiment that Mendel conducted to verify his theory. Figure 4 lists the results of this experiment for all seven traits. You can tell which trait is the dominant one by consulting the column headed by "first generation." The trait that we singled out for illustration is the third one. Notice that not exactly three-fourths of the second-generation plants are red; instead, it is 75.9 percent. (That is, $705/929 \approx 0.759$.)

Why do you think Mendel's results did not *exactly* match the theoretical prediction? Why were there about 76 percent red-flowered plants

FIGURE 4

Characters	First generation	Second generation				
		Number of plants			Percentage	
		Dominant	Recessive	Total	Dominant	Recessive
1. *Seeds:* round vs. wrinkled	All round	5,474	1,850	7,324	74.74	25.26
2. *Seeds:* yellow vs. green	All yellow	6,022	2,001	8,023	75.06	24.94
3. *Flowers:* red vs. white	All red	705	224	929	75.89	24.11
4. *Flowers:* axial vs. terminal	All axial	651	207	858	75.87	24.13
5. *Pods:* inflated vs. constricted	All inflated	882	299	1,181	74.68	25.32
6. *Pods:* green vs. yellow	All green	428	152	580	73.79	26.21
7. *Stem length:* tall vs. dwarf	All tall	787	277	1,064	73.97	26.03
Totals		14,949	5010	19,959	74.90	25.10

in his garden rather than precisely 75 percent? This is a crucial issue that pervades all of probability and statistics. The answer is not easy, and we will discuss it in detail in later sections.

Summary

The title of this section is doubly suggestive. The term "genetic approach" refers to the origin and development of a subject. In this sense, we presented Mendel's theory as part of the origin of probability theory, although the roots of the subject lay in sixteenth- and seventeenth-century gambling. The other sense refers to genetics, the branch of biology dealing with hereditary traits.

Mathematically, the importance of Mendel's achievement is that it illustrates the basic principle that if an experiment that consists of two equally likely outcomes is performed twice, then four possibilities arise, each of which has the same likelihood of occurring.

Mendel's results underlie a basic division in probability, for they show the distinction between theoretical and empirical probability. The theoretical probability that a second-generation flower is red is 3/4; the empirical probability is 705/929, which is slightly greater than 3/4. Which one do you think is correct?

Publication

Mendel's approach to genetics was revolutionary, and, like many revolutionary scientific advances, it met with an indifferent, if not hostile, reception. First, there was the difficulty of publication. Although his results were published in 1866, only an obscure European journal would accept for publication this newcomer's highly mathematical work. Even though he tried to have his paper published in a more respected journal, he met with no success. The biologists were not prepared for unconventional methods.

Second, there was the lack of realization of the importance of Mendel's theory. Most of the basic knowledge of chromosomes was known by 1884, the year in which Mendel died, yet his work continued to be ignored. In fact, the poor monk spent much of his later years embroiled in some rather bitter wrangling within the church. Finally, in 1900, three other experimenters independently rediscovered the Mendelian principle: Correns (in Germany), DeVries (in the Netherlands), and Von Tschermak (in Austria). Two years later an American biologist (Sutton) published the first analysis of Mendel's results in relation to the behavior of chromosomes. This event marked the birth of genetics as a theory. This theory offered abundant predictions, and the testing of them led to the rapid growth of genetics as an exact science. Most of the subsequent success was due to experiments that were conducted on the common fruit fly.

Mendel laid the groundwork for mathematical biology in his manuscript on genetics. His clear exposition of statistical reasoning and his

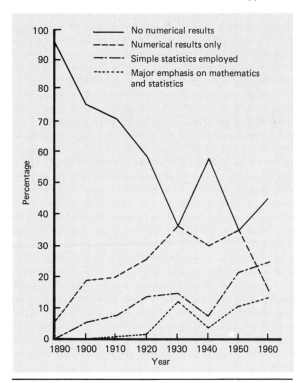

FIGURE 5 Percentage of articles involving numerical and statistical work in decennial issues of *The American Naturalist.*

use of algebra and geometry to clarify his experimental procedures and predictions changed the course of biological research forever. It might have taken biologists awhile to follow Mendel's lead, but once they did there was no turning back.

The book "Biometry," by R. R. Sokal and F. J. Rohlf, documents this ever-increasing application of probability and statistics to biological phenomena. The authors show the results of a survey of eight volumes of *The American Naturalist* from 1890 to 1960. Their analysis is reprinted in Figure 5. *The American Naturalist* has a very wide spectrum of coverage and is thus a good indicator of trends in biological research. Another measure of the importance of statistical reasoning in the biological sciences is the dramatic influx of new biological journals introduced in the last 10 years that are devoted solely to probabilistic and statistical studies. Of the new journals published, about 50 percent are statistically oriented.

Soon after the biological researchers picked up on Mendel's lead, the other physical and social sciences began to use statistical reasoning too. The same measure of the significance of this revolution that Sokal and Rohlf used in biology can be applied to sociology, psychology, political science, and various other disciplines.

The power of statistics and probability lies in its ability to establish order to seemingly haphazard or chaotic behavior in nature. Our goal is to investigate the technique of this type of scientific study and see how it can help us obtain information about certain natural phenomena.

The next section will be devoted to defining the language and notation used in probability theory. We will also begin the attack on the problem of why Mendel's results did not agree exactly with the theoretical predictions.

EXERCISES

1 Which one of the two probabilities—theoretical or empirical—do you think is correct? (See the last paragraph in the section summary.)

2 The term "genetics" comes from the word "gene," which is derived from the Greek *genos,* meaning birth, race, or kind. The words "gender," "gendarme," "genocide," and "general" all have the same root. Can you name any others?

Problems 3 to 10 are designed to give you further insight into Mendel's theoretical explanations of why three-fourths of the second-generation plants have red flowers and one-fourth have white flowers.

3 Start with 8 red and 8 white marbles. Put them in eight placeholders, each holding 2 red marbles or 2 white marbles. (Of course, you don't need marbles; two different-colored chips or two different types of coin will do. You also don't need real placeholders; you can just pair them off.) You now have four placeholders with 2 red marbles and four placeholders with 2 white marbles. Consider four empty placeholders into which we are going to put two marbles. As in the accompanying figure, put the placeholders with 2 red marbles on the left, the blank ones in the middle, and the ones with white marbles on the right. If you select 1 marble at random from each placeholder on the left and 1 from the right, and then put them in the empty placeholders, you have a red and a white marble in each. Explain how this demonstration is analogous to the first generation of hybrid plants in Mendel's experiment.

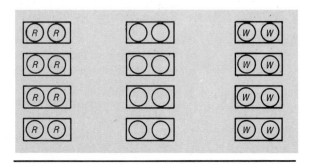

4 Take two sets of four placeholders, each placeholder having 1 red and 1 white marble. Put four empty placeholders between the two sets. Consider the experiment of drawing 1 marble at random from one placeholder on the left and 1 from the right and putting them into an empty placeholder in the middle. Do this for the placeholders on the top row; then proceed to the second, third, and fourth rows. What are the possibilities for the middle placeholders? Explain how this experiment is analogous to the second generation of Mendel's experiment.

5   Which arrangement of the marbles in the middle placeholders of Problem 4 best represents the outcome of the second generation of Mendel's experiment?

6   What percentage of the red-flowered plants in the first generation of Mendel's experiment are purebred and what percentage are hybrid?

7   Do Problem 6 for the second generation.

8   Consider a bag with 200 red and 200 white marbles. If you draw the marbles from the bag 2 at a time until all are taken from the bag, what percentage of the pairs drawn would you expect to be both red, both white, red and white?

9   Would the result of Problem 8 be the same if the bag contained 600 red and 600 white marbles?

10   Would the result of Problem 8 be the same if the bag contained 600 red and 300 white marbles?

11   If you flip a coin and record whether a head or tail appears, there are two possibilities: heads or tails. What are the possibilities if you flip two coins and record the two faces?

12   Suppose one of your enemies offers to play the following game: "We'll flip two pennies; you give me a dollar if a head and a tail appears and I'll give you a dollar if two heads appear. We'll flip again and double the bet if two tails appear." Do you think the game is fair or are you being hoodwinked? If you think the game is unfair, how would you change the rules to make it fair? In the long run, how much more money would your enemy win than you, if any?

13   Draw a card from the deck of ordinary playing cards and record its color. There are two possibilities, red or black. Now draw another card. What are the four equally likely possibilities?

14   In Problem 13, if you consider only the three possibilities—2 red cards, 2 black cards, 1 black and 1 red card—would you consider them equally likely?

15   Begin with 800 pure red-flowered and 800 pure white-flowered plants.
   (*a*)   When they are cross-fertilized, how many first-generation plants are produced?
   (*b*)   How many of them will be red?
   (*c*)   When the first-generation red-flowered plants are inbred, how many second-generation plants are produced?
   (*d*)   How many of them will be red?

16   Do Problem 15, beginning with 1856 pure plants of each color.

17 Do Problem 15, beginning with 1858 pure plants of each color. How does this exercise differ from Problem 16?

18 (a) Mendel also studied the color of the seeds of the pea plant, which were either yellow (*Y*) or green (*G*). Which color do you think is dominant? (See Figure 4.)

   (b) The pea plants are also *Y* or *G*. Which color do you think is dominant?

   (c) Do you find this contrast surprising?

19 If a tall-stemmed plant is cross-fertilized with a dwarf-stemmed plant, what do you think the length of the offspring will be? (See Figure 4.)

20 Mendel arrived at 705 red-flowered plants and 224 white-flowered plants in the second generation. How many plants in the parent generation did he begin with?

21 Mendel arrived at 6022 plants whose seeds were yellow (*Y*) and 2001 whose seeds were green (*G*) in the second generation. How many plants in the parent generation did he begin with?

22 If 16,046 first-generation plants are inbred, how many of their offspring will have the recessive trait? How many actually did, according to Figure 4?

23 Do Problem 22 for the number 1160.

24 If neither Mendel nor any of his successors had discovered a theory of genetics, what would be the chances of a second-generation plant having a red-colored flower, on the basis of the empirical evidence?

25 Do Problem 24 for (a) a yellow seed and (b) a yellow pod.

26 The three-card game, sometimes referred to as the "problem of the three chests," illustrates the $\frac{3}{4}$ versus $\frac{1}{4}$ problem in a little more deceptive setting. Three cards are put into a hat. One is white on both sides; another is red on both sides; the third is red on one side and white on the other. The dealer reaches into the hat and draws a card, placing it on a table so the side up is the only one seen by the player. Suppose the side up is white (the problem is the same if it is red; just reverse the colors). The dealer says: "This cannot be the red-red card, so it is either the white-white or the white-red card. The other side is either red or white. I'll bet even money that the other side is white." Is the bet fair?

6-2 THEORETICAL PROBABILITY EXPERIMENTS

Mendel's breakthrough in the field of genetics was both a theoretical and an empirical milestone. In this section we will study some more theoretical experiments and defer our discussion of empirical experiments for awhile. However, as we consider these simple experiments, ones that you can easily do yourselves, we will try to keep in mind their empirical base.

The simple experiments that we will investigate in this section are:

1 Flip a coin and record either heads or tails.

2 Roll an ordinary die* and record the number on the top face.

3 Flip two coins and record the number of heads observed.

4 Draw a marble from an urn that contains some red and some white marbles and record the color of the marble.

5 Draw two marbles from an urn and record the color of each.

6 Draw a playing card and record its suit.

7 Draw two playing cards and record the suit of each.

These experiments will enable us to introduce the language and elementary concepts of probability theory.

What We Mean by an Experiment

It is easy to classify Mendel's work in the garden as an experiment. Medical, scientific, and social science research consists of conducting many serious experiments, such as the search for a vaccine, the psychological experiments by Pavlov, safety tests for new drugs and inventions, and the determination of whether a new social program will be cost-effective. When we study statistics, we will concern ourselves with similar types of experiments. For now, however, we will consider simpler versions of these more serious experiments. The reason for this approach, of course, is that it will be much easier for us to explain their theoretical foundation. Once we master the underlying ideas of these experiments, we will be able to tackle the more complicated ones with greater facility and confidence.

Thus by the word "experiment" we mean an act whose possible outcomes are foreseeable and which can be repeated under given conditions.

Equally Likely Outcomes

One of the most fundamental and intuitive ideas in probability theory is that of *equally likely outcomes*. When we flip a fair coin, we usually assume that the two possible outcomes—observing heads and observing tails—have an equal chance of occurring. If an ordinary die is tossed, it is reasonable to assume that any one of the six faces—numbered 1 to 6— is just as likely to turn face-up as the others. Hence the six outcomes are assumed to be equally likely.

The question of equally likely outcomes is an intuitive one, but nevertheless it is a key issue. Why do we feel free to assume that the two outcomes of tossing a coin, and the six outcomes of tossing a die, are equally likely? Is it obvious?

---

*Most of us are accustomed to rolling a pair of dice instead of just one. The word "die" might seem awkward to you. Remember that the singular of "dice" is "die."

It is the symmetry and the homogeneity of the coin and the die that convince us that the outcomes have an equally likely chance to occur. The coin is just a bit less symmetric than the die, but the die is used almost exclusively for games of chance and hence it is constructed to ensure equally likely outcomes. Did you ever notice, for instance, how the dice are delivered to the gaming tables in casinos? They come wrapped in cellophane and are sparkling new. This ensures the gambler that all is fair (where, unfortunately, the term "fair" here means that the odds are weighted in the house's favor).

Willard W. Langcor* attempted to confirm his intuitive supposition that the outcomes of tossing a die were equally likely. He tossed precision-made dice 2 million times, using a new die for each 20,000 tosses. He recorded whether the number of the top die was even or odd. What do you think was the percentage of the outomes that were even? Langcor found that 50.045 percent were even. He repeated the experiment for inexpensive dice. The percentage of even outcomes was only a trifle larger, 50.725 percent.

We do not expect you to actually conduct such laborious experiments yourself. It is sufficient to recognize that when we *assume* that outcomes are equally likely, we have some facts to back up that assertion. Probability theory has its roots in the assumption that the outcomes of certain experiments are equally likely; yet, it is just as important to keep in mind that, whenever possible, theoretical assumptions must be examined to see if they fit the facts.

When we say that the two outcomes of flipping a coin are equally likely, do we have any facts to test our hypothesis? You have probably tossed a coin many times and your gut feeling is that heads comes up just about as often as tails.† As further evidence that this assumption is correct, we refer you to J. E. Kerrich's book, "An Experimental Introduction to the Theory of Probability" (Belgisk Import Co., Copenhagen), where the author describes the coin-tossing experiment that he designed while interned during World War II. He tossed a coin 1000 times and recorded the number of heads observed. He conducted the experiment 10 times and the outcomes were

502, 511, 497, 529, 504, 476, 507, 528, 504, 529

Note that each number is close to 500, which is *exactly* one-half of the 1000 tosses, but none is actually equal to 500. The numbers are sufficiently close to the theoretical assumption that we are confident in assuming that heads and tails are equally likely.

---

*See F. R. Mosteller et al., "Probability with Statistical Applications," Addison-Wesley, Reading, Mass., 1961.
†Even the great moguls of the National Football League determine which team gets first choice in the college draft by flipping a coin.

As one more example of equally likely outcomes, consider an ordinary deck of 52 playing cards. If the cards are adequately shuffled, they are indistinguishable from one another; so each card has an equal chance to be drawn as any other. A quick perusal of the many thorough books on strategy in card games such as poker, bridge, and blackjack shows that the accepted tactics are based on the assumption that each card has a 1-in-52 chance of being drawn.

Thus we see that the basis for the assumption of equally likely outcomes rests on the belief that if we were to conduct the experiment very many times in exactly the same manner, we would observe that each outcome would occur with about the same frequency as the others.

Probabilities: Concrete Measure of Chance Variability

We have used the words "probability" and "chance" almost synonymously. We will now assign an explicit meaning to probability. Its definition will be derived from the previous discussion of equally likely outcomes. It will hinge on the idea of percentages of outcomes if an experiment is repeated very many times.

For example, if we consider the experiment of tossing a coin, the two outcomes are heads and tails, which we will abbreviate by H and T, respectively. Since they are assumed to be equally likely, one-half the time an H will occur and one-half the time a T will occur. We express this by saying that "the probability of H occurring is $1/2$." In symbols we write

$$\Pr(H) = \frac{1}{2}$$

Similarly, we write $\Pr(T) = 1/2$, which is read "the probability of observing T, or tails, is $1/2$."

In general, we use the notation

$$\Pr(\_) = \_$$

to denote that "the probability of __ is equal to __," where the first placeholder represents an outcome of an experiment and the second is a number.

Let us look at two examples of experiments in order to get practice at using this new notation.

> **Example 1**
> If we consider the experiment of tossing a die, the six outcomes, observing 1, 2, 3, 4, 5, and 6, are equally likely, and so each has the probability $1/6$ assigned to it. For example, we say that the probability of observing 3 is $1/6$, and we write
>
> $$\Pr(3) = \frac{1}{6}$$

Example 2

If we draw a card from an ordinary deck of cards, all 52 cards are equally likely to appear, and hence the probability of drawing any one of them, say, the king of spades, is 1/52. Symbolically we write

$$\Pr(\text{king of spades}) = \frac{1}{52}$$

We can now state our first general principle for assigning probabilities to outcomes of experiments. *If an experiment has n equally likely outcomes, then the probability that any one of the outcomes will occur is 1/n.*

Elusive Equally Likely Outcomes

Sometimes it is not apparent what the equally likely outcomes of an experiment are. For example, suppose you flipped a coin twice. At first it appears that there are three possible outcomes: heads-heads, heads-tails, and tails-tails. If this were the case the probability of two heads would be 1/3. To see that this reasoning is incorrect let us picture two different coins, say a nickel and a quarter, as in Figure 1. The four distinct possibilities are: (1) heads on both coins; (2) tails on both coins; (3) heads on the nickel, tails on the quarter; (4) heads on the quarter, tails on the nickel.

If you flipped the same two coins many times, what proportion of the times would you expect to observe each of the four possibilities? If the coins are fair, then each possibility, or outcome, is as equally likely to occur as another. Hence you would expect to observe each outcome about 1/4 of the time. For example, if you flipped the coins 100 times, you would expect to observe 2 heads about 25 times, 2 tails about 25 times, heads on the quarter and tails on the nickel about 25 times, tails on the quarter and heads on the nickel about 25 times.

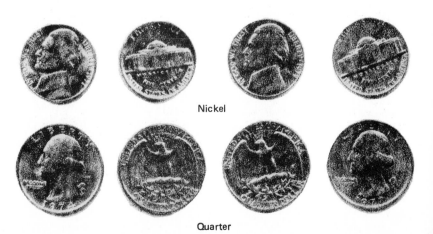

Nickel

Quarter

FIGURE 1

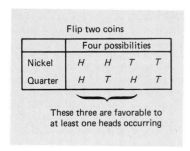

FIGURE 2

Using the notation for probabilities, if we let $2H$ represent the outcome of observing 2 heads, $2T$ represent 2 tails, HT represent first a heads and then a tails, and TH first a tails and then a heads, we have that

$$\Pr(2H) = \frac{1}{4} \qquad \Pr(HT) = \frac{1}{4}$$

$$\Pr(2T) = \frac{1}{4} \qquad \Pr(TH) = \frac{1}{4}$$

If we let "H and T" represent getting a heads and getting a tails, in either order, then

$$\Pr(H \text{ and } T) = \frac{1}{2}$$

Now let us see how the exercise of flipping two coins parallels Mendel's experiment. Suppose we flipped two coins and recorded the proportion of times that *at least* 1 heads occurred. Can you predict what that percentage is? Three of the four possibilities are favorable: TH, HT and HH (remember that at least 1 heads occurs if 1 heads or 2 heads are observed). Hence one would expect to observe at least 1 heads 3/4 of the time. We express this by writing $\Pr(\text{at least } 1H) = 3/4$. See Figure 2.

In Mendel's experiment, there were two possibilities for the color of the flower of the pea plant, red or white. Hence when the reproductive cells, each carrying the red trait or the white trait, joined to make the initial cell of the plant, there were four possibilities, three of which produced red flowers. Hence 3/4 of the time red flowers would be expected to be observed. See Figure 3.

What is the probability of observing a red-flowered plant? Since there are four equally likely outcomes and three of them are favorable to observing a red-flowered plant, the probability is 3/4. We express this by

$$\Pr(\text{red-flowered plant}) = \frac{3}{4}$$

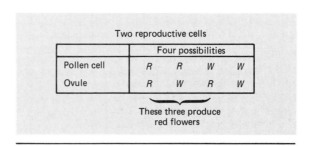

FIGURE 3

We can now generalize our notion of probability in terms of equally likely outcomes to the case where there is more than one equally likely outcome favorable to the desired possibility. *If an experiment has n equally likely outcomes and if an event is described by m equally likely outcomes, then the probability of the event occurring is m/n.*

In other words, we can describe the probability of an event having *m* favorable equally likely outcomes out of a total of *n* equally likely outcomes as

$$\frac{\text{Number of favorable outcomes}}{\text{Total number of outcomes}} = \frac{m}{n}$$

Let us look at an example of this definition.

Example 3

Problem   A card is drawn from an ordinary deck of cards. Determine the probability of observing a

(*a*) spade   (*b*) king   (*c*) picture card

Solution   There are 52 equally likely outcomes so the denominator in each problem is 52. We must find the numerator for each event.

(*a*) There are 13 spades, so

$$\text{Pr(spade)} = \frac{13}{52} = \frac{1}{4}$$

(*b*) There are 4 kings, so

$$\text{Pr(king)} = \frac{4}{52} = \frac{1}{13}$$

(*c*) There are 12 picture cards—the jack, queen, and king of each of the four suits—so

$$\text{Pr(picture card)} = \frac{12}{52} = \frac{3}{13}$$

Urn Problems

Let us consider one more type of experiment, one which is easy to visualize and similar in its simplicity to the others presented so far, but one that can be expanded easily to describe more complicated problems.

An urn problem refers to the experiment of drawing one or more marbles from an urn that contains a certain number of marbles that have distinguishing characteristics. For example, an urn might contain 3 marbles, 1 red, 1 white, and 1 blue. The experiment would be to draw 1 marble at random from the urn. The term "at random" means that each

FIGURE 4   This urn contains 3 marbles: 1 red ⓡ, 1 white ⓦ, and 1 blue ⓑ. Each has the probability of 1/3 to be drawn.

marble is as likely to be drawn as any other. Then each has the probability of 1/3 to be drawn. As we can see from Figure 4, it is easy to visualize an urn problem. An urn problem can be expanded by adding more marbles to the urn and by drawing more than one marble. We will consider both these cases in Examples 4 to 7.

It should be noted here that the importance of urn problems does not stem from the fact that urns and marbles are important in themselves. It is that they are simple and easy to visualize, and many complex probability problems can be restated so they are equivalent to an urn problem. Thus urn problems serve as models of seemingly more complicated problems. For instance, whenever a selection process is done at random, the problem is similar to an urn problem. The urn can also be visualized as a bag or a box containing numbered slips of paper, a bingo machine containing ping-pong balls with various letters and numbers, or a phone book out of which various people will be selected.

Let us look at a few simple urn problems and determine the probabilities involved.

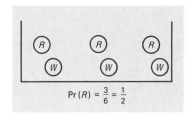

FIGURE 5   An urn containing 6 marbles: 3 red and 3 white.

Example 4

An urn contains 3 red marbles and 3 white marbles. A marble is drawn at random.

Problem   What is the probability that the marble is red?

Solution   There are six possibilities. Each is as likely to occur as the other. The event of drawing a red marble has three favorable outcomes, and hence the probability of observing a red marble, denoted by R, is

$$\Pr(R) = \frac{3}{6} = \frac{1}{2}$$

See Figure 5.

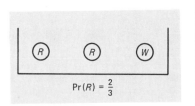

FIGURE 6   An urn containing 3 marbles: 2 red and 1 white.

Example 5

An urn contains 2 red marbles and 1 white marble. A marble is selected at random.

Problem   What is the probability that the marble is red?

Solution   There are three possibilities, two of which are favorable to the outcome of drawing a red marble. Hence the probability of drawing a red marble is

$$\Pr(R) = \frac{2}{3}$$

See Figure 6.

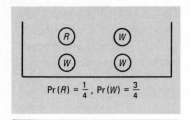

$$\Pr(R) = \tfrac{1}{4}, \; \Pr(W) = \tfrac{3}{4}$$

FIGURE 7 An urn containing 4 marbles: 1 red and 3 white.

Example 6

An urn contains 1 red marble and 3 white marbles. A marble is drawn at random.

Problem What is the probability that the marble is (*a*) red, (*b*) white?

Solution (*a*) There are four equally likely possibilities. There is one possibility that is favorable to drawing a red marble; so $\Pr(R) = 1/4$.

(*b*) There are three possibilities favorable to drawing a white marble; so $\Pr(W) = 3/4$. See Figure 7.

The next urn problem entails drawing 2 marbles from an urn. Notice that there are two distinct ways to draw 2 marbles from an urn, either with or without replacing the first one drawn. If you draw 2 marbles at once, or, equivalently, draw 1 and then another without replacing the first, it is called drawing *without replacement*. In Example 7 the first marble is replaced before selecting the second; this is called drawing *with replacement*. In a later section we will learn how to handle drawing without replacement.

Example 7

An urn contains 1 red and 1 white marble. One of the marbles is drawn at random, and its color is recorded. Then it is placed back in the urn and again 1 of the 2 marbles is drawn at random and its color is recorded.

Problem What is the probability that both marbles are white?

Solution There are two choices for the first marble drawn, *R* or *W*. There are also two choices for the second marble. The following tree diagram describes the possibilities.

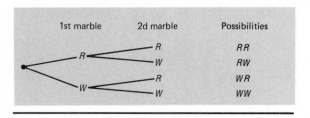

Hence there are four distinct possibilities. Only the last one, *WW*, is favorable to drawing 2 white marbles. Therefore

$$\Pr(WW) = \frac{1}{4}$$

See Figure 8.

Notice that the reasoning used to solve the problem in Example 7 is exactly the same used by Mendel to describe the heredity process. It is

FIGURE 8

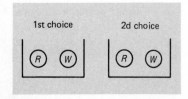

also exactly the same reasoning used earlier in this section to determine the probabilities of the experiment of tossing two coins.

Summary

In order to emphasize that the definition of probability has an empirical base, we described what is meant by the word "experiment." The meaning of the word is expanded from its usual, more restricted context of referring to experiments in the physical or social sciences, to include elementary exercises that are easy to do and understand, such as those involving tossing coins or dice.

One of the underlying principles of probability theory is that of equally likely outcomes. While we treat this as an intuitive notion, the idea of an experiment having equally likely outcomes has an empirical base. Thus it would be incorrect to assume that when two coins are tossed there are three equally likely outcomes (no heads, one heads, two heads) because empirical evidence shows that this is not true. Experience with tossing coins shows that there are four equally likely outcomes.

EXERCISES

In Problems 1 to 8 an experiment is described. List the possible outcomes and determine whether they are equally likely. What is the probability of each equally likely outcome?

1 A card is drawn from an ordinary deck of cards and the color of the card is observed.

2 A card is drawn from an ordinary deck of cards and the suit of the card is observed.

3 A card is drawn from an ordinary deck of cards and whether or not it is a picture card is observed.

4 A die is tossed and the top face is observed to be either even or odd.

5 A die is tossed and the top face is observed to be a prime number, a composite, or neither.

6 The four aces of an ordinary deck of cards are placed face down on a table. One card is chosen at random and its suit is observed.

7 In Problem 6 one of the four cards is chosen at random and its color is observed.

8 Three aces, the ace of clubs, hearts, and diamonds, are placed face down on a table. One card is chosen at random and its color is observed.

In Problems 9 and 10 an experiment is described that is very similar to the experiment of tossing two coins. List the four equally likely outcomes.

9   Two cards are drawn from an ordinary deck of cards and their color is observed. They are drawn with replacement, meaning that one card is drawn, its color is observed, and then it is replaced into the deck before the second card is drawn.

10 A roulette wheel has 38 slots, 36 of which are numbered in random order from 1 to 36. The latter are alternately colored red and black. The other two slots are called 0 and 00 and are colored green. The wheel is spun and a white ball is rolled in the opposite direction and bounces until it lands in a winning slot. Do not count the two slots 0 and 00. Play the game twice and bet only on red or black.

In Problems 11 to 15 an experiment is described whose outcomes are equally likely. Find the probability of the given event.

11 A cube has six sides each painted a different color, one of which is red. If a side is chosen at random and the color is observed, find the probability that it is red.

12 In roulette there are 38 slots into which the ball can come to rest. Find the probability that the slot labeled 00 is observed. (See Problem 10.)

13 In the game of bingo usually ping-pong balls are drawn at random and the letter (either B, I, N, G, or O) and a number are called out. Assume that there are the same number of balls having each letter. What is the probability of observing a B?

14 You are trying to remember your license plate number. You can remember the first five digits but you are drawing a complete blank as to the last digit. You choose one of the 10 digits 0 to 9. What is the probability that you will choose 0?

15 In Problem 14 suppose you remember that the number is even. What is the probability you will choose 0?

In Problems 16 to 23 assume that an urn contains the given number of red and white marbles. A marble is selected at random and its color is observed. Find the probability that the marble is red.

16   3 red and 1 white.   20   4 red and 2 white.

17   4 red and 1 white.   21   2 red and 4 white.

18   3 red and 2 white.   22   5 red and 2 white.

19   5 red and 1 white.   23   4 red and 3 white.

Problems 24 to 29 have a little twist to them. Assume an urn has an unknown number of marbles in it. Some are red and some are white. You know how many white marbles there are and you know $\Pr(R)$, the probability that a marble drawn at random is red. Find the number of marbles that are in the urn. We will start with an example.

Example
Three white and $\Pr(R) = 1/2$.

Solution One-half the total number of marbles are red because $\Pr(R) = 1/2$. Hence one-half the marbles must be white. There must be 6 marbles, 3 red and 3 white.

24 4 white and $\Pr(R) = \dfrac{1}{2}$.

25 6 white and $\Pr(R) = \dfrac{1}{2}$.

26 3 white and $\Pr(R) = \dfrac{1}{4}$.

27 6 white and $\Pr(R) = \dfrac{1}{4}$.

28 6 white and $\Pr(R) = \dfrac{2}{5}$.

29 6 white and $\Pr(R) = \dfrac{1}{7}$.

6-3   SAMPLE SPACES AND ADDITION RULES

Most people are familiar with the game of Monopoly. It is a common board game that requires each player to move a marker around the perimeter of the board, by moving the number of spaces equal to the sum of the top faces of two dice. A diagram of the Monopoly board is given below. Consider the following familiar problem that occurs often in the course of the game. (See Figure 1.)

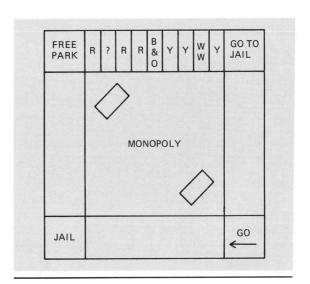

FIGURE 1

MONOPOLY PROBLEM   You are on Free Parking and your opponents have monopolies on the red and yellow spaces with hotels (and their exorbitant rents). If you land on one of these spaces, you will go bankrupt and lose the game. You will be safe if you land on any one of the other spaces. What is the probability that you will land on a safe square on your next turn?

The problem is more difficult than it may appear to you at first. It would be incorrect to use the following reasoning: There are 11 possible outcomes (the numbers 2 to 12 that are the sums on the top faces of the dice). There are 5 spaces that are disastrous (3 yellow and 2 red; the first red, Kentucky, is not possible because you cannot roll 1), and so there are 6 safe possibilities; hence the probability of landing safely is 6/11. Do you see why this reasoning is faulty? It is because the 11 outcomes are not equally likely. A 7 occurs much more often than a 2 or a 12. Maybe you would like to take a stab at solving this problem before we move on. It will take us a few more pages to discuss the ideas necessary to give you the solution.

Sample Spaces

In this section we will study a few more theoretical experiments in order to give you more experience at deriving probabilities and using the notation. Each of the experiments may appear to be concrete in that we will describe real objects—coins, dice, and cards—but the probabilities will be theoretical. We will logically piece the possibilities together rather than actually toss coins and dice or draw cards from a deck. The first few concepts will be derived via the following example.

Example 1

If three coins are tossed and the top faces are observed, what are the possibilities? You might be tempted to say that there are four: 3 heads (and no tails of course), 2 heads, 1 heads, and no heads. Are these four possibilities equally likely? If so, we would conclude that the probability of observing any one of them, say, of observing 3 heads, is 1/4. From our previous discussions you should have at least an inkling that they are not equally likely. To see this, we assume that the three coins are distinguishable. As before, even if they look alike, we can call them the first coin, the second coin, and the third coin.

Suppose the coins are a penny, a nickel, and a dime. There is only one way to get 3 heads, but there are three ways to get 2 heads. Observing 2 heads is the same as getting 1 tails, and there are three ways of getting 1 tails—a tails on one coin and a heads on the other two. Hence there are three ways to get 2 heads. Similarly, there are three ways to get 1 heads and one way to get no heads. These eight possibilities are depicted in Figure 2.

Note that the eight possibilities listed in Figure 2 are equally likely so the probability of observing any one of them is 1/8. If we let *H* stand for

Possibility	Penny	Nickel	Dime
1	*H*	*H*	*H*
2	*H*	*H*	*T*
3	*H*	*T*	*H*
4	*H*	*T*	*T*
5	*T*	*H*	*H*
6	*T*	*H*	*T*
7	*T*	*T*	*H*
8	*T*	*T*	*T*

FIGURE 2   The experiment of tossing three coins. (Let *H* stand for heads and *T* for tails.)

heads and T stand for tails, then we can label the eight possibilities by referring to the penny as the first coin, the nickel as the second, and the dime as
the third. Then HTH refers to the third possibility in Figure 2: heads on the
penny, tails on the nickel, and heads on the dime. We can express the probabilities as

$$\Pr(HHH) = \frac{1}{8} \qquad \Pr(THH) = \frac{1}{8}$$

$$\Pr(HTH) = \frac{1}{8} \qquad \Pr(TTH) = \frac{1}{8}$$

$$\Pr(HHT) = \frac{1}{8} \qquad \Pr(THT) = \frac{1}{8}$$

$$\Pr(HTT) = \frac{1}{8} \qquad \Pr(TTT) = \frac{1}{8}$$

The following definition lists several important terms in probability
theory. It shows how set theory forms the basis for probability. The necessary ingredients from set theory were described in Section 2-5.

DEFINITION   A *sample space S* of an experiment is a set of all
possible outcomes of the experiment such that any performance of
the experiment results in one and only one element in S. An *event*
is a subset of the sample space. A *simple event* consists of one outcome in S; a *complex event* consists of more than one outcome in
S.

In Example 1 the sample space is the set of eight outcomes {*HHH,
HTH, HHT, HTT, THH, TTH, THT, TTT*}. Each one of the outcomes
can be regarded as a simple event. For instance, {*HHH*} is the simple
event of tossing all heads. The next example will illustrate complex
events.

Example 2
Problem   If three coins are tossed, what is the probability of observing
exactly 1 heads?
Solution   In the experiment of tossing three coins presented in Example 1,
the event of observing exactly 1 heads is a complex event that consists of the
three simple events

$$HTT \qquad THT \qquad TTH$$

Since this event can occur by observing one of these three simple events, the probability of its occurring is 3/8. We express this in our notation by letting $1H$ stand for the event of observing exactly 1 heads (and 2 tails, of course). $Pr(1H)$ is the probability of observing $1H$, which is

$$Pr(1H) = \frac{3}{8}$$

An alternate way of describing the probability of 1 head is to notice that it is the sum of the three probabilities of the three simple events *HTT*, *THT*, and *TTH*. That is,

$$Pr(1H) = Pr(HTT) + Pr(THT) + Pr(TTH)$$
$$= \frac{1}{8} + \frac{1}{8} + \frac{1}{8} = \frac{3}{8}$$

A word of caution is in order here. In Example 2 we found $Pr(1H)$ by *adding* the probabilities of the simple events that define the complex event. This is always possible if simple events are used to describe the complex event. We will discover later when it is justified to use this addition procedure to find probabilities of complex events, and indicate when it leads to the wrong answer.

We first listed four possibilities when we conducted the experiment of tossing three coins: 3 heads, 2 heads, 1 heads, no heads. We will denote them by $3H$, $2H$, $1H$ and $0H$. The first and fourth are simple events, while the second and third are complex. We list the probabilities of the four events.

$$Pr(3H) = \frac{1}{8}$$

$$Pr(2H) = \frac{3}{8}$$

$$Pr(1H) = \frac{3}{8}$$

$$Pr(0H) = \frac{1}{8}$$

Note that the sum of the probabilities of the eight simple events is 1, and that the sum of the probabilities of the four events first examined in the discussion is also 1. The probability of "at least 1 heads" occurs when any one of the eight equally likely outcomes is observed except the last one, *TTT,* that is, the event $0H$. Hence Pr(at least 1 heads) = 7/8. Another way of looking at this is the following. Note that

$$\text{Pr}(0H) = \frac{1}{8} = 1 - \frac{7}{8}$$

$$= 1 - \text{Pr(at least 1 heads)}$$

Let us return for a moment to our definition of a sample space of an experiment. Note that we defined *a* sample space rather than *the* sample space. This is because there usually is more than one way to define a sample space for an experiment. It depends upon what one is interested in recording that determines the sample space. When tossing three coins, we could record each outcome as a sequence of three H's and T's. There would then be eight equally likely simple events. On the other hand, we could decide to record the number of heads that appear, so the sample space would have four outcomes: $0H$, $1H$, $2H$, and $3H$.

Thus we have examined two different sample spaces for the experiment of flipping three coins—one with equally likely outcomes and one without equally likely outcomes. They are contrasted in Figure 3. There are actually many more sample spaces that can be used to describe this experiment. Can you find some?

Now we will consider another experiment that has two natural, distinct sample spaces. Each sample space depends upon what outcomes are recorded in the statement of the experiment.

Equally likely outcomes			Record the number of heads	
Simple events	Pr		Simple events	Pr
HHH	$\frac{1}{8}$		3H	$\frac{1}{8}$
HHT	$\frac{1}{8}$		2H	$\frac{3}{8}$
HTH	$\frac{1}{8}$		1H	$\frac{3}{8}$
HTT	$\frac{1}{8}$		0H	$\frac{1}{8}$
THH	$\frac{1}{8}$			
THT	$\frac{1}{8}$			
TTH	$\frac{1}{8}$			
TTT	$\frac{1}{8}$			

FIGURE 3 Two sample spaces for the experiment of flipping three coins.

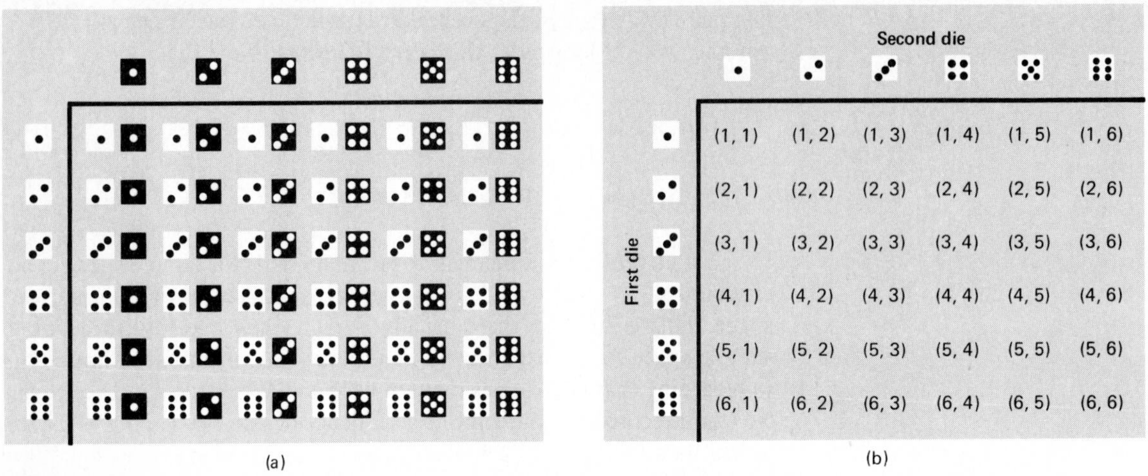

(a)

(b)

FIGURE 4

Example 3

Consider the experiment of tossing two ordinary dice and recording the numbers on the two top faces. There are six numbers on each die and they are equally likely to occur, providing the die is fair, and so the probability of observing a specific number on one die is $1/6$. When two dice are tossed, what are the possibilities? You might be tempted to say that there are 11 possibilities: 2, 3, 4, 5, 6, 7, 8, 9, 10, 11, and 12. These numbers, however, are the *sums* of the numbers on the two top faces, and our experiment is to record the numbers themselves, not their sum.

As with the coins, it is helpful to distinguish the two dice. Picture them to be different colors, or simply refer to one as the first die and the other the second. Figure 4 shows the 36 possibilities. We distinguish the dice by representing a throw by an ordered pair of numbers, with the first number in the ordered pair representing the number on the first die and the second number corresponding to the second die. Thus (5, 6) represents the throw of 5 on the first die and 6 on the second. Note that (5, 6) $\neq$ (6, 5).

The 36 combinations are equally likely, and so the probability of observing each one of them is $1/36$. Frequently the sum of the numbers is the desired outcome. As was mentioned above, there are 11 sums. They are not equally likely. Throwing a 2 (which is the vernacular for observing a sum of 2) can occur in only one way, i.e., getting a 1 on the first die and a 1 on the second; so it is considered a simple event. [We refer to it as (1, 1). Often it is called "snake-eyes."]

Similarly, throwing a 12 can occur only when we observe (6, 6), and it too corresponds to a simple event. If we let Pr(2) represent the probability of throwing 2 and Pr(12) represent the probability of throwing 12, we have

$$\text{Pr}(2) = \frac{1}{36} = \text{Pr}(12)$$

The other sums are complex events.

FIGURE 5 A sample space for the experiment of tossing two dice and recording the sum of the top faces.

Sum	2 12	3 11	4 10	5 9	6 8	7
Probability	$\frac{1}{36}$	$\frac{2}{36}$	$\frac{3}{36}$	$\frac{4}{36}$	$\frac{5}{36}$	$\frac{6}{36}$

In how many ways can we roll 3? We can throw (1, 2) or (2, 1) so there are two ways. Thus the probability of throwing 3 is $2/36 = 1/18$, so $\Pr(3) = 1/18$. There are three ways to roll 4: we can throw (1, 3), (2, 2), or (1, 3). Hence $\Pr(4) = 3/36 = 1/12$. In a similar way one computes the remaining probabilities. We record them all below, and list them in Figure 5.

$$\Pr(2) = \frac{1}{36} = \Pr(12)$$

$$\Pr(3) = \frac{2}{36} = \frac{1}{18} = \Pr(11)$$

$$\Pr(4) = \frac{3}{36} = \frac{1}{12} = \Pr(10)$$

$$\Pr(5) = \frac{4}{36} = \frac{1}{9} = \Pr(9)$$

$$\Pr(6) = \frac{5}{36} = \Pr(8)$$

$$\Pr(7) = \frac{6}{36} = \frac{1}{6}$$

Example 3 shows that a sample space for the experiment of tossing two dice can be defined in at least two ways. If the description is to observe the sum on the dice, then the simple events are 2, 3, 4, 5, 6, 7, 8, 9, 10, 11, and 12, as recorded in Figure 5. They are not equally likely and their probabilities are not equal. If, on the other hand, one observes the two faces, then there are 36 simple events. The 36 simple events in the latter description are equally likely, and each has a probability of 1/36. Nine of the eleven events obtained by summing the numbers on the faces are then complex events.

Let us look at another example of a complex event.

Example 4

Problem If two dice are tossed, what is the probability that the sum is at least 11?

Solution First we must make clear what is meant by "at least." It is equivalent to the expression "greater than or equal to." Thus the expression "at least 11" is equivalent to "greater than or equal to 11," which is the same as "11 or 12." We want to find Pr(11 or 12).

A hasty answer to the problem might be 2/11, based upon the following reasoning. The sample space consists of the 11 sums

$$S = \{2, 3, 4, 5, 6, 7, 8, 9, 10, 11, 12\}$$

Two of them are favorable, so Pr(11 or 12) is 2/11. But this is wrong, because the outcomes (the 11 sums) are not equally likely.

Earlier we emphasized the fact that an experiment can have more than one sample space. For this experiment it is easier to consider the sample space of 36 equally likely outcomes shown in Figure 4. There are two ways of throwing 11, so

$$Pr(11) = \frac{2}{36}$$

and one way of throwing 12, so

$$Pr(12) = \frac{1}{36}$$

Thus

$$Pr(\text{at least } 11) = Pr(11) + Pr(12)$$

$$= \frac{2}{36} + \frac{1}{36} = \frac{3}{36} = \frac{1}{12}$$

This same result can be obtained from the sample space $S = \{2, 3, \ldots, 11, 12\}$ too, even though the outcomes are not equally likely. You assign $Pr(11) = 2/36$ and $Pr(12) = 1/36$, and then proceed as above.

It is important for you to understand what is meant by the expressions "at least" and "at most" because you will encounter both of them frequently. Example 4 describes "at least." What do you think the probability is that the sum is at most 11?

The Addition Rule

Notice that in Example 4 we have added probabilities again. We separated the complex event of rolling at least an 11 into two other events, the simple event of rolling a 12 and the complex event of rolling an 11. We are allowed to add the probabilities of complex events to obtain the probability of the union of these events as long as they do not intersect, i.e., if there is no event that is common to any two of them.

A natural question arises. When is it not valid to add probabilities? Let us look at the example of rolling just one die. The sample space consists of the simple events, 1, 2, 3, 4, 5, and 6, each with a probability of 1/6. What is the probability of rolling at least 5 or an even number? To roll at least 5 is a complex event. There are two ways to roll at least 5—that is, rolling 5 or 6—so

$$\text{Pr(at least 5)} = \text{Pr(5 or 6)} = \text{Pr(5)} + \text{Pr(6)}$$

$$= \frac{1}{6} + \frac{1}{6} = \frac{2}{6} = \frac{1}{3}$$

You can roll 2, 4, or 6 to get an even number so

$$\text{Pr(even number)} = \frac{3}{6} = \frac{1}{2}$$

Can we add these numbers, $\frac{1}{3} + \frac{1}{2} = \frac{2}{6} + \frac{3}{6} = \frac{5}{6}$, to get the probability of the event of observing at least 5 or an even number? To find the probability, we count the simple events favorable to its occurrence. It occurs if a 5 or a 6 is rolled, or if a 2, 4, 6 occurs. But 6 is mentioned in both descriptions and it only makes sense to list it once. So the favorable outcomes are 2, 4, 5, and 6. Four simple events are favorable, so the probability is

$$\text{Pr(at least 5 or an even number)} = \frac{4}{6} = \frac{2}{3}$$

Notice that we get the wrong answer if we add the probabilities of the two events that describe the desired event. The difficulty arises when the event 6 is present in both lists of complex events. If we add the probabilities we are equivalently adding the probability of the simple event 6 twice; this is the cause of the error.

We now need a definition to explain when it is legal to add probabilities.

DEFINITION: Two events E and F are called *mutually exclusive* if $E \cap F = \varnothing$. Otherwise they are called *compatible*.

In other words, two events are mutually exclusive if they cannot occur simultaneously. Some examples will clarify these definitions.

Example 5

If we toss a die, the events of rolling an even number and rolling an odd number are mutually exclusive. Rolling an even number is compatible with rolling a number greater than 4 (that is, a 5 or 6), since both events will occur if a 6 turns up.

Often the description or definition of an event is too long or complicated to fit comfortably in the context of a paragraph. In the next example we adapt the notation

E: description of E

to define the event E.

> **Example 6**
> If we roll two dice, consider the three events:
>
> > A: observing an even number
> >
> > B: observing 2, 3, 5, or 7
> >
> > C: observing 7 or 11
>
> Then A and C are mutually exclusive; A is compatible with B since they both occur if a 2 is thrown; B is compatible with C since they both occur if 7 is thrown.

The Addition Rule Revisited

We can now state the general principle that governs when probabilities can be added.

> RULE 1 (THE ADDITION RULE): If an event A can be regarded as the union of two other events that are mutually exclusive, then the probability that A will occur is the sum of the probabilities of the other two events. In other words, if $A = B \cup C$ where $B \cap C = \emptyset$, then
>
> $$\Pr(A) = \Pr(B) + \Pr(C).$$
>
> On the other hand, it is not valid to add the probabilities if the events are compatible, that is, if there is at least one outcome common to both events. The rule also holds if the event is the union of any finite number of events.

In the next example we compute the probabilities of the events in Example 6 and use them to help explain the application of the addition rule.

Example 7

In Example 6, the three events A, B, and C of the experiment of tossing two dice were defined as:

A: observing an even number

B: observing 2, 3, 5, or 7

C: observing 7 or 11

The probability of A occurring can be derived by adding the simple events favorable to the event A. Thus

$$\Pr(A) = \Pr(2, 4, 6, 8, 10, \text{or } 12)$$
$$= \Pr(2) + \Pr(4) + \Pr(6) + \Pr(8) + \Pr(10) + \Pr(12)$$
$$= \frac{1}{36} + \frac{3}{36} + \frac{5}{36} + \frac{5}{36} + \frac{3}{36} + \frac{1}{36} = \frac{18}{36} = \frac{1}{2}$$

The probability of B can be derived by noting that

$$\Pr(B) = \Pr(2, 3, 5, \text{or } 7)$$
$$= \Pr(2) + \Pr(3) + \Pr(5) + \Pr(7)$$
$$= \frac{1}{36} + \frac{2}{36} + \frac{4}{36} + \frac{6}{36} = \frac{13}{36}$$

The probability of C occurring can be determined by noting that

$$\Pr(C) = \Pr(7 \text{ or } 11) = \Pr(7) + \Pr(11)$$
$$= \frac{6}{36} + \frac{2}{36} = \frac{8}{36} = \frac{2}{9}$$

Let D be the event of either A or C occurring. In other words D is defined as

D: observing an even number, or a 7, or an 11

Since A and C are mutually exclusive, we can use the addition rule to find $\Pr(D)$. Hence

$$\Pr(D) = \Pr(A \text{ or } C) = \Pr(A) + \Pr(C)$$
$$= \frac{1}{2} + \frac{2}{9} = \frac{9}{18} + \frac{4}{18} = \frac{13}{18}$$

What happens when the two events that comprise another complex event are compatible, that is, their intersection is not the empty set? We investigate an example first and then derive the general rule.

Example 8

In the experiment of Example 6, let the event E be defined by

E: observing B or C

In other words, $E = B \cup C$. Can we find $\Pr(E)$ by using $\Pr(B)$ and $\Pr(C)$? First of all note that we can compute $\Pr(E)$ directly because we can find the simple events that comprise it. In other words, E is defined by

E: observing 2, 3, 5, 7 or 7, 11

Note that the simple event of observing 7 is mentioned twice. It is just this redundancy that must be accounted for. Note that 7 is the single element in $B \cap C$. We can state the definition of E in terms of simple events by

E: observing 2, 3, 5, 7, or 11

Hence we can compute $\Pr(E)$ directly:

$$\Pr(E) = \Pr(2) + \Pr(3) + \Pr(5) + \Pr(7) + \Pr(11)$$

$$= \frac{1}{36} + \frac{2}{36} + \frac{4}{36} + \frac{6}{36} + \frac{2}{36} = \frac{15}{36} = \frac{5}{12}$$

If we had tried to compute $\Pr(E)$ by adding $\Pr(B)$ and $\Pr(C)$, we would have obtained the wrong answer:

$$\Pr(B) + \Pr(C) = \frac{13}{36} + \frac{8}{36} = \frac{21}{36} = \frac{7}{12}$$

The reason we get the wrong answer is that the probability of observing 7 was used twice in the above sum, once when we computed $\Pr(B)$ and once when we computed $\Pr(C)$. Just as it makes sense to mention 7 only once in the definition of E, it makes sense to add its probability only once when computing $\Pr(E)$. The outcome 7 is an element of $B \cap C$; thus if an element is in both sets, its probability gets added twice when computing the probability of each set, so it must be subtracted once. Hence we get the correct probability for $\Pr(E)$ by computing

$$\Pr(E) = \Pr(B) + \Pr(C) - \Pr(7)$$

$$= \frac{13}{36} + \frac{8}{36} - \frac{6}{36} = \frac{15}{36} = \frac{5}{12}$$

We can now express a rule that allows us to find the probability of the union of two events. Rule 1 states that if the sets are mutually exclusive we can simply add their probabilities to compute the probability of their union. Rule 2 extends this to sets that are not mutually exclusive.

RULE 2 (UNION OF TWO EVENTS): If A, B, and C are events such that A is the union of B and C—that is, A is defined by $A = B \cup C$—then

$$Pr(A) = Pr(B \cup C)$$
$$= Pr(B) + Pr(C) - Pr(B \cap C)$$

Note that if B and C are mutually exclusive then the event $B \cap C$ is an impossible event, and so $Pr(B \cap C) = 0$. In this case Rule 2 reduces to Rule 1. In other words Rule 2 includes Rule 1 in the sense that Rule 1 is a special case of Rule 2. We mentioned Rule 1 first because it is so important. Let us look at one more example to explain how to use Rule 2.

Example 9
In the experiment of tossing two dice and recording the sum of the top faces, consider the following events

A: observing an even number

B: observing a number less than 7

C: observing 6, 7, or 8

Problem Determine $Pr(A)$, $Pr(B)$, $Pr(C)$, $Pr(A \cup B)$, and $Pr(A \cup C)$.
Solution The probabilities of events A, B, and C can be calculated directly from Figure 4. They are

$$Pr(A) = \frac{18}{36} = \frac{1}{2} \qquad Pr(B) = \frac{15}{36} = \frac{5}{12} \qquad Pr(C) = \frac{16}{36} = \frac{4}{9}$$

Directly from the definition of the events we see that

$A \cap B$: observing 2, 4, or 6

$A \cap C$: observing 6 or 8

Hence

$$Pr(A \cap B) = \frac{1}{36} + \frac{3}{36} + \frac{5}{36} = \frac{9}{36} = \frac{1}{4}$$

$$Pr(A \cap C) = \frac{5}{36} + \frac{5}{36} = \frac{10}{36} = \frac{5}{18}$$

Now we can use Rule 2 to compute $Pr(A \cup B)$ and $Pr(A \cup C)$ by reducing them to probabilities that have already been calculated.

$$Pr(A \cup B) = Pr(A) + Pr(B) - Pr(A \cap B)$$

$$= \frac{18}{36} + \frac{15}{36} - \frac{9}{36} = \frac{24}{36} = \frac{2}{3}$$

$$Pr(A \cup C) = Pr(A) + Pr(C) - Pr(A \cap C)$$

$$= \frac{18}{36} + \frac{16}{36} - \frac{10}{36} = \frac{24}{36} = \frac{2}{3}$$

Complementary Probabilities

Often in games of chance, like backgammon or Monopoly, you are interested in rolling one of two or more numbers. For example, suppose rolling a 5, 6, or 7 would be disastrous. What are your chances of escaping, that is, rolling another number, any number but those three? Since there are 11 sums, you want the probability of rolling any one of eight of the others, namely 2, 3, 4, 8, 9, 10, 11, or 12. Hold on though! It would be easier to find the probability of rolling 5, 6, or 7.

$$Pr(5, 6, \text{ or } 7) = Pr(5) + Pr(6) + Pr(7)$$

$$= \frac{4}{36} + \frac{5}{36} + \frac{6}{36} = \frac{15}{36} = \frac{5}{12}$$

In other words, there are 15 ways of rolling 5, 6, or 7. Then there must be 21 (= 36 − 15) ways of not rolling a 5, 6, or 7; i.e., the probability of throwing one of the other eight numbers is 21/36 = 7/12. Thus

$$Pr(\text{not rolling 5, 6, or 7}) = 1 - \frac{15}{36} = \frac{21}{36} = \frac{7}{12}$$

Recall that the complement of a set S is denoted by S'. Thus if E is an event in a probability experiment, E' is the event that E will not happen. It is called the *complement* of E. For example, if E is the event that a 5, 6, or 7 will appear when two dice are rolled, then E' is the event that 5, 6, or 7 will not occur, so that 2, 3, 4, 8, 9, 10, 11, or 12 will occur. The fact that we just saw, namely,

$$Pr(5, 6, \text{ or } 7) = 1 - Pr(\text{not 5, 6, or 7})$$

is an example of a more general rule.

RULE 3 (COMPLEMENTARY PROBABILITIES): If E is an event and E' is its complement, then $Pr(E) = 1 - Pr(E')$.

It is instructive to see why Rule 3 is true. Let E be any event in a sample space S. Then either E occurs or it does not. Thus

$$\text{Pr}(E \cup E') = 1$$

By using Rule 1 we can rewrite the left-hand side of this formula as

$$\text{Pr}(E) + \text{Pr}(E') = 1$$

Now subtract $\text{Pr}(E')$ from both sides. You obtain

$$\text{Pr}(E) = 1 - \text{Pr}(E')$$

This is precisely what Rule 3 states.

Rule 3 is frequently used in problems involving expressions such as "at least" and "at most." The next example illustrates one such instance.

Example 10
Problem If two dice are tossed, what is the probability that the sum is less than 11?
Solution Let E be the event

E: the sum is less than 11

We want to calculate $\text{Pr}(E)$. Notice that the complement of E is the event

E': the sum is at least 11

It is easier to calculate $\text{Pr}(E')$ than $\text{Pr}(E)$ because it involves fewer simple events. In fact, in Example 4 we found $\text{Pr}(E') = 1/12$. By means of Rule 3 we obtain

$$\text{Pr}(E) = 1 - \text{Pr}(E') = 1 - \frac{1}{12} = \frac{11}{12}$$

Notice that the following two expressions are equivalent:

less than 11

at most 10

Earlier we suggested that you find Pr(at most 10). This is equivalent to Pr(less than 11), which, according to Example 10, is equal to 11/12.

Ode to a Mathematical Urn: More Examples

In order to provide more examples of the concepts contained in this section we will consider some urn problems.

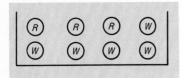

FIGURE 6

Example 11

Suppose there are 3 red and 5 white marbles in an urn. See Figure 6. There are 8 equally likely outcomes, each with probability 1/8. Since there are 3 simple events corresponding to the event of drawing a red marble, the event R, and 5 corresponding to drawing a white marble, the event W, we have the probabilities $\Pr(R) = 3/8$ and $\Pr(W) = 5/8$.

Problem What is the probability of not drawing a red marble?

Solution By a slight revision of Rule 3,

$$\Pr(R') = 1 - \Pr(R)$$

$$= 1 - \frac{3}{8} = \frac{5}{8}$$

Note that in this problem the event R' is the same as drawing a white marble.

Example 12

Suppose there are 3 red, 4 white, and 5 green marbles in an urn. See Figure 7. We will let R stand for observing a red marble, W a white marble, and G a green marble.

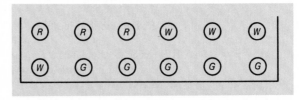

FIGURE 7

Problem Compute $\Pr(R \cup W)$ and $\Pr(R \cup G)$.

Solution The probabilities are

$$\Pr(R) = \frac{3}{12} = \frac{1}{4}$$

$$\Pr(W) = \frac{4}{12} = \frac{1}{3}$$

$$\Pr(G) = \frac{5}{12}$$

We see that $R \cup W$ is composed of the mutually exclusive events of "red" and "white." Hence we can add their probabilities to obtain

$$Pr(R \cup W) = Pr(R) + Pr(W)$$

$$= \frac{1}{4} + \frac{1}{3} = \frac{3}{12} + \frac{4}{12} = \frac{7}{12}$$

Similarly, we get

$$Pr(R \cup G) = Pr(R) + Pr(G)$$

$$= \frac{1}{4} + \frac{5}{12} = \frac{3}{12} + \frac{5}{12} = \frac{8}{12} = \frac{2}{3}$$

There are alternate ways of computing these numbers. Note that $R \cup W = G'$ and $R \cup G = W'$. Thus we could have used Rule 3 to find these two probabilities. That is

$$Pr(R \cup W) = Pr(G') = 1 - Pr(G) = 1 - \frac{5}{12} = \frac{7}{12}$$

and

$$Pr(R \cup G) = Pr(W') = 1 - Pr(W) = 1 - \frac{1}{3} = \frac{2}{3}$$

as we saw above.

All Hands on Deck: More Examples

We close this section with one more type of experiment that is familiar to you, choosing a card from an ordinary deck of cards. This example is fruitful because most people have used playing cards a great deal so that cards are familiar objects, yet the deck is complicated enough so that many intricate examples from probability theory can be generated.

Example 13

A card is drawn from an ordinary deck.

Problem What is the probability of drawing each of the following?

   (*a*) A heart
   (*b*) A picture card
   (*c*) A red card
   (*d*) A red picture card
   (*e*) A red card or a picture card

Solution   (*a*) There are 13 cards of each suit so there are 13 hearts. Hence Pr(heart) = 13/52 = 1/4.

Another way to proceed is to view the sample space as 4 suits. Then the probability of selecting 1 suit is 1/4.

(*b*) There are 3 picture cards in each suit: jack, queen, and king. Thus there are 12 picture cards, so Pr(picture card) = 12/52 = 3/13.

Another way is to let the sample space be the 13 different kinds of cards. Since 3 of them are picture cards, the probability is 3/13.

(c) Half the cards are red and half are black, so Pr(red card) = 1/2. Equivalently, there are 26 red cards in the deck, so Pr(red card) = 26/52 = 1/2.

(d) We solve this problem by counting the number of red picture cards and dividing that number by 52. There are 2 red suits, hearts and diamonds, and each suit has 3 picture cards, so there are 6 red picture cards. Hence Pr(red picture card) = 6/52 = 3/26.

(e) We can solve this problem in two ways, by counting the number of cards defined in the event or by using Rule 2. We choose the latter for two reasons—first, because it again illustrates the use of Rule 2, but mainly because the leg work has already been done in previous parts of the example. Let the two events in parts (b) and (c) be defined by

P: observing a picture card

R: observing a red card

Then our problem is to find $Pr(P \cup R)$. In part (d) we found $Pr(P \cap R)$ = 6/52. From Rule 2 we have

$$Pr(P \cup R) = Pr(P) + Pr(R) - Pr(P \cap R)$$

$$= \frac{12}{52} + \frac{26}{52} - \frac{6}{52}$$

$$= \frac{32}{52} = \frac{8}{13}$$

Summary

There is a lot of material in this section, so we will review the essentials.

The important definitions include the following terms: sample space, event, simple event, complex event, compatible events, and mutually exclusive events.

We covered three rules for combining probabilities. Rule 1 deals with the addition of probabilities of events that are mutually exclusive. Rule 2 tells us how to find the probability of a complex event that is the union of any two events. Rule 3 reveals the connection between the probability of an event and its complement.

There are a number of ideas concerning the fundamentals of probability theory that we have thus far treated intuitively. We feel that they are elementary enough that to have singled them out earlier would have interrupted our discourse. We record them now.

1   All numbers that represent probabilities are between 0 and 1.

2   The probability of an event that is certain to occur is 1.

3 The probability of an impossible event is 0.

4 The sum of the probabilities of all the outcomes in a sample space is 1.

Epilogue: How Likely Are Your Chances of Throwing the Board in Disgust

Not everyone heaves the Monopoly board into the air to vent frustration upon going bankrupt. Only those with tender egos. But now you can readily calculate your probability of escaping and remaining in the game. (At which point you will probably throw the board anyway.)

We started this section with a problem from Monopoly. Your marker was on Free Parking and your opponent had hotels on the three red properties and the three yellow properties. If you land on any of the five properties, you go bankrupt. What is the probability of escaping on your next roll of the dice?

If we assume the sample space consists of the integers 2 to 12, then the event E of escaping on your next turn is defined as

E: rolling 2, 5, 8, 10, 11, or 12

Remember that even if you go to jail (a roll of 10) you are still in the game. The simplistic reasoning would state that there are 11 possible rolls and 6 are favorable to escape; so your chances are better than even. What do you think? Let us compute $\Pr(E)$ by using Rule 1.

$$\Pr(E) = \Pr(2, 5, 8, 10, 11, 12)$$

$$= \Pr(2) + \Pr(5) + \Pr(8) + \Pr(10) + \Pr(11) + \Pr(12)$$

$$= \frac{1}{36} + \frac{4}{36} + \frac{5}{36} + \frac{3}{36} + \frac{2}{36} + \frac{1}{36} = \frac{16}{36} = \frac{4}{9} \approx 0.44$$

Your actual chances of escaping are less than even. Do you see why that is the case even though more than half the possible spaces are safe? It is because the spaces that have the hotels have higher probabilities than the other spaces. Thus when you lose you will not throw the board, simply because now you recognize that the odds are against you.

EXERCISES

In Problems 1 to 8 three coins are tossed. What is the probability of observing the given event?

1 No heads

2 Exactly 1 heads

3 At least 1 heads

4 At most 1 heads

5 3 tails

6 No tails

7 At least 2 tails

8 No heads or 3 heads

In Problems 9 to 24 two ordinary dice are tossed and the sum of the top faces is recorded. What is the probability of observing each event?

9 7 or 11

10 2 or 12

11 2, 3, or 4

12 A sum less than 5

13 A sum greater than 5

14 A sum less than 7

15 A sum greater than 7

16 An even number

17 An odd number

18 An even number greater than 7

19 Doubles or a number greater than 9

20 Doubles or a number less than 6

21 At least 10

22 At most 10

23 At most 4

24 At least 4

In Problems 25 to 28 an urn contains 10 marbles and 1 is drawn at random. What is the probability of observing a red marble if the urn has:

25 5 red and 5 white marbles

26 6 red and 4 white marbles

27 7 red and 3 white marbles

28 2 red and 8 white marbles

In Problems 29 to 35 an urn contains red, white, and green marbles, and 1 marble is drawn at random. What is the probability of observing a red marble if the urns contain the following?

29 3 red, 2 white, and 5 green marbles

30 3 red, 1 white, and 1 green marble

31 5 red, 2 white, and 7 green marbles

32 2 red, 5 white, and 7 green marbles

33 7 red, 1 white, and 2 green marbles

34 9 white and 6 green marbles

35 5 red marbles

In Problems 36 to 43 a card is drawn at random from an ordinary deck. What is the probability that it is:

36 A club

37 A spade

38 A black picture card

39 A red king

40 A card that is not a club

41 A card that is not a spade

42 A black card or a picture card

43 A heart or a card less than 5

Problems 44 to 49 refer to the game Monopoly.

44 If your marker lies on Free Parking and your opponent has a monopoly on the red spaces, what is your probability of avoiding them on your next throw?

45 If your marker lies on Free Parking and your opponent has a monopoly on the yellow spaces, what is your probability of avoiding them on your next throw?

46 Answer the Monopoly problem for the marker on New York Avenue, the space before Free Parking.

47 Answer the Monopoly problem for the marker on Tennessee Avenue, two places before Free Parking.

48 Are your chances of surviving the Monopoly problem better if the marker lies on New York Avenue or Free Parking?

49 Are your chances of surviving the Monopoly problem better if the marker lies on Tennessee Avenue or Free Parking?

Odds

Racetracks use a form of probabilities called *odds* to state the chances of a horse to win. If Judy's Boy has odds of 2 to 1 and Anita's Boy has odds of 40 to 1, it is much more likely that Judy's Boy is going to win. The odds given at racetracks are *odds against* an event happening. It is easy to convert odds against to probabilities. If the odds against an event are n to m, then the probability of an event occurring is $m/(n + m)$. For example, the probability of Judy's Boy winning is $1/(2 + 1) = 1/3$ while the probability of Anita's Boy winning is $1/(40 + 1) = 1/41$. As another example, the probability of rolling a 7 when two dice are tossed is $1/6$; so to compute the odds against rolling a 7 we let $1/6 = 1/(5 + 1)$ and hence the odds against are 5 to 1. In general, if the probability of an event is a/b, then the odds against the event are $(b - a)$ to a. In Problems 50 to 55, convert the odds against to probabilities. In Problems 56 to 61 convert the probabilities to odds against.

50 3 to 1

51 3 to 2

52 3 to 7

53 10 to 1

54 10 to 7

55 1 to 5

56	1/3		59	0.5
57	3/5		60	0.01
58	3/20		61	9/10

6-4   CONDITIONAL PROBABILITY

Probability theory is often used to make predictions about the future, using past performance as a guide. Sometimes the likelihood that an event will take place is altered by new information. Often we get a different perspective of a problem by looking at it from a slightly different angle. In this section we will discuss how the theory of probability handles the type of situation where a new or different condition is included in the problem. Often the probability of an event occurring will change if we are given that this new piece of information has occurred. This altered probability is called *conditional probability*. We will start this section by presenting three situations from everyday life that illustrate, from an intuitive point of view, how conditional probabilities are used.

1   *To picnic or not to picnic.* Last night the weather forecast indicated that the probability of rain today was 30 percent, and so we optimistically went ahead with our plans for a picnic. It is now noontime here on the blanket in the park and ominous black clouds are forming in the western sky. Mary and Bill just arrived, driving from a point due west, in a very wet car. With this additional information the probability of rain today is assigned a number much greater than 30 percent.

2   *To bet or not to bet.* Oddsmakers in sports constantly deal with probabilities. When they list the odds that a certain team will win a future game, they are relying on their knowledge and experience to assign a probability that the team will win or lose. Sometime the probabilities change when new information is uncovered. One of the most interesting examples of this occurred in a 1965 National Football League game between the New York Giants and the Baltimore Colts. Johnny Unitas, one of the greatest quarterbacks of all time, had led the Colts all year. The bookmakers made the Colts 2-1 favorites to win; in other words the probability that the Colts would win was 2/3 (see the explanation following Problem 49 in Section 6-3 to see how to translate odds into probabilities). During the week before the game Unitas was hurt and it was determined that he could not play. The Colts' only player with any experience at quarterback was a halfback named Tom Matte. The odds in favor of the Colts dropped drastically. The probability that the Colts could win, given that Unitas would not play, was set at 1/6. However, the Colts defied the odds and won the game.

3   *To jury or not to jury.* The law has minimal appreciation for mathematics, and it appears to be especially suspicious of statistical reasoning. By tradition each case on trial is considered on its own merits. It is only compared with previous trials (called *precedents*) if the cases are remarkably similar. However, slowly but surely, the precepts of quantification are gaining a foothold in the legal world. Here is a particular example.

A fundamental question confronting most defendants is whether to select a trial by jury or by judge alone. This perennial debate centers on whether the defendant's chances of acquittal are better with a jury or with the trial judge. Of course, if in most cases the jury and the judge's decisions agree, the jury system is not only superfluous but unnecessarily expensive. On the other hand, if they rarely agreed one would have to question the entire judicial system and the meaning of justice. One of the first thorough studies of this question was done by Kalvern and Zeisel,* who analyzed 3576 criminal jury trials in which the presiding trial judges agreed to report how they would have decided the case if there were no jury. The results of the survey are presented in Table 1. The numbers in the table are percentages. For example, in 14 percent of the cases (i.e., $14\% \times 3576 \approx 501$ cases) the jury acquitted the defendant and the judge agreed with the verdict. In 64 percent of the cases the jury convicted the defendant and the judge agreed. Therefore in 78 percent ($= 14 + 64$) of the cases the jury and the judge agreed. This is about what one would expect if the trial by jury method of justice is to be meaningful in the sense that the percentages do not approach either extreme. The jury is not superfluous (almost total agreement) nor is there a wide range of disagreement.

TABLE 1

Percent Agreement and Disagreement between Jury and Trial Judge

	Jury	
	Acquitted	**Convicted**
Judge:		
Acquitted	14	3
Convicted	19	64

There is an additional way to look at the data, one that illustrates what we mean by conditional probabilities. It is important to note that each of the percentages is a probability. For example, the probability that the judge and the jury agree is 78 percent or 0.78. There are two other probabilities that can be derived from the table that are very interesting for most defendants. They are:

(i) The probability that the judge would have agreed with the jury, *given that the jury had convicted the defendant*

(ii) The probability that the judge would have agreed with the jury, *given that the jury had acquitted the defendant*

To find (i) we note that since we are given the information that the jury has convicted we are interested only in the cases recorded under "Jury, Convicted," which is a total of 67 percent ($= 3 + 64$), or 2396 ($= 0.67 \times 3576$)

Jury

Convicted
3
64

---

*H. Kalvern, Jr., and H. Zeisel, "The American Jury," Little, Brown, Boston, 1966.

```
┌─────────────────────────┐
│  ┌───────────────────┐   │
│  │    Jury           │   │
│  ├───────────────────┤   │
│  │ Acquitted         │   │
│  │    14             │   │
│  │    19             │   │
│  └───────────────────┘   │
│                          │
└─────────────────────────┘
```

cases. To determine (i) we must answer the question: Of these 2396 cases, what percentage did the judge agree with, that is, convict? The answer is: In 64 percent of the total number of cases—that is, in 2288 (= 0.64 × 3576) cases—the judge agreed with the decision to convict; so 96 percent (= 64/67) of the time the judge agreed with the jury, given that the jury had convicted.

Can you calculate (ii) now? If we are given that the jury has acquitted, then we are concerned with only 33 percent (= 14 + 19) of the cases. The percentage of these cases in which the judge agreed is 42 percent (= 14/33). More importantly, from the defendant's point of view, given that the jury acquitted, the judge would have convicted in 58 percent (= 19/33) of the cases.

The main point of this discussion is that a person on trial is much better off choosing a trial by jury unless there are extenuating circumstances (for example, the lawyer's case might hinge on a subtle but compelling legal point understood by a judge but beyond the capabilities of a jury). This is because all too often (58 percent of the time) a judge would have convicted the defendant when the jury gave an acquittal.

Mathematical Treatment

We will now tackle conditional probability from a more rigorous mathematical standpoint. Once the necessary notation has been introduced the definition will be presented.

In Example 1 we will discuss a simple problem that uses conditional probability. It is a new way of looking at a familiar experiment. We will use the example to introduce our notation.

> **Example 1**
> If a fair die is rolled the probability of observing a 6 is 1/6. Suppose we are informed that an even number has been rolled. Is the probability of observing 6 changed? Is it increased or decreased? Since we know the number rolled is even it must be
>
> 2, 4, or 6
>
> These three outcomes are equally likely to occur. Instead of being one of six equally likely outcomes, the event of observing 6 is now one of three equally likely outcomes. Hence the probability of observing 6, given the additional knowledge that an even number has occurred, is 1/3.

Example 1 is a bit wordy even though the ideas presented in it are relatively simple. To express the concepts more succinctly we introduce some new notation. Let E and F be the events defined by

E: observing 6

F: observing an even number

Then the relevant probabilities can be expressed as follows:

$$\Pr(E) = \frac{1}{6} \quad \text{while} \quad \Pr(E \text{ given that } F \text{ has occurred}) = \frac{1}{3}$$

In addition, we can abbreviate the phrase

E given that *F* has occurred

by the notation

$E \mid F$

Hence $\Pr(E \mid F)$ is called the *conditional probability of E given F.* It is often shortened to simply "the probability of *E* given *F*."

We will now look at an example that is a bit more complex than Example 1. It will provide us with further insight into the definition of conditional probability.

Example 2

Consider the experiment of rolling two dice and recording the sum of the top faces. Define

E: observing an even number

F: observing 4, 7, or 10

Problem Compute $\Pr(E \mid F)$. This is, what is the probability of *E* given that *F* has occurred?

Solution A hasty approach to this problem would use the following reasoning. Since *F* is given, the only possibilities are 4, 7, and 10. Two of them are even. Thus $\Pr(E \mid F) = 2/3$. But this reasoning is faulty. Do you see why?

A better approach is to use the sample space of 36 possible outcomes as seen in Figure 4 of Section 6-3.. Since we know that *F* has occurred, we know that 1 of 12 ordered pairs, (1, 3), (2, 2), (3, 1), (1, 6), (2, 5), (3, 4), (4, 3), (5, 2), (6, 1), (4, 6), (5, 5), or (6, 4), has been rolled. Event *E* will occur if we get a roll of 4 or 10, each of which will occur in three cases. Thus 6 ordered pairs are favorable. Hence

$$\Pr(E \mid F) = \frac{6}{12} = \frac{1}{2}$$

In order to define $\Pr(E|F)$ let us look more closely at the solution of Example 2. The denominator was obtained by counting the 12 ordered pairs favorable to event F. In other words, the denominator for $\Pr(E|F)$ is equal to $\Pr(F)$. The numerator was obtained by counting the 6 ordered pairs favorable to E that were *also* favorable to F. In other words, the numerator of $\Pr(E|F)$ is equal to $\Pr(E \cap F)$. In Example 2, if we had started with $\Pr(E \cap F)$ and $\Pr(F)$, we would have obtained

$$\frac{\Pr(E \cap F)}{\Pr(F)} = \frac{6/36}{12/36} = \frac{6}{12} = \frac{1}{2} = \Pr(E|F)$$

This leads us to the definition for $\Pr(E|F)$.

DEFINITION: If A and B are events and $\Pr(B) \neq 0$, then the conditional probability of A given B is

$$\Pr(A|B) = \frac{\Pr(A \cap B)}{\Pr(B)}$$

As an immediate application of the definition we solve the problem posed in Example 1, where E was the event of tossing a 6 on the roll of a die and F was the event of observing an even number. By the definition,

$$\Pr(E|F) = \frac{\Pr(E \cap F)}{\Pr(F)} = \frac{\Pr(E)}{\Pr(F)} = \frac{1/6}{3/6} = \frac{1}{3}$$

which is the answer given in Example 1.

It is sometimes said that mathematical formulas are merely bunches of meaningless symbols. The formula given in the definition above, however, shows how unfair such a criticism is. An analysis of this formula offers an alternate way of viewing conditional probability, one that corresponds to our intuitive notion of it. For $\Pr(E \cap F)/\Pr(F)$ is the proportion of cases where both events E and F occur, if it is already known that F must occur. This is precisely what the notation $\Pr(E|F)$ means.

Another way of looking at the conditional probability $\Pr(E|F)$ is that by restricting our attention to only those outcomes favorable to F, and then determining how many of these are favorable to E, we are considering a new, restricted sample space. For instance, in Example 1 the original sample space consisted of the six outcomes 1, 2, 3, 4, 5, 6. When we computed $\Pr(E|F)$ we restricted our sample space (because we were given that F occurred) to the three outcomes 2, 4, 6. Therefore, since the simple events are equally likely, we can view the conditional probability of E given F in this alternate but equivalent way:

$$\Pr(E\,|\,F) = \frac{\text{number of outcomes favorable to } E \text{ and } F}{\text{number of outcomes favorable to } F} = \frac{1}{3}$$

This suggests an equivalent definition of conditional probability, where *the notation n(E) means the number of outcomes favorable to event E.*

ALTERNATE DEFINITION: If A and B are events in which B occurs at least 1 time, and the simple events of the sample space are equally likely, then

$$\Pr(A\,|\,B) = \frac{n(A \cap B)}{n(B)}$$

Return now to Example 2, where two dice were thrown and

E: the sum is even

F: the sum is 4, 7, or 10

If it is known that the sum is 4, 7, or 10, then the sample space is reduced to the 12 equally likely outcomes

(1, 3), (2, 2), (3, 1)

(1, 6), (2, 5), (3, 4), (4, 3), (5, 2), (6, 1)

(4, 6), (5, 5), (6, 4)

Now just count the outcomes with an even sum. There are 6 altogether, the first 3 and the last 3. Thus

$$\Pr(E\,|\,F) = \frac{n(E \cap F)}{n(F)} = \frac{6}{12} = \frac{1}{2}$$

Let us consider two more examples to help clarify the definition and its alternate form.

Example 3

A psychology class has 80 students of whom 53 are male and 22 are seniors. In addition, 10 of the seniors are female.

Problem   If a student is chosen at random, what is the probability that the person is a senior, given that a male was chosen?

Solution   We adopt the following notation concerning the events in question:

M: the student is male

S: the student is a senior

We are given $\Pr(M) = 53/80$, $\Pr(S) = 22/80$, and $\Pr(S \cap M') = 10/80$. Note that $\Pr(M \cap S) = 12/80$ because if there are 22 seniors, 10 of whom are female, the number of male seniors is 12. We have

$$\Pr(S|M) = \frac{\Pr(S \cap M)}{\Pr(M)} = \frac{12/80}{53/80} = \frac{12}{53}$$

or, using the alternate definition,

$$\Pr(S|M) = \frac{n(S \cap M)}{n(M)} = \frac{12}{53}$$

Example 4

It is often the case that real data are presented by means of a table. The information in a table can be accurately ascertained by using the thought processes inherent in the concept of conditional probability, as we saw in the introduction to this section. As an illustrative example we will discuss an important study conducted during World War II. The question raised was whether ships that are attacked by Japanese suicide planes should take violent maneuvers to avoid being hit or should continue on a steady course and trust their antiaircraft fire to repel the attack. It was important to determine whether the tactic of violent maneuvers would spoil the aim of the attacking kamikazes more than they would disrupt the aim of the antiaircraft guns.

To answer this question data were gathered on 365 ships that experienced kamikaze attacks in the summer of 1943.* The ships were categorized as "large," such as cruisers, battleships, and aircraft carriers, and "small," such as destroyers and auxiliaries.

The data are presented in the following table.

	Large Ships (L)		Small Ships (S)		Total	
Maneuvering (M):						
Number of attacks	36		144		180	
Number hit		8		52		60
Nonmaneuvering (N):						
Number of attacks	61		124		185	
Number hit		30		32		62
Total						
Number of attacks	97		268		365	
Number hit		38		84		122

*This application was adapted from P. M. Morse and G. E. Kimball, "Methods of Operations Research," Office of the Chief of Naval Operations, Navy Department, Washington, D.C., 1946, chapter 5. This was the first self-contained exposition of operations research and it contains a wealth of fascinating applications. However, there are very few copies of it available because it was classified by the Navy until about 1960.

Can you determine the conclusions reached by the Navy from the table? To decipher the implications contained in the table let us define the following events:

L: large ship

S: small ship

M: ship maneuvered to avoid attack

N: ship held a steady course

H: ship was hit by the kamikaze

Problem 1 Compare $\Pr(H|M)$ and $\Pr(H|N)$.
Solution We have $\Pr(H \cap M) = 60/365$, which is the number of ships that maneuvered and were hit (60) divided by the total number of ships. Also, $\Pr(M) = 180/365$, and hence

$$\Pr(H|M) = \frac{\Pr(H \cap M)}{\Pr(M)} = \frac{60/365}{180/365} = \frac{60}{180} = \frac{1}{3}$$

Therefore one-third of the ships that maneuvered were hit. What about those that did not manuever? We have $\Pr(H \cap N) = 62/365$, $\Pr(N) = 185/365$, and hence

$$\Pr(H|N) = \frac{\Pr(H \cap N)}{\Pr(N)} = \frac{62/365}{185/365} = \frac{62}{185}$$

The number is very close to $1/3$. Therefore, for all practical purposes, one-third of the ships that did not take violent maneuvering were hit. It seems like a stalemate, but this data included both large and small ships.

Now the Navy compared the large ships with the small ones.

Problem 2 Compare $\Pr(H|L \cap M)$ and $\Pr(H|S \cap M)$. In other words, compare the probability that a ship was hit given that it maneuvered and was large, versus the probability that it was hit given that it maneuvered and was small.
Solution We get

$$\Pr(H|L \cap M) = \frac{\Pr(H \cap L \cap M)}{\Pr(L \cap M)} = \frac{8/365}{36/365} = \frac{8}{36} = \frac{2}{9}$$

and

$$\Pr(H|S \cap M) = \frac{\Pr(H \cap S \cap M)}{\Pr(S \cap M)} = \frac{52/365}{144/365} = \frac{52}{144} = \frac{13}{36}$$

Therefore about 36 percent ($\approx 13/36$) of the small ships that maneuvered were hit, whereas only about 22 percent ($\approx 2/9$) of the large ships that maneuvered were hit.

Problem 3   Compare $\Pr(H|L \cap N)$ and $\Pr(H|S \cap N)$.
Solution   We get

$$\Pr(H|L \cap N) = \frac{\Pr(H \cap L \cap N)}{\Pr(L \cap N)} = \frac{30/365}{61/365} = \frac{30}{61}$$

and

$$\Pr(H|S \cap N) = \frac{\Pr(H \cap S \cap N)}{\Pr(S \cap N)} = \frac{32/365}{124/365} = \frac{32}{124} = \frac{8}{31}$$

Therefore about 50 percent ($\approx 30/61$) of the large ships that did not maneuver were hit, whereas only 26 percent ($\approx 8/31$) of the small ships that held a steady course were hit.

Problem 4   What was the Navy's recommendation?
Solution   Easy! When attacked by kamikazes, large ships should employ radical maneuvers and small ships should hold a steady course.

Tree Diagrams

The formula for conditional probability can be used in a modified form in order to solve certain types of problems via *tree diagrams*. If an experiment occurs in stages, or if it can be viewed as happening in two or more steps, then usually the analysis of the experiment can be given in terms of tree diagrams. The tree diagram begins with the possible outcomes at the first stage of the experiment; then for each of these outcomes the possibilities at the second stage are listed. For instance, suppose at the first stage there are two possibilities A and B, each with probability $1/2$. (Think of them as the two possibilities for your grade in this course. Choose two other letters if you are less optimistic.) We express this in the first part of a tree diagram as:

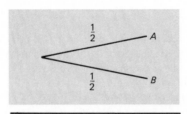

Suppose for each of these there are two possible outcomes at the second stage, say D and E (think of D as representing making the dean's list and E as representing everyone's list). Suppose that if outcome A occurs then the chances of D occurring are very good, say 90 percent, but if B occurs then there is only a 30 percent chance of D occurring. These probabilities are conditional probabilities. In other words, the probability that D occurs given that A has occurred is 0.9. Therefore

$$\Pr(D|A) = 0.9 \quad \text{and} \quad \Pr(D|B) = 0.3$$

Similarly, since E must occur if D does not occur,

$$\Pr(E \mid A) = 0.1 \qquad \text{and} \qquad \Pr(E \mid B) = 0.7$$

We express this in terms of a tree diagram as in **Figure 1**.

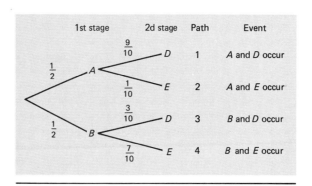

FIGURE 1

Suppose we are interested in the probability of D occurring. There are two paths in the tree that lead to the occurrence of D, one through A and one through B. Schematically they are:

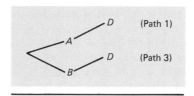

Path 1 represents the event $A \cap D$ and path 3 represents $B \cap D$. Since they are mutually exclusive, $\Pr(D) = \Pr(A \cap D) + \Pr(B \cap D)$. To compute $\Pr(A \cap D)$ we use a modified form of the rule for conditional probabilities, which we will derive now.

Recall that for any events E and F we have

$$\Pr(E \mid F) = \frac{\Pr(E \cap F)}{\Pr(F)}$$

If we multiply each side of the equation by $\Pr(F)$, we have

$$\Pr(E \cap F) = \Pr(F) \cdot \Pr(E \mid F)$$

This is a very important formula called the *product rule* of probability and so we single it out. Note that $E \cap F = F \cap E$, and make this minor adjustment in the statement of the rule.

RULE 4 (THE PRODUCT RULE):   For two events E and F,

$$\Pr(F \cap E) = \Pr(F) \cdot \Pr(E|F)$$

Here is where the tree diagram makes life a little simpler for us. To calculate $\Pr(D)$ from the tree diagram, note that we get the same computation by multiplying the probabilities on the branches of the tree along each path leading to D. That is, along path 1 through A we encounter $\Pr(A) = 1/2$ on the first branch and $\Pr(D|A) = 9/10$ on the second branch. These are the probabilities used to compute $\Pr(A \cap D)$. Similarly, along path 3 we encounter $\Pr(B) = 1/2$ and $\Pr(D|B) = 3/10$, which was used to calculate $\Pr(B \cap D)$. Figures 2 and 3 present two tree diagrams. The one in Figure 2 lists all the pertinent information so that you can more easily follow the discussion, while the tree diagram in Figure 3 contains only the information necessary to solve this type of problem. Henceforth our tree diagrams will look like that in Figure 3.

FIGURE 2

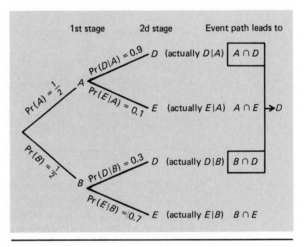

We can now use the product rule to find the probability that D will occur in our example. We have $\Pr(A) = 1/2$ and $\Pr(D|A) = 9/10$, and hence from the product rule

$$\Pr(A \cap D) = \Pr(A) \cdot \Pr(D|A)$$

$$= \frac{1}{2} \cdot \frac{9}{10} = \frac{9}{20}$$

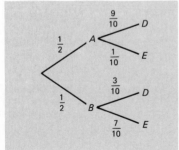

FIGURE 3

Similarly, from the product rule

$$\text{Pr}(B \cap D) = \text{Pr}(B) \cdot \text{Pr}(D|B)$$

$$= \frac{1}{2} \cdot \frac{3}{10} = \frac{3}{20}$$

Therefore

$$\text{Pr}(D) = \text{Pr}(A \cap D) + \text{Pr}(B \cap D)$$

$$= \frac{9}{20} + \frac{3}{20} = \frac{12}{20} = \frac{3}{5}$$

(Thus you have a 60 percent chance of making the dean's list!)

Tree Diagrams: Selection without Replacement

We are now able to tackle the type of problem, mentioned earlier, that involves a selection procedure without replacement. The solution of such problems involves tree diagrams. Let us look at an example.

Example 5

An urn contains 6 marbles: 2 red and 4 white; 2 marbles are selected at random without replacement.
Problem What is the probability that both marbles are white?
Solution There are two ways of interpreting the physical procedure of selecting the marbles without replacement. We could either reach in the urn and take 2 marbles in one hand or we could select 1 marble, record its color, and then select a second marble without replacing the first. Theoretically, the two procedures amount to exactly the same problem. As far as the computation is concerned, it is usually easier to make the selections one at a time.* Thus we will picture the problem in two steps, selecting 1 marble and then another. At each stage there are two possibilities, a red marble, R, or a white one, W. At the first stage the probabilities are $\text{Pr}(R) = 2/6 = 1/3$ and $\text{Pr}(W) = 4/6 = 2/3$. At the second stage the probabilities are altered, depending upon whether a red or white marble was chosen in the first step. Hence the second stage entails conditional probabilities.

It is convenient now to introduce a tree diagram. Figure 4 contains the diagram for this problem. At stage 2 there are two possibilities, R and W, associated with each of the two outcomes at stage 1. The conditional probabilities are computed from the makeup of the urn after the first marble has

---

*This is an important distinction to recognize because often a problem is stated in terms of the former procedure (for example, select two cards from a deck or choose two people from a group), but to solve the problem, you must picture the selection procedure in the latter way—the selection of the first entity followed by the selection of the second, without replacement.

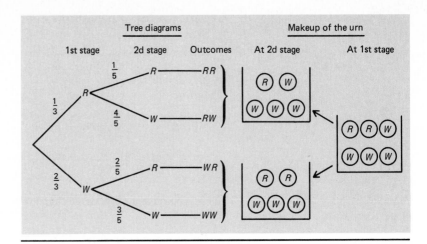

FIGURE 4

been removed. That is, if a red marble was drawn first, then the urn contains 5 marbles; 1 red and 4 white. If a white marble was drawn first, then 2 red and 3 white remain. Therefore the probabilities at the second stage depend upon what color marble was drawn at the first stage. The probabilities for the second draw are:

If a red marble was drawn first: $\quad \Pr(R) = \dfrac{1}{5} \quad \Pr(W) = \dfrac{4}{5}$

If a white marble was drawn first: $\quad \Pr(R) = \dfrac{2}{5} \quad \Pr(W) = \dfrac{3}{5}$

We insert these probabilities on the branches of the tree, along with the makeup of the urn given the outcome of the first draw in Figure 4. There are four possible outcomes—*RR, RW, WR, WW*—where the first letter corresponds to the color of the marble on the first draw and the second letter corresponds to the color of the marble on the second draw.

The problem is to find the probability that both marbles are white. This event is described by the bottom path. On the first draw we have $\Pr(W) = 2/3$ and on the second we have $\Pr(W \text{ on second} | W \text{ on first}) = 3/5$. Hence by the product rule the probability that they are both white is

$$\Pr(WW) = \frac{2}{3} \cdot \frac{3}{5} = \frac{2}{5}$$

Example 6
Consider the following two events, when cards are drawn from an ordinary deck:

F: one card is drawn, and it is a heart

S: two cards are drawn, and the second is a heart

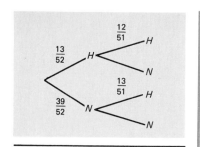

FIGURE 5

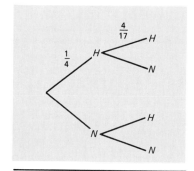

FIGURE 6

Problem Which event has a greater chance of occurring, F or S?

Solution We will find the probability of each event. The first one is obvious: $\Pr(F) = 13/52 = 1/4$. The second one is not so easy.

Draw a tree diagram of the possible outcomes and assign probabilities along the relevant branches of the tree. Let H denote "heart" and N denote "not a heart." See Figure 5.

There are two possible paths for S to occur, and we add their probabilities to obtain $\Pr(S)$. Then

$$\Pr(S) = \frac{13}{52} \cdot \frac{12}{51} + \frac{39}{52} \cdot \frac{13}{51} = \frac{13 \cdot 12 + 39 \cdot 13}{52 \cdot 51}$$

$$= \frac{13(12 + 39)}{52 \cdot 51} = \frac{13 \cdot 51}{52 \cdot 51} = \frac{13}{52} = \frac{1}{4}$$

Notice that $\Pr(F) = \Pr(S)$. Both events have the same chance of occurring.

The next two examples will give you an indication of how to compute the probabilities of poker hands. They also show that sometimes the full tree diagram is not needed to answer the specific problem.

Example 7

Problem Two cards are drawn (without replacement) at random from an ordinary deck. What is the probability that they are both hearts?

Solution At first it might seem as though there are 52 possibilities at the first stage of the experiment, but since we are only interested in whether the card is a heart or not, there are only two outcomes at each stage: a heart, denoted by H, or a card from another suit, denoted by N in the tree diagram in Figure 6. At the first stage there are 52 cards in the deck and 13 are favorable to H; so $\Pr(H) = 13/52 = 1/4$. At the second stage there are 51 cards in the deck. If a heart was drawn at the first stage, then $\Pr(H$ on second$|H$ on first$) = 12/51 = 4/17$. The problem asks for $\Pr(HH)$, and by the product rule, or, equivalently, by multiplying the probabilities on the branches of the tree diagram, we get

$$\Pr(HH) = \Pr(H) \cdot \Pr(H \text{ on second}|H \text{ on first})$$

$$= \frac{1}{4} \cdot \frac{4}{17} = \frac{1}{17}$$

The next example demonstrates how a tree diagram may require more than two stages to solve the problem. The experiment in Example 8 has five stages.

Example 8

In some games of poker five cards are dealt. A *flush* is a hand of five cards such that all the cards are of the same suit.

Problem What is the probability of being dealt a flush?

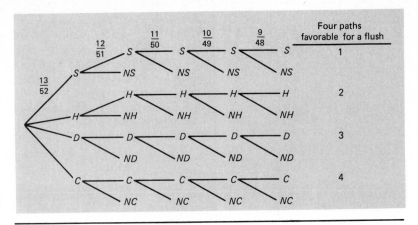

FIGURE 7 Tree diagram for the probabilities of dealing five spades.

Solution The process of dealing five cards can be thought of as a five-stage process, the first card through the fifth card. The tree diagram given in Figure 7 reflects the fact that there are four possibilities for the first card: a spade, S; a heart, H; a diamond, D; or a club, C. At the second stage we are interested only in whether the second card is the same suit as the first card. Thus if the first card is a spade, we are concerned with whether the second card is a spade, S, or not a spade, NS. Similarly, for the other three suits there are two possibilities at the second stage. Hence there are eight possibilities altogether at the second stage.

The third, fourth, and fifth stages are similar. Again we are interested in whether the card dealt agrees or does not agree in suit with the previous card. But note that if the card at stage 2 does not agree with the card at stage 1 (so either NS, NH, ND, or NC occurred), we are not interested in the next card because a flush cannot occur. Hence any part of the tree to the right of any of these outcomes is ignored.

We inserted the probabilities for drawing five consecutive spades in the tree diagram. They are computed by noticing that at each subsequent stage after the first there is one less card in the deck (so the denominators decrease by 1), and that card was a spade; so there is one less spade in the deck (and thus the numerators decrease by 1 also). We did not include the probabilities on the other three paths of the tree that comprise a flush being dealt in order to keep the diagram as uncluttered as possible. They are exactly the same as the probabilities that five successive spades are dealt. (The three other paths correspond to dealing five successive hearts, diamonds, and clubs.) Therefore once we have computed the probability of dealing five successive spades, we can get the probability of a flush by multiplying it by 4 (there are four identical paths). Therefore, by multiplying the probabilities on the branches of the tree we get

$$\Pr(\text{flush}) = 4 \cdot \Pr(\text{five successive spades})$$

$$= 4 \cdot \frac{13}{52} \cdot \frac{12}{51} \cdot \frac{11}{50} \cdot \frac{10}{49} \cdot \frac{9}{48}$$

$$= 4 \cdot \frac{1}{4} \cdot \frac{4}{17} \cdot \frac{11}{50} \cdot \frac{10}{49} \cdot \frac{3}{16}$$

$$= \frac{33}{16,660} \approx 0.002 \qquad \text{or about } 0.2\%$$

Hence a flush is very rare. It will be dealt, on the average, about twice every 1000 hands. If you are ever dealt one, bet your heart out. But beware that other hands, such as a full house (three of a kind and two of a kind) and four of a kind, have smaller probabilities and thus will beat your prized flush.

Summary

In this section we covered two important concepts in the theory of probability: conditional probability and tree diagrams. Intuitively, the conditional probability of an event E takes into account the knowledge that some other event has already occurred which may or may not affect the chances of E occurring. For two events E and F we define the conditional probability of E given F by

$$\Pr(E|F) = \frac{\Pr(E \cap F)}{\Pr(F)}$$

If a probability experiment can be interpreted as being conducted in stages, then usually a tree diagram can be used to describe the experiment. The probabilities on the tree are conditional probabilities, except for those at the first stage. Events are described by various paths through the tree. The probability of an event described by a path is the product of the probabilities on the branches of the path. This is derived from the product rule, which states that for any two events E and F,

$$\Pr(F \cap E) = \Pr(F) \cdot \Pr(E|F)$$

EXERCISES

In Problems 1 to 4 a single die is rolled. Find the probability of each event.

1   A 2, given that an even number was observed.

2   A 2, given that a number less than 4 was observed.

3   A 6, given that a number greater than 4 was observed.

4   A 5, given that a prime number was observed.

In Problems 5 to 10 two dice are rolled and their sum observed. Find the probability of each event.

5   A sum of 11, given that an odd number was observed.

6   A sum of 7, given that an odd number was observed.

7   A sum of 8, given that an even number was observed.

8   A sum of 2, given that a number less than 7 was observed.

9   A sum of 6, given that a double (both numbers the same) was observed.

10   A sum of 7, given that a double was observed.

In Problems 11 to 18 two cards are drawn from an ordinary deck, without replacement. Find the probability of observing the event.

11   The second card is black, given that the first is black.

12   The second card is a club, given that the first is a club.

13   The second card is a club, given that the first is black.

14   The second card is black, given that the first is red.

15   They are both clubs.

16   The second is a jack, given that the first is black.

17   They are both jacks.

18   The second is a picture card, given that the first is a picture card.

In Problems 19 to 28 an urn contains 10 marbles: 5 red, 3 white, and 2 blue. Two marbles are selected at random without replacement. Find the probability of each of the following events.

19   Both are red.

20   Both are white.

21   Both are blue.

22   Neither is red.

23   Neither is white.

24   Neither is blue.

25   One is red and the other is white.

26   One is red and the other is blue.

27   They are the same color.

28   They are not the same color.

Problems 29 to 34 refer to an urn containing 2 red, 3 white, and 5 blue marbles. Two marbles are selected from the urn.

29   What is the probability that at least one of the marbles is white, if the selection is made with replacement?

30   What is the probability that at least one of the marbles is white, if the selection is made without replacement?

31   Which has the greater chance of selecting at least one white marble, with or without replacement?

32 Repeat Problem 29 for a red marble.

33 Repeat Problem 30 for a red marble.

34 Repeat Problem 31 for a red marble.

In Problems 35 to 38 compute the stated probability.

35

37

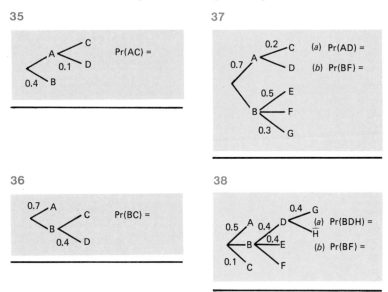

36

38

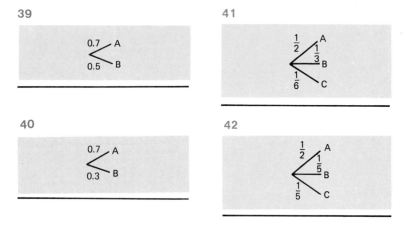

In Problems 39 to 42 determine whether the stated numbers can refer to a probability experiment.

39

41

40

42

In Problems 43 to 50 refer to the table and definition of *L*, *S*, *M*, *N*, and *H* from Example 4. Find the probability of observing the event.

43   $H|L$ 47 $(H \cap M)|L$

44   $H|S$ 48 $(H \cap N)|L$

45   $M|L$ 49 $(H \cap M)|S$

46   $M|S$ 50 $(H \cap N)|S$

In Problems 51 to 53 a hand of five cards is dealt. Find the probability of observing the event. (Draw tree diagrams to aid your solutions.)

51   Four aces.

52   Four of a kind.

53   Three aces.

54   Explain how the two Venn diagrams below can be used to describe the definition of conditional probability.

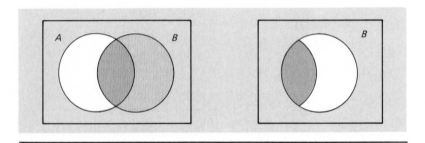

6-5   INDEPENDENT EVENTS

Some of the most successful gamblers today are mathematicians. Of course, most of them do not call themselves by that name. In fact, the casinos from whom they win their money refer to them contemptuously as "card counters," and try to bar them from their gaming tables, especially the blackjack table. One such gambler, Edward O. Thorp, who by day was a mathematics professor, has written a fascinating book that describes his system of beating the house. His method involves counting cards and memorizing the odds that result after some cards are played. By doing this he was able to tip the balance from the house's small advantage to his own.

Thorp reminds us of a seventeenth-century gambler, Antoine Gombaud, the Chevalier de Méré, who was also concerned with small advantages. Let us take a look at the situation that confronted Gombaud. He was concerned with two events:

A: at least one 6 will turn up on four throws of a single die

B: at least one pair of 6s will turn up on 24 throws of a pair of dice

He wanted to know the answers to two questions:

Q1: Which event is more likely to occur?

Q2: Is either one in the gambler's favor?

First let us examine a rule of thumb that Gombaud used to solve these questions. Since a 6 will occur once in six tosses of a single die, and a pair of 6s will come up once every 36 tosses of a pair of dice, the 6 is six times more likely (6/36) to occur than the pair of 6s (1/36). But the pair of dice is thrown six times more than the single die. Thus, he concluded, these two conditions will negate each other; so the two events, A and B, are equally likely to occur.

This answered Q1 (at least according to Gombaud's reasoning): Neither event is more likely to occur. They are equally probable. The Chevalier should have been content with this answer, but he wasn't. You see, gamblers rely more on experience (what we have called empiricism) than pure reason (theory), and Gombaud's experience had convinced him that event A occurred slightly more often than event B.

What do you do when you find yourself stuck in such a situation? Why, you call your local mathematician. And that is just what Gombaud did. He asked his friend Blaise Pascal if he could settle the matter. Pascal did. As a matter of fact, he wrote to one of his friends, Pierre Fermat (the lawyer who dabbled in mathematics), and he too settled the problem. Interestingly, each man attacked it from an entirely different approach; yet both arrived at precisely the same conclusion. The ensuing correspondence between them came to form the basis of modern probability theory.

The solution of this problem rests on a topic called *independent events*. As we learn how to handle events that are independent, we will develop a rule that governs when it is allowable to multiply probabilities. Before we tackle the solution of the problem posed by Antoine Gombaud, we will discuss independent events, first from an intuitive point of view and then from a more rigorous viewpoint.

Some Everyday Examples of Independent Events

There are many ways of comparing two events. One way is to determine if they are independent, in the sense that the occurrence of one affects the chances that the other will occur. Let us look at some examples.

1   If we flip two coins, say a nickel and a quarter, the two events "heads on the nickel" and "heads on the quarter" do not have any impact on each other. The occurrence of one does not affect the other, so we say they are independent.

2   Reggie Jackson is known as a "streak" hitter, which means he usually gets his hits in bunches and suffers long slumps with few hits in between. Thus if Reggie gets a hit in his first time at bat, it is much more likely that he is in a streak rather than a slump, so the likeli-

hood of his getting a hit in his second time at bat increases. Therefore the two events "get a hit the first time up" and "get a hit the second time up" are not independent for Reggie.

3   Calvin Murphy set an all-time National Basketball Association record in 1980 to 1981 by making 95 percent of his free-throw attempts. The man is a machine! He has perfected his skill to such a high degree of accuracy that very little affects him. Most basketball players do not shoot as well in the beginning of the game, when they are not warmed up, and at the end, when they are very tired, as in the middle of the game. Murphy, on the other hand, is the personification of consistency. One would say that his chances of successfully making a free throw remain the same no matter what time in the game it is attempted. Each attempt is independent of the others.

How can we translate these last two examples into our more rigorous language of probability theory? Consider the two events

A: Reggie gets a hit in his first time at bat in a game.

B: Reggie gets a hit in his second time at bat in a game.

Suppose Reggie is hitting 0.333, so that he has gotten a hit in one-third of his previous times at bat. We can then say that $\Pr(A) = 0.333$ and $\Pr(B) = 0.333$. If he gets a hit in his first time at bat, then he is more likely to get a hit in his second time at bat. In terms of the language of probability, if we know that A has occurred, then the probability that B will occur is increased. Thus $\Pr(B|A) \neq \Pr(B)$ for these two events. Intuitively, if two events are dependent, then $\Pr(B|A)$ is not equal to $\Pr(B)$. Knowing that A has occurred alters the likelihood that B will occur.

In contrast, we now consider the two events

E: Calvin Murphy makes his first free-throw attempt.

F: Calvin Murphy makes his second free-throw attempt.

If we assume that $\Pr(E) = \Pr(F) = 0.95$—in other words we assume that Murphy makes 95 out of 100 free throws and he is just as likely to make any attempt as any other—then the knowledge that E has occurred does not affect the probability that F will occur. That is, $\Pr(F|E) = \Pr(F) = 0.95$. The events E and F are independent in the sense that neither affects the likelihood that the other will occur.

The Definition of Independent Events

The introduction to this section was meant to provide you with an intuitive grasp of what is meant by independent events. We are now ready to present the mathematical definition of this concept.

DEFINITION: Two events E and F are said to be *independent* if $\Pr(F|E) = \Pr(F)$. If they are not independent, they are said to be *dependent*.

We can couple the definition of independence with the product rule to obtain a very useful rule that governs when we can multiply probabilities of independent events. The product rule states that for any two events E and F

$$\Pr(E \cap F) = \Pr(E) \cdot \Pr(F|E)$$

If the two events E and F are independent, then $\Pr(F|E) = \Pr(F)$. Substituting the latter equation into the former yields

$$\Pr(E \cap F) = \Pr(E) \cdot \Pr(F)$$

This is an important rule so we single it out. We have proved only one-half of the rule. We will leave the other half to the exercises.

RULE 5 (THE PRODUCT RULE FOR INDEPENDENT EVENTS): Two events E and F are independent events if and only if

$$\Pr(E \cap F) = \Pr(E) \cdot \Pr(F)$$

We have used the expression "independent events" as though it might imply a one-sided relation. For instance, if

$$\Pr(F|E) = \Pr(F)$$

then the very wording of this equality implies that F is independent of E. Is E independent of F too? The answer is "yes" because, from Rule 5, if $\Pr(E \cap F) = \Pr(E) \cdot \Pr(F)$, then

$$\Pr(E|F) = \frac{\Pr(E \cap F)}{\Pr(F)} = \frac{\Pr(E) \cdot \Pr(F)}{\Pr(F)} = \Pr(E)$$

This means that E is independent of F. Thus independent events occur in pairs.

We would like to emphasize that the product rule can be used in two ways:

1   If E and F are known to be independent, then we can compute the probability that both events will occur by multiplying their respective probabilities.

2   If the three numbers $\Pr(E)$, $\Pr(F)$, and $\Pr(E \cap F)$ are known, then we can determine whether E and F are independent by testing whether the equation $\Pr(E) \cdot \Pr(F) = \Pr(E \cap F)$ is true.

Example 1

It is known that approximately 4 percent of all people on earth are red-headed. Suppose two people are selected at random. Since the number of people is so large it is reasonable to assume that the choice of the first person and the second are independent events. In other words, the outcome of the first selection does not affect the outcome of the second selection.

Problem   What is the probability that both people are red-headed?

Solution   The probability that the first person chosen is red-headed is 0.04. Since the events are independent, the probability that the second person is also red-headed is 0.04. Hence, by the product rule for independent events,

$$\Pr(\text{both red-headed}) = \Pr(\text{first red-headed} \cap \text{second red-headed})$$

$$= \Pr(\text{first red-headed}) \cdot \Pr(\text{second red-headed})$$

$$= (0.04)(0.04) = 0.0016$$

Here is a similar example.

Example 2

A company determines that 2 percent of the items produced by a particular machine are defective. Suppose two items are chosen at random. If there is a large number of items from which to select, then it is reasonable to assume that the selection of the first and second items is independent. That is, the outcome of the first selection (defective or not) in no way affects the outcome of the second item selected.

Problem   What is the probability that both are defective?

Solution   Since the events are independent, the probability of the first choice being defective is 0.02 and the probability of the second being defective is also 0.02. Hence, by the product rule for independent events,

$$\Pr(\text{both defective}) = \Pr(\text{first defective}) \cdot \Pr(\text{second defective})$$

$$= 0.02 \cdot 0.02 = 0.0004$$

Rule 5 can be extended to any finite number of events. Thus if A, B, and C are three independent events, then

$$\Pr(A \cap B \cap C) = \Pr(A) \cdot \Pr(B) \cdot \Pr(C)$$

We will now look at some examples that extend the applicability of the product rule for independent events to experiments with more than two stages.

Example 3

If we toss two coins, we have already determined that the probability of observing two heads is $1/4$. We can use the product rule to compute this probability, and then we can easily extend the notion to the experiment of tossing more coins. Think of the experiment of tossing two coins as conducting the experiment of tossing a single coin twice. We say that each time we toss a coin it is a *trial* since we are conducting exactly the same experiment. The trials are independent because the outcome of one does not affect the other. On each trial $\Pr(H) = \Pr(T) = 1/2$. If we let "*H* on 1st" represent observing heads on the first toss, "*H* on 2d" represent observing heads on the second toss, and *HH* represent the intersection of the two, that is, heads on the first and on the second toss, then we can use the product rule to compute

$$\Pr(HH) = \Pr(H \text{ on 1st} \cap H \text{ on 2d})$$
$$= \Pr(H \text{ on 1st}) \cdot \Pr(H \text{ on 2d})$$
$$= \frac{1}{2} \cdot \frac{1}{2} = \frac{1}{4}$$

This is precisely what we expected.

We now use this line of reasoning to compute the probability of observing three heads, *HHH,* when three coins are tossed. If we consider the experiment as consisting of three independent trials of tossing one coin with $\Pr(H) = 1/2$ on each trial, then

$$\Pr(HHH) = \frac{1}{2} \cdot \frac{1}{2} \cdot \frac{1}{2} = \frac{1}{8}$$

Similarly, if four coins are tossed, then

$$\Pr(HHHH) = \left(\frac{1}{2}\right)^4 = \frac{1}{16}$$

In general, if *n* coins are tossed, then

$$\Pr(\text{all heads}) = \left(\frac{1}{2}\right)^n = \frac{1}{2^n}$$

The previous example demonstrated one of the fundamental uses of the concept of independent events and the product rule. Often an experiment can be viewed as consisting of a number of stages where the experiment at each stage is identical with the other stages. These are called *independent trials,* and the product rule for independent events allows us to multiply the probabilities computed at each stage to find the probability of an event that is the intersection of these events at the separate stages.

As another example consider an urn problem that is done with replacement. Since the marble that is chosen is replaced in the urn before the next selection, the makeup of the urn is the same for each selection. Hence the selections are independent trials. (Of course, if an urn problem is done without replacement, then the selections are dependent.)

Let us look at an urn problem that demonstrates how to use the product rule.

Example 4
An urn contains 6 marbles: 1 red, 2 white, and 3 blue.

Problem 1   If 2 marbles are drawn at random, with replacement, what is the probability that both are red?

Solution   The experiment consists of two independent trials, each a selection of a marble at random. At each stage the probability of observing a red marble, is $1/6$. Hence $\Pr(RR) = 1/36$.

Problem 2   If 3 marbles are drawn at random, with replacement, what is the probability that all 3 are blue?

Solution   The experiment consists of three independent trials, and on each trial $\Pr(B) = 3/6 = 1/2$. Hence $\Pr(BBB) = (1/2)^3 = 1/8$.

Problem 3   If 10 marbles are selected at random, with replacement, what is the probability that all are white?

Solution   Since on each of the 10 trials $\Pr(W) = 2/6 = 1/3$, we have

$$\Pr(\text{all 10 are } W) = \left(\frac{1}{3}\right)^{10} = \frac{1}{3^{10}} = \frac{1}{59{,}049}$$

$$\approx 0.000017$$

Systems Arranged in Parallel and in Series

One of the most useful applications of the concept of independent events arises in the study of systems that operate via two or more subsystems or components, each of which operates independently of the other components in the whole system. Most complex machines, computer systems, and integrated networks, such as power lines for major cities, are examples of such systems.

Associated with each system is the probability that it will operate satisfactorily (over a specific amount of time). This is called its *reliability.* Let us look at an example to help clarify the point.

Example 5
In order to run a large, complex payroll system a company uses two computers *in parallel,* which means that they work simultaneously. Normally the system runs with both computers operating, but the system can run if one computer malfunctions. The system is "down" only if both computers are down (which is computer talk for "not operating"). Experience shows that each computer is "up" (operating) 97 percent of the time.

Problem   If we assume that the computers run independently, what is the probability that the payroll system will operate satisfactorily? That is, what is its reliability?

Solution The payroll system will not run if both computers fail to operate. The probability that one will be down is 0.03; so the probability that both are down, from the product rule for independent events, is Pr(both down) = 0.03 · 0.03 = 0.0009. Hence the probability that the payroll system will run is 1 − 0.0009 = 0.9991. It is instructive to point out that the system has a much higher reliability than either of its components.

Example 5 demonstrates the need for what is often called a *backup system.* One computer was backing up the other in the sense that it could handle the load if one failed. A concrete example of such a system is the computer facility aboard the space shuttle, first launched in April 1981. The shuttle was equipped with four computers that operated independently. They had a dual purpose—each had its own tasks to accomplish during the flight but it also served as a backup for the others. If two did not agree on a calculation, then the third and fourth were consulted. The historic first flight of the space shuttle *Columbia* was delayed for several days during the last few minutes of the countdown because the reliability of the computer system was too low.

The system that works in parallel is much different than the type that operates *in series,* which is a system whose subsystems are dependent upon one another. For example, the Apollo mission to land on the moon necessitated the creation of an environment similar to that here on earth so that the astronauts could sustain themselves in a foreign and hostile environment for about 2 weeks. For the mission to be successful all the components had to function without flaw. If any one of the major systems failed, the mission would have to be aborted.

In fact the *Apollo 14* mission experienced just such a catastrophic failure while en route to the moon. The subsequent effort to return the crew safely to earth involved many long, grueling hours of work on the part of the NASA flight control group, and it marks a high point in space technology.

A simplified model of the major components of the Apollo spacecraft is given in Figure 1. The five main subsystems were: (1) the main thruster engine, (2) the service propulsion system, (3) the command service module, (4) the lunar excursion module (LEM), and (5) the LEM engine. Suppose each subsystem has a probability of success of 0.99 (this is the reliability of the subsystem). What is the reliability of the whole system? We let R stand for the reliability of the whole system, and let

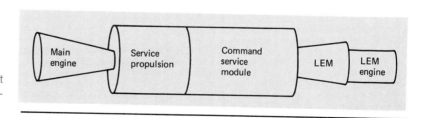

FIGURE 1 Apollo spacecraft with five independent components.

the subsystems be denoted by A, B, C, D, and E. Then we let $R(A)$ represent the reliability of A. In other words, $R(A) = \Pr(A$ operates successfully$)$. We similarly define $R(B)$, $R(C)$, $R(D)$, and $R(E)$. Thus $R(A) = R(B) = R(C) = R(D) = R(E) = 0.99$. If we assume that each subsystem operates independently of the others, then the product rule states that

$$R = R(A)R(B)R(C)R(D)R(E)$$

$$= (0.99)^5 \approx 0.95$$

Notice that for a system whose components work in parallel the reliability is greater than when the components work in series.

Let us pursue this last point with a simpler example.

Example 6

Suppose a system is composed of three components, A, B, and C. If they work in parallel, they can be described by the diagram in Figure 2. To get an idea of how the diagram depicts the situation, think of electricity flowing from left to right through the wires linking A, B, and C. Think of A, B, and C as having switches so that if any one of them fails its switch is turned off and no electricity can flow through that wire. The current will flow from left to right as long as at least one switch is open.

However, if the components are in series, as in Figure 3, then the current will flow only if all three switches are on.

Suppose the reliability of each component is given by

$$R(A) = 0.9 \qquad R(B) = 0.8 \qquad R(C) = 0.7$$

Then the probability of each component's failing can be obtained by Rule 3:

$$\Pr(A \text{ fails}) = 0.1 \qquad \Pr(B \text{ fails}) = 0.2 \qquad \Pr(C \text{ fails}) = 0.3$$

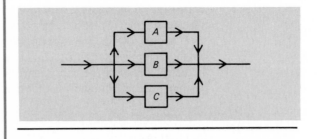

FIGURE 2   A system with three subsystems in parallel.

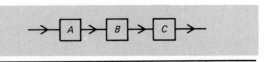

FIGURE 3   A system with three subsystems in series.

Let us compare the reliability of the entire system in parallel with the reliability of the entire system in series.

We will begin by computing the reliability of the system in parallel. Rather than proceed directly, it is better to determine what the probability of failure is, because the system will fail if and only if all three components fail. Thus

$$\text{Pr(system fails)} = \text{Pr}(A \text{ fails} \cap B \text{ fails} \cap C \text{ fails})$$

Under the assumption that the components operate independently, Rule 5 yields

$$\text{Pr}(A \text{ fails} \cap B \text{ fails} \cap C \text{ fails}) = \text{Pr}(A \text{ fails}) \cdot \text{Pr}(B \text{ fails}) \cdot \text{Pr}(C \text{ fails})$$

Putting these two formulas together and substituting their values yields

$$\text{Pr(system fails)} = (0.1)(0.2)(0.3) = 0.006$$

The reliability of any system is given by

$$R = 1 - \text{Pr(system fails)}$$

Hence $R = 1 - 0.006 = 0.994$.

Next consider the entire system in series (see Figure 3). We get

$$\text{Pr(system operates)} = \text{Pr}(A \text{ operates} \cap B \text{ operates} \cap C \text{ operates})$$
$$= \text{Pr}(A \text{ operates}) \cdot \text{Pr}(B \text{ operates}) \cdot \text{Pr}(C \text{ operates})$$

But reliability corresponds to the probability of operating, so

$$R = R(A)R(B)R(C)$$

Hence the reliability of the system in series is

$$R = (0.9)(0.8)(0.7) = 0.504$$

We present a tree diagram that is intended to graphically depict the reasoning used to determine how to compute the reliability of the two different systems. In Figure 4 if we let A, B, and C represent the three components of a system, then we place an S or an F immediately after the A, B, or C to indicate the event that the component operated successfully or failed, respectively. There are eight different paths. (This tree is very similar to the experiment of tossing three coins. Can you describe the comparison?) Only path 1, where all three operate successfully,

$$AS \text{———} BS \text{———} CS$$

is favorable to a system in series operating successfully. Only path 8, where all three fail,

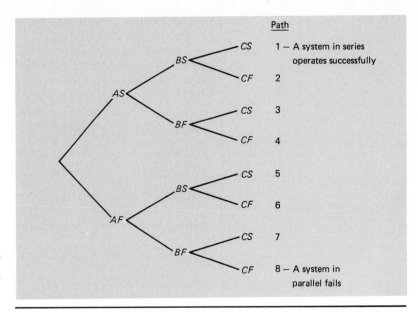

FIGURE 4   The possibilities of success (*S*) and failure (*F*) of a system with three subsystems *A*, *B*, and *C*.

$$AF \text{———} BF \text{———} CF$$

describes the possibility that a system in parallel will fail. That is, a system in parallel will operate successfully if any of the seven paths numbered 1 to 7 occurs. Rather than find each of their probabilities and add them, we make use of complementary probabilities.

When contrasting these two types of systems, you might want to keep the following guidelines in mind. If the system is in parallel, then it is easier to work with its probability of failure, because the whole system will fail when only one of its components fails. However, you can obtain the reliability of a system in series by multiplying the reliabilities of its components, because the system will operate only when all of its components operate. In short:

System	Diagram	Approach
Parallel	$\begin{bmatrix} A \\ B \\ C \end{bmatrix}$	Failure
Series	-*A*-*B*-*C*-	Success

The Test for Independence

The next few examples will describe how to use the product rule for independent events to determine whether or not two events are independent. In Examples 7 and 8 it is somewhat apparent that the events are independent, but they will demonstrate the use of the rule. In Example 9 it is not so clear beforehand whether the events are independent.

Example 7

Consider the experiment of tossing two dice and the sample space consisting of the 36 equally likely outcomes. We center our attention on the two events

> E: 6 on the first die

> F: an even number on the second die

Problem   Are E and F independent?

Solution   From an intuitive point of view it is apparent that they are independent because any outcome on the first die does not affect the outcome of the second die. Let us see if they are independent according to the product rule. We need to see if $\Pr(E) \cdot \Pr(F) = \Pr(E \cap F)$. We have $\Pr(E) = 1/6$ and $\Pr(F) = 1/2$. Hence $\Pr(E) \cdot \Pr(F) = 1/6 \cdot 1/2 = 1/12$. Thus we expect that $\Pr(E \cap F)$ will turn out to be $1/12$. To determine how many of the 36 ordered pairs in the sample space are in $E \cap F$, we need to determine how many numbers on the second die will yield an even sum together with a 6 on the first die. See Figure 4 in Section 3. There are three: 2, 4, and 6. (The others yield odd sums.) Hence $E \cap F = \{(2, 6), (4, 6), (6, 6)\}$, and so $\Pr(E \cap F) = 3/36 = 1/12$, as expected.

Example 8

Toss three coins and consider two events:

> E: all three are heads or all three are tails

> F: heads on the first coin

Problem   Are E and F independent?

Solution   This problem is a little harder to answer from an intuitive point of view. If we are given that F has occurred, is $\Pr(E)$ altered? There are eight outcomes in the sample space, and four are favorable to F (they are *HHH, HHT, HTH, HTT*) and two are favorable to E (they are *HHH, TTT*). Therefore $\Pr(E) = 2/8 = 1/4$, and if F has occurred only one of the four outcomes is favorable to E; thus $\Pr(E|F) = 1/4$. Hence according to the definition of independence, E and F are independent since $\Pr(E) = \Pr(E|F)$.

Let us check the product rule. We have $\Pr(E) = 1/4$, $\Pr(F) = 1/2$, and, since $E \cap F$ contains the single event *HHH*, $\Pr(E \cap F) = 1/8$. Therefore $\Pr(E) \cdot \Pr(F) = 1/4 \cdot 1/2 = 1/8 = \Pr(E \cap F)$ and they are indeed independent.

Example 9

Toss two dice, form their sum, and consider the three events

> E: 2, 7, or 11

> F: an odd number

> G: 4, 7, or 10

Problem   Are E and F independent? Are F and G independent?

Solution We have $\Pr(E) = 9/36 = 1/4$, $\Pr(F) = 18/36 = 1/2$, and $\Pr(G) = 12/36 = 1/3$. We note that

$$E \cap F = \{7, 11\}$$
$$F \cap G = \{7\}$$

and so $\Pr(E \cap F) = 8/36 = 2/9$ and $\Pr(F \cap G) = 1/6$. Since

$$\Pr(E) \cdot \Pr(F) = \frac{1}{4} \cdot \frac{1}{2} = \frac{1}{8}$$

and

$$\Pr(E \cap F) = \frac{2}{9}$$

it follows that

$$\Pr(E) \cdot \Pr(F) \neq \Pr(E \cap F)$$

Therefore E and F are dependent. But

$$\Pr(F) \cdot \Pr(G) = \frac{1}{2} \cdot \frac{1}{3} = \frac{1}{6}$$

and

$$\Pr(F \cap G) = \frac{1}{6}$$

Hence F and G are independent.

WARNING!

The two terms "independent events" and "mutually exclusive" are very often confused. The source of this confusion is the English-language phrase "has nothing to do with." From an intuitive point of view, when we say that the occurrence of one event has nothing to do with the occurrence of another, we mean that the events are independent. An equivalent phrase would be "does not affect the probability of." But when the phrase is applied to *sets,* as in one set has nothing to do with another, the intuitive implication is that they have no element in common. Thus there are two completely different interpretations of the phrase, and it is important that you avoid confusing them. In Example 9, notice that no pair of sets is mutually exclusive. But two are independent (F and G), while two are dependent (E and F). (See also Problem 31.)

The concept of independence occurs in many diverse fields. In many statistical studies the crux of the issue centers on whether two events are independent or whether one affects the other. For example, many people still cling to the belief that smoking cigarettes and suffering lung cancer are independent despite the fact that extensive and thorough research has yielded convincing evidence to the contrary. A major debate in economics is concerned with the rate of inflation and the rate of unemployment and whether they are dependent upon each other. We will cover more on this important topic when we study statistics.

Epilogue: The de Méré Problem

We opened this section with a description of the classic problem that the seventeenth-century rogue Antoine Gombaud, the Chevalier de Méré, posed to his local mathematician, Blaise Pascal. Let us review the question raised by Gombaud. He was concerned with two events:

A: at least one 6 will turn up on four throws of a single die

B: at least one pair of 6s will turn up on 24 throws of a pair of dice

The two questions confronting him were:

Q1: Which event is more likely to occur?

Q2: Is either one in the gambler's favor?

We attack the problems using the complements of the two events. We first look at the complement of A, which is defined by:

A': no 6 will turn up on four throws of a die

We can consider the experiment of four throws of a die as four independent trials of the experiment of tossing one die. In order for A' to occur we must observe any one of the other five numbers on each trial. The probability of that event, on each trial, is 5/6. The trials are independent so

$$\Pr(A') = \left(\frac{5}{6}\right)^4 = \frac{5^4}{6^4} = \frac{625}{1296} \approx 0.48$$

Hence

$$\Pr(A) = 1 - \Pr(A') = 1 - \frac{625}{1296} \approx 0.52$$

In the same way we can compute $\Pr(B)$ by first computing $\Pr(B')$, using the fact that throwing a pair of dice 24 times is equivalent to conducting 24 independent trials. The event we are concerned with at each trial is observing any pair of numbers except (6, 6). The probability of that event is 35/36. Hence

$$\Pr(B') = \left(\frac{35}{36}\right)^{24} = \frac{35^{24}}{36^{24}}$$

At this juncture we have a distinct advantage over Pascal because he did not have a calculator. How long do you think it took him to carry out these computations? Using our calculator we find

$$\Pr(B') \approx 0.51$$

and so

$$\Pr(B) = 1 - \Pr(B') \approx 1 - 0.51 = 0.49$$

Armed with $\Pr(A)$ and $\Pr(B)$ we can answer Gombaud's two questions. Since $\Pr(A) > \Pr(B)$, event A is more likely to occur. Since $\Pr(A) > 0.5$ event A is in the gambler's favor, but not by very much. How would you assess the Chevalier de Méré's intuitive ability to determine empirical probabilities?

EXERCISES

In Problems 1 to 6 determine whether the given events A and B are independent.

1   Mary has not studied for her history exam, which is a 10-question true-false test. She must guess at each answer.

   A: Mary gets the first question correct.
   B: Mary gets the second question correct.

2   John McEnroe is one of the top-ranked tennis players in the world. He has a "killer instinct" in that he often crushes weaker opponents. Suppose he is playing in a tournament that requires him to win two out of three sets to win a match. The experiment consists of the first round of the tournament, in which he is pitted against a weaker opponent.

   A: McEnroe wins the first set.
   B: McEnroe wins the second set.

3   A: It rained last Christmas.
   B: It will rain next Christmas.

4   A: The leaves of the oak on the right side of my house have started to change color.
   B: The leaves of the oak on the left side of my house have started to change color.

5   A: Jean is a registered Republican.
   B: Jean voted for Ronald Reagan in 1980.

6   *A*: Bill owns a Rolls Royce.
    B: Bill voted for Ronald Reagan in 1980.

In Problems 7 to 10 assume that *A* and *B* are independent. Find $\Pr(A \cap B)$.

7   $\Pr(A) = 0.5, \Pr(B) = 0.2$ 9 $\Pr(A) = 0.9, \Pr(B) = 0.8$

8   $\Pr(A) = 0.7, \Pr(B) = 0.1$ 10 $\Pr(A) = 0.9, \Pr(B) = 0.99$

In Problems 11 to 14, *n* coins are tossed. Find the probability of observing *n* heads, no heads, at least 1 heads.

11   $n = 4$ 13 $n = 8$

12   $n = 5$ 14 $n = 10$

In Problems 15 to 18 an urn contains 3 red, 4 white, and 5 blue marbles. Two marbles are drawn, with replacement. Find the probability of the event.

15   Both are red.   17   One is red and one is white.

16   Both are blue.   18   They are not the same color.

In Problems 19 to 22 an urn contains 3 red, 5 white, and 7 blue marbles. Three marbles are drawn, with replacement. Find the probability of the event.

19   All 3 are red.   21   All 3 are the same color.

20   All 3 are blue.   22   All 3 are different colors.

In Problems 23 to 26 two cards are drawn, with replacement, from an ordinary deck. Find the probability of the event.

23   Both are clubs.

24   Neither is red.

25   Both are picture cards.

26   One is a picture card and the other is not.

27   Large computer facilities usually have a backup system that copies all the work being done by the computer on a device that is external to the machine because when the computer "crashes," that is, breaks down, information in the computer is lost. A computer center determines that its computer has a reliability of 0.95 and its backup system has a reliability of 0.99. If you are running a program, what is the probability that it will be lost due to malfunction of both the computer and its backup system? Assume that they operate independently.

28   Dome houses are constructed primarily of wooden frames in the shape of triangles that are fit together in pentagonal shapes. Since these pentagons are in turn fit together, the tolerance of the size of the triangles that are to be shipped to the construction site is rather strict. A company has two assembly lines that manufacture the triangular-shaped pieces and they operate independently. One assembly line produces defective triangles at

the rate of 1 in every 20. The other produces defective triangles at the rate of 1 in every 25. What is the probability that a triangle chosen at random will be defective?

29   If a key component in a traffic controller's radar equipment has a reliability of 0.95, how many components must be used as backups to ensure a reliability of at least 0.999?

30   We proved one part of Rule 5. Prove the other part of Rule 5; namely, if $\Pr(E \cap F) = \Pr(E) \cdot \Pr(F)$, then E and F are independent.

31   Suppose two events, E and F, are mutually exclusive. Use Rule 5 to show that if neither E nor F is the empty set [or the impossible event, which implies that $\Pr(E) \neq 0$ and $\Pr(F) \neq 0$], then E and F are dependent.

32   Calvin Murphy is a 95 percent free-throw shooter. If he shoots two foul shots, what is the probability that he will make (*a*) both, (*b*) neither, (*c*) at least one?

33   (*a*)   George and Lillian Egan's first 14 children were all boys. What is the probability of such an event occurring?
(*b*)   What is the probability that their fifteenth child was a girl? (Jennifer Egan was born in November 1981, ending her parent's streak at 14.)

6-6   COUNTING TECHNIQUES

Hundreds of thousands of people daily plunk down a dollar or so, and for all too many of them it is a significant investment, in the hopes of "hitting the big one" in the state lottery. Each day, as their ticket chunks out of the computer, visions of expensive cars and wall-to-wall video monstrosities dance in their heads. The odds against their winning are monumental, but on they play. In most states the game is played by selecting a three-digit number. The winning number is then selected at random in the evening. The usual payoff is 500 to 1; that is, for every dollar bet on the winning number, the state pays $500. What do you think the probability of winning is?

If you *knew* what the number was going to be, actually knew it by some foolproof method, would you only bet $1? Of course not! You would gather all available cash and purchase tickets in a flurry. But of course this is impossible because the state governments have taken great precautions to prevent even the slightest cheating.

Why then, on April 24, 1980, did a white Cadillac pull up in front of the Dew Drop Inn in South Philadelphia and the driver purchase 904 lottery tickets? And that was just the beginning. Later in the day 1700 tickets were purchased at C and T Cold Cuts by the same man. By the end of the day he had purchased over 10,000 tickets, all on the numbers that are combinations of 4 and 6. Throughout the state accomplices had purchased another 10,000 tickets. If they bet on all the combinations of 4 and 6, how much did they improve their chances of winning? Does this seem like irrational behavior?

The Pennsylvania lottery is run each day. At 7:00 in the evening a 1-minute TV show, produced in Pittsburgh and aired throughout the state, is used to select a three-digit number. Every night a senior citizen selects one ping-pong ball from each of three large urns. Each urn contains 10 balls numbered 0 to 9.

On April 24, 1980, Violet Lowery stepped up to the platform and proceeded to draw three ping-pong balls. Each was numbered 6. The winning number was 666, thus making a great many of those 10,000 tickets purchased by the man in the white Caddie big winners. They had tickets worth over three-quarters of a million dollars. Their cohorts in crime (you didn't think they were just *lucky,* did you?) across the state had another $1 million in winning tickets.

The next day many of the sellers of the tickets, most of whom are local shop owners who also pay out the winnings for a fee from the state, complained that the payoffs they were making were enormous. Could there have been a fix? The number 666 was very suspicious. The Bible calls it "the number of the beast." But how suspicious was it?

Here are four questions raised by the possible rigging of the "daily number." What is the probability of the number 666 being drawn? Is it less likely, the same, or more likely to be drawn than any other three-digit number? If the state pays $500 for every dollar wagered on the winning number, how much money would the state expect to pay out on a given day? What facts would the state need in order to launch a full-scale investigation to determine if the fix was on?

The answers to the first three questions can be given by probability theory. We can only give hints to the answer of the fourth. Perhaps you can make an educated guess at the answers now. By the end of this section we will supply the answers.

The Multiplication Principle

To answer the first question, about the probability of drawing 666, we use a fundamental rule of counting called the *multiplication principle.* We split the problem into three stages: the first stage is to select one of the 10 numbers 0 to 9; the second stage is to select another digit, where in this problem repetitions are allowed; and the third stage consists in selecting another digit, again allowing repetitions. At each stage there are 10 ways to select a number. The multiplication principle states that there are

$$10 \cdot 10 \cdot 10 = 1000$$

total ways of selecting a three-digit number. (Notice that a three-digit number can begin with 0. Thus 007 is one such number.)

There is another method of reasoning that shows that this result is correct. Each three-digit number is simply a number between 000 and 999, and there are 1000 such numbers. You probably even find this latter line of reasoning easier than the former. The reason we first solved the problem in that manner was to introduce the multiplication principle. It can be applied to many other problems that cannot be solved as easily as this one by using the second line of reasoning. Before presenting the multiplication principle in its general form, let us cover an example whose solution requires its use.

> **Example 1**
> In the Pennsylvania lottery, three distinct urns are used to select the daily numbers, so repetitions are allowed. Suppose a lottery uses just one urn containing 10 balls, each numbered with a digit from 0 to 9. Suppose three balls are selected in order without replacement.
> **Problem** How many ways are there to select a three-digit number from the 10 digits 0 to 9 without repeating any digit?
> **Solution** We divide the problem into three stages and then determine how many ways are possible at each stage. At stage 1 there are 10 numbers that can be chosen. At stage 2 there are only 9 because one number, the one on the first ball selected, is not possible. At stage 3 there are only 8 balls left in the urn, so there are 8 ways of selecting the third digit. The multiplication principle then says we multiply the number of possibilities available at each stage. The result is
>
> $$10 \cdot 9 \cdot 8 = 720$$
> first second third
> stage stage stage
>
> Hence there are 720 total ways of selecting a three-digit number with no repetitions.

We are now ready to present the multiplication principle in its general form.

RULE 6 (MULTIPLICATION PRINCIPLE): If a counting problem can be divided into two stages such that at stage 1 there are n_1 ways to make a selection and at stage 2 there are n_2 ways, then the total number of ways to make the two selections is

$$n_1 \cdot n_2$$

Similarly, if the problem requires a third stage where there are n_3 ways to make the third selection, then the total number of ways to make the three selections is

---

$$n_1 \cdot n_2 \cdot n_3$$

In general, if there are m stages and at the ith stage there are n_i ways to make a selection, then the total number of ways to make the m selections is

$$n_1 \cdot n_2 \cdot n_3 \cdot \cdots \cdot n_m$$

The multiplication principle is applied to problems that can be visualized in stages, where at each stage a number of different selections can be made. The first question you must ask yourself, especially if the same entities are being selected at each stage, is whether the selection process is done with replacement or without. For example, in a lottery that selects a three-digit number, we saw that if repetitions are allowed the total number of possibilities is $10 \cdot 10 \cdot 10 = 1000$, whereas if repetitions are not allowed the total number of possibilities is $10 \cdot 9 \cdot 8 = 720$.

Permutations and Combinations

If the process is done with replacement, the multiplication principle suffices as the tool to use. However, if the process is done without replacement, we need a further refinement to ensure correct solutions. Let us first look at an example to see the two different types of problems possible when the selection is done without replacement.

Example 2

The Scholarship Committee has been given $1200 to award to two individuals. They have narrowed the list of candidates to five. The committee is unsure as to how to split the money, whether to give awards of $800 and $400 or to grant two $600 awards.

Problem 1 In how many ways can the committee select two of the five individuals and award one $800 and one $400 scholarship?

Problem 2 In how many ways can the committee select two of the five individuals and award two equal $600 scholarships?

Solution Before reading the solution, convince yourself that the answers to the two problems will be different. Can you see why? Can you see that some of the pairings in Problem 1 are duplicates when considered as possibilities for Problem 2?

To solve Problem 1, we can use the multiplication principle. Consider the problem as consisting of two stages, the first one to select the $800 award recipient and the second to select the $400 award recipient. There are five choices at the first stage and four at the second. Hence there are

$$5 \cdot 4 = 20$$

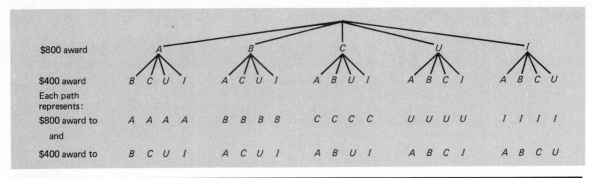

FIGURE 1   There are 20 possible ways to arrange 5 people 2 at a time.

total pairings. One can visualize this result by considering a tree diagram as in Figure 1. We denote the five individuals by *A*, *B*, *C*, *U*, and *I*. We emphasize that there is a big difference if *U* will get $800 and *I* will get $400 versus if *I* will get $800 and *U* will get $400.

Problem 2 cannot be solved by using only the multiplication principle. Note that each of the possibilities in the solution to Problem 1 appears in the list twice. For example if *U* and *I* are two of the five individuals, then *U-I* is one possibility, implying that *U* will get $800 and *I* $400, and *I-U* is another of the 20 possibilities. The order of the individuals is important in Problem 1. In Problem 2 the order is not relevant because the awards are identical, so we simply want to select two people. In the listing of the 20 possibilities in Problem 1, each choice of two people is mentioned twice, once in each ordering. Therefore there are 10 pairings for Problem 2.

It is the concept of *order* that separates the two problems in Example 2. In the first problem it mattered whose name came first and whose came second. In the second problem the order in which the names were arranged was immaterial. The following definition is meant to clarify this point.

DEFINITION: Given a set of distinct objects, an arrangement of these objects without repetition in a specific order is called a *permutation* of the objects. If a certain number of the objects are selected and the order of the objects is not considered, it is called a *combination* of the objects.

Let us look at some examples that demonstrate the definition.

Example 3
Problem How many permutations are there of the letters A, P, E?

Solution We can list them:

APE PEA EPA

AEP PAE EAP

Thus there are six permutations of the three objects.

Example 4

Problem Given the four letters A, P, E, and T, how many combinations consisting of two letters are there?

Solution There are three combinations which contain the letter A:

AP AE AT

There are also three combinations containing P, but the AP combination has been included above. The other two are:

PE PT

Again, there are three combinations containing E, but AE and PE have already been accounted for. The remaining one is

ET

Finally, there are three combinations containing T. However, all of them have already been included in the list. Hence, altogether there are six combinations of two letters that can be formed from the set of four letters. They are

AP AE AT PE PT ET

The next example describes the difference between permutations and combinations.

Example 5

Problem A sorority has an eight-member executive board. The board must submit two slates of candidates for the whole sorority to vote upon. They must elect:

(*a*) a president, a vice president, and a secretary on one slate,

(*b*) a three-member committee that will serve on the all-university sorority council on another slate.

In how many ways can the executive board choose each type of slate from among their eight members?

Solution The first question you must ask yourself is whether the selection process is with or without replacement (or repetitions). Each of these problems is done without replacement, and thus the next question is whether the problem entails permutations or combinations.

In part (*a*) the order of the listing of the three people chosen is important. Therefore each arrangement of three individuals is a permutation. In this case the multiplication principle applies. We can view the problem in three stages: first choose a president, then a vice president, and next a secretary. There are 8 choices at the first stage, 7 at the second, and 6 at the third, so there are

$$\underline{8} \;\cdot\; \underline{7} \;\cdot\; \underline{6} = 336$$

pres. vice pres. sec.

total ways of permuting the 8 people taken 3 at a time.

In part (*b*) the order is not important. To determine how many distinct combinations there are we must find out how many times each combination of three objects appears as a distinct permutation in the list of 336 permutations above. Let us look at one particular combination, say "Adams-Berry-Carter," which we abbreviate as A-B-C. In part (*a*), the permutation A-B-C implied that A is slated as president, B as vice president, and C as secretary. The permutation A-C-B is different from A-B-C; yet both represent the same combination. We must determine in how many ways the three letters are permuted. There are six ways, which can be ascertained by listing

A-B-C	B-A-C	C-A-B
A-C-B	B-C-A	C-B-A

or by using the multiplication principle. The latter method implies that there are three choices for the first stage (choosing a president), two choices for the second stage (vice president), and one for the third stage (secretary). Hence there are $3 \cdot 2 \cdot 1 = 6$ ways to permute three objects. Therefore we can conclude that in the list of 336 permutations, each distinct combination appears 6 times, so there are $336/6 = 56$ total ways of forming (combining) a three-person committee from the eight people.

Rules for Handling Permutations and Combinations

We now want to extract from the previous examples the general rules needed to handle problems involving permutations and combinations. In the examples in this section there is a similarity in the expressions formed by multiplying successive numbers. This leads us to the next definition.

DEFINITION: The expression *n*!, read "*n* factorial," is defined by

$$n! = n \cdot (n - 1) \cdot (n - 2) \cdot \cdots \cdot 3 \cdot 2 \cdot 1$$

and

$$0! = 1$$

According to folklore the exclamation point was chosen because $n!$ increases at an astonishing rate as n gets larger. Example 6 will illustrate its dramatic growth. Actually the symbol was introduced in 1808 in France for the sole purpose of circumventing printing difficulties with the notations that had been used before then.

> **Example 6**
> Calculate $n!$ for the first few integers.
>
> $$1! = 1 \qquad\qquad 5! = 5 \cdot 4 \cdot 3 \cdot 2 \cdot 1 = 120$$
>
> $$2! = 2 \cdot 1 = 2 \qquad 6! = 6 \cdot 5 \cdot 4 \cdot 3 \cdot 2 \cdot 1 = 720$$
>
> $$3! = 3 \cdot 2 \cdot 1 = 6 \qquad 7! = 7 \cdot 6! = 7 \cdot 720 = 5040$$
>
> $$4! = 4 \cdot 3 \cdot 2 \cdot 1 = 24 \qquad 8! = 8 \cdot 7! = 40{,}320$$

We will often have occasion to evaluate expressions involving two or more factorials. The next example demonstrates the method for handling such expressions.

> **Example 7**
> **Problem** Evaluate the expressions
> $$(a)\ \frac{5!}{3!} \qquad (b)\ \frac{7!}{4!} \qquad (c)\ \frac{10!}{7!}$$
> **Solution** To evaluate the expressions we expand each factorial and then cancel the common factors.
>
> $$(a)\ \frac{5!}{3!} = \frac{5 \cdot 4 \cdot \cancel{3} \cdot \cancel{2} \cdot \cancel{1}}{\cancel{3} \cdot \cancel{2} \cdot \cancel{1}} = 5 \cdot 4 = 20$$
>
> $$(b)\ \frac{7!}{4!} = \frac{7 \cdot 6 \cdot 5 \cdot \cancel{4} \cdot \cancel{3} \cdot \cancel{2} \cdot \cancel{1}}{\cancel{4} \cdot \cancel{3} \cdot \cancel{2} \cdot \cancel{1}} = 7 \cdot 6 \cdot 5 = 210$$
>
> $$(c)\ \frac{10!}{7!} = \frac{10 \cdot 9 \cdot 8 \cdot \cancel{7} \cdot \cancel{6} \cdot \cancel{5} \cdot \cancel{4} \cdot \cancel{3} \cdot \cancel{2} \cdot \cancel{1}}{\cancel{7} \cdot \cancel{6} \cdot \cancel{5} \cdot \cancel{4} \cdot \cancel{3} \cdot \cancel{2} \cdot \cancel{1}} = 10 \cdot 9 \cdot 8 = 720$$

Perhaps you recognized the expression in Example 7(c). It is the number of ways of selecting a three-digit number with no repetitions. It was calculated in Example 1 by using the multiplication principle. We can view this expression in a more general setting because there is nothing significant about a three-digit number in the calculation. Whenever we start with 10 distinct objects and permute them 3 at a time, the multiplication principle tells us that there are $10 \cdot 9 \cdot 8$ total possible permutations. Because we can express it succinctly in a formula we use the notation

$$\frac{10!}{7!} = \frac{10!}{(10 - 3)!}$$

to represent the number of ways to permute 10 distinct objects 3 at a time. In the same way the expressions in Example 7(*a*) and (*b*) represent the number of ways to permute (*a*) 5 distinct objects 2 at a time and (*b*) 7 distinct objects 3 at a time. This leads to the following rule concerning permutations:

RULE 7 (PERMUTATIONS):   The number of ways of permuting *n* distinct objects taken *r* at a time, denoted by $P(n, r)$, is given by

$$P(n, r) = \frac{n!}{(n - r)!}$$

Let us look at an example that illustrates the definition.

Example 8

Problem   Find the number of ways to permute 25 objects 2 at a time.

Solution   We must calculate $P(25, 2)$, which is

$$P(25, 2) = \frac{25!}{(25 - 2)!} = \frac{25!}{23!}$$

$$= \frac{25 \cdot 24 \cdot 23!}{23!} = 25 \cdot 24$$

$$= 600$$

How can we derive the formula for combinations from the rule governing permutations? The clue is in Example 5. The number of permutations of 8 objects taken 3 at a time was calculated to be $P(8, 3) = 8 \cdot 7 \cdot 6 = 336$. The number of combinations of 8 objects taken 3 at a time was derived from $P(8, 3)$ by noting that each combination of 3 objects appeared in the list $3! = 6$ times, which is the number of ways of permuting 3 objects 3 at a time. Hence the number of combinations was $P(8, 3)/3! = 8!/(5! \cdot 3!) = 336/6 = 56$. This motivates the following rule.

RULE 8 (COMBINATIONS):   The number of combinations of *n* distinct objects taken *r* at a time, denoted by $C(n, r)$, is

$$C(n, r) = \frac{n!}{(n - r)!r!}$$

Let us look at two examples that use Rule 8.

Example 9

Quinella wagering at racetracks requires that the bettor choose two horses. The bettor wins if the horses finish first or second, where the order of their finish is immaterial.

Problem If there are 10 horses in the race, how many distinct quinella tickets are there?

Solution Since the order is not considered, the solution is $C(10, 2)$.

$$C(10, 2) = \frac{10!}{8!2!} = \frac{10 \cdot 9}{2 \cdot 1} = 45$$

Hence there are 45 distinct tickets. Would it be advantageous to bet \$2 on each combination in order to be sure to win?

Example 10

A history professor wants to assign a take-home exam consisting of five essay questions. She has 50 students in her class and she wants to ensure that no two students have exactly the same questions.

Problem What is the smallest number of questions that she must devise in order to have at least 50 distinct exams, in the sense that two exams are distinct if at least one question is different?

Solution We must first recognize that the order of the questions is immaterial. In other words, a different permutation of the same five questions will result in the same exam. Therefore we must find the smallest number of questions, say n, such that the number of combinations of n things taken 5 at a time is at least 50. That is, we must find the smallest number n such that $C(n, 5) \geq 50$. It should be clear that $n = 5$ and $n = 6$ are too small. Let us try 7.

$$C(7, 5) = \frac{7!}{5!2!} = \frac{7 \cdot 6}{2 \cdot 1} = 21$$

This is too small also. Let us try 8.

$$C(8, 5) = \frac{8!}{5!3!} = \frac{8 \cdot 7 \cdot 6}{3 \cdot 2 \cdot 1} = 8 \cdot 7 = 56$$

Hence the instructor can devise 56 distinct exams consisting of 5 questions taken from 8 different questions.

**Epilogue:
The Lottery Fix
Revisited**

We can now answer the questions posed in the introduction to this section. When the number 666 was drawn on April 24, 1980, there should not have been any alarm that the lottery was fixed because 666 is simply one of the 1000 possibilities that are equally likely. Hence the probability of each is $1/1000 = 0.001$.

At first, state officials vehemently denied that there could have been any foul play associated with the drawing. They had made large payoffs before.

There are certain numbers that people choose more often than others. For example, any number whose second digit is 1 or 2 has a large frequency of play because it is in the form of a birthdate. For instance, 827 can be interpreted as 8-27, or August 27, and people with that birthdate will often play that number. The payoff for 666 was extremely large, but it could be accounted for because of the unusual character of the number.

Within a few days, a number of investigative reporters were on the scent. One significant fact that they uncovered was that there were a few other three-digit numbers that had very similar waging patterns on that fateful day, but most of them did not have the obvious aesthetic appeal as 666. It was soon discovered that all the numbers that consisted of 4s and 6s were bet very heavily. How many such numbers are there? Using the multiplication principle, we see there are three selections to be made and at each stage there are two ways to make the selection, either 4 or 6. Hence there are

$$2 \cdot 2 \cdot 2 = 8$$

total ways of arranging the two numbers in a three-digit number. The gamblers were confident that one of these eight numbers would be drawn. Thus they placed heavy bets on all eight numbers. This provided the authorities with very good evidence of a conspiracy, but it was still not known how they pulled it off.

How much money would the state expect to pay out? This is a little harder to assess but it is still within our grasp. If each of the 1000 numbers were bet with the same frequency (a lousy assumption, as we mentioned earlier, but more on that shortly) and the state takes in about \$1 million each day (a valid assumption), so that about \$1000 is bet on each number, then the winning number would pay out \$500 $\cdot$ 1000 = \$500,000. Hence the state pays out about half of what it takes in.

What about the bogus assumption that the numbers are bet with equal frequency? To be sure, some numbers are bet more than others. If a number with a higher frequency is chosen, the state pays out more than the \$500,000 calculated above. But if a number with a lower frequency wins, the payout is smaller. In the long run things average out so that the state keeps about half of what it takes in. The largest payoff before April 24, 1980, was about \$900,000 and the amount wagered rarely fluctuated much from the \$1 million mentioned earlier.

How did the state determine that the fix was on? On April 24, 1980, over \$1.4 million was wagered and the payoff exceeded \$2.4 million. The state lost a cool \$1 million. Was this recognized immediately? Are you kidding? Of course not. In fact the State Revenue Secretary Howard Cohen, whose department was responsible for running the lottery, stated emphatically for months that it was "impossible" to rig the lottery. Six months later when the facts were uncovered, he was forced to resign.

In 1982 two men were sentenced to prison terms of up to 7 years for their roles in the lottery fix. Ironically, the day on which they went to prison Delaware's daily number came up 555. There was no unusually heavy betting on it, however, and no investigation was begun.

Can you guess how they rigged the lottery? In each urn they weighted those balls numbered 4 and 6, ensuring that the tubes that pull one ball from each urn would pull a 4 or a 6. The nonweighted balls continued to whirl around, thus giving the appearance that the choice was random and the lottery was fair.

EXERCISES

In Problems 1 to 10 use the multiplication principle to answer the question.

1   Most license plates for cars in Illinois consist of two letters followed by four digits. Given that the four digits are 1111, how many distinct license plates can be made?

2   In Problem 1, given that the two letters are AB, how many distinct license plates can be made?

3   How many license plates can be made that consist of two letters followed by four digits?

4   How many license plates can be made that consist of three letters followed by three digits?

5   If a coin is flipped (and *H* or *T* is recorded) and then a die is tossed (and the number on the top face is recorded), how many outcomes are possible?

6   If two coins are flipped (and either *H* or *T* is recorded for each) and two dice are tossed (and the numbers on the top faces are recorded), how many outcomes are possible?

7   A combination lock has a dial with 10 digits. Three numbers are chosen at random to form the combination that opens the lock. [For example, if the combination is 3-5-3, most locks require you to turn the dial to the right to 3, then to the left to 5 (sometimes you must bypass the first digit on the way to the second), and then right to 3.] If the numbers can be repeated, how many lock combinations* can be formed?

8   In most locks it is not possible to have a lock combination (see footnote) consisting of three digits such that the first and second digits are the same. It is possible that the second and third digits are equal. Given this restriction, how many lock combinations are possible if the lock has 10 digits?

9   If the dial of a lock has 30 numbers and repetition of the digits is allowed, how many lock combinations consisting of 3 numbers are possible?

10   If the dial of a lock has 30 numbers, how many lock combinations consisting of 3 numbers are possible if the first and second numbers are not allowed to be equal?

---

*The use of the word "combination" when referring to a sequence of numbers that opens a lock, as in this problem, is a little different than the mathematical term. A "combination lock" requires a permutation of numbers to open it. The order of the digits in the lock combination is obviously important. In mathematics a combination refers to a set of objects where the order is immaterial. We hope this helps to clarify the distinction between permutations and combinations. We will refer to a combination that is meant to open a lock as a lock combination.

In Problems 11 to 16 determine how many permutations there are of the given number of objects.

11 5 objects taken 2 at a time.

12 5 objects taken 3 at a time.

13 6 objects taken 2 at a time.

14 10 objects taken 5 at a time.

15 12 objects taken 4 at a time.

16 20 objects taken 3 at a time.

In Problems 17 to 22 determine how many combinations there are of the given number of objects.

17 5 objects taken 2 at a time.

18 5 objects taken 3 at a time.

19 10 objects taken 2 at a time.

20 10 objects taken 7 at a time.

21 20 objects taken 5 at a time.

22 25 objects taken 5 at a time.

In Problems 23 to 35 a counting problem is given. First determine whether the solution involves permutations or combinations, and then use the appropriate formula to answer the question.

23 A club consisting of 10 members must elect a president and a vice president. In how many ways is it possible?

24 In how many ways can five people line up at a supermarket checkout line?

25 If the call letters of a radio station must begin with the letter W, how many different stations could be designated by using four letters, such as WIND, without using repetitions?

26 The Coast Guard uses flags raised on a flagpole to signal ships concerning such matters as weather conditions. If three different flags are available, how many signals can be sent using all three flags if the order of the arrangement is important?

27 The number of signals available in Problem 26 can be increased if signals consisting of just two flags are considered. How many signals are possible using either two flags or three flags, where the order of the flags is important?

28 An encyclopedia consists of 10 volumes. If they are placed on a shelf at random, in how many ways can they be arranged?

29 If 10 women compete in a marathon, in how many ways can the first three places be taken?

30 A classroom has 10 seats. In how many ways can four students be assigned to the seats?

31 An agricultural experimenter has designed an experiment to test the effects of 2 types of insecticide and 4 types of fertilizer on 6 different types of crops. She is going to subdivide the plot of land and use a different combination of the three entities on each subplot. How many subplots does she need?

32 A veterinarian wishes to test 3 different levels of dosage of a vaccine on 4 types of animals at 2 different age levels of the animals. How many classes of vaccinations is he going to study?

33 There are 20 candidates for the all-star basketball game, and 5 will be selected for the first team. In how many ways can 5 of the 20 be chosen?

34 A total of 12 people work as programmers in a computer center, and 3 are to be chosen to work a 2-hour shift as consultants. In how many ways can 3 be chosen from the 12?

35 Seven astronauts have reached the final training level for a particular space flight. (*a*) In how many ways can 2 be chosen from the 7? (*b*) In how many ways can a captain and a copilot be chosen?

36 (*Pigs-get-fat-hogs-get-slaughtered department*) It has been conjectured that the Pennsylvania lottery scam team could have milked the state for much more money by placing about one-half the number of bets and repeating the procedure often. Why?

6-7 SOLVING PROBABILITY PROBLEMS

In the early eighteenth century two men wrote books on probability that improved the pioneering efforts of Pascal and Fermat. Their books have become classics in the field because they established the form and content of the subject as we know it today. Yet their lives were almost entirely different. Both were targets of religious persecution, but one, who came from a family of geniuses, prospered in spite of the experience and lived a comfortable life as a professor of mathematics. The other suffered from discrimination and never held a full-time job.

James Bernoulli (1654–1705) was born into an amazing family. His grandfather had fled Antwerp because of the religious persecution of the Protestant sect of Huguenots. The family eventually settled in Basel, Switzerland.

James was the first in a long line of Bernoullis to achieve worldwide eminence in mathematics. His two younger brothers, four of his nephews, and three of their sons also achieved fame. Within a century eight Bernoullis had amassed numerous awards for their contributions to mathematics. While James remained in Basel as professor of mathematics, his relatives held prestigious posts as far away as Leningrad, where one was held in such high esteem by Empress Catherine that she paid for his public funeral. The Bernoulli family is to mathematics what the Bach family is to music.

Abraham De Moivre (1667–1754) was also a Huguenot. His family fled France for England. Unlike Bernoulli, who had a brilliant family to rely on for inspiration and knowledge, De Moivre depended upon his acquaintance with other British scientists to keep him abreast of current mathematical developments. Two in particular were champions of his cause, Sir Isaac Newton and Edmund Halley (of comet fame). Unfortunately, because he was not British, De Moivre was never able to obtain a university teaching position; so he spent his life conducting long hours of private tutoring lessons, in spite of the fact that he carried out original research in several fields.

The books that they wrote reflect their different circumstances. Bernoulli's book, "The Art of Conjecturing," was written in Latin and presents some of his original results on infinite series. De Moivre's book, "Doctrine of Chances," deals with the type of problems that we have been considering in this chapter: those involving dice, cards, and urns. His intent was to develop an "algebra of probability" in which probability problems could be solved by certain rules and procedures similar to the methods in algebra. This led him to base his work on permutations and combinations, which also forms the basis for Bernoulli's work.

De Moivre derived the rules for permutations and combinations from the principles of probability, but today it is customary to proceed in the opposite direction. In the preceding section we developed the rules for permutations and combinations. In this section we will present a series of examples to illustrate how they can be applied to probability problems.

Examples

The examples in this section will make use of the counting techniques discussed in Section 6-6. In all the problems the outcomes will be equally likely, so the formula that will be applied to all cases is

$$\text{Pr(event)} = \frac{n(\text{event})}{n(\text{sample space})}$$

Whenever possible we will express the solution as a number instead of a form involving $P(n, r)$ or $C(n, r)$.

Example 1
A lottery number is a three-digit number obtained by selecting three ping-pong balls from an urn. The urn contains 10 balls, each of which is numbered with a digit from 0 to 9.
Problem If you have bought 5 tickets with different three-digit numbers, what is your probability of winning?
Solution The general formula for probability becomes

$$\text{Pr(winning)} = \frac{\text{number of ways of winning}}{\text{number of possible lottery numbers}}$$

Example 1 in Section 6-6 showed that there are $P(10, 3)$ possible lottery numbers. Since you will win if 5 of these are selected, the probability becomes

$$\text{Pr(winning)} = \frac{5}{P(10, 3)} = \frac{5}{720} = \frac{1}{144}$$

Example 2
Problem If 10 runners compete in the 1-mile run in the Olympics, and 3 are Americans, what is the probability that the Americans will win all three medals? (Assume that all runners have the same chance of winning.)

Solution Since we are not interested in the order of finish, the number of possible finishes is $C(10, 3)$. Of these, there is one way for the three Americans to win the medals. Thus

$$\text{Pr(win 3 medals)} = \frac{1}{C(10, 3)} = \frac{1}{120}$$

An alternate way to approach this problem is to regard the order of finishes. The Americans can win the medals in 3! ways, depending on who wins the gold, silver, and bronze medals. The number of possible finishes becomes $P(10, 3)$ when the order of finish matters. Then

$$\text{Pr(win 3 medals)} = \frac{3!}{P(10, 3)} = \frac{6}{720} = \frac{1}{120}$$

Example 3

Five coins are tossed.

Problem (*a*) What is the probability that 2 of them will land heads?

(*b*) What is the probability that at least 2 of them will land heads?

Solution The number of ways in which 5 coins can land is determined from the multiplication principle to be 2^5. This forms the denominator of both parts of the problem.

(*a*) The number of ways in which 5 coins can land with 2 of them heads is $C(5, 2)$. Thus

$$\text{Pr}(2H) = \frac{C(5, 2)}{2^5} = \frac{10}{32} = \frac{5}{16}$$

(*b*) Define the event

E: at least two coins will land heads

Then its complement is

E': no coins land heads or one coin lands heads

The 5 coins can land with 0 heads in $C(5, 0)$ ways. They can land with 1 heads in $C(5, 1)$ ways. Thus

$$\text{Pr}(E') = \frac{C(5, 0) + C(5, 1)}{2^5} = \frac{1 + 5}{2^5} = \frac{6}{32} = \frac{3}{16}$$

According to Rule 3 on complementary probabilities,

$$\text{Pr}(E) = 1 - \text{Pr}(E') = 1 - \frac{3}{16} = \frac{13}{16}$$

Example 4

Problem If a two-digit number is formed by selecting two numbers from the set $\{1, 2, 3, \ldots, 9\}$, what is the probability that both digits are even?

Solution The order of the digits is important here because, for example, 12 is even but 21 is not. Thus we are dealing with permutations. There are 4 even numbers in the set $\{2, 4, 6, 8\}$. There are $P(4, 2)$ ways of arranging 2 of them. The total number of possible two-digit numbers is $P(9, 2)$. Hence

$$\text{Pr(both even)} = \frac{P(4, 2)}{P(9, 2)} = \frac{6}{36} = \frac{1}{6}$$

Another way to solve this problem is by means of the tree diagram below, where E = even and O = odd.

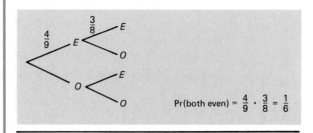

$$\text{Pr(both even)} = \frac{4}{9} \cdot \frac{3}{8} = \frac{1}{6}$$

Example 5

A field-hockey coach has 20 girls on her team. A starting team consists of 11 players.

Problem If the coach selects the starting team at random, what is Amy's probability of being one of the starters?

Solution The number of possible starting teams is $C(20, 11)$. That is the easy part. If Amy is on the starting team, then the remaining 10 places can be chosen from the other 19 girls. There are $C(19, 10)$ ways in which this can be done. Thus

$$\text{Pr(starting)} = \frac{C(19, 10)}{C(20, 11)} = \frac{19!/10!9!}{20!/11!9!} = \frac{19!\,11!}{10!\,20!} = \frac{11}{20}$$

Do you see how to generalize Example 5 to teams with other numbers of players and starters? See Problem 24 in the exercises.

Example 6

An urn contains 7 red marbles and 5 blue marbles, and 4 marbles are selected at random.

Problem What is the probability of selecting 2 red marbles and 2 blue marbles?

Soution Think of the 7 red marbles as having numbers painted on them, so that it is possible to distinguish one red marble from another. Then the number of ways to select 2 of these marbles is $C(7, 2)$, where combinations are used instead of permutations because, for example, the selection of marbles 1 and 2 does not depend upon the order of selection.

Similarly, there are $C(5, 2)$ ways of selecting 2 blue marbles from the original 5. By the multiplication principle there are $C(7, 2) \cdot C(5, 2)$ ways to select the 4 marbles so that 2 are red and 2 blue.

Altogether there are $C(12, 4)$ ways of selecting 4 marbles from a set of 12. The probability of selecting 2 red marbles and 2 blue marbles is

$$\Pr(2 \text{ red and } 2 \text{ blue}) = \frac{C(7, 2) \cdot C(5, 2)}{C(12, 4)}$$

EXERCISES

Express the answers as numbers whenever possible.

1 A lottery number is a three-digit number obtained by selecting 3 balls from an urn with 10 balls numbered 0 to 9. If you have bought 10 tickets with different three-digit numbers, what is your probability of winning?

2 Repeat Problem 1 for a lottery with four-digit numbers and 100 tickets.

3 A novice chugger boasts that he can distinguish the tastes of beers. If glasses of Miller, Schlitz, and Budweiser are imbibed and he guesses randomly, what is his probability of guessing all three correctly?

4 If Alex, Barb, and Zeke sit randomly on three seats in the same aisle, what is the probability that Zeke will sit beside Barb?

5 If the letters a, e, and t are typed at random, what is the probability that they will spell a word?

6 If Deb, Ken, Paula, and John sit randomly on four seats in the same aisle, what is the probability that Deb and Ken will sit beside each other but Paula and John will be separated?

7 In the general election for three positions on City Council there are eight candidates, including three Democrats. What is the probability that all three Democrats will win?

8 An urn contains 3 red marbles, 5 white marbles, and 2 blue marbles. If 3 marbles are selected from it, what is the probability that all 3 will be red?

9 If the 1-mile run in the Olympics includes 12 runners, of whom 4 are European, what is the probability that all three medals will be won by Europeans?

10 If the 100-meter run includes 10 runners, of whom 3 are American, what is the probability that exactly 2 of the Americans will win a medal?

11 Flip a coin 4 times. Compute the probabilities of obtaining 0, 1, 2, 3, and 4 heads. Which outcome has the highest probability of occurring?

12 Repeat Problem 11 for a coin tossed 5 times. What is the essential difference between these two problems?

13 Flip a coin 10 times and compute the following probabilities.
(*a*)   Exactly 5 heads
(*b*)   Exactly 4 heads
(*c*)   Exactly 6 heads
(*d*)   At least 9 heads
(*e*)   At most 2 heads

14 Why are the solutions to parts (*b*) and (*c*) in Problem 13 equal?

15 Flip a coin 12 times and find the probability of obtaining 6 heads. Is this number greater than the solution to Problem 13(*a*)?

16 Which has the greater probability, getting 4 heads in 8 flips of a coin or getting 5 heads in 10 flips of a coin?

17 A two-digit number is formed by selecting two numbers from the set {1, 2, 3, 4, 5, 6, 7}. What is the probability that both digits are odd?

18 If two letters are chosen at random from the alphabet, what is the probability that both are vowels?

19 If three letters are chosen at random from the alphabet, what is the probability that all three are vowels?

20 A three-digit number is formed by selecting three numbers from the set {1, 2, 3, 4, 5, 6, 7}. What is the probability that all three are odd?

21 If you are 1 of 12 players on a basketball team, and the starting 5 players are chosen at random, what is the probability that you will be a starter?

22 A basketball team of 12 players includes two stars whose contracts specify they must start every game. If you are one of the remaining players, and the starting 5 players are chosen at random, what is the probability that you will be a starter?

23 A doubles team in tennis is chosen at random from 6 available players, including you. What is the probability that you will be selected?

24 A team is composed of p players, and s of them are starters. If you are a team member and the starting team is selected randomly, what is the probability that you will be a starter?

25 In Example 6, what is the probability of selecting at most 1 red marble?

26 In Example 6, what is the probability of selecting at least 1 red marble?

27 In Example 6, what is the probability of selecting 3 red marbles and 1 blue marble?

28 In Example 6, what is the probability of selecting 4 red marbles?

29 Three marbles are selected from an urn containing 6 red and 4 white marbles. Find the probability that (*a*) all 3 are red, (*b*) none is red, (*c*) at least 1 is red.

30 If 4 marbles are selected from the urn in Problem 29, what is the probability that at most 1 of them is white?

31 For a family of four, find the probability that (*a*) no two of them have the same birthday, (*b*) at least two of them have the same birthday.

32 Repeat Problem 31 for a family of 10.

Courtesy of Francoise Ulam.

7

Transition from Probability to Statistics

The division between probability and statistics is so subtle that it becomes murky. At first we felt that one section—one on empirical probability—would be adequate to explain the connection and the difference between the two. Then we noticed that the section on the Monte Carlo method—which generates probabilities by simulating experiments many times—also uses topics from statistics to determine probabilities. Hence the first two sections in this chapter explain how probabilities are determined from lists of data, first by generating the data by simulation (Monte Carlo) and then by gathering the data from actual experience (empirical probability).

The next two sections lean more toward statistics. Section 7-3, ''Design of an Experiment,'' explains how important it is to set up an experiment carefully so that the results are acceptable. In this section you will see how percentages are used, just as we used probabilities, to interpret results of experiments. The last section discusses how statistics are gathered and how there often is more than one way to interpret the data.

7-1 THE MONTE CARLO METHOD

This section provides another side to probability. It shows how experimentation can be used to suggest an answer in cases where there is little or no hope of finding a theoretical solution. Let us consider three typical problems.

1 Two sharpshooters engage in a duel in which the first one to hit the target wins. They shoot alternately. The probability of each one hitting the target is 1/2. What is the probability that the first person to shoot wins the duel?

2 Baseball players exude warmth and confidence during hitting streaks and grouse during slumps. What is the probability that a .300 hitter will get at least one hit in 44 consecutive games?

3 Families of five people frequently comment that two of them were born in the same month. Is this a mere coincidence?

These three problems appear to be similar to the kind of probability problem encountered in the previous chapter. They can, in fact, be solved by the methods introduced there. However, the first involves an infinite sum, while the others reduce to expressions that are difficult to evaluate.

In this section we will introduce an entirely different way to solve such problems, called the *Monte Carlo method*. It is used to determine the probability of an event when either

(i) there is no theory that adequately describes that event, or

(ii) there is a theory that describes the event, but the computations involved become unwieldy if not downright impossible.

It was developed by a Polish mathematician, Stanislaw Ulam, in the 1940s after he had fled his homeland because of the threat of Nazi Germany. He was an integral cog in the Manhattan Project at Los Alamos, New Mexico, where he helped develop the atomic bomb.

The quotation below presents a good introduction to the Monte Carlo method. It is taken from Ulam's autobiography.*

The idea for what was later called the Monte Carlo method occurred to me when I was playing solitaire during my illness. I noticed that it may be much more practical to get an idea of the probability of the successful outcome of a solitaire game by laying down the cards, or experimenting with the process and merely noticing what proportion comes out successfully, rather than try to compute all the combinatorial possibilities which are an exponentially increasing number so great that, except in very elementary cases, there is no way to estimate it. This is intellectually surprising, and if not exactly humiliating, it gives one a feeling of modesty about the limits of rational or tradi-

---

*S. M. Ulam, "Adventures of a Mathematician." Copyright © 1976 S. M. Ulam. Used with permission of Charles Scribner's Sons.

tional thinking. In a sufficiently complicated problem, actual sampling is better than an examination of all the chains of possibilities.

The idea was to try out thousands of such possibilities and, at each stage, to select by chance, by means of a "random number" with suitable probability, the fate or kind of event, to follow it in a line, so to speak, instead of considering all branches. After examining the possible histories of only a few thousand, one will have a good sample and an approximate answer to the problem. All one needed was to have the means of producing such sample histories.

It seems to be that the name Monte Carlo contributed very much to the popularization of this procedure. It was named Monte Carlo because of the element of chance, the production of random numbers with which to play the suitable games.

Sampling

Ulam's quotation contains the seeds of the two essential ingredients in the Monte Carlo method. The first answers the question: "When do you use it?" You use it whenever your problem is so complicated that the theory alone cannot solve it. In spite of the many principles elaborated in Chapter 6, some occasions demand an empirical approach. The other aspect of the Monte Carlo method is the means of obtaining samples. The examples will illustrate several ways of producing them.

Example 1

Advertisers like promotions that encourage consumers to collect parts of the package of products that include the trademark. In particular, soft drink companies favor games that require saving bottlecaps that have letters or numbers stamped on them.

Problem   Seven-up is running a promotion that requires the consumer to save bottlecaps with the letters in "Seven-up." For each set of the six letters—s, e, v, n, u, p—you get a free bottle. About how many bottles of Seven-up would you have to purchase in order to obtain all six letters?

Solution   This is not a "solution" in the usual sense. It is an indication of a method by which you could approximate an answer rather than a principle that would allow you to compute it precisely.

At this point it is up to you to experiment. This does not mean that you must run out and buy Seven-up, of course. Instead, it means that you must find some implement that has six equally likely outcomes. An ordinary die will do just fine.

Now you toss the die and record the top number. Do it again and record the second number if it is different from the first. Continue this until all six numbers have appeared. How many tosses were required altogether? Record this number; then repeat the entire process several more times. What do you think the answer to the problem is?

We performed this task 100 times and obtained the following results, where N represents the number of tosses required to obtain all six numbers.

	N		N		N		N		N
1	13	21	10	41	19	61	13	81	13
2	15	22	23	42	30	62	19	82	9
3	14	23	9	43	9	63	24	83	10
4	10	24	10	44	11	64	13	84	19
5	10	25	6	45	17	65	21	85	19
6	8	26	12	46	13	66	14	86	8
7	11	27	11	47	15	67	17	87	18
8	12	28	11	48	12	68	8	88	11
9	18	29	19	49	11	69	7	89	20
10	11	30	9	50	24	70	13	90	11
11	8	31	20	51	9	71	9	91	19
12	20	32	24	52	10	72	9	92	25
13	11	33	10	53	15	73	6	93	19
14	30	34	15	54	29	74	11	94	19
15	8	35	12	55	21	75	15	95	10
16	10	36	8	56	11	76	15	96	14
17	6	37	10	57	9	77	13	97	11
18	9	38	20	58	11	78	26	98	14
19	6	39	10	59	20	79	9	99	10
20	16	40	29	60	12	80	15	100	19

This presented us with quite a bit of raw, undigested data. To obtain an answer we found the average of these numbers:

$$\frac{1407}{100} = 14.07$$

Thus our sample indicates that, on the average, about 14 tosses will be required to get all six numbers. How does this compare with your sample?

Example 1, which we will refer to as the soda-pop example, typifies the kind of problem that we will encounter here. Even though a theoretical answer is known, we suggest an empirical approach to finding it. In fact, strictly speaking, this example is concerned with an "expected number," a concept we did not introduce, so our only approach at this point is an empirical one.

Four Steps

The Monte Carlo method consists of four steps. First, a means must be found for producing random outcomes. The die was tossed in the soda-pop example. Second, a sample must be generated to model the given problem. This involved tossing the die and counting the number of throws required for getting all six outcomes. Third, numerous samples must be taken. Usually 100 is sufficient. Finally, the desired answer is derived from the samples. We merely found the average of the 100 sample numbers.

Step 1, a means for producing random numbers, was satisfied by a die for the soda-pop example. You might also have used a spinner which has six equally likely outcomes. But what would you have done if the set consisted of 10 letters instead of 6? For this, you might even use an ordinary telephone book. Problems 1 to 6 of the exercises will introduce you to some of these ways of producing random numbers, but several other alternate means are available. For instance, there are entire books of random numbers. Such books may become extinct soon, however, because some hand calculators have random-number keys and most computers contain commands for generating random numbers.

The three examples below will illustrate the Monte Carlo method further. They will also describe how random-number generators can be used.

Applications

Remember that the aim of the examples below is to get you actively involved in experimenting and deriving answers based upon your experiments. In each one, you should follow the four steps listed above.

> **Example 2**
> Two sharpshooters engage in a duel in which the first one to hit the target wins. They shoot alternately. The probability of each one hitting the target is 1/2.
> **Problem**   What is the probability that the first person to shoot wins the duel?
> **Solution**   First you need a model for producing two equally likely outcomes. A coin can serve as a model. Let *H* stand for "hitting the target." Appropriately, *H* can stand for "heads" too. Flip the coin. If an *H* appears, the first sharpshooter wins; if not, flip again. If the second toss is *H*, the second sharpshooter wins; if not, flip again. Continue flipping until an *H* appears and record the winner of the duel.
>
> Repeat this entire process nine more times. How many times did the first sharpshooter win? What is the probability of the first sharpshooter winning?

In Example 2, as with the soda-pop example, a theoretical answer is known. It is 2/3. How does your empirical probability compare with this one?

The next example is somewhat different. It illustrates a very effective way to model numerous situations.

> **Example 3**
> Pete Rose is a .300 hitter. This means roughly that over his entire career he batted safely 30 percent of the time.

Problem   Construct a model for Pete Rose's performance during a season in which he bats four times per game for 150 games.

Solution   Produce a list of 600 random digits (that is, numbers from 0 to 9). Divide the list into 150 sets of four digits, where each set will denote the four at-bats during that game. Let the three digits 0, 1, 2 represent hit (*H*) and the seven digits 3, 4, 5, 6, 7, 8, 9 represent an out (*O*). This provides a model for one such season.

Once you have produced such a model you can examine it and answer several questions. For instance:

1   What was Rose's batting average that season? Remember that theoretically it should be .300.

2   In how many games did he have no hits?

3   What was his longest hitting streak? (This means the number of consecutive games in which he got at least one hit. Rose holds the NL record for this: he had a 44-game hitting streak in 1978.)

4   What was his worst slump? (We will define this as the number of consecutive at-bats in which he did not get a hit.)

5   How many perfect 4-for-4 days did he have? (What is the theoretical probability of having such a day?)

We wrote a computer program to produce 150 random numbers between 0 and 9999, inclusive. It is given in Table 1. Interpret 0–2 as *H* and 3–9 as *O,* as in Example 3. We have underlined the numbers that correspond to hits. Our answers to the five questions above, on the basis of this particular simulation, are:

TABLE 1

7406	6377	2551	7243	1341	4192	9592	4262	0640	1006
2297	5536	3633	9820	5907	4133	0673	1960	2705	5892
0682	8865	1003	7507	5845	0509	2737	6860	6822	4155
5165	3725	8103	8919	4531	0362	2512	2940	8915	1945
9041	6748	4006	9827	3207	7706	4827	8326	8890	8303
0023	7838	0502	0891	0102	1853	0411	1597	9578	8950
5518	8059	3691	7577	6001	4102	6951	3638	6503	7122
2813	9440	3937	1776	6397	5828	5835	0583	6021	4507
8555	5355	0898	9255	6150	3306	3514	6333	9284	1555
3676	6927	1018	0219	0208	7325	4805	6831	4597	3201
2606	6371	3132	8374	1241	7542	0286	9568	7003	6169
2011	8670	7143	2806	4675	2949	7373	9153	0462	9515
3142	5091	5914	9388	5849	6527	5034	3097	0537	8484
5996	6246	7063	5759	0371	0594	0661	5860	1119	2573
9463	9874	1085	3594	5785	3574	3742	0075	5045	3664

1 .312

2 33

3 16

4 17

5 2

Example 3 is important even if you do not like baseball. In fact, it is important even if you like baseball but not Pete Rose. Its importance lies in the way that random numbers can be used to model real-life situations. Take the soda-pop example. Instead of tossing dice or plucking a spinner, you could just produce a list of random digits, ignore 0, 7, 8, and 9, and observe how many of the other outcomes were required to produce the digits 1 to 6. A phone book would accomplish this same task if you use the last four digits of each phone number.

Before leaving this example, however, we should point out that we have not answered the question posed at the beginning of this section about the probability of a .300 hitter getting a hit in 44 consecutive games. Such a question can be answered by a rather easy theoretical analysis. Instead, we have seen how to simulate various kinds of seasons that such a hitter could have, because this kind of model can be applied in numerous situations.

Now we turn to our third problem posed earlier.

Example 4

A family consists of five people.

Problem What is the probability that at least two of them were born in the same month?

Solution To simulate the months of the year we must find some means of producing 12 equally likely outcomes. Some spinners are divided into 12 sections. There are even dodecahedral dice, which have 12 faces. Another suggestion is to use the telephone book, but you will have to exercise caution when using random digits in this manner (or from a calculator or computer).

Consider the last two digits of the telephone numbers, but regard only 01 to 96. Ignore 00, 97, 98, and 99. Divide 12 into each number, and record only the remainder after you have performed the division. Here are a few examples, where the remainder has been circled:

$$
\begin{array}{ccc}
2 & 4 & 7 \\
12\overline{)29} & 12\overline{)59} & 12\overline{)84} \\
\underline{24} & \underline{48} & \underline{84} \\
\textcircled{5} & 11 & \textcircled{0} \\
 & \textcircled{E} &
\end{array}
$$

For the sake of clarity, let us denote the remainders 10 by T and 11 by E.

In this way, the phone book will produce 12 equally likely outcomes (the remainders): 0, 1, 2, 3, 4, 5, 6, 7, 8, 9, *T, E*.

Now divide the list of these numbers into sets of five. Each set will model the birth months of the family members. Produce 100 families (500 numbers altogether). In how many sets does some number appear at least 2 times? (For instance: 10125, 21*EE*5, and 23232 do, but *TE*970 and 012*T*3 do not.) On the basis of your sample, what do you conclude the probability to be? The probability can be calculated theoretically, and Problem 19 will supply it.

The wording in Example 4 differs slightly from the problem as it was stated in the introduction to this section. There we asked, "Is this a mere coincidence?" The solution to Problem 19 shows that the probability is approximately 0.62, so it is no coincidence at all. Would you be surprised if the probability for four-member households is also greater than 1/2? (See Problem 19.)

Summary

The Monte Carlo method is used to compute probabilities in cases where either the theory is missing or the computations are extremely difficult. Roughly, this method produces a model of the given specific problem by using random events to produce numerous samples. The final answer is drawn from these samples. It makes use of methods of statistics that will be introduced in the next chapter.

The Monte Carlo method presents one side of empirical probability, a side in which there is no previous history to draw conclusions from. The samples are taken in order to provide such a history. In the next section we will introduce another side of empirical probability, one in which the probabilities are based upon previous performance or upon samples drawn from part of an entire population.

EXERCISES

Problems 1 to 7 present experiments that you can readily conduct in order to obtain empirical data. You will be asked to compare your data with the corresponding theoretical probability whenever possible.

1   Shuffle an ordinary deck of 52 playing cards and then deal them out 2 at a time recording whether 2 red cards, 2 black cards, or 1 red and 1 black card appear. Compute the percentage for each occurrence. Does this agree with your theoretical analysis of the probabilities? (Compare with Example 2 in Section 6-1.)

2 Toss two coins 50 times, keeping careful tally of all of the four equally likely possibilities. Compute the percentage of times that each occurs. Does this agree with your theoretical analysis of the probabilities?

3 Choose a page in a phone book at random and select the first 100 names. Record the last digit in the phone numbers. Compute the percentage that each digit 1, 2, . . . , 9 occurs. Does this agree with your theoretical analysis of the probabilities?

4 Choose a page in a phone book. Record the first digit in the phone numbers, skipping those phone numbers that start with a letter. Record 100 numbers and compute the percentage that each digit 1, 2, . . . , 9 occurs. Are your results different from those of the previous problem?

5 Most people have a definite preference for either Coca-Cola or Pepsi-Cola. You can use these two brands for this experiment, or choose two other brands of cola, or two brands of any similar drink, perhaps even lemonade or orange juice. Be sure that you are convinced you can tell the difference. Have someone else pour 10 glasses with the colas in any sequence of brands. Record the number of times you guess the correct brand. What percentage of correct responses would convince you that you are an "expert"? What percentage of correct responses would indicate that you were simply guessing?

6 Select a sports article in the newspaper and record the length of each of the first 100 words. Compute the percentage of times that each word length 1, 2, 3, . . . occurs. Do the same experiment for the first 100 words in an article on the editorial page. Do you expect the results to agree? Do they?

7 Flip a bottlecap 50 times from a height of at least 3 feet and let it land on a hard surface. Record whether it lands upright or upside down. Compute the percentage of upright landings. Do the same experiment with the bottlecap landing in your hand. Do the results agree with the first experiment?

In Problems 8 to 21 you should make use of the Monte Carlo method in order to obtain an approximate answer. In some cases you will have to devise a way of simulating the problem with an appropriate model. The answers to the odd numbered problems are the theoretical answers.

8 How many soda bottles would you expect to buy in order to obtain a complete set of 10 bottle caps?

9 A hockey coach chooses players for her team at random. If each player can play only one of the six positions, about how many girls will she have to pick before being able to field an entire team?

10 How many times would you expect to throw a single die before all six numbers appear?

11 Two sharpshooters engage in a duel in which the first one to hit a target wins. They shoot it alternately. Suppose the probability of the first sharp-

shooter's hitting the target is 1/2, and the second's is 3/4. Design an experiment to model the duel. Conduct the experiment to obtain the empirical probability that the first sharpshooter wins the duel.

12 (*a*) Repeat Problem 11, with Pr(first) = 0.4 and Pr(second) = 0.7.
   (*b*) Repeat Problem 11, with Pr(first) = 0.4 and Pr(second) = 0.6.
   (*c*) What is the essential difference between parts (*a*) and (*b*)?

13 (*a*) What is the theoretical probability that Pete Rose will get no hits in four at-bats?
   (*b*) What is the empirical probability based upon the model given in Table 1?

14 (*a*) What is the theoretical probability that Pete Rose will bat a perfect 4-for-4?
   (*b*) What is the empirical probability based upon the model given in Table 1?

15 What is the theoretical probability that Pete Rose will (*a*) get at least one hit in four at-bats; (*b*) get at least one hit in 44 consecutive games in which he gets four at-bats?

16 A baseball team that wins 60 percent of its games will usually win its division. Construct a model for such a team over a 162-game season, using random digits. What was the team's longest losing streak? What is the theoretical probability of this happening?

17 George Brett almost hit .400 during the 1980 baseball season. Construct a model for one such season of 162 games in which Brett had four official at-bats each game. What is his batting average in this model?

18 Could you use the sum of two dice in order to model Example 4 for a family of 11 members? Why?

19 (*a*) What is the theoretical probability that at least 2 members in a family of 5 will have the same birth month?
   (*b*) Repeat part (*a*) for a family of 4.
   (*c*) Repeat part (*a*) for a family of 10.

20 All athletes are interested in personal records (PRs). This exercise will help you predict how many times a PR will be set when each outing is independent of the others. This occurs, for instance, when an athlete is not following a particular training regimen.

Choose a random number between 0 and 9999, inclusive. This is a PR. Choose a second number. If it is greater than the first, you have set a new PR. Choose a third number. If it is greater than the numbers that preceded it, then this number becomes the new PR.

Choose 100 numbers altogether. How many PRs were set?

21 Refer to a meteorological table of temperatures in the city you live in. Consider the highest temperature recorded each year for the last 100 years. In how many of these years was a PR set? (See Problem 20.) How does this result compare with the one in Problem 20?

7-2   EMPIRICAL PROBABILITY

How do you know whether that coin in your pocket is fair?

Theoretically your coin is fair if $\Pr(H) = 1/2$. But how do you test this in practice? If you flip it 10 times and get 4 heads, what do you conclude? Remember that J. E. Kerrich tossed a coin 10,000 times and obtained 5087 heads. Is this enough evidence to conclude that the coin is fair?

The question about the fairness of a coin is a difficult one to answer, but it illustrates the basic division of probability into two parts, theoretical probability and empirical probability. For Kerrich's coin, the empirical probability is $\Pr(H) = 0.5087$, whereas theoretically $\Pr(H) = 0.5$. Formally an *empirical probability* is a probability that is based upon past experience. Thus, for Mendel, the theoretical probability of a plant being red is $3/4 = 0.75$, but the empirical probability is $705/929 = 0.7589$.

The Monte Carlo method would seem to fall into the realm of empirical probability, because it is based upon models that simulate real events. The models then provide a "history" for the event and the probability is determined from this history. However, certain theoretical assumptions about the event are made in order to construct a model. For instance, in the problem about the birth months of five-member families we assumed that births occur evenly throughout the year, so that the probability of being born on a given day is $1/365 = 0.0027$. That this is not the case in practice can be seen in Figure 1, which is a graph of births in 1977 in New York. Thus the Monte Carlo method represents a mix between theoretical and empirical probability.

We have seen that probability theory attempts to establish order to seem-

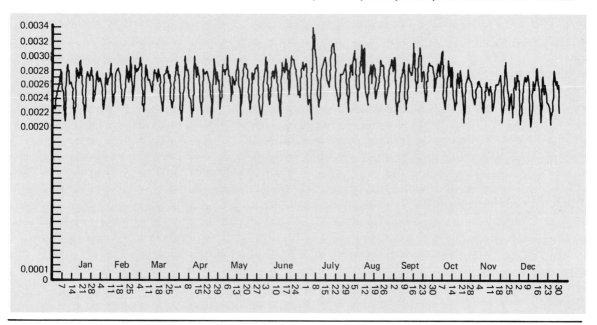

FIGURE 1   Probability of birth in New York State in 1977.

ingly haphazard occurrences. There are many instances when no such theoretical considerations can describe the fluctuations and irregularities in processes governed by chance. In such instances, the variability must be analyzed solely with empirical data. By so analyzing past experience we hope to discover certain regularities that help us understand the processes and predict the future with greater confidence.

Order in Nature

In this section we will relate three anecdotes about individuals who made important decisions based upon empirical probabilities. We will see how a small weather-forecasting company grew to national prominence, how a baseball manager shuffled his lineup to pull off a stunning victory, and how the Salk vaccine became part of our nation's immunization plan. You, too, are often confronted by problems that can be solved by using empirical probability. We provide an example of rival claims about the efficiency of a pregnancy test to illustrate this.

Much of the work of science is to find order in the seemingly haphazard occurrences in nature. Kepler found order in the heavens and Galileo discovered basic principles about the earth. Newton found laws that united their theories. Einstein found laws that govern the universe from a different perspective.

Often, however, it is not possible with the knowledge at hand to find such laws. Before the laws are found, nature must be observed, data collected and analyzed, and regularity discerned. If the future cannot be predicted absolutely by universal laws, then it must be expressed in terms of probabilities.

Meteorology is an important field of scientific research that relies on probability, rather than universal laws, to predict the future. Weather prediction is a relatively young scientific discipline. Recent technological innovations, most notably computers and satellites, have provided meteorologists with powerful tools to analyze weather patterns. Let us consider one example of a meteorologist's struggle to predict the future.

Example 1

Joel Myers looked at his computer printout in disbelief. A winter storm hitting the eastern seaboard with such magnitude is not too rare. It would level its wintry fury in about one month's time, somewhere in late January. The snow would be deep—many feet in Buffalo, perhaps 2 feet in New York and Philadelphia, close to that much in Washington. It would wreak havoc; businesses would close, heating bills would rise, but they could cope. They are used to handling rough winters. This one looked like it would be more difficult, but no undue concern would be called for.

It was the latter part of the printout that had him stunned. In all his years of weather reporting he had never seen anything like it. He felt that the prediction made by his computer was faulty. His data had to be checked and rechecked. No one would believe him unless he had conclusive evidence. If he were correct, if he could predict such a phenomenon so far in advance, not only would it enhance his firm's reputation, but there was big money to be made on the side.

Dr. Joel N. Myers is the founder and president of a small weather-forecasting firm called Accu-Weather, located in central Pennsylvania. He and his associates have contracts primarily with private industry, notably radio and television stations. Up and down the east coast you hear the familiar lead-in, "Now it's time for Accu-Weather and Joel Myers." Their forecasts are informative as well as entertaining. They are highly respected because of their high degree of accuracy.

Weather forecasting is based upon empirical probabilities. Weather patterns consisting of high and low pressure areas, cloud formations, temperature fluctuations, and various other data form the basis for determining probabilities. The statement that "there is a 30 percent chance of rain today" means, intuitively, that 30 percent of the times that this particular weather pattern has been observed, rain followed. It also implies, of course, that 70 percent of the times it did not.

Accu-Weather's computer printout predicted just such a pattern forming. The probability was high that the storm would hit in about one month, and most likely its ferocity would not be blocked by the Appalachians from moving southward, as is usually the case. The stunning part of the report, recognizable only to the astute eyes of a seasoned meteorologist, showed clearly that subfreezing temperatures and snowstorms would reach Florida. The sunshine state would be blanketed by cold and snows. Citrus crops and tourist trade would be devastated.

Joel Myers did check and recheck his data. It showed conclusively that in one month, if past weather patterns in the north could be applied with equal veracity in the south, Florida would suffer its worst winter storm on record. He had such faith in his data that he bought heavily in citrus futures, believing that the damage done would drive prices up and earn him a handsome profit. Very few others followed suit. It was a tale bordering on the unbelievable.

In January 1977, Miami suffered its worst winter storm on record. Temperatures reached the low 20s and 5 inches of snow fell. Accu-Weather was a success: meteorologically *and* financially.

Today Accu-Weather is the nation's largest and best-known private weather-forecasting service. It has provided weather forecasts for the 1980 Winter Olympics in Lake Placid, presidential inaugurations, NASA launchings, and even for the Penn State football program.

Conditional Probabilities

In some problems we consider only those outcomes that are related to a specific set of circumstances, rather than all the outcomes. The probabilities so described are called *conditional probabilities*. The next example illustrates how one must occasionally analyze certain segments of the past to develop more significant indicators of the future.

Example 2

Danny Ozark looked down the bench. If Bowa got a hit, he would have to pinch hit for the pitcher, Steve Carlton. It was the bottom of the ninth, two out and the score was 4-0. If the Phillies lost this second game of the doubleheader, they would fall four games behind the Pittsburgh Pirates, with only 2 weeks of the season to go. If they could salvage this second game, they would be only two games back and a come-from-behind victory could give them a great deal of momentum.

The bases were loaded so it was all up to Bowa. But Ozark had to think ahead. If Bowa did get on base, he could bat Luzinski, Boone, Unser, or Gross. Was the probability of one of them getting a hit greater than the other three? Luzinski had a batting average of .316, which means that he had 111 hits in 351 times at bat (i.e., .316 = 111/351). Boone was hitting .297 (92 for 310), Unser was .305, and Gross was .291.

Bowa lined a bullet over Stargel's head at first base. Two runners scored and men were left on second and third. The score now was 4-2. There were two outs in the bottom of the ninth, and a hit could tie the game, whereas an out would just about end the season.

Did the batting averages adequately reflect the chances of each player getting a hit in this crucial situation? No. There was much more to it than that. Ozark had done his homework. He had to. A split-second decision was called for, so the pertinent percentages had to be at his fingertips.

The pitcher had to be taken into account. Grant Jackson, a left-hander, was on the mound. Left-handed pitchers usually do better against left-handed batters than right-handed batters. For instance, Luzinski was batting .330 against left-handers (63 for 191) and .300 (48 for 160) against right-handers, and Boone was .307 (62 for 202) against lefties and .278 (30 for 108) against righties. Both Luzinski and Boone were right-handed batters. Unser and Gross were left-handed batters who were hitting .324 and .310, respectively, against righties and .245 and .238, respectively, against lefties.

So the choice seemed clear. The pitcher was left-handed and Luzinski had the best average against lefties; so he should send up Luzinski.

It was not as easy as that, however. Ozark knew that as soon as he sent up Luzinski, or even Boone, the Bucco manager, Chuck Tanner, would replace Jackson with a righty, probably Blyleven, because their ace right-hander, Tekulve, had been used earlier. So if he sent up Luzinski (the superstar who was not starting only because a muscle pull prevented him from playing in the field but not from batting), he would be wasted because Tanner would put in a righty, which would force Ozark to counter with a left-handed batter. It appeared to be better to send up Boone, and save Luzinski in case the game went into extra innings.

Ozark did send up Boone. Tanner called Blyleven, so Ozark was confronted with another decision, Unser or Gross. It might seem clear that the choice would be Unser because his record against righties was much better than Gross's. But Ozark had done his homework very well. Gross "owned" Blyleven. His lifetime average against him was .462 (6 for 13), whereas Unser's was .267 (4 for 15). With no hesitation Ozark summoned Gross. He drilled the 2-and-1 count slider to left center, driving home two runs and tying the score. The Phils went on to score another run for a stunning ninth-inning victory.

Ozark's knowledge of past records gave him the insight to judge which player had the best choice of success in a particular situation. This brief fictional episode gives you a glimpse into the intricate maneuvering that is inherent to major-league baseball. Big-league managers constantly juggle past records. They ask: Is he in a slump? How does he hit in June? Does this park give him a nightmare? Can he handle the pressure in September?

All these questions are answered by past records and are given in terms of percentages, which are precisely the type of entities we call empirical probabilities. It is interesting to observe how dependent our national pastime is on these probabilities.

Drug Studies

Our next example comes from the drug industry. It illustrates how empirical probabilities helped to determine the effectiveness of the Salk vaccine some 30 years ago. Further details about this historic study are given in Exercise E. A related study that you might also find interesting investigates the relationship between depression and the use of marijuana. It is described in Exercise. F.

Example 3

Jonas Salk developed his polio vaccine in the early 1950s. The dreaded disease had struck many hundreds of thousands of people, mostly children, since 1916. Salk demonstrated the effectiveness of the vaccine in his laboratory, but it could not be deemed safe and effective until a large-scale trial was conducted to see if the vaccine protected children without causing any major detrimental side effects.

In 1954 the Public Health Service organized an experiment involving nearly 1 million children. It is a very difficult and complex task to devise such an important and large test so that bias, that is, unwarranted favoritism, either for or against, is all but eliminated. One method used to reduce bias is to make the test a *double-blind* test. That is, some of the children were given the vaccine while others were given a placebo. None of them knew which they had received. Also, the doctors treating the patients were not told who was in the treatment group (those who were administered the vaccine) or who was in the control group (those who were given a placebo). The results are given below.

The Double-Blind Randomized Controlled Experiment

	Size	Polio Cases
Treatment	200,000	56
Control	200,000	142
No consent	350,000	161

SOURCE: Thomas Francis, Jr., *American Journal of Public Health*, 1955.

Altogether, 750,000 parents were asked to allow their children to participate; 350,000 declined and the 400,000 others were randomly selected for the treatment group or the control group. The table shows that 56 children in the treatment group and 142 in the control group contracted polio, while 161 in the nonconsent group caught the disease. Since the sizes of the three groups differ, it is best to consider percentages of cases rather than the actual data. The table then becomes:

	Polio Cases, %	Rate per 100,000
Treatment	0.028	28
Control	0.071	71
No consent	0.046	46

It was evident that the treatment group had a much smaller rate than the control group and the nonconsent group. This offered conclusive evidence for the effectivness of the Salk vaccine, and the "miracle drug" has been used ever since.

Not-What-It-Seems Department

The next example provides another illustration of the use of conditional probabilities by giving an instance where there are two sides to an issue, and the two conflicting parties use their own conditional probabilities to prove their claims. The topic is the safety of a pregnancy test, a delicate issue if there ever was one.

Example 4

Pharmaceutical companies are required by strict federal government regulations to conduct exhaustive tests on their drugs before they are allowed to market them to the general public. There are many reasons for exerting tight control over the drug industry, one of the most important being the control of side effects. Often a drug will cause the beneficial function for which it was designed, but it will also produce harmful effects. In addition, the effectiveness of a drug must be ascertained and explained to the public. Such information greatly affects the usefulness to the consumer.

The Warner-Lambert Company, a large pharmaceutical firm, has marketed a do-it-yourself pregnancy test. This is a major development, for it eases the burden of making an appointment with a physician every time a woman suspects she is pregnant. This will often help financially as well as psychologically, as the cost of the test is around $15, while an appointment with a physician could cost $75 or more.

There is one drawback to the test, however. Experiments have shown that there is a 20 percent chance of an error being made by the test if the woman is pregnant. How significant is this drawback? Is the percentage of error high enough to discourage its use? Only time will tell. It should be clearly explained to the public what this percentage of 20 percent really means. There is a growing controversy over how it is reported in the media.

The American Medical Association (AMA), the organizing body and political arm of physicians, wants to emphasize the 20 percent figure, while the drug industry takes the wholly different stand that the actual number of women for whom the test records an error is far less than 20 percent.

The surprising part of the controversy is that *both* sides are correct! Each is considering the problem from a perspective that displays its argument in a favorable light. The AMA wants the public to decide that the 20 percent error rate is too high for comfort, that a little extra money spent on a doctor's appointment is well worth it. The drug industry wants the public to see the error rate as much less than 20 percent, perhaps as little as 2 percent. They claim that if doubt still persists then the test can be administered twice, reducing the risk to a negligible figure (see Exercise I).

To see why each side can claim that its argument is valid, we will consider some numbers and scrutinize more closely the description of the error. Suppose 1000 women were tested. That is, each was given the test to administer. They were chosen at random, and so it was unknown, at the time of testing, whether they were pregnant. Suppose that after nine months it was determined that 100 had been pregnant at the time of testing. What does the 20 percent error rate mean to you?

It means that about 20 of the 100 women who had been pregnant concluded from the results of the test that they were not then pregnant. It does not mean that the test erred for 200 of the total of 1000 women. Thus 20 percent of the women who were pregnant observed an error, not 20 percent of the total population. Hence the drug company can claim that the test failed for only 20 of the 1000 women, or 2 percent.

It makes a great difference to the public when they are spending their hard-earned money for a product whether it has a probability of failure of 0.2 or 0.02.

The alert reader might have noticed that there is another type of error that the test could have. A woman might not be pregnant, yet the test may indicate that she is. This is generally referred to as a type II error as opposed to the type I error described above. For this kind of test the type II error is very small and, for all practical purposes, can be neglected.

Example 4 shows one way in which probability is used in the drug industry. It also points out that the buyer must remain skeptical of advertised claims (2 versus 20 percent) until further information regarding the product is published. *(Caveat emptor!)*

Summary

This section presented several examples of probabilities based on past performances. We saw that an examination of the empirical probabilities led Joel Myers to expand Accu-Weather into a forecasting service of national renown, helped the Phillies win a game, and supplied Jonas Salk with evidence attesting to the effectiveness of his vaccine. Another example illustrated how competing special interest groups used empirical probability to "prove" their claims about the efficacy of a pregnancy kit.

The kind of examples and exercises given in Sections 7-1 and 7-2 should be contrasted with those in Chapter 6, which were concerned with theoretical considerations only. The foundation has now been built for the transition to the related subject of statistics.

EXERCISES

We use letters in addition to numbers to order the exercises in this section because each one has several parts.

A. The London Aerial Bombardment

This exercise examines a critical moment in history during World War II. Great Britain had its back to the wall and found it necessary to apply rigorous analysis to its precarious position. They not only solved many pressing problems that threatened their own as well as all of western Europe's very existence, but in so doing the British researchers opened a new dimension in the development of probability theory. We will look at one aspect of the dramatic turn of events that changed the course of history.

The German bombardment of London was reaching catastrophic proportions. More antiaircraft artillery was being manufactured, but where to locate the guns to yield more maximum effectiveness was a critical problem. The distribution of bomb hits seemed haphazard (i.e., random) at first glance. To determine whether the aerial bombardment was centered on a specified list of targets, the terrain of the greater London area was divided into 576 small regions, each $\frac{1}{4}$ square kilometer. The number of bomb hits landing in each area was recorded. There were 537 bomb hits during the time interval of the study. The data are given in the accompanying table.

Number (n) of Hits	Number (N) of Regions with n Hits	Bomb Hits, nN
0	229	0
1	211	211
2	93	186
3	35	105
4	7	28
5	0	0
6	0	0
7	1	7
	576	537

For example, there were 93 regions that had 2 bomb hits, accounting for 186 ($= 93 \cdot 2$) bomb hits.

1 Pick one of the 576 regions at random. What is the probability that the region suffered (*a*) no hits, (*b*) 1 hit, (*c*) at least 1 hit, (*d*) at most 1 hit, (*e*) 5 hits, (*f*) fewer than 5 hits, (*g*) more than 5 hits?

2 The British concluded from this data that the spatial distribution of the aerial bombardment was not haphazard. Further analysis demonstrated that positioning artillery at specific locations would prove very effective. (*a*) Can you explain how the British reached this conclusion? (*b*) Do you think that certain regions were singled out for extra bombing? (*c*) About what percentage of the regions required no antiaircraft artillery?

B. Sex Discrimination Suit

The accompanying table gives the 1973 admission data of the graduate school of a large university on the west coast. The figures were used in a sex discrimination case against the school in 1975. The data are broken down into six majors, which were identified in the suit but not published.

| Major | Men | | Women | |
	Number of Applicants	Percent Admitted	Number of Applicants	Percent Admitted
A	825	62	108	82
B	560	63	25	68
C	325	37	593	34
D	417	33	375	35
E	191	28	393	24
F	373	6	341	7

1 Which major appears most biased against women? Against men?

2 What is the probability that a man would be admitted to major (*a*) A, (*b*) B, (*c*) C?

3 What is the probability that a woman would be admitted to major (*a*) A, (*b*) B, (*c*) F?

4 If the suit charged sex discrimination against women, which of the above questions would be the focal point of the suit?

Side Effects of Drugs

A great deal of research is being done on side effects of many drugs. One of the most common types of study is one that compares age with the use of the drug. A study on women using oral contraceptives is given in the following table. Users and nonusers of the pill were compared according to their age and their blood pressure.

The following table gives distribution of systolic blood pressure, cross-tabulated by age and pill use, for women in the Contraceptive Drug Study, excluding those who were pregnant or taking hormonal medication other than the pill. Class intervals include the left endpoint, but not the right. Percents may not add to 100 because of rounding.

Blood Pressure, Millimeters	Age 17–24		Age 25–34		Age 35–44		Age 45–58	
	Nonusers, %	Users, %	Nonusers, %	Users, %	Nonusers, %	Users, %	Nonusers, %	Users, %
<90	....	1	1	....	1	1	1	....
90–95	1	....	1	....	2	1	1	1
95–100	3	1	5	4	5	4	4	2
100–105	10	6	11	5	9	5	6	4
105–110	11	9	11	10	11	7	7	7
110–115	15	12	17	15	15	12	11	10
115–120	20	16	18	17	16	14	12	9
120–125	13	14	11	13	9	11	9	8
125–130	10	14	9	12	10	11	11	11
130–135	8	12	7	10	8	10	10	9
135–140	4	6	4	5	5	7	8	8
140–145	3	4	2	4	4	6	7	9
145–150	2	2	2	2	2	5	7	9
150–155	....	1	1	1	1	3	2	4
155–160	....	....	....	1	1	1	1	3
≥160	....	....	....	....	1	2	2	5
Total percent	100	98	100	99	100	100	99	99
Total number	1206	1024	3040	1747	3494	1028	2172	437

1 What percentage of users have a blood pressure of at least 140 and are aged (*a*) 17 to 24, (*b*) 25 to 34, (*c*) 35 to 44, (*d*) 45 to 58?

2 What percentage of nonusers have a blood pressure of at least 140 and are aged (*a*) 17 to 24, (*b*) 25 to 34, (*c*) 35 to 44, (*d*) 45 to 58?

3 Repeat Problems 1 and 2 with blood pressure of 140 replaced by 120.

4 (*a*) In your estimation, does this study provide evidence that age is a significant factor in causing high blood pressure? (*b*) Is the use of the pill a significant factor in causing high blood pressure? (*c*) Is age combined with the use of the pill, taken together, a significant factor?

D. Salk Vaccine Field Test

1 Another experiment designed to determine the effectiveness of the Salk vaccine was conducted in 1954 by the National Foundation for Infantile Paralysis (NFIP). All second-graders whose parents would consent were vaccinated. This became the treatment group. Children in grades 1 and 3 would become the controls. Discuss the flaws in this experiment that could produce bias. You will want to consider the facts that polio is a contagious disease, that those in grade 3 were older than those in grade 2, that parental consent was not required for the control group, and that doctors treating the patients would know whether the child received the vaccine. Can you find an advantage of this test over the one discribed in the text?

2   The results of the NFIP experiment are given in the accompanying table.

	Size	Cases
Grade 2 (treatment)	225,000	56
Grades 1, 3 (control)	725,000	390
Grade 2 (no consent)	125,000	55

Translate the data into percentages of cases of polio per size of the group. Express the percentages in terms of probabilities. Is this experiment more or less conclusive than the test described in the text?

E.   A More Complex Drug Study

Research concerning the relationship between drug use and psychological conditions is a very important and current topic. An example of this type of study is one done by Paton, Kessler, and Kandel (1977). Questionnaires were given to a random sample of high school students in New York State. Two separate surveys were conducted in order to test the validity of the results. That is, if the data from the first survey agreed with the second, they were considered more reliable. The study centered on the feelings of depression accompanying drug users and nonusers. The students were divided on the basis of their use of marijuana during the past 30 days. Feelings of depression were recorded, and the data are presented in the accompanying table.

The data are arranged in 16 entries. There are four subblocks, each containing 4 entries. The upper left-hand block, containing the numbers

255   110
78   243

corresponds to respondents who were current users in both surveys. The upper right-hand block, containing the numbers

First Survey (Fall 1971) / Second Survey (Spring 1972)	Current Marijuana User		Not Current Marijuana User		
	Depressed, N	Not Depressed, N	Not Depressed, N	Depressed, N	Total N
Current marijuana user:					
Depressed, N	255	110	78	45	(488)
Not depressed, N	78	243	15	77	(413)
Not current marijuana user:					
Depressed, N	159	58	1019	482	(1718)
Not depressed, N	44	135	401	1586	(2166)
Total N	(536)	(546)	(1513)	(2190)	

SOURCE: Paton, Kessler, and Kandel, 1977.

$$\begin{array}{cc} 78 & 45 \\ 15 & 77 \end{array}$$

corresponds to respondents who were current users during the first survey and not during the second survey. Thus there were 686 (= 255 + 110 + 78 + 243) respondents who were current users in both surveys. The figure 255 refers to the number of respondents who (1) were current users in both surveys and (2) were depressed in both surveys. Thus the probability that a respondent who was a current user in both surveys was depressed in both surveys is 255/686 = 0.37.

1 What is the probability that a respondent who was a current user in both surveys reported no depression in both surveys?

2 What is the probability that a respondent who was a current user in both surveys reported depression in one survey and no depression in the other?

3 How many respondents reported in both surveys that they were not current users?

4 What is the probability that a respondent who reported in both surveys that he or she was not a current user also reported depression in both surveys? Reported no depression in both surveys? Reported depression in one survey and no depression in the other?

F. Big Bang Theory of Killer Innings

This theory asserts that in a majority of baseball games, the winning team scores more runs in one inning than the loser does in nine innings. An examination of the 20 World Series before 1980 reveals that the theory holds in 69 of the 121 games.

1 What is the empirical probability that the big bang theory holds, on the basis of this evidence?

2 What is the corresponding probability for your favorite team over the course of a 162-game season?

G. Fair Coins and Dice

In Section 6-2 we presented the results of experiments conducted by Kerrich and Langcor on coins and dice, respectively. On the basis of this evidence, what is the probability that:

1 A flipped coin lands with tails up?

2 A tossed die lands with an odd number facing up?

H. Sex

We hope that the title of this exercise attracted your attention but did not titillate you. Anyway, it does concern sex. You will have to make use of your reference library to answer parts (a) to (c) in Problem 2.

1 In two species of seal it has been found that females giving birth early in the season produce 12 males for every 10 females. Those who pup late produce a male-female ratio of 8 to 10.

(*a*) What is the probability that an early-producing female will deliver a male?

(*b*) What is the corresponding probability for late-producing females?

(*c*) The study that led to these figures was published in the February 13, 1981, issue of *Science* magazine. *The New York Times* account (April 17) based on it concluded that the figures evened out. Is this conclusion valid from this evidence?

2   (*a*)   What is the probability that a child born during 1983 is a girl?
    (*b*)   What is the probability for a child born in the United States?
    (*c*)   What is the probability for a child born in your local hospital?
    (*d*)   Do you expect any of the answers to these three questions to be 1/2?

I. Pregnancy, Revisited

We thought it natural to follow our "sex" exercise with one involving the Warner-Lambert Company's do-it-yourself pregnancy test.

1   Recall that the Company reported a 0.02 probability of error if the woman actually was pregnant. What will the probability of error be if a pregnant woman has taken the test twice?

J. Cycles

1   Notice the obvious cyclical period in Figure 1 of this section. What is the reason for this pattern?

7-3   DESIGN OF AN EXPERIMENT

There is a fine line between probability and statistics. Sometimes it is difficult to determine where one ends and the other begins. Example 3 in the previous section, which dealt with the Salk vaccine for polio, illustrates the close relationship between these two subjects. The final result, which led to compulsory immunization in the United States, was based upon a probabilistic argument. But the means for constructing the experiment and verifying its conclusion belong to the realm of statistics.

This section is concerned with one aspect of the transition from probability to statistics, the design of an experiment. Many a study has been negated only because of the fact that the experiment was not designed properly.

When an investigator designs an experiment, it is often the case that time or financial restrictions are imposed, intentionally or unintentionally, so that the study cannot be done as completely as possible. For example, if you want to ascertain the behavior of a very large population (such as the voters in the United States during a presidential election or all people who are eligible for a vaccination), it is not feasible to contact everyone; so a small sample is selected. Then, on the basis of the evidence from the sample, a decision is made about the whole population, such as who will win the election or whether the vaccine is successful. The basic questions are: (1) How should the sample be chosen to guarantee that it is an accurate representation of the whole population? (2) What safeguards can be built into the experiment to ensure that the results are valid?

Can Life Be Prolonged?

We will pursue some of the answers to these questions through anecdotes. The first anecdote will show how an intricately designed experiment can shed light on a tantalizing "mind over matter" question. Keep the outcome of Anecdote 1 in mind as you study statistics, because it will show you one important reason why the safeguards mentioned in the previous paragraph are necessary.

Often in the movies the hero is seen cradling the slain villain-turned-good-guy in his arms, pleading with his nemesis to hang on just long enough to confess the true identity of "Mr. Big." Didn't Ronald Reagan once catch up with Jesse James in this manner? Do such feats occur in real life? Do some individuals have a will that is strong enough to postpone death in order to experience some anticipated event? It hardly seems possible, you say?

Anecdote 1

David Phillips, a sociologist at the State University of New York, Stony Brook, tried to answer this question in an article entitled "Deathday and Birthday: An Unexpected Connection."* His thesis rested on the power of the human brain. This is a wide open, active field of research requiring expertise in various fields such as psychology, biology, medicine, and sociology.

Phillips' conjecture was that human beings have much more control over their physiological processes than popularly thought. For example, he believed that an individual with a strong mental capacity could effectively control many physical aspects of life such as disease, pain, and even longevity. Thus someone with a strong mental makeup is much less likely to suffer pain and disease than others.

Phillips sought analytical proof. He decided to center on one conjecture: people can prolong life if the will and motivation is strong enough.

When confronted with such a statement, most of us would term it too far-fetched to have any connection with reality. Even if it were true, it would seem impossible to conduct an "experiment" to test a hypothesis involving deceased people who lived beyond their "due date." What well-defined criteria could be used? If ever a problem seemed far removed from mathematical analysis, this was it!

Phillips did come up with an argument, however. We hope you will be convinced that it carries considerable merit. Phillips decided to base his study on influential individuals. He wanted to see whether a significant proportion of them were able to prolong their death for an appreciable amount of time. First he had to find a motivation, preferably common to all, that would cause them to exercise their mental powers to stave off death.

Phillips reasoned that famous people were invariably honored on their birthday, especially if they were in the twilight of their careers. Anticipation of such honor would be great. Is it possible that someone nearing death could

---

*See J. M. Tanur et al. (eds.), "Statistics: A Guide to the Unknown," Holden-Day, San Francisco, 1978, pp. 71–85.

put it off in order to experience the guaranteed recognition on one's birthday?

It is impossible to determine whether a particular individual thwarted death to live beyond his or her due date. The question is meaningful only when a large sample is studied.

Phillips assumed that under normal circumstances the death dates of a large segment of the population should be distributed uniformly throughout the dates of the year. In other words, there should be approximately the same number of deaths in each month and on each day of the month. If one initially assumed that there would be no connection between the birthdate and the death date and if a sample that is representative of the whole population were chosen at random, one would expect that about 1/12 of the sample would die in each of the 12 months of the year.

The next step was to formulate the hypothesis that the month in which an individual dies, the death month, is related to (is dependent upon) the birth month. That is, some people postpone death in order to celebrate their birthdays.

It was decided to measure the time in months between a subject's birth and death. Each person was assigned a whole number between -6 and 5; if the subject dies 6 months before the month of birth, the number -6 was assigned, while it was 0 if the person died in the same month that he or she was born. Similarly, if a person dies 1, 2, 3, 4, or 5, months after the birth month, the assigned, number was 1, 2, 3, 4, or 5 respectively. For example, John Adams, the second U.S. President, was born on October 30, 1735, and died on July 4, 1826; so, since July is 3 months before October, he was assigned -3. Albert Einstein was born on March 14, 1879, and died on April 18, 1955; so he was assigned the number 1.

A total of 1251 names were selected from the reference books "Four Hundred Notable Americans" and "Who Was Who in America." If the deaths of these famous people were independent of their birthdates, one would expect to observe about 1/12 of the deaths occurring in each of the 12 designated time periods, from 6 months before the birth month until 5 months after the birth month. Since one-twelfth of 1251 is 104.25, one would expect, if birth and death months are independent, about 104 in each category.

The following table presents the results of Phillips' study. Figure 1 gives a graphical depiction of the data in the table.

Number of Deaths before, during and after the Birth Month

x	-6	-5	-4	-3	-2	-1	The birth month 0	1	2	3	4	5
Number of deaths	90	100	87	96	101	86	119	118	121	114	113	106

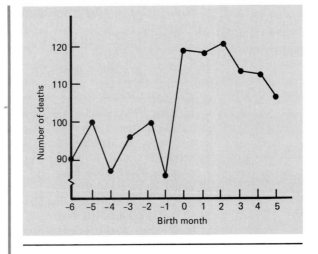

FIGURE 1 Number of deaths before, during, and after the birth month.

The data in the table exhibit a distinct dip in the number of deaths in the month before the birth month: 86 deaths were recorded, whereas about 104 were expected. Could this dip be attributed to chance? One would not expect *exactly* 104 in each category, but is 86 significantly fewer deaths? Consider the next four categories corresponding to $x = 0, 1, 2$, and 3; the birth month and the 3 months following. Each figure is much greater than 104. If independence were assumed, one would expect about $1/3$ of the deaths to occur in this 4-month period, that is, about $417 = 1/3 \times 1251$ deaths. Phillips observed $472 (= 119 + 118 + 121 + 114)$, which is 55 more deaths than expected. That total is about 13 percent more than expected ($13\% \approx 55/417$).

Phillips concluded his paper by studying a smaller number of famous people who, by an interesting measure depending upon children's biographies, were judged to be more famous than the others in the study. The size of the death rise in the birth month and the following 3 months was even larger than the 13 percent observed for the whole group. This subgroup of more famous individuals had a death rise of 58 percent.

Do Phillips' results prove that humans have the ability to prolong life? Certainly not. It does not prove the thesis in the sense of a mathematical proof. But it does provide interesting evidence to support the claim. Whether the evidence is strong enough to convince you is an individual matter. One thing is certain, we believe: the credibility of the assertion is much more acceptable with Phillips' results than without.

The major lesson to be learned from Phillips' experiment is that when a researcher wants to study a phenomenon often the key to a successful experiment lies in the choice of measure used. In designing the experiment to test his death-dip hypothesis, Phillips chose to study the measure relating the time of the death month before, during, or after the birth month of famous people. Once the measure was determined, gathering the data was a menial and time-consuming task.

Another lesson that we can learn from Phillips' study is that the power of the brain is often underestimated. We must never take it for granted. In the next example we will see how the brain can even mask pain if the individual wants to be cured. If the science of statistics is to accurately judge performance in an experiment, safeguards must be taken so that bias cannot affect the outcome.

Can Ulcers Be Frozen to Death?

There is one very compelling question concerning research using statistical analysis: When is the evidence gathered by the researcher strong enough to convince the reader of the validity of the thesis? It is usually easier to judge when the evidence is not strong enough. Often it is not the results that are unconvincing, but it is the design of the experiment that produced the results that is lacking. Even a seemingly well-conceived experiment may yield faulty results. We will illustrate this with an example from medicine.

When a new technique is devised to combat an illness, tests must be run to document its effectiveness. If only two variables are studied, such as birth and death dates in Anecdote 1, or treatment and cure for ulcers in Anecdote 2, then every conceivable precaution should be taken in the experiment to minimize the effects of other variables which are not considered. This is called *bias*. When a variable that is not controlled affects the outcome of the experiment, than it is said that bias is introduced and the relationship between the primary variables is not uncovered.

Anecdote 2

In 1962 Wangensteen announced a startling new method for treating ulcers.* He reported that his new technique of freezing the lining of the stomach was not only very effective but it was a vast improvement over the normal treatment that required tedious and risky surgery.

The success of Wangensteen's technique was dramatic. It replaced the usual complicated surgical treatment with a far less drastic procedure. It stopped the digestive process by freezing the stomach lining for about an hour, thereby allowing the ulcer to start the healing process. All 24 patients in the study were cured.

Even though the procedure was widely adopted, many doctors were skeptical. The doubt did not arise from Wangensteen's results, but from the design of his experiment. No one questioned the fact that all 24 patients were cured; it was simply a case of determining what was the cause of the cure. As we saw in Anecdote 1 the brain is a powerful force. Perhaps it was not the procedure but the belief in the procedure that caused the cure.

For this reason a well-designed experiment will include a *control group,* which is a set of people who have the same salient properties as the people in the treatment group but who are not treated. Usually the control group is given what appears to be the same treatment but in reality is a sham.

Journal of the American Medical Association, vol. 180, 1962, pp. 439–444.

Many other factors may also enter the picture. Their effects may first appear negligible but are often critical. For example, Wangensteen himself was the examining physician determining the success of the experiment. He was certainly biased toward the success of the trial. The diagnostician should not know which patients have undergone the new procedure. In this way the judgment is blind in the sense that it is not biased toward those who have undergone the treatment. An experiment is double blind if it (1) has a treatment group and a control group chosen at random, and (2) the diagnosticians who measure the success of the treatment do not know who is in either group when the judgment is made. Thus a double-blind experiment avoids two kinds of bias, that on the part of the subject and that of the diagnostician.

Wangensteen's experiment had neither of these controls. Could these biases have made a difference in the outcome of the experiment? At first it would seem not to matter, as the evidence appears conclusive. In 1963 Ruffin decided to find out for himself. He devised a double-blind experiment to evaluate Wangensteen's technique. At five separate hospitals, 160 patients with ulcers were chosen and randomly assigned to two groups: 82 were in the treatment group and 78 were in the control group. Those in the treatment group were given the gastric freezing technique. The control group was given the same procedure except, unknown to them, the coolant was removed from the stomach before freezing could take place. Since all the patients were anesthetized, they could not tell whether their stomachs were actually subjected to the gastric freezing. Periodic checkups were given to the patients over a 2-year period by doctors who did not know who had been given the true treatment.

Ruffin's results can be summarized as follows:

1   In the first 6 weeks after the application of the treatment about 30 percent in both groups were "cured"; that is, they were symptom-free. Most others were "improved," about 50 percent in the treatment group and about 40 percent in the control group. Thus, after 6 weeks, Ruffin's results were similar to Wangensteen's, as 80 percent (= 30% + 50%) in the treatment group and 70 percent (= 30% + 40%) in the control group were on the mend.

2   After 2 years most patients suffered a relapse. Significantly large percentages in both groups were clinically worse: about 45 percent in the treatment group and about 40 percent in the control group.

3   At no point during the 2-year period of evaluation was there a significant difference between the two groups. Improvement and relapse occurred in almost exact percentages in the two groups. As Ruffin claimed in his article published in 1969 in the *New England Journal of Medicine,* "The results of this study demonstrate conclusively that the freezing procedure was no better than the sham in the treatment of duodenal ulcer."

It should be clear from Anecdote 2 that whenever a researcher designs an experiment it must, whenever possible, be double blind.

**Vitamin C and
the Common Cold**

Another important ingredient in designing a valid experiment involves the selection of the sample when it is impossible or inconvenient to consider the whole population. We will present one more anecdote on this subject, but since it is such an important aspect of statistics, we will spend an entire section in Chapter 8 on one important aspect of sampling—sample surveys.

> **Anecdote 3**
>
> One of today's most fashionable crazes, even as measured in our age of fads, is the almost religious reliance on vitamin C to ward off various inflictions, most notably the common cold. The common cold is, at its most innocent, a terrible nuisance and, at its worst, the precursor of serious illness. Cures have been sought by medical researchers for many years. In 1970 the esteemed scientist Dr. Linus Pauling wrote a revealing and sensational report* stating that regular massive doses of vitamin C will prevent the common cold and other maladies as well.
>
> Many people latched onto the idea, and pharmaceutical companies churned out vitamin C pills in assorted sizes, colors, textures, and potencies. By and large the medical community looked on Pauling's pronouncement with skepticism. His book, while convincing from a lay point of view, had very little scientific evidence to back up his controversial hypothesis.
>
> One of the shortcomings of Pauling's argument was the lack of statistical evidence in the studies that claimed that massive doses of vitamin C prevented the common cold. One of the investigators quoted in his book was Dr. Edme Regnier, whose paper, "The Administration of Large Doses of Ascorbic Acid in the Prevention and Treatment of the Common Cold," published in *Review of Allergy* (October 1968), received praise by Pauling and was used in many of his conclusions.
>
> Regnier's article begins with a description of his own history of infections from which he had suffered since childhood—50 to 60 bouts over the next 20 years from age 7. He tried many treatments and finally was successful with massive doses of vitamin C. From this personal experience he hypothesized that large doses of vitamin C could be used to almost completely suppress colds, leaving the patient completely symptom-free and preventing secondary bacterial infection. (Common colds are caused by viruses which can lower your immunity to bacteria and thus increase the likelihood of infection.)
>
> At the start of his study Regnier had good intentions to ensure that his experiment was designed well. One early fault, however, was the size of the sample and the manner in which it seems to have been chosen. Regnier's study was limited to 22 subjects (who were acquainted with him, either as friends or patients). He set out to conduct a double-blind experiment. However, the investigator's blind, that is, Regnier himself, was removed because he felt that a cold was a tenuous activity. Its treatment required a high degree of informality to encourage the subjects. Later, the single remaining

*L. Pauling, "Vitamin C and the Common Cold," W. H. Freeman, San Francisco, 1970.

blind was removed when some subjects recognized that they were being treated with a placebo and refused to proceed. The results of the study were therefore biased since the subjects and the investigator were convinced of the effectiveness of the treatment they were testing.

The treatment of the subjects was not consistent from subject to subject, nor even from cold to cold. A statistical analysis of Regnier's data cannot be made, primarily because he reported no numerical data, only observations.

Here are the main faults of the study from a statistical point of view.

1 Lack of a random sample

2 Lack of a double-blind experiment

3 Biased investigator and subjects

4 Inconsistency of treatment

5 Lack of statistical data

Another study prominently quoted by Pauling was a double-blind experiment involving 279 students at a ski school in Switzerland.* The period of observation was only 1 week for each individual.

Even though the studies on which he rested his case had little scientific evidence, the scientific eminence of Linus Pauling, a Nobel Prize winner in biochemistry, and the large public interest in the matter prompted other investigators to test the hypothesis with properly controlled trials.

One of the most thorough studies was done by Anderson, Reid, and Beaton† of the University of Toronto's School of Hygiene. They were highly skeptical of Pauling's claims and so were surprised when their results partially substantiated some of the assertions in his book. The study involved 1000 subjects who were instructed to follow a regimen similar to Regnier's and Pauling's: 1000 milligrams per day and an increase to 4000 milligrams per day at the onset of a cold. Vitamin C and placebo tablets were identical in all obvious aspects. A computer-generated coding system created well-matched treatment and control groups. The study ran for a period of 4 months.

Some of the conclusions of the study were:

1 The number of subjects with no colds was 105 (26 percent) for the treatment group and 76 (18 percent) for the placebo group. This shows a significant difference between the two groups, but it does not support the greater claims of reduction in frequency of colds asserted by Pauling.

2 The number of days of symptoms per cold was 5 percent lower for the vitamin group.

3 The number of days confined to home was 30 percent lower for the vitamin group. The symptoms alleviated by vitamin C were found to be con-

---

*G. Ritzel, "Ascorbic Acid and Infections of the Respiratory Tract," *Helvetica Medica Acta,* vol. 28, 1961, p. 63.
†"Vitamin C and the Common Cold: A Double-Blind Trial," *Canadian Medical Association Journal,* September 1972.

stitutional (chills, fever, malaise) rather than local (nose, throat, chest congestion). There is a connection between this and the number of days confined to home because constitutional symptoms are more likely to cause sufficient discomfort to warrant a day off.

4 Seventy-seven precent of the vitamin group said that they had an increased sense of well-being while taking the drug. This is one of Pauling's key arguments supporting the treatment. But 78 percent of the placebo group said the same thing! Thus the claim that vitamin C increases the sense of well-being appears to be a mental condition induced not by the vitamin but by the impression that one *should* feel better while taking it.

5 The number of subjects who reported side effects was 12 percent for the treatment group and 11 percent for the control group.

One of the most striking differences between the two experiments—Regnier's and the Anderson study—is one of attitude. Regnier approached his study convinced of the validity of his thesis and allowed his confidence to bias the design of his experiment. It is not that his results were alarmingly wrong. The key point is that the design of his experiment was faulty so his results cannot be accepted as statistically valid.

On the other hand, the University of Toronto investigators took great pains to ensure that their study was statistically valid by eliminating as much bias as possible in the design of the experiment—even though they themselves were biased toward the lack of validity of the hypothesis. They were not satisfied with one study but conducted two more very similar ones. In addition, they concluded their article with a caution that not enough research had been done to make recommendations concerning the use of vitamin C, especially with respect to the safety of prolonged use of large dosages.

Summary

In this section we found that an investigator must do more than just gather data. An experiment that is meant to test the validity of a hypothesis must be well designed to avoid bias that may render the conclusions worthless.

We have described what statisticians call the *method of comparison.* An investigator sets out to determine the effects of a *treatment* on a *response,* for example, the effects of vitamin C on the response of not getting colds. He or she compares the effects on a treatment group, one that gets the treatment, to a control group, one that does not. If the subjects in the control group are given a placebo, then the experiment is called a *blind* experiment. This avoids the bias of the subjects experiencing the response because they believed they should experience it, not because they actually did.

A well-defined experiment should be double blind, whenever possible. A double-blind experiment avoids bias on the part of the subjects as well as the evaluators because the investigator does not know which subjects are in the treatment group.

A topic that was just touched upon in this section is the method of choosing the subjects for the experiment. To avoid bias they should be chosen in a random way that avoids any judgment of the investigator as to which subjects are chosen for the experiment and for the treatment and control groups. We will pursue this idea more thoroughly in Section 8-6.

EXERCISES

1 What is wrong with basing a study on people chosen at random from a telephone book?

2 What is wrong with basing a study on people selected at random from voter lists?

3 Is an experiment faulty if it does not include the number of nonrespondents, that is, the number of people who did not respond to a questionnaire?

4 In an experiment to determine people's preferences about sensitive issues, is it faulty to get them by personal interviews?

5 Design an experiment to determine whether beer drinkers can distinguish between different brands of beer.

6 Design an experiment to determine whether nonsmoking spouses of smokers are more likely to contract cancer than the smoker.

7 Many creatures have the ability to point themselves toward home even after being taken to a distant unfamiliar place. Design an experiment to determine whether humans have this ability. (See *Science,* May 29, 1981.)

8 Roughly two-thirds of all humans are right-handed, while the rest are either left-handed or ambidextrous. Design an experiment with newborn infants to suggest a reason for this preference. (See *Science,* May 1, 1981.)

9 In the Salk polio vaccine trial the children of parents who refused to participate contracted polio at the rate of 46 per 100,000. If you lump together the control and treatment groups—who consist of children whose parents participated—the rate of those who got polio was 49 per 100,000. Suppose some parents say that, on the basis of these figures, they refuse to allow their children to take the vaccine. Is this a good argument from a statistical point of view?

10 The summer after the volcano at Mount St. Helens erupted, the corn crop in Iowa was very poor. Many seeds failed to germinate. Does this implicate the fallout of volcanic ash as the cause of the poor crop?

11 In the October 14, 1976, issue of the *San Francisco Chronicle* it was reported that conventional treatment had failed to alleviate pain for 31 patients suffering from severe headaches as a result of spinal punctures. The patients were then given five acupuncture treatments, and 30 of the 31 patients experienced "complete and permanent relief." Compare this study with the Wangensteen report and comment on whether there is convincing statistical evidence as to the effectiveness of acupuncture.

12 People who are arrested and charged with a criminal act are held in jail until they can be brought to trial unless they are released either by posting bail or on their own recognizance (that is, they either pay the court bail money, which is returned if they appear in court, or they are simply allowed to promise to appear in court). Studies have shown that those detained in jail are 2 to 3 times as likely to be convicted as those who are released. Does this prove that a person who is detained is more likely to be convicted than a person who is released?

7-4 DESCRIPTIVE AND INFERENTIAL STATISTICS

The world of statistics often seems baffling to the beginning student. At times it baffles the experts, too. In today's society it seems that there are statistics concerning every facet of our lives, but most people still hold a view of statistics similar to Disraeli's: "There are three kinds of lies: lies, damned lies, and statistics." As a branch of science, the discipline of statistics is indispensable but, as Disraeli suggests, its methods can also be used to mislead. Thus an understanding of the basic concepts of the subject is essential.

Descriptive Statistics

Statistics is concerned with the collection, analysis, and interpretation of data. It can be separated into two general areas, *descriptive* and *inferential,* which we will explain by means of examples drawn from sports, warfare, sociology, and political science.

Often a set of data is too large to handle without resorting to certain types of numbers that describe the data. One of these numbers is probably familiar to you, the *average,* or more properly, the *arithmetic mean.* If you have taken three exams and received scores of 85, 90, and 95, then your average is

$$\frac{85 + 90 + 95}{3} = \frac{270}{3} = 90$$

The average describes how you have done on the three exams. It is one of the measures of a set of data that we refer to as descriptive statistics, and we will study it in more detail in Chapter 8. In this section we will encounter many such measures.

Inferential Statistics

When the statistics gathered about a particular event are used to make a decision, that is, when the data are interpreted so that a judgment can be made, we say that inferential statistics is being used. Let us take a look at our national pastime to illustrate this concept.

Example 1

Suppose it is July 1, 1980, and you are out at the old ballpark on a sunny summer Sunday, enjoying a game between, say, the Yankees and the Angels. You are handed a ballot for the All-Star game. You immediately turn to the outfield section of the ballot to vote for your favorites: Reggie Jackson, Fred Lynn, and Don Baylor. But wait! A little voice reminds you that our national pastime thrives on statistics. Are your three favorite outfielders actually playing better than the rest? You grab the Sunday sports section and check the league statistics. To judge how well someone is hitting, three measures are regarded as most important: batting average (number of hits divided by number of at-bats), home runs, and runs batted in (basically, you get a run batted in, or RBI, if you get a hit or make an out and a base runner scores as a result). Not one of your three is in the top 10 in hitting or the top five in RBIs, and only Jackson is among the leaders in home runs. Then you notice a name you've never seen before—Oglivie. He plays for Milwaukee. He is sixth in batting average, having 83 hits in 249 at-bats for .333 average (83/249), and he is first in both home runs and RBIs. Even though you have never heard of him, seen him play, or know how well he fields, his statistics convince you he deserves your vote for the All-Star game. You have used descriptive statistics to infer that Oglivie is having the best year of all the outfielders in the American League. See Figure 1.

AMERICAN LEAGUE*
(Based on 165 at bats)

	G	AB	R	H	Pct.
Molitor Mil	47	190	41	68	.358
Carew Cal	68	263	33	90	.342
Brett KC	45	169	30	57	.337
Orta Cle	61	230	38	77	.335
Cooper Mil	63	258	34	86	.333
Oglivie Mil	66	249	52	83	.333
Yount Mil	59	250	53	83	.332
Buddy Bell Tex	52	205	32	68	.332
Bumbry Bal	70	276	51	91	.330
Hurdle KC	59	180	24	59	.328

HOME RUNS: Oglivie, Milwaukee, 20; Reggie Jackson, New York, 18; Thomas, Milwaukee, 15; Armas, Oakland, 14; Rice, Boston, 13; Mayberry, Toronto, 13; Nettles, New York, 13.

RUNS BATTED IN: Perez, Boston, 55; Oliver, Texas, 55; Oglivie, Milwaukee, 55; Hebner, Detroit, 50; Cooper, Milwaukee, 49; Armas, Oakland, 49.

PITCHING (7 Decisions): Stone, Baltimore, 10-3, .769; John, New York, 10-3, .769; Gura, Kansas City, 10-3, .769; Cleveland, Milwaukee, 6-2, .750; McGregor, Baltimore, 8-3, .727; Rainey, Boston, 8-3, .727; Farmer, Chicago, 5-2, .714.

*As of June 30, 1980.

FIGURE 1

Example 1 demonstrates how descriptive statistics can be used in making judgments or inferences. The process of using statistics to solve problems and make decisions is called *inferential statistics*. The methods used in inferential statistics are usually much more difficult to apply than in the example above. We will cover more on this subject in Chapter 8. The rest of this section will be concerned with three examples of where and how statistics can be used to help solve complicated human problems.

Comparative Effectiveness— World War II

In the previous example we used certain measures—batting average, home runs, RBIs—to compare one ballplayer's performance with that of his peers. Even though we might not have known that 20 home runs is a large number in July, we compared Oglivie's total with the rest of the players in the league and saw that he was the best home run hitter at the time. In studying statistics you will often be required to compare certain numbers with others and draw a conclusion. Obtaining or calculating the numbers is in the realm of descriptive statistics. Drawing the conclusion is part of inferential statistics. The next example presents a moment in history where inferential statistics was used to effect a change in a government's policy during World War II.

Example 2

In World War II the methods of conducting war had undergone a phenomenal growth in complexity. (Recall the ship-maneuvering example from Section 6-4.) Important decisions could no longer be left up to the experienced veteran officers. The British government was quick to realize that nonmilitary personnel could be utilized to evaluate the effectiveness of many strategies. Consequently the government enlisted the help of statisticians, mathematicians, psychologists, and many other types of talented civilians who knew little about war tactics but could provide clues to strategies in specific areas never before conceived of by the military.

An interesting example is presented in a study done by the Director of Naval Operations Research, Admiralty, on the relative importance of different types of armor on British cruisers. It was always assumed that the highest priority for the placement of armor was on deck because the attacks came overwhelmingly from bombs from airplanes and shells from ships. For example, during the study there were 21 casualties due to shells, 65 due to bombs, only 10 due to mines, and 30 due to torpedoes. These numbers seemed to confirm the tacit assumption of the shipbuilders.

The Operations Research group (or the OR group, for short) took a different tack on the problem. In many cases it is necessary to compare relative effectiveness of two different tactics. It is often difficult to find a common unit of measure. Care must be taken in choosing the unit of measure so that bias, whether intended or not, does not enter into the decision. If certain important aspects of the problem are omitted or deemed negligible, a wrong conclusion might be reached. Thus it is important that the unit of measure not prejudge the outcome; it must be chosen in such a way that it provides an objective, equitable, and usable standard of comparison. We will find that many of the methods of statistics are designed to ensure objectivity in drawing conclusions.

The OR group noticed that the simple unit of measure of counting numbers of casualties prejudiced the problem in favor of the attacks above water by shells and bombs. This is because there were far more planes and ships than submarines. It does not take into account the severity of the damage. It is necessary to compare the relative importance of ship damage to ship sinking. The problem was to find a measure that would relate the two in a meaningful way. The OR group chose the amount of time a dockyard will take to make up the loss. That is, a damaged ship requires a certain amount of time to repair and a sunk ship requires a specific amount of time to replace by building another. In either case, the time is lost and cannot be made up. Therefore the unit of measure that the group adopted was ship-months lost.

In the Admiralty study, the number of months lost due to the time that a damaged cruiser was out of service was tabulated for each casualty, and the value of 36 ship-months was given to a ship sunk because that is the time needed to build a new cruiser. The data are compiled in Table 1.

Several conclusions can be drawn from the table. Let us compare the bomb casualties with the torpedo casualties. As we said before, there were more than twice as many bomb as torpedo casualties, 65 to 30. But note the percentage of torpedo casualties that resulted in sunk ships was 37 percent (11/30) as compared with just 14 percent (9/65) for bomb casualties. When translated into ship-months lost, these figures indicate 400 by torpedo casualties as compared with 320 by bomb casualties. Also note that torpedoes damaged ships much more severely than bombs: only 19 casualties caused 180 ship-months lost compared with 56 ships damaged by bombs causing only 90 ship-months lost. The last row in the table is the total number of cruiser-months lost divided by the number of casualties. Thus the entry 19 comes from $19 \approx 580/30$. It clearly shows that torpedo damage is much more severe and serious than other war-related damage. Because of the results of this study and many others like it, ships were built with more antisubmarine armor, and antisubmarine tactics were studied and perfected.

TABLE 1

Casualties to Cruisers by Enemy Action

	Shell	Bomb	Mine	Torpedo	Total
Ships sunk	3	9	1	11	24
Ships damaged	18	56	9	19	102
Total casualties	21	65	10	30	126
Cruiser-months lost:					
By sinking	110	320	40	400	870
By damage	30	90	60	180	360
Total	140	410	100	580	1230
Percent	11	34	8	47	100
Cruiser-months per casualty	7	6	10	19	10

The result was that it became unprofitable for the Germans to attack in the North Atlantic and the U-boats were forced to seek combat elsewhere, thereby negating their effectiveness. This was the turning point of the Battle of the Atlantic.

Graphs—Do They Give the Whole Picture?

In studying statistics we will often use graphical means to display the data. You should be aware at the outset, however, that one of the primary ways that people misuse and mislead via statistics is by misrepresenting data and causal relationships in graphs. We now present an example from applied sociology where a graphical depiction of the data had a profound effect on driving habits of motorists in the northeastern section of the United States.

Example 3

In 1955 the Connecticut Motor Vehicle Department issued an announcement stating that 324 people had been killed in automobile accidents on Connecticut highways. This was an incredibly large increase over the previous year, when only 242 deaths were recorded. The 1955 total was a record high. It was early December when the trend was noted, and the hazardous Christmas and New Year's holidays were approaching. Just before Christmas Governor Abraham Ribicoff announced what would become a highly controversial campaign against highway fatalities by cracking down on speeding in the state.

Governor Ribicoff firmly believed that if speed were adequately controlled the number of fatalities would plummet. It appeared that the existing methods of handling speeders by court hearings and a "point system" were obviously not stringent enough. Consequently, on December 23, 1955, the governor announced that anyone convicted of speeding would have his or her license suspended for 30 days on the first offense, 60 days on the second, and indefinitely on the third.

The furor from the public was deafening. License suspension for speeding increased a whopping 1800 percent in the first 3 months of 1956, from about 160 to 2860. In the first 6 months the number went from 210 to 4560, an increase of over 2100 percent.

Neighboring states, the Teamsters, bus companies, and lobbyists for sales companies claimed they were paying an exorbitant price for a dubious benefit. It was incumbent upon the governor to demonstrate the usefulness of the controversial program.

By the end of the year, Connecticut had recorded 284 automobile fatalities. The governor stated, "With the saving of 40 lives in 1955, a reduction of 12.3 percent from the 1955 motor vehicle death toll, we can say the program is definitely worthwhile." Accompanying this statement was the graph in Figure 2.

There are two points of interest on the graph. The top left point represents the number of fatalities, marked on the vertical line on the left, namely 324, in 1955, which is marked directly below the point. The lower point on the right represents the number of fatalities, 284, recorded in 1956. The solid line is meant to reinforce the idea that the total number of deaths decreased from 1955 to 1956. This graph is very convincing. In fact, Governor Ribicoff

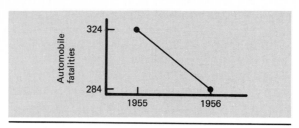

FIGURE 2

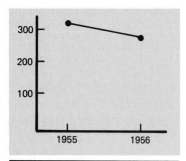

FIGURE 3

received a telegram of commendation for the program from the National Safety Council.

Are you convinced? Do you see anything wrong with the graph? Is anything missing? If you study the graph for awhile, you should note that there are two important ingredients missing.

First of all, the vertical axis that measures the number of fatalities is missing the interval of numbers from 0 to 280. Does this make a difference? Consider Figure 3, where we have plotted the same two points but included the entire vertical axis. The graphical representation of the decrease is much less spectacular in Figure 3 than in Figure 2.

The second significant missing portion of relevant data are the numbers corresponding to the years before 1955. Could it be that the dramatic decrease from 1955 to 1956 was due to a natural fluctuation? Of course, we cannot be sure until we consider this data.

In their classic paper on the Connecticut crackdown on speeding,* Campbell and Ross studied the causal relationship between the crackdown and the decrease in the number of fatalities. Their findings produced far-reaching ramifications in diverse fields. The study raised the question of the effectiveness of massive government intervention in society. While the answer was far from clear, it prompted sociologists, psychologists, and scientists to study the methods of measuring effectiveness of large sociological programs, both in government and the public sector.

In Figure 4 you see the graph depicting not only the two points plotted in Figures 2 and 3, but also those corresponding to the 4 years before 1955 and the 3 years after 1956. The vertical line labeled "Treatment" indicates that the crackdown, or the treatment, took place in 1956, and thus during the time span between the two points labeled 1955 and 1956 on the horizontal axis. (This type of graph will be discussed in more detail in Section 8-1.)

Campbell and Ross raised the question of whether the decrease was actually caused by the crackdown, or whether there were other factors not mentioned in the state's analysis that contributed as well. We will point out just a few of their findings here and in the exercises. For more details we refer you to their very interesting paper.

1 The oscillatory behavior of the data before 1956 (that is, the totals go down-up, down-up) would lead one to predict a drop in 1956 even without a crackdown.

*P. T. Campbell and H. L. Ross, "The Connecticut Crackdown of Speeding: Time Series Data and Quasi-Experimental Analysis," *Law and Society Review,* vol. 3, 1968, pp. 33–53.

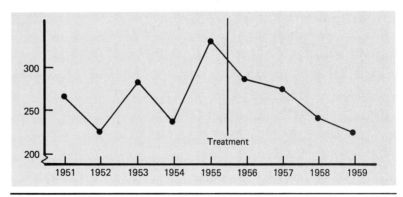

FIGURE 4

2 Clearly the largest change from one year to the next occurs in the striking rise in the death toll from 1954 to 1955. Perhaps merely a simple announcement of the fact to the public would cause a sober awareness of the problem and hence contribute to a subsequent decrease.

3 When compared with the results in neighboring states, the data from Connecticut are very similar. New York, Massachusetts, and Rhode Island all had increased fatalities in 1955, although not quite so dramatic as in Connecticut. All three had very similar declines in their highway death tolls in 1956.

The important point in this example is that when an investigator uses inferential statistics to draw conclusions, every effort must be made to uncover other outside forces that may have a bearing on the causal relationships. Many of the methods of statistics that we will study are designed to negate the effect of possible unseen or unmeasurable factors. Such factors are called *confounding effects,* and they can really confound your conclusions.

Who's Voting for Whom?

The final example in this section will demonstrate how another type of graph, called a *bar graph,* conveys the meaning of the data in a visual way. It is a classic study of voting behavior, and it provided the first large-scale detailed analysis of the behavioral characteristics of a socially diverse population.*

Example 4
We all belong to one or more societal groups. They may be religious, racial, ethnic, or political. Members of societal groups recognize their membership and this often affects their behavior. It has long been believed that certain generalizations could be made about specific characteristics of members of a group, such as occupational tendencies, places of residence, educational goals and attainments, and many other facets of their lives. Of course, judg-

*B. Berelson, P. Lazarsfeld, and W. McPhee, "Voting: A Study of Opinion Formation in a Presidential Election," University of Chicago Press, Chicago, 1954.

ments concerning the whole group can be subject to error when applied to an individual case.

One important type of study that sheds light on many diverse social interactions is voting behavior. The first large-scale study of the voting tendencies of societal groups was conducted by Berelson, Lazarsfeld, and McPhee during the 1948 presidential election between Harry Truman and John Dewey. As we will see in Section 8-6, the election itself dramatically affected the polling agencies, whose job it was to predict the outcome, and the discipline of statistics itself. Why the outcome of the election took such a drastic turn can be partially explained by the Berelson study.

The study was conducted in Elmira, New York. By concentrating their efforts on a small geographical area, the investigators could survey the largest number of people within their limited financial constraints. Do you see anything wrong with limiting the survey to the populace of one city? We will answer that question at the end of the section.

There are a number of reasons why Elmira was chosen. It was small enough so they could question the majority of the voting public. It was large enough so that most ethnic and societal groups were represented in sufficient numbers. Besides, who could accuse them of spending their grant money for a vacation in Elmira?

We will present a number of their findings, some of which you would have been able to predict, but some of which will surprise you. In each case we will pose a question, ask you to predict the outcome, and then present the conclusions of the study. Each conclusion will be accompanied by a graphical representation of the data in the form of a bar graph. This type of graph will have bars of equal width so that the heights of the bars, representing percentages, can be used immediately to compare the data.

Problem 1   What percentage of the following groups voted Republican: white native-born Protestant, Catholic, Jewish, black? (See Figure 5.)

Solution.   More than four-fifths (81 percent) of the white native-born Protestants voted Republican, whereas about one-third of the Catholics and the Jews and only about one-fifth (19 percent) of the Elmira blacks voted Republican. Politicians have always been concerned with organizing the "black vote" or the "Jewish vote," since the constituents are often numerous,

FIGURE 5   The minorities and the "majority" vote differences (percentage Republican of two-party vote).

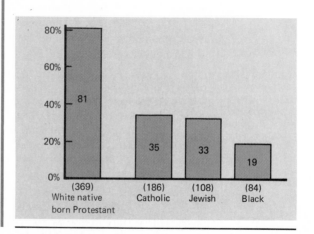

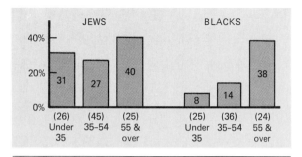

FIGURE 6   Percentage Republican of two-party vote.

have a strong group identification, and share common characteristics that the politician can identify and direct an appeal to.

Problem 2   Does age make a difference? In particular, for the Jewish and the black votes, is the younger generation more likely (or less likely, or about equally likely) to support the Democrats? (See Figure 6.)

Solution   About the same percentage of Jews and blacks 55 years of age and older voted Democratic, but the percentage increased dramatically for younger voters, especially among blacks. Thus the more-educated youth were thought to lean much more closely to the Democratic party than their elders, and it is toward that group that the Democratic party looked for its standard-bearers.

Problem 3   Does a strong group identification affect voting trends? The investigators wanted to know if those who were more involved in their societal group voted differently than those less involved. They chose to consider Catholics and Protestants. They decided that the number of years of residence in Elmira would be an adequate measure of group identification. It was believed that in the long run those who lived in the same community the longest time would tend to have the strongest ties to the group. (See Figure 7.)

Solution   In splitting the voting populace into three groups (those who had lived in Elmira over 20 years, from 5 to 19 years, and less than 5 years) it

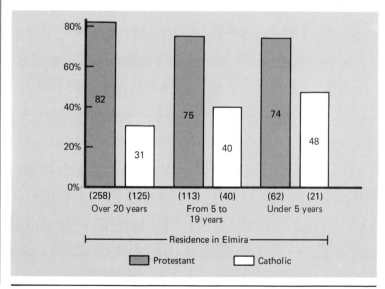

FIGURE 7   Percentage Republican of two-party vote.

was found that just about the same percentage of Protestants voted Republican (82, 75, 74 percent, respectively) in each category. However, the Catholic vote was strongly tied to religious affiliation. The longer that Catholics lived in Elmira, the more likely they were to vote Democratic. In the chart you can see that the percentage that voted Republican decreased from 48 percent (and hence 52 percent voted Democratic) to 40 to 31 percent as length of residence in Elmira increased.

The conclusions of the Berelson study are significant, but the actual figures must be taken with a grain of salt. The conclusion from this study that societal group affiliation affects voting behavior is inescapable, but clearly Elmira is not a perfect cross section of the whole country. Certainly a Democrat in the north (say, Ted Kennedy) is different from a Democrat in the south (say, Jimmy Carter). Therefore it would be incorrect to try to project onto the whole country the actual percentages in the study. The study was a pioneering classic in its design as far as its detail and large-scale effectiveness, however. It was conducted at an opportune time, when the very roots of the effectiveness of polling agencies were shook by the 1948 presidential election. It provided the groundwork for many of the methods and procedures now used in the industry to ensure valid prediction.

Summary

In this section we presented four examples to illustrate and contrast descriptive and inferential statistics. The example from baseball indicated that several measures can be used as standards for comparison, and it showed how you can make inferences based upon the descriptive data. The second example, which was concerned with the placement of armor on battleships in World War II, showed that the choice of a standard unit of measure is not always clear-cut, but that statistics can be used both to help make this decision and to infer conclusions based upon it.

The strict enforcement of speed limits by Governor Ribicoff illustrated how graphs of statistics can be used (and misused) to prove a point. It also showed, however, that you have to exercise extreme caution in designing your experiment so that the biases are minimized.

The example on voting patterns in Elmira indicated how bar graphs are used to depict the findings of a study. It further showed how inferences can be made from these graphs. Section 8-1 is devoted to additional ways of graphing sets of data.

EXERCISES

1 What is the difference between descriptive and inferential statistics?

2 (*a*) If we use a ballplayer's batting average as the probability that he will get a hit in his next at-bat, what is the probability that Oglivie will get a

hit on his next at-bat on June 30, 1980? (*b*) If he gets two hits in four at-bats on June 30, what is the probability that he will get a hit in his next at-bat?

3   Why was the simple unit of measure of counting numbers of casualties first employed by the Admiralty ineffective in assessing damage done to British cruisers? What unit was adopted by the Operations Research group? What factor did the new unit of measure take into account that the former one did not?

4   Use Table 1 to give the following probabilities. What is the probability (*a*) that a ship sunk was sunk by a bomb, (*b*) that a ship sunk was sunk by a torpedo, (*c*) that a ship sunk was not sunk by a torpedo, (*d*) that a ship damaged was damaged by a torpedo, (*e*) that a ship that was sunk or damaged was sunk or damaged by a torpedo, (*f*) that a ship that was sunk or damaged was not sunk or damaged by a torpedo?

5   Why is the graph in Figure 2 somewhat misleading?

6   Compare Figure 2 with Figure 3. Which graph more accurately reflects the data?

7   In Figure 4, why is the solid black vertical line included?

8   Can you think of outside forces that could have contributed to the drop in the number of deaths on Connecticut highways from 1955 to 1956?

9   Use the graphs depicting the voting trends of the electorate of Elmira, New York, in the 1948 Presidential election to answer these questions. What is the probability that (*a*) a Catholic voted Republican, (*b*) a black voted Republican, (*c*) a Jew under 35 years of age voted Republican, (*d*) a black under 35 years of age voted Republican, (*e*) a Protestant who lived in Elmira for less than 5 years voted Republican?

8 Statistics

The study of statistics as a mathematical science is a relatively new discipline. Statistics were gathered as mere facts, like those in an almanac, soon after people learned to count. However, it was not until the late seventeenth century, when an unlikely character studied London death records in his leisure time, that the science of statistics was born. It started when John Graunt showed that the deaths due to certain illnesses followed a regular pattern. Graunt was not a scientist and did not follow up his studies. Following the lead of Graunt there were many individuals who helped develop the framework of statistics. Among the most notable were Adolph Quetelet, who improved upon Graunt's mortality tables, which provided the foundation for the first life insurance companies. Gregor Mendel applied probability and statistical methods to biology. Sir Francis Galton (1822–1911), the precocious cousin of Charles Darwin, extended Mendel's work on the mathematical study of heredity. Then in 1900 Karl Pearson (1857–1936), often called the father of statistics, applied statistical methods to other areas of biology, and statistical methods were popularized. Thus statistics is still in the embryonic stages of its evolution.

We open this chapter with the rudiments of statistics—histograms, the average, and the standard deviation—in order to introduce you to the tools of the trade. The last three sections on the normal curve, its approximation, and sample surveys, give you a glimpse at how statistical methods apply to everyday problems. Section 8-6 will demonstrate how dramatic the evolution of statistics has been over the last 50 years as a predictor of human behavior. To bring the point across we focus our attention on presidential elections because one gets immediate feedback, on election day, of whether the poll was correct. Major improvements in the methods of statistical studies, and the subsequent improvement in survey practice, have come about because of the colossal mistakes made by polling agencies in presidential elections.

8-1 HISTOGRAMS, OR A PICTURE IS WORTH 1000 WORDS

Are your height and weight above average or below? Perhaps you are "just average." How do we know what average is? The U.S. government provides statistics that help us answer these types of questions, as well as other more important ones. The Current Population Survey (CPS) reports on the vital statistics of a representative cross section of the country each month. It contacts about 40,000 households in order to keep abreast of important trends in the populace. These statistics are used to predict inflationary trends, unemployment forecasts, birthrates, and various other demographic variables. Table 1 gives the CPS data on height of men and women in the United States.

TABLE 1

Men		Women	
Height, Inches	Number in Thousands	Height, Inches	Number in Thousands
Under 60	01	Under 53	0.4
60	01	53	0.5
61	05	54	0.5
62	09	55	0.2
63	17	56	0.2
64	37	57	10
65	35	58	13
66	70	59	38
67	63	60	45
68	94	61	83
69	54	62	104
70	63	63	73
71	32	64	90
72	28	65	47
73	11	66	44
74	06	67	14
75	01	68	12
Over 75	01	69	01
		70	01
		Over 70	0.5
	528		577.3

Endpoint Convention

In Table 1 the heights are actually intervals. The first entry in the table for men is "Under 60" inches and the corresponding frequency is 1, meaning approximately 1000 of the 528,000 men surveyed were under 60 inches, or under 5 feet tall. The next entry is labeled "60 inches." It represents the interval of numbers between 60 and 61 inches.

A convention must be adopted to include either the left-hand endpoint of the interval (in this case 60) or the right-hand endpoint (61). The convention arbitrarily adopted in this survey is to include the left-hand endpoint. Thus the entry "60" actually means all those heights x such that $60 \leq x < 61$. If someone's height is 5 feet exactly, he or she is tallied in the "60" column, and if someone's height is 5 feet 1 inch (or 61 inches), that person is tallied in the "61" column. This is called an *endpoint convention*.

Histograms

As presented in the table the data do not make much sense. They present too much information, and a glut of information can actually cause a bigger problem than not having enough. They must be summarized. The first method that we present is called a histogram. We call Table 1 a *frequency distribution*. When we graph this information using a bar graph, it is called a *histogram*.

Let us draw a histogram for the height distribution for women. The interval sizes correspond to the widths of the bars. If the interval sizes are equal, then the heights of the bars are the frequencies. If the interval sizes are not equal then the areas of the bars must be considered. Since the interval sizes in Table 1 are equal (except for the first and last ones, which we will ignore for the time being) we can draw a histogram, using the frequencies as the heights of the bars, as in Figure 1.

FIGURE 1   Heights of women.

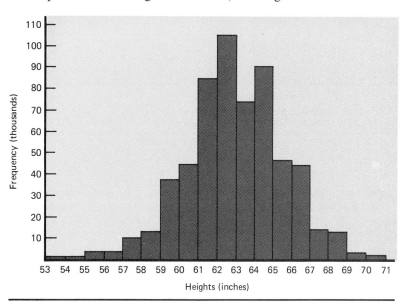

Figure 1 shows how the data pile up in the middle. The distribution is somewhat symmetric about the interval from 62 to 63 inches. This gives us a good, but rough, indication that the average height for women is about 63 inches, or 5 feet 3 inches. In the next section we will calculate the exact average of the data.

This behavior of the data for heights, bunching up around the middle and then spreading out thinly at either end, is a very important feature. Many sets of data behave this way. If a distribution has this property, it is called a *normal distribution*. Examples of data that are normally distributed are height and weight of large populations, IQs, sizes of species of animals, and lengths of snowflakes.

Skewed Distribution

An example of something that is not normally distributed is income. Consider the data supplied by the current population survey on income of families of the United States in 1979 (Table 2).

We can plot the income intervals on the horizontal axis. Since they are of equal length (except for the last one; more on it a little later), we can plot the percentages of families having those incomes as the heights of the bars. We then get the histogram in Figure 2.

We again note that the statisticians who compiled the data in Table 2 established the endpoint convention that the left-hand endpoint would be included in the interval, and so the right-hand endpoint would not. Thus, the entry "5000–10,000" in Table 2 means all those families whose incomes x satisfied $5000 \leq x < 10,000$.

TABLE 2

Distribution of Income of Families in the United States (1979)

Income Level, Dollars	Percent
0– 5000	4
5000–10,000	13
10,000–15,000	17
15,000–20,000	20
20,000–25,000	16
25,000–30,000	12
30,000–35,000	7
35,000–40,000	4
40,000–45,000	3
45,000–50,000	2
50,000–55,000	1
Over 55,000	1

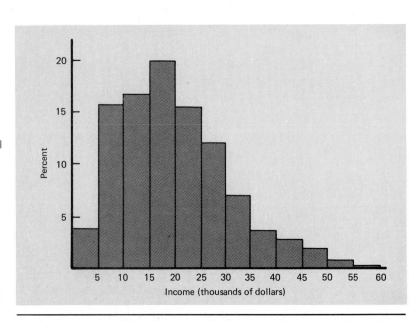

FIGURE 2 Income of families in the United States (1979).

This presents an example of a skewed distribution. Since the long tail, where the frequencies are small, is toward the higher numbers on the right of the horizontal axis, it is said to be *skewed to the right*.

Frequency Polygons

When you look at a histogram and consider it as a picture of the distribution of the data, your eye naturally focuses on the tops, or heights, of the bars. In Figure 1 we pointed out the small heights to the extreme left and right and the gradual increase in heights as you move to the middle. If we connect the midpoints of the tops of the bars in the histogram with straight segments, we get what is called a *frequency polygon*. The frequency polygon is another form of graph that emphasizes the heights of the bars of a histogram. See Figure 3, where we have drawn a frequency polygon from the histogram in Figure 1, and Figure 4, which is the frequency polygon for the histogram in Figure 2.

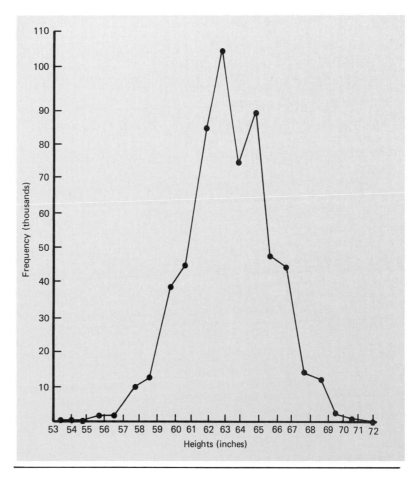

FIGURE 3 Frequency polygon for heights of women.

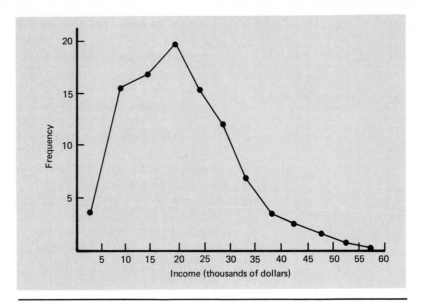

FIGURE 4 Frequency polygon for the income of families in the United States (1979).

If we compare the frequency polygon for a normally distributed set of data with one that is skewed, we get a graphical illustration of the difference between the shapes of the two types of distribution. See Figure 5.

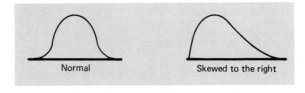

FIGURE 5   A comparison of normal and skewed distributions.

Unequal Intervals

We now tackle the question of unequal intervals, which we already promised you twice in this section. In Table 2 you can see that the last interval is open-ended in the sense that it includes all incomes greater than $55,000. It obviously does not have the same length as the other intervals, namely $5000. Its length would be from $55,000 to the largest salary in the United States, which, at the time you read this, will be earned by some 9-foot-tall basketball player making well over $1 million per year. The percentage of those making $50,000 to $55,000 is 1, the same as for those earning over $55,000. It would present an unrealistic picture if the heights of the two bars were the same. The proper approach is to consider the *area* of the bar when determining the height. Since the lengths of the first 11 bars are the same, by assigning frequencies as their heights, we automatically have their areas in the same proportion. For example, the first bar has height 4 (percent) and the fourth has height 20 (percent);

the fourth height is 5 times larger than the first. Their corresponding areas are $4 \cdot 5000 = 20,000$ and $20 \cdot 5000 = 100,000$, and again the area of the fourth is 5 times larger than the first.

What is the length of the last interval? It would be unreasonable to assume it is as large as we flippantly alluded to before, but, on the other hand, it must be larger than 5000. For the sake of expedience, let us assume it has length 20,000, in which case the interval becomes 55,000 to 75,000. What height should we choose? Since the previous interval had the same percentage, 1 percent, the last bar should have the same area, namely 5000×1 (percent) $= 5000$. Its length is 20,000, so its height, say h, should satisfy $20,000h = 5000$. Thus $h = 5000/20,000 = 1/4$. Hence in Table 2 we let the height of the last bar be 1/4. Notice that we chose, somewhat arbitrarily, the length of the last interval to be 20,000. If we had determined that 30,000 would have been a more reasonable "guesstimate," what height would we have chosen?

In general, if the data are given in terms of unequal intervals, the heights are determined by letting the frequencies be the areas of the bars. Then the height of a bar is the area (or frequency) divided by the length of the interval.

Randomness and the Normal Distribution

One reason why heights and weights of large populations are normally distributed and income is not is due to the fact that heights and weights occur in a random way, whereas income is not distributed randomly at all. This concept of *randomness* is as essential in the study of statistics as it is in probability. It is linked very closely with the idea of the normal distribution. We will spend more time on each concept later, but we would like to introduce these key ideas at this early stage.

Let us tie these concepts together with an example from probability theory with which you are now familiar. Suppose we toss a coin 1000 times and each time record if we get heads or tails. Providing the coin is fair, each toss corresponds to a random, or equally likely, choice of heads or tails. Earlier we expressed this by saying the probability of getting heads is 1/2, as is the probability of getting tails. How many heads would you expect to observe in our experiment of 1000 tosses? The number 500 comes to mind. But you would not expect to get exactly 500 every time you toss the coin 1000 times. If you conducted the experiment very many times (we really do not expect you to do this), recorded the number of heads observed, and then plotted these values in a histogram, you would expect to see a distribution very close to the normal distribution, with the outcomes bunching up around 500, as in Figure 6.

In other words, because we believe that the outcome in tossing a fair coin is a random event, we expect some totals of the number of heads to be less than 500, about an equal amount greater than 500, and not very many far away from 500. In short, we expect the data to be normally distributed.

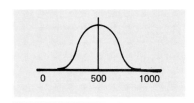

FIGURE 6   Hypothetical distribution: number of heads recorded in 1000 tosses of a coin.

Uniform Distribution

What happens when a small sample of a population is selected? If the process of selection is not random, can we be sure that certain characteristics that are normally distributed in the whole population, such as height and weight, are still normally distributed in the sample?

Let us consider a particular example. The Pittsburgh Steelers were the National Football League champions in 1979. How are their weights distributed? Obviously their average weight is larger than the whole population, so even if the weights are bunched up around the average, the distribution will be farther to the right on the horizontal scale. They are not selected randomly by the football team; they are chosen because they have specific skills. On the other hand, they do have many characteristics of the whole population; some are heavier (the linemen) and some are lighter (the backfield). What do you think? Will the distribution be normal, skewed right or left, or of some other type? Data are given in Table 3.

The histogram compiled from the data in Table 3 is given in Figure 7. You see immediately that the distribution is not like a normal or a skewed distribution. The frequencies are virtually the same for all the values on the horizontal axis. This is called a *uniform distribution*.

Why did this distribution of weights fail to exhibit a normal distribution? It is not simply because too few people were chosen. In fact, every professional football team will have a similar uniform weight distribution (see Problem 14). So even if we surveyed all the players in the NFL, which is a fairly large sample of about 1300 individuals, we would still see a uniform weight distribution. The reason for this type of distribution was hinted at before. The players on the teams are specifically chosen for particular skills. A certain number must be fast and light, a specific number of others must be stronger and heavier. Thus the sample is very selective and does not resemble the whole population. Obviously it is not a random selection from the U.S. population.

TABLE 3

Weight Distribution of the Pittsburgh Steelers (1979)

Weight, Pounds	Number
170–179	1
180–189	6
190–199	5
200–209	6
210–219	6
220–229	6
230–239	2
240–249	7
250–259	6
260–269	4

SOURCE: NFL Yearbook, 1979.

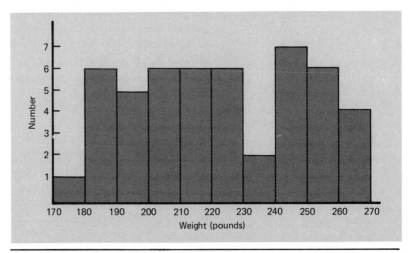

FIGURE 7 Histogram of the weight distribution of the Pittsburgh Steelers.

Grouping Data

In the examples presented thus far the data were grouped in order to make the analysis manageable. If the data are given in an ungrouped fashion, the investigator must determine how to arrange the entries into groups so that the distribution can be determined. In the previous example an alphabetical listing might have been given, in which case the weights of the individuals would have to be grouped. (See Table 4.)

Here are a few procedural guidelines. The number of groups, or intervals, is an arbitrary choice for the investigator. A few simple rules must be followed. The number of intervals should not be so small as to prevent a good spread of the data, and the number should not be so large that too few entries are in most of the groups. In general, for small sets of data, 4 to 10 groups can be used, while for larger sets, as many as 15 or 20 can be used.

In the previous example, we found it convenient to choose 10 intervals. The length of each interval was 10 pounds because the spread of weights, from the smallest at 177 to the largest at 265 (see Table 4), could be covered by going from 170 to 270. After selecting the intervals, go through the data and tally each entry in its corresponding interval. In

TABLE 4

Name	Pos.	Ht.	Wt.	Name	Pos.	Ht.	Wt.
Anderson, Fred	DE-DT	6-5	236	Johnson, Ron	CB	6-10	200
Anderson, Larry	CB	5-11	177	Kolb, Jon	T	6-2	252
Banaszak, John	DE-DT	6-3	244	Kruczek, Allen	QB	6-1	205
Beasley, Tom	DT	6-6	253	Lambert, Jack	LB	6-4	220
Bell, Theo	WR	5-11	180	Mendich, Jim	TE	6-2	214
Bieler, Rocky	RB	5-11	210	Moser, Rick	RB	6-0	210
Blount, Mel	CB	6-3	206	Mullins, Gerry	G	6-3	244
Bradshaw, Terry	QB	6-3	215	Oldhern, Ray	S	5-11	192
Brown, Larry	T	6-4	245	Petersen, Ted	C-T	6-5	244
Cole, Robin	LB	6-2	220	Pinney, Ray	T-C	6-4	240
Colquill, Craig	P	6-2	182	Reutershan, Randy	WR	5-10	182
Courson, Steve	G	6-1	250	Shell, Donnie	S	5-11	180
Cunningham, Bennie	TE	6-5	247	Smith, Jim	WR	6-2	205
Davis, Sam	G	6-1	255	Smith, Laverne	RB	5-10	193
Delophane, Jack	RB	5-10	206	Stallworth, John	WR	6-2	183
Dungy, Tony	S	6-0	190	Stoudt, Cliff	QB	6-4	216
Dunn, Gary	DT	6-3	247	Swann, Lynn	WR	6-0	180
Furness, Steve	DT-DE	6-4	255	Thomas, J. T.	CB	6-2	196
Garcia, Roy	K	5-10	185	Thornton, Sidney	RB	5-11	230
Greene, Joe	DT	6-4	264	Tooms, Laren	LB	6-3	222
Greenwood, L. C.	DE	6-7	250	Wagner, Mike	S	6-2	200
Grossman, Randy	TE	6-1	215	Webster, Mike	C	6-2	250
Ham, Jack	LB	6-1	225	White, Dwight	DE	6-4	265
Harris, Franco	FB	6-2	225	Winston, Dennis	LB	6-0	228
Hicks, John	G	6-2	260				

Table 3, we tallied the first weight listed in Table 4, 236 pounds (Fred Anderson), in the interval 230 to 240. The numbers in Table 3 are the sums of the tallies for each interval.

The entire process, beginning with a list of the data and ending with graphs to picture it, is summarized below. The example is drawn from the weights of the Pittsburgh Steelers as given in Table 4.

---

1 Begin with the raw data.

236	177	244	253	180
210	206	215	245	220
182	250	247	255	206
190	247	255	185	264
250	215	225	225	260
200	252	205	220	214
210	244	192	244	240
182	180	205	193	183
216	180	196	230	222
200	250	265	228	

2 Identify the highest and lowest weights.

Highest weight = 265
Lowest weight = 177

Find the range.

Range = 88

Choose an interval.

Interval = 10 pounds

3 Tally the weights in each interval.

4 Write the frequencies. (The sum of the frequencies should be the total number of players, which yields a good way to check your work.)

Weight	Tally	Freq.
170–179	I	1
180–189	JHI II	7
190–199	IIII	4
200–209	IIII I	6
210–219	IIII I	6
220–229	JHI I	6
230–239	II	2
240–249	JHI II	7
250–259	IIII II	7
260–269	III	3

5 Draw the histogram (see Figure 7).

6 Draw the frequency polygon (see Exercise 18).

Summary

In this section we might not have discovered that a histogram is worth a thousand words, but maybe that it takes a thousand words (or so!) to explain histograms. In any case, a histogram is a picture—a graphical depiction of the data presented in a statistical study to help determine how the data are distributed.

The amassing and grouping of data are usually not sufficient to give a good picture of the situation, especially if the data are large in number. To draw a histogram, first group the data in intervals and record the frequencies in each group. Label the horizontal axis with the intervals. If the intervals are of equal length, then the frequencies are the heights of the bars and the bars all have equal length. If the lengths are unequal, then the areas of the bars must be determined from the frequency so that the height of each bar is equal to the frequency divided by the length.

The endpoint convention refers to the decision, usually aribtrary, that takes place when the data are large in number. It is made by the statistician to specify whether to include the left or right endpoints in the intervals.

A histogram is a bar graph. A frequency polygon is another form of graph that gives a picture of the frequency distribution. The polygon is formed by joining consecutive midpoints of the tops of the histogram bars with straight line segments.

We covered three basic types of distributions. Figure 8 gives the shape of the three respective distributions: normal, skewed to the right and skewed to the left, and uniform.

Of the three types of distributions the normal distribution is of paramount importance. One of the key issues that we will cover later is the close connection between a random event and the normal distribution.

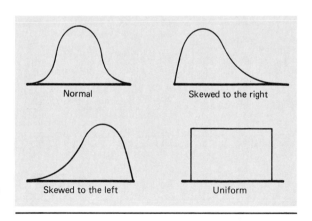

FIGURE 8 A comparison of distributions.

EXERCISES

1 We have said that there should usually be about 10 intervals in a histogram. What problems are encountered if fewer than 5 intervals are used or if more than 20 are used?

2 What advantage is gained if equal intervals are chosen? Why should one opt for unequal interval sizes on occasion?

3 To choose an interval size when equal intervals are desired, one usually takes the range of the data, that is, the largest value minus the smallest value, and divides this number by the number of intervals desired, usually about 10. If the range divided by the number of intervals is a fraction (which it frequently is), the nearest whole number is the natural choice. Assume that each of the pairs of numbers below represents the smallest and largest values in a set of data. If 10 equal intervals are desired, what should the interval length be?

(a)	10, 30	(d)	−6, 50	(g)	−6.1, 11.2
(b)	10, 37	(e)	7.2, 28.2	(h)	701, 1200
(c)	0, 42	(f)	7.2, 50.8		

4 Comment on the statement: If data are given in a grouped fashion, then there exists one histogram for the data, but if the data are not grouped, it is wrong to talk about *the* histogram for the data.

5 The ages of the students in a first-year statistics course are given below. Draw a histogram for the data.

20	18	20	24	20
19	18	19	18	18
22	26	19	18	22
19	24	25	24	23
18	19	19	18	
27	20	30	20	

6 Scores on a history midterm examination are given below as they would appear in the professor's grade book, where names are in alphabetical order. In order to determine how the class did as a whole on the exam group the data in intervals 5 units long, starting with 60–64, 65–69, 70–74, ... , 95–99. Find the frequency for each interval and draw the corresponding histogram.

84	88	84	70	62
68	90	88	83	90
80	61	96	83	83
87	68	62	95	78
64	82	72	91	74
77	78	87	67	80
91	68	71	78	78
90	72	90	88	90
92	98	66	84	75
81	76	65	80	80

5	16	27	18	9
34	22	11	7	2
36	7	13	8	10
8	10	12	12	17
27	11	2	2	19
26	2	8	10	14
20	3	3	10	17
23	6	5	18	20
−1	12	71	4	−6
37	3	22	11	11
				1

7 The table to the left gives the percentage change in population for each of the 50 states in the United States and the District of Columbia from 1960 to 1970 as reported by the U.S. Census Bureau. The percentages are excerpted from an alphabetical listing of the states. Thus the first entry, 5, corresponds to an increase of 5 percent in the population of the first state in the list, Alabama. The smallest value is −6 and the largest is 71, but the next largest is 37, so choose an interval size of 5, starting with −10 and going up to 40, and let the last interval be "over 40." Record the frequencies and draw the histogram.

4	27	24	12	20
8	22	32	23	33
22	31	19	8	26
34	10	2	42	27
28	16	5	5	44
23	16	16	22	16
13	21	17	49	23
8	19	23	29	20

8 The seasonal snowfall in inches (rounded off) for Philadelphia from October through April from 1930–1931 to 1969–1970 is given in the table to the left. Choose an interval size so that you have about 10 intervals. Record the frequency for each interval. Draw the histogram. Guess what the average annual snowfall is for Philadelphia.

10	12	14	16	12
19	12	19	13	13
11	15	12	16	14
18	14	16	15	16
13	14	13	11	11
10	17	14	13	14
14	10	10	18	13
14	14	13	24	13

9 The table to the left is excerpted from the "Official Baseball Guidebook, 1973." It gives the number of wins recorded by the top 40 pitchers in the National Baseball League in 1973 as they are listed for earned run average. Choose an interval size so that you have about 10 intervals and record the frequency for each interval. Draw the histogram for the data.

2	5	3	33	246
<	5	54	9	54
7	51	<	20	21
10	60	<	<	
9	17	<	34	
8	9	<	8	
1	10	13	6	
14	<	4	36	

10 The table to the left gives a listing of populations of the 35 largest countries in Europe in 1973 taken from an alphabetical listing. The numbers have been rounded off to the nearest million and < means that the country's population was less than 500,000. Choose a size for the intervals so that you get about 10 intervals, but let the last interval be "over 60" since the largest value, 246, is much larger than the other values. Let the first interval be "less than 1."

When doing Problems 11 and 12 remember that when data are presented in unequal intervals, special care must be taken when determining the heights of the rectangles in the histogram.

11 The distribution by age for men and women in the United States in 1970 as reported by the U.S. Census Bureau is given in the table below. The left endpoint is included in the interval. The intervals are unequal. Interpret "75 and over" as "75–85" since the number of people over 85 was negligible. Draw the histogram.

Age	% of Population	Age	% of Population
0–5	7	35–45	11
5–14	19	45–55	10
14–18	8	55–65	10
18–21	5	65–75	6
21–25	6	Over 75	4
25–35	12		

12 Sometimes histograms will have large peaks when data bunch up about more than one value. An example of such data is years of education. The data below give the distribution of the number of years of education completed by persons 25 years old or more in the United States. One set of data is for 1960 and the other for 1970. Draw a histogram for each set of data. Where are the peaks and why do they occur at those particular values? The left endpoint is included.

	Percent of People	
Years of Education	1960	1970
0–5	8	5
5–8	14	9
8–9	18	13
9–12	19	17
12–13	25	34
13–16	9	10
16 or more	8	11

SOURCE: *Statistical Abstract,* 1976, table 199.

13 One hundred students were asked what their grade point averages (GPA) were, where the GPA is determined by assigning 4 for every A, 3 for every B, and so on. The results are presented in the frequency polygon below. Interpret why there are peaks at 2 and 3 and why the curve touches the horizontal axis at 0 and 4.

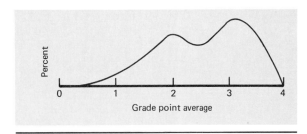

14   Select your favorite football team (pro or college) and draw a histogram of the weights of the players. (You will have to go to a local bookstore or library to find the appropriate references.)

15   Draw a histogram for the heights of the players on the Pittsburgh Steelers.

16   Draw histograms for the heights and weights of a professional baseball team. What do you expect them to look like beforehand?

17   Uniform distribution is (choose one):
     (*a*)   An army handout of battle fatigues
     (*b*)   Done when a baseball team breaks from spring training
     (*c*)   A level concept in statistics

18   Draw a frequency polygon for the weights of the Pittsburgh Steelers corresponding to the intervals given in the box immediately preceding the section summary.

19   Use the histogram below to insert either "increase" or "decline" into the newspaper headline of the article that contained the histogram: "1960 Census Finds Sharp _____ in Size of American Households."

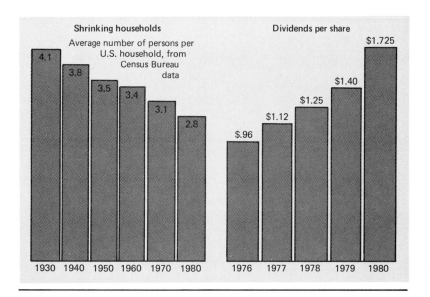

20   InterNorth Company used the histogram of dividends per share (as given in the figure on page 503) in a newspaper advertisement to attract investors. Does the company seem like a sound investment to you, or do you think the statistics lie?

21   (*a*)   Is either of the graphs below a frequency polygon?
   (*b*)   About how many pounds has U.S. peanut production per year increased from 1964 to 1980?
   (*c*)   Milk prices were 3 times higher in 1980 than in 1965. Does the histogram reflect this accurately?

Not just peanuts

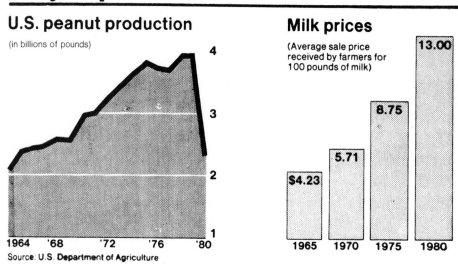

U.S. peanut production

(in billions of pounds)

1964   '68   '72   '76   '80

Source: U.S. Department of Agriculture

Milk prices

(Average sale price received by farmers for 100 pounds of milk)

$4.23   5.71   8.75   13.00

1965   1970   1975   1980

22   The graph on the next page is a slick way of depicting the trend of record and tape sales from 1975 through 1980.
   (*a*)   Do you think that such sales have peaked and are now on the permanent decline?
   (*b*)   Are the relative heights accurate?
   (*c*)   What kind of graph is it?

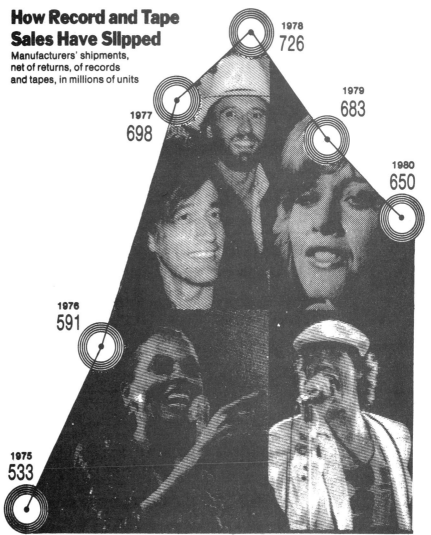

How Record and Tape Sales Have Slipped
Manufacturers' shipments,
net of returns, of records
and tapes, in millions of units

1978
726

1979
683

1977
698

1980
650

1976
591

1975
533

Source: Recording Industry Association of America

8-2 MEASURES OF CENTRAL TENDENCY

If viewed as the mere gathering of data, statistics is not a new phenomenon. In fact, the Bible and older documents contain lengthy lists of data. However, it was not until the seventeenth century that statistics was used as a major weapon on problems in the humanities and social sciences.

The first person to use statistics in this creative way was an unlikely candidate indeed. He was John Graunt (1620–1674), an English haberdasher who had no formal training in mathematics. But he did have a very curious avocation—he studied the death records of English cities. He found that the percentage of deaths from accidents, suicides, and various diseases remained constant over given time periods. Thus events such as deaths that appear to occur haphazardly or randomly actually turn out to exhibit surprising regularity.

It is somewhat ironic to say that something is born of death, but that is exactly what happened with statistics, for it was conceived from Graunt's deathly conclusions. His book, which was published in 1662, founded the science of statistics and launched the trend toward the scientific method in the social sciences. Studies of population and income followed soon thereafter, and the first life insurance companies were formed around 1700.

It was the social ills brought on by the industrial revolution that prompted social scientists to seek solutions of major problems through statistical methods. The task of statistics became to summarize, digest, and extract meaningful information from such large quantities of data.

Today the statistical approach dominates most of the social sciences and many of the humanities. Many students who major in psychology, sociology, political science, and other areas find that they must master the fundamentals of statistics as they begin to pursue their chosen area of interest.

The aim of this chapter is to supply the basic tools that you will need in further courses. We have already introduced what we call the "geometric approach," where graphs are used to depict the data. There is an "algebraic approach" too, and it will be covered in this and the next section. After that we will combine the two approaches and obtain an even clearer understanding of the problems to be studied.

The Mean

Remember that the mere accumulation of statistics accomplishes very little in itself, for only in the very simplest of cases will it be possible to draw conclusions from the data. The extraction of facts from large masses of data, such as those given earlier about heights in the United States, is accomplished by means of two basic mathematical concepts, an average and the dispersion from that average. The goal of this section is to describe what we mean by average and to illustrate how it can be obtained most effectively. In the next section we will study how to measure the way in which data vary from the average.

The simplest and most common mathematical device for distilling knowledge from data is the *arithmetic average,* also known as the *mean.* We have already used this concept intuitively when we spoke of the various baseball averages. In fact, the fourth step in the Monte Carlo method consisted entirely of finding the mean of a set of numbers, though we did not call it by that name. You are certainly aware of the fact that your grade in this course will be based on the average of your individual exam grades, and that your GPA in turn depends on your course grades.

So you have encountered the term "average" many times. Let us consider two examples in order to motivate our formal definition of this concept.

Example 1

If your exam grades are 82, 84, 78, and 76, your average is determined by adding the four numbers and dividing that sum by the number of numbers you summed. Thus your average would be

$$\frac{82 + 84 + 78 + 76}{4} = \frac{320}{4} = 80$$

Example 2

If six of your classmates have weights 160, 182, 196, 143, 150, and 129, can you determine what the average weight of these six people would be? Their average weight is

$$\frac{160 + 182 + 196 + 143 + 150 + 129}{6} = \frac{960}{6} = 160$$

These examples lead us to the following definition:

DEFINITION: The arithmetic average, or mean, of a list of n numbers $x_1, x_2, \ldots, x_n$, is their sum divided by how many numbers there are, namely n. We will denote it by $\bar{x}$. Therefore, the formula for the mean is

$$\bar{x} = \frac{x_1 + x_2 + \cdots + x_n}{n} \qquad (1)$$

In Example 1, we had $n = 4$, $x_1 = 82$, $x_2 = 84$, $x_3 = 78$, $x_4 = 76$, and $\bar{x} = 80$. In Example 2, $n = 6$, $x_1 = 160$, $x_2 = 182$, $x_3 = 196$, $x_4 = 143$, $x_5 = 150$, $x_6 = 129$, and $\bar{x} = 160$.

In a certain intuitive sense the arithmetic average measures the "center" of a set of data. In Example 1 the four exam grades are clustered about the number 80, even though none of them is in fact equal to 80. At first glance it is harder to see that the numbers in Example 2 are clustered about the number 160 because they are more scattered than those in Example 1 even though one of them is equal to 160. Notice that the difference between the data and the mean balances out; in other words, the sum of the differences is always 0. For instance, in Example 1 we see that

$$(82 - 80) + (84 - 80) + (78 - 80) + (76 - 80) = 0$$

The Median

Notice that we have been using the term "arithmetic average" rather than just "average." This is because there are three different concepts that are often referred to by the term average. They are the mean, median, and mode. The mean is the concept that we have been discussing thus far in this section, and it is what is usually meant by average. However, in some cases the other measures are more informative. The *median* of a list of numbers is the middle number when the list is arranged in ascending order, provided that there is an odd number of entries in the list. If there is an even number, then the median is the mean of the middle two numbers.

> ### Example 3
> *Problem* What are the mean and median of the 10 numbers 1, 1, 2, 3, 3, 3, 5, 5, 6, 7?
> *Solution* The mean is
>
> $$\frac{(1 + 1 + 2 + 3 + 3 + 3 + 5 + 5 + 6 + 7)}{10} = \frac{36}{10} = 3.6$$
>
> whereas the median is the fifth plus the sixth number divided by 2; that is, $(3 + 3)/2 = 3$.

The Mode

The *mode* is the number that appears most often. Thus in Example 3 the mode is 3. The mode is not used as a measure of central tendency as much as the mean and the median, and we will not pursue it here.

The Mean versus the Median

It is very important to recognize the difference between the mean and the median, especially since many times they are both referred to as the average. You must therefore know which idea is being presented when the term average is used, since average may mean either mean or median.

Let us look at an example of where the two concepts tend to obfuscate rather than explain the data.

> ### Example 4
> Recently there was a long and bitter strike by the public school teachers. The city was in financial straits and wanted the teachers to forgo a pay increase or at least accept only a small raise and thereby help the city weather its financial crisis. The teachers saw inflation dwindling the buying power of their present salaries and so were demanding a raise at least equal to the cost-of-living increase. Each side had a stake in convincing the public of the validity of its stand.

On successive days an administrative official from the city said in an interview that the average salary of the teachers was $18,000 and a teachers' union official was quoted as saying that the average teachers' salary was $15,000. Was one of them lying? If not, could they possibly be talking about the same data?

Neither was lying, and yes, they were using the same data. The apparent contradiction is that the two officials were using two different meanings of average.

In order to explain how there can be such a large discrepancy between the two measures, let us look at a simpler example.

Example 5

Suppose a small company has 10 employees, 9 of whom make $15,000 each, whereas the vice president (probably the brother-in-law of the president) makes $45,000.

Problem What is the average salary?

Solution The median is the fifth plus the sixth largest salaries divided by 2, which is $15,000. The mean is the sum of the nine salaries plus the $45,000 salary, divided by the total number of salaries, 10. Hence the mean is

$$\bar{x} = \frac{9 \cdot 15,000 + 45,000}{10} = \frac{135,000 + 45,000}{10}$$

$$= \frac{180,000}{10} = 18,000$$

You might argue that Example 5 is far-fetched. Nevertheless it makes the point that the mean and the median can be significantly different for a particular set of data. This is the case if the entries in the list are not clustered evenly about the mean but are strung out, with some entries being either much larger or much smaller than the mean (but not both, for then they would cancel each other out and the mean and the median would be close). This is the situation with many familiar collections of data, especially income. In Figure 1 we consider three types of distributions that are symmetric. The mean and the median are very close for these kinds of distribution.

If a set of data is skewed to the right, then the relatively large values have a greater effect on the mean than on the median, and so the mean is larger than the median. This was the case in Example 5. If the data are skewed to the left, then the mean is less than the median. (See Figure 1 on page 510.)

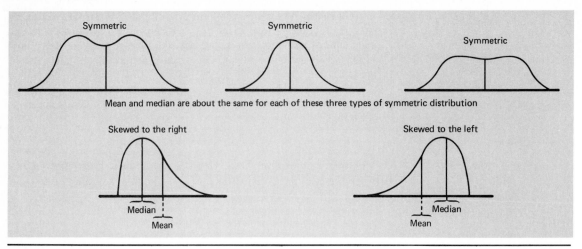

Mean and median are about the same for each of these three types of symmetric distribution

FIGURE 1

An Alternate Formula for the Mean

If the set of data contains numbers that are repeated, then the formula for the mean can be expressed in a more manageable form. This is especially convenient if the data are expressed in grouped form.

We will first consider a very simple example in which one would easily use Formula (1) to find the mean. We will show how we can develop the more general formula from this relatively short list of data, and then use it to handle some examples of larger sets of data.

> **Example 6**
> A fisherman hauls in a catch of 10 fish weighing 5, 6, 7, 6, 6, 7, 8, 5, 5, and 8 pounds, respectively.
> **Problem** What is the mean of the catch?
> **Solution** The mean $\bar{x}$ of the weights, according to Formula (1), is
>
> $$\bar{x} = \frac{5 + 6 + 7 + 6 + 6 + 7 + 8 + 5 + 5 + 8}{10} = 6.3$$
>
> We could group the data as follows:
>
> | Weight, pounds | 5 | 6 | 7 | 8 |
> |---|---|---|---|---|
> | Number of fish | 3 | 3 | 2 | 2 |
>
> If we take the 5-pound fish together, we have a total weight of $5 \cdot 3 = 15$ pounds. Taking the 6-pound fish we have a total weight of $6 \cdot 3 = 18$ pounds. Similarly, the 7- and 8-pound fish yield weights of $7 \cdot 2 = 14$ and $8 \cdot 2 = 16$ pounds. Thus the total weight is $5 \cdot 3 + 6 \cdot 3 + 7 \cdot 2 + 8 \cdot 2 = 15 + 18 + 14 + 16$. Hence the mean is $\bar{x} = 63/10 = 6.3$ pounds. Therefore this method of getting $\bar{x}$ agrees with the first way.

In Example 6 we see that the mean can be found from grouped data by taking each repeated value in the list (that would be 5, 6, 7, and 8, respectively) and multiplying it by the number of times it is repeated, which is called its *frequency*. In Example 6 the respective frequencies would be 3, 3, 2, and 2, corresponding to the values 5, 6, 7, and 8.

We can express the procedure of Example 6 in a general formula as follows:

ALTERNATE FORMULA FOR THE MEAN: If a set of data is grouped so that there are n entries, and the values are x_1, $x_2, \ldots, x_r$ with frequencies $f_1, f_2, \ldots, f_r$, respectively, then the mean $\overline{x}$ is given by

$$\overline{x} = \frac{f_1 x_1 + f_2 x_2 + \cdots + f_r x_r}{n} \qquad (2)$$

where $f_1 + f_2 + \cdots + f_r = n$.

Look back at Example 4. The mean of the salaries was found to be $18,000. Which formula was used? You should see that the alternate formula is the same as the original formula, except that the common data are grouped together.

The following method is particularly effective when you have to find the mean of data with certain frequencies.

1 List the data on a row.

2 List the corresponding frequency under each datum.

3 Find the total of the frequencies.

4 List the products of each datum and its frequency. Put them beneath the data.

5 Find the total of these products.

6 Form the mean by dividing the total in step 5 by the total in step 3.

We now consider an example where the alternate formula for the mean is the more reasonable choice.

Example 7

Many introductory psychology courses are large, and often the tests are multiple-choice. A Psych 1 instructor has 100 students. The grades on a 10-question test are entered into a grade book alphabetically according to the students' last names. In order to find the mean the grades are grouped as follows:

Number correct	10	9	8	7	6	5	4	3	2	1	0
Number of students	11	16	20	15	13	10	10	4	1	0	0

Problem What is the mean?

Solution Using the alternate formula for the mean, with $n_1 = 10$, $n_2 = 9$, $n_3 = 8, \ldots, n_{11} = 0$, $f_1 = 11$, $f_2 = 16$, $f_3 = 20, \ldots, f_{11} = 0$, we get

$$\bar{x} = \frac{1}{100}[(11 \cdot 10) + (16 \cdot 9) + (20 \cdot 8) + (15 \cdot 7) + (13 \cdot 6) +$$

$$(10 \cdot 5) + (10 \cdot 4) + (4 \cdot 3) + (1 \cdot 2) + (0 \cdot 1) + (0 \cdot 0)]$$

$$= \frac{1}{100}(110 + 144 + 160 + 105 + 78 +$$

$$50 + 40 + 12 + 2 + 0 + 0)$$

$$= \frac{701}{100} = 7.01$$

Hence the (arithmetic) average grade is about 7.

If you follow the six-step method described above, you get

Number correct (x)	10	9	8	7	6	5	4	3	2	1	0	Total
Number of students (f)	11	16	20	15	13	10	10	4	1	0	0	100
fx	110	144	160	105	78	50	40	12	2	0	0	701

Once again

$$\bar{x} = \frac{701}{100} = 7.01$$

Grouped Data in Intervals

As we saw in the previous section, often the data are grouped into blocks corresponding to intervals rather than to specific numbers. To find the mean and the median of such data sets we use the midpoints of the intervals. The next two examples will illustrate this.

Example 8

The weights of the Pittsburgh Steelers as grouped in Table 3 of Section 8-1 indicated that the distribution was uniform. What would you guess the mean of the weights to be? We grouped the data so they would be easier to handle. The mean could be found by adding the 49 separate entities and then dividing that sum by 49, according to Formula (1). (See Problem 36.)

Problem 1 What is the mean weight of the Steelers?

Solution Using the grouped data, we note that there are six players weighing between 180 and 189 pounds. The midpoint of that interval is 185 (assuming that the number 189 represents those weights in the interval that are less than 190 pounds). We could approximate that there are six Steelers weighing 185 pounds. Similarly, the midpoint of the interval from 190 to 199 is 195 pounds; so we can approximate that there are five players who weigh 195 pounds. In this way we transform the data in Table 3 into the data in the table below:

Wt. (x)	175	185	195	205	215	225	235	245	255	265	Total
No. (f)	1	6	5	6	6	6	2	7	6	4	49
fx	175	1110	975	1230	1290	1350	470	1715	1530	1060	10,905

Using the alternate formula for the mean, we get

$$\bar{x} = \frac{175 + 1110 + 975 + 1230 + 1290 + 1350 + 470 + 1715 + 1530 + 1060}{49}$$

$$= \frac{10,905}{49} \approx 222.6$$

Problem 2 Would you expect the median of this set of data to be close to the mean?

Solution The distribution of the data is uniform, and so it is not skewed; hence, the two measures should be close. To find the median we note that there are 49 entries, so the middle entry is the twenty-fifth. The data are presented in ascending order in the preceding table as we read from left to right. Thus we start from the left and add the number of entries until we get to the twenty-fifth. There are 24 entries in the first five columns of the table, so the twenty-fifth is in the sixth column, or 225 pounds.

Example 9

In the previous section we approximated the "average height for women" to be about 63 inches, or 5 feet 3 inches. We did not distinguish between the mean or median because the data are normally distributed, and so the two measures are very close. We can now calculate the mean from the data list given in Table 1 of Section 8-1. We will calculate the mean of the height distribution for men, 18 to 79 years old, and let you calculate the precise mean for women as an exercise (see Problem 33).

Problem 1 What is the mean height of men in the United States?

Solution We first locate in Table 1 the midpoint of each interval because the data are grouped into intervals. Then we list the frequencies f_i (numbers in thousands), and the products $f_i \cdot x_i$, where x_i is the midpoint of the ith interval and f_i is its corresponding frequency. The sum of the products is then divided by the sum of the frequencies to find the mean.

TABLE 1

Height Distribution for Men, 18 to 79 Years

Height x_i Inches	Frequency f_i	$x_i f_i$
59.5	1	59.5
60.5	1	60.5
61.5	5	307.5
62.5	9	562.5
63.5	17	1,079.5
64.5	37	2,386.5
65.5	35	2,292.5
66.5	70	4,655.0
67.5	63	4,252.5
68.5	94	6,439.0
69.5	54	3,753.0
70.5	63	4,441.5
71.5	32	2,288.0
72.5	28	2,030.0
73.5	11	808.5
74.5	6	447.0
75.5	1	75.5
76.5	1	76.5
	528	36,015.0

$$\overline{x} = \frac{36,015}{528} \approx 68.2$$

What do we do about the first and last midpoints, that is, the ones corresponding to "under 60" and "over 76"? We arbitrarily choose them to be 59.5 and 76.5. Since their corresponding frequencies are relatively small, other numbers could have been chosen without greatly affecting the value for the mean. Hence the mean is $\overline{x} = 36,015/528 \approx 68.2$, or about 5 feet 8 inches. Thus the average man is about 5 feet 8 inches tall.

Problem 2   Find the median height of men in the United States.

Solution   The total of the frequencies is 528. Half of this is 264; so the median is the mean of the two-hundred-sixty-fourth and the two-hundred-sixty-fifth entry. Both these entries are 68.5; so the median is 68.5.

Summary

There are three measures of central tendency that are referred to as averages; so you must be careful to distinguish between them. They are the mean, the median, and the mode. The first two are the most useful ones.

If the data are distributed symmetrically, then the mean and median are close. If the distribution is skewed, then the mean is affected by the extreme values and will be significantly different from the median.

If the data set is given in intervals, then the midpoints of the intervals can be used to approximate the mean and the median.

EXERCISES

In Problems 1 to 10 find the mean of the list of numbers.

1   2, 5, 7, 10

2   7, 2, 1, 0, 5, 4

3   −1, 0, 2, −2, 1, 0

4   −8, 10, 20, 16, 1, −1, 2, 4

5   6.1, 0.1, −3.2, 6.7, 3.3

6   2, 5, 7, 9, 2, 1, 13, 12, 15, 20

7   2, 5, 7, 9, 2, 1, 13, 12, 50, 60

8   2, 5, 7, 9, 2, 1, 13, 12, −50, −60

9   0.06, 0.07, 0.02, 0.10, 0.05, 0.02

10   5.2, 6.1, 10,8, 11.2, 6.3, 0.5, 8.7, 7.2

In Problems 11 to 14 use the alternate formula to find the mean of the sets of data.

11
Value	0	1	2	3
Frequency	2	5	3	4

12
Value	1	2	3	4
Frequency	8	2	1	5

13
Value	3	5	7	9
Frequency	2	1	0	4

14
Value	11	12	13	14	15
Frequency	3	5	2	0	1

In Problems 15 to 20 find the mean and the median of the sets of data.

15   0, 2, 4, 3, 5, 6, 8, 6, 4, 2

16   0, 2, 4, 3, 5, 6, 8, 8, 8, 8

17   0, 2, 4, 3, 5, 0, 1, 8, 8, 9

18   0, 1, 0, 1, 9, 8, 11, 1, 8, 1

19   12, 15, 5, 18, 17, 8, 13, 22, 2, 8

20   92, 95, 85, 98, 97, 88, 93, 102, 82, 88

In Problems 21 to 24 the data are given in grouped fashion. To find the mean find the midpoints of the intervals; use them as the value x_i in the alternate formula for the mean.

21

Interval	Frequency
0–2	4
2–4	5
4–6	3
6–8	0
8–10	1

23

Interval	Frequency
0–5	16
5–10	14
10–15	10
15–20	8
20–25	2
25–30	1

22

Interval	Frequency
0–4	2
4–8	4
8–12	7
12–16	1
16–20	8

24

Interval	Frequency
12–15	2
15–18	1
18–21	8
21–24	12
24–27	15
27–30	20

In Problems 25 to 28 find the mode of the sets of data.

25 2, 1, 2, 8, 1, 2, 3, 2, 7, 7

26 16, 15, 14, 10, 15, 10, 15, 16, 15

27 91, 90, 92, 78, 91, 90, 91, 88, 72, 75

28 6, 7, 6, 8, 7, 7, 7, 6, 8, 2, 5

In Problems 29 to 32 a sketch of a curve is given. Mark where the mean and the median lie on the horizontal axis. Are they closer to 5, 10, or 15?

29

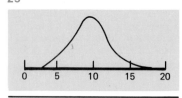

31

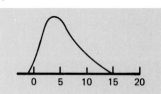

30

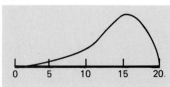

32

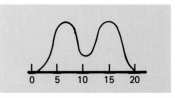

33 Approximate the (arithmetic) average height of women, aged 18 to 79 years, from Table 1 in Section 8-1.

34 In Problem 1, compute $(2 - \overline{x}) + (5 - \overline{x}) + (7 - \overline{x}) + (10 - \overline{x})$. Do you think that this will always be the case?

35 The alternate formula for the mean contains two variables, r and n. Which one will always be greater than or equal to the other?

36 (a) Use the precise weights of the 49 Pittsburgh Steelers to determine the team's mean weight. How does this compare with the mean weight as computed in Example 8?
    (b) Do part (a) for the median.

37 In Section 7-2 we described J. E. Kerrich's experiment of tossing a coin 1000 times. He performed it 10 times and observed the following number of heads:

    502, 511, 497, 529, 504, 476, 507, 528, 504, 529

    What is the mean number of tosses? What is the median?

38 The Boston Celtics won the 1980–1981 National Basketball Association (NBA) championship. The heights of their players were:

    6-1, 6-9, 6-6, 6-2, 6-10, 6-5, 6-2, 6-8, 6-10, 7-$\frac{1}{2}$, 6-10

    What was the Celtics' mean height? What was the median height?

39 The chart on the following page was taken from a newspaper account of the average (i.e., mean) salaries of each of the 26 professional baseball teams for 1979 and 1980. The overall average (mean) salary was obtained by adding up the 26 entries and dividing by 26. That's how the figures $144,000 and $114,000 were obtained. Is this method valid? (*Hint:* Do all teams pay the same number of players?)

40 A set of data has five entries. Each entry is either 0, 1, or 2. What is the set if the mean is 0? If the mean is 2? Can the mean be 3?

41 A set of data has five entries. Each entry is either 10, 11, or 12. If the median is 10 and the mean is 10.8, what is the set?

42 Ten employees in a company have an (arithmetic) average salary of $15,000. Another person is hired at a salary of $15,000. What is the mean salary of the 11 employees? A twelfth person is hired at a salary of $50,000. What is the mean salary of the 12 employees?

43 Some businesses have high demand periods during the year and hire and fire employees during particular seasons on a fairly regular basis. Examples would be farming and construction enterprises and toy companies. During periods of peak hiring mean salaries usually go down and when high layoff periods occur the mean salaries typically increase. Can you explain the phenomenon?

The Salary Standings
A Breakdown of the Average Salary per Player Paid by Each Major League Team

1980* Rank	1979* Rank	Club	Average Salary, Thousands	
			1980	1979
1	1	N.Y. Yankees	$243	$199
2	2	Philadelphia	221	198
3	3	Pittsburgh	199	174
4	5	California	191	155
5	6	Boston	185	145
6	9	Los Angeles	183	134
7	21	Houston	177	74
8	12	St. Louis	173	117
9	4	Cincinnati	163	165
10	13	Chicago Cubs	160	104
11	8	Milwaukee	159	137
12	7	Montreal	158	143
13	10	Texas	149	129
14	11	San Francisco	148	120
15	18	Atlanta	148	90
16	14	San Diego	139	104
17	16	Cleveland	127	98
18	17	N.Y. Mets	126	93
19	15	Baltimore	116	101
20	19	Kansas City	100	91
21	24	Detroit	87	63
22	25	Seattle	82	62
23	22	Minnesota	80	70
24	20	Chicago White Sox	72	74
25	23	Toronto	67	67
26	26	Oakland	55	41
		Avg. salary	$144	$114

---

*Rankings in terms of salary.
This information has been supplied by the Major League Baseball Players Association.

44   Criticize the assertion, "Obviously there must be as many people with above-average intelligence as there are with below-average intelligence."

45   Is it safe for you to step into a swimming pool whose mean depth is 4 feet?

8-3 STANDARD DEVIATION

Let us recall some events from the American Revolution. In the 1760s the British government adopted the policy of taxing the American colonies. The most famous and most important of these measures was the Stamp Act, which was later repealed and replaced by the Townshend Acts, both of which provided taxes to maintain the British army in America. This led to the Boston Tea Party, and the decision of King George III that the colonies should be subdued. Revolt followed. Hostilities broke out in April 1775, and on July 4 of the following year the delegates to the second Continental Congress approved the declaration that "These United Colonies are, and of right ought to be, free and independent states." The war came to an official end with the Treaty of Paris of 1783, in which Britain recognized the independence of the United States.

That sequence of events answered one of the two great questions that were posed in the revolutionary era: how should power be distributed between Britain and America? But the other question—How should power be divided among the Americans—remained unresolved. It is instructive to think not of *the* revolution, but of two revolutions, one external and the other internal. We outlined the external one above; now let us recall some events from the internal revolution.

The Continental Congress functioned as a national government until 1781, when the Articles of Confederation, the first national constitution, were adopted. However, the articles contained one major defect: they did not provide for a strong central government. This is understandable; American heads of state had no intention of replacing one form of despotism with another. As a result, the central government's lack of power to tax led to financial disaster, and its lack of power to regulate commerce led to grave commercial problems. The situation was so serious that historians have come to call the period 1783 to 1787 not *a* critical period but *the* critical period in American history.

A constitutional convention was convened in hot, steamy Philadelphia from May to September 1787, where the members hammered out a document that would prove to be the cornerstone of our government—the Constitution. The 13 colonies were even less united than are the states today. Yet politicians from all ideologies recognized the importance of the passage of the Constitution. "The Federalist" papers were published by Alexander Hamilton, John Jay, and James Madison to persuade the citizens of New York State to vote for ratification. The problem today arises from the fact that the papers were published anonymously, over the signature "Publius," in New York newspapers and in book form in 1788 and republished extensively. They have earned an important niche in political philosophy.

Of the 85 papers there is general agreement on the authors of 70 of them: 51 were penned by Hamilton, 14 by Madison, and 5 by Jay. Of the remaining 15 papers, 3 appear to be of joint authorship and the other 12 are in dispute between Madison and Hamilton. The usual means for determining authorship—political content and writing style, for example—have failed to solve the problem. At various times during the last two centuries one author was favored over the other only to suffer disrepute in the next change in the political climate.

The first seemingly conclusive evidence was provided by Mosteller and Wallace* in 1964. Since each of the three original authors consciously used the same formalistic legal writing style, the usual linguistic methods of studying sentence length and word usage proved debatable. Mosteller and Wallace turned to what are called *marker words,* or words that are exceptionally common to one author's style and not the other's. The difficulty is in finding such words and proving that the difference in usages is significant. The nagging uncertainty never disappears, as perfect markers can never be found, but sometimes nearly conclusive cases can be made using marker words. Statistics enters the picture to yield evidence and assess its strength in the support of a hypothesis in favor of one of the authors.

The question of the authorship of "The Federalist" papers forged an important link between statistics and history. The particular concept that was fundamental to this determination is the standard deviation. In this section we will define it in two ways (one intuitive, the other formal), show how to compute it, and illustrate it with several examples. Then we will return to the revolutionary era.

Definitions

The mean of a set of data measures, from an intuitive point of view, the "center" of the data. It depends on all of the data but does not tell anything about how the data are distributed. To get a more meaningful description of the data we must determine how the entries are spread around the average. One such measure is the range of the data: the largest number minus the smallest number. (We encountered this concept in Section 8-2.) Of course, the range tells us nothing about the spread of the numbers in between. Some sets of data are bunched up close to the mean (think of a normal distribution, for example, the heights of women given earlier) and some are spread out evenly (think of a uniform distribution, for example, the weights of the Pittsburgh Steelers), whereas other sets of data have various types of spread.

In this section we will study the most important measure of dispersion, called the *standard deviation.* We abbreviate the standard deviation by the letter σ, the Greek letter sigma. It will provide us with a good measure of how the data are dispersed about the mean. In order to introduce this concept we will consider the following problem.

PROBLEM: On a 10-point quiz, five students score a total of 30 points. List three ways in which their scores could be distributed and discuss the dispersion of the data in each case.

---

*F. Mosteller and D. L. Wallace, "Inference and Disputed Authorship: The Federalist," Addison-Wesley, Reading, Mass., 1964.

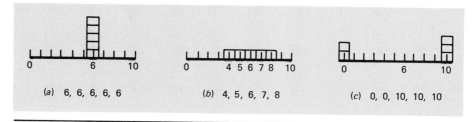

(a) 6, 6, 6, 6, 6   (b) 4, 5, 6, 7, 8   (c) 0, 0, 10, 10, 10

FIGURE 1

There are many ways to choose from. We list three: (*a*) 6, 6, 6, 6, 6, (*b*) 4, 5, 6, 7, 8, and (*c*) 0, 0, 10, 10, 10. Their histograms are given in Figure 1.

Since the scores total 30, no matter what the distribution of the 5 scores is the mean of each set of scores is $\overline{x} = 30/5 = 6$. There is no variation among the scores in (*a*), so its standard deviation should be 0. The spread in (*c*) is about as wide as possible. Since standard deviation measures the dispersion about $\overline{x} = 6$, the value for (*c*) should be large. Clearly the scores in (*b*) vary a little from the mean, so its standard deviation should be greater than the one in (*a*) but less than the one in (*c*).

Now let us state an intuitive description of standard deviation and then illustrate it with an example.

INTUITIVE DESCRIPTION OF THE STANDARD DEVIA-
TION: The standard deviation σ of a set of data is a measure of how far away from the mean the entries are. Most entries will be within one σ of the mean; that is, they will be within one σ above or below the mean. Very few will be more than two σ's away from the mean and only rarely will there be any entries more than three σ's away.

In the problem above the data were presented geometrically. Usually, however, they are presented as a list of numbers. Consider the following three sets of data:

Set *A*:   6, 5, 5, 4, 5, 3, 5, 7, 5, 5

Set *B*:   6, 9, 7, 6, 5, 5, 2, 1, 4, 5

Set *C*:   6, 2, 10, 9, 4, 1, 0, 1, 7, 10

The mean for each set is 5, but the spread about the mean is much different for each set. Set *A* has very little dispersion, set *B* has a little more spread, and set *C* has the largest spread about 5. What would you predict the σ for each set to be? Perhaps at this juncture you would not even want to guess a particular number for their respective σ's, but from the intuitive description of the σ you should be able to compare their relative sizes. The σ for set A will be relatively small, since the spread about the mean is small; the σ for set B will be larger, and for set C larger yet. As is suggested by its name, the σ will be a measure of the deviations from the mean. If we calculate the deviations of each entry from the mean for set A, since $\overline{x} = 5$, we get the 10 numbers:

$$6 - 5 = 1 \qquad 5 - 5 = 0 \qquad 7 - 5 = 2$$
$$5 - 5 = 0 \qquad 3 - 5 = -2 \qquad 5 - 5 = 0$$
$$5 - 5 = 0 \qquad 5 - 5 = 0 \qquad 5 - 5 = 0$$
$$4 - 5 = -1$$

If we add the deviations, we get 0 because the negative numbers cancel the positive ones. In fact, this will always happen. Can you see why? By the very definition of the mean, some members will be less and some greater, so their total effect will be one of cancellation. Thus we must get rid of the negative signs. First, consider the following approach. Both 3 and 7 are 2 units away from the mean, so their cumulative deviation is 2 $+ 2 = 4$. Thus one way to measure the total deviation is to sum the absolute values of the deviations. For set A we would have:

$$1 + 0 + 0 + 1 + 0 + 2 + 0 + 2 + 0 + 0 = 6$$

There is another way to get rid of the negative signs in the deviations. We can square each. Remember that the square of a positive or a negative number is always a positive number. This has two primary advantages over taking absolute values: (1) it is much easier to express and calculate with squared numbers than it is with absolute values, and (2) as numbers get larger in absolute value their squares are much larger than their absolute values [provided the numbers are greater than 1 (what happens if the numbers are between 0 and 1?)]; so the entries that are far away from the mean will have large deviations and thus will provide a heftier contribution to the σ than if absolute values were taken.

Thus the first step in finding the σ for a set of data entails calculating the square of the deviations from the mean. For set A we have:

$$(6 - 5)^2 = 1^2 = 1 \qquad\qquad (3 - 5)^2 = (-2)^2 = 4$$
$$(5 - 5)^2 = 0^2 = 0 \qquad\qquad (5 - 5)^2 = 0^2 = 0$$
$$(5 - 5)^2 = 0^2 = 0 \qquad\qquad (7 - 5)^2 = 2^2 = 4$$
$$(4 - 5)^2 = (-1)^2 = 1 \qquad\qquad (5 - 5)^2 = 0^2 = 0$$
$$(5 - 5)^2 = 0^2 = 0 \qquad\qquad (5 - 5)^2 = 0^2 = 0$$

We then find the mean of these numbers by adding them and dividing the sum by the number of data. Since there are 10 numbers, we divide their sum by 10 and get

$$\frac{1 + 0 + 0 + 1 + 0 + 4 + 0 + 4 + 0 + 0}{10} = \frac{10}{10} = 1$$

This number is called the *variance*. The standard deviation is the square root of the variance; that is,

$$\sigma = \sqrt{\text{variance}} = \sqrt{\text{mean of the squared deviations}}$$

The square root is taken to offset somewhat the squaring performed earlier. It is easy to see that for set A, $\sigma = \sqrt{1} = 1$.

Can you now find the σ for set B? We first find the deviations and square them. We will arrange the data in a form that is a little more useful than the method used above in the table to the left. Then we get:

$$\text{Variance} = \frac{48}{10} = 4.8$$

Hence $\sigma = \sqrt{4.8} \approx 2.2$. This agrees with our earlier discussions in that the σ for set B is larger than the σ for set A. Also notice that most of the entries are within one σ of the mean; that is, most are between $5 - \sigma \approx 5 - 2.2 = 2.8$ and $5 + \sigma = 5 + 2.2 = 7.2$. None of the entries is beyond three σ's, that is, less than $5 - 3\sigma \approx 5 - 3(2.2) = 5 - 6.6 = -1.6$ or more than $5 + 3\sigma \approx 5 + 3(2.2) = 5 + 6.6 = 11.6$. (See Figure 2.)

Data x	$x - \bar{x}$	$(x - \bar{x})^2$
6	1	1
9	4	16
7	2	4
6	1	1
5	0	0
5	0	0
2	−3	9
1	−4	16
4	−1	1
5	0	0
		48

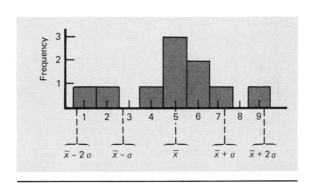

FIGURE 2   Set *B* histogram.

We can now give the general formula for σ in the following formal definition.

DEFINITION: The standard deviation σ of a data set of n entries $x_1, x_2, \ldots, x_n$ with mean $\bar{x}$ is defined to be

$$\sigma = \sqrt{\frac{(x_1 - \bar{x})^2 + (x_2 - \bar{x})^2 + \cdots + (x_n - \bar{x})^2}{n}}$$

The expression inside the square root sign is called the variance; that is, $\sigma = \sqrt{\text{variance}}$.

Notice that the σ has the same units as the entries in the data set. The variance is in squared units. For example, if the data set were in meters, the variance would be in squared meters while the σ would be in meters.

Previously we asked you to guess what the σ for set C is, but we recognized that it was perhaps difficult for you to respond confidently, except to note that it would be larger than the σ for set B. Do you now feel confident enough to approximate the σ for set C?

The formula given in the definition of the standard deviation is appropriate in the sense that it corresponds to our intuitive sense of the grouping of data about the mean. But in a practical sense it is inappropriate. That is, when you are confronted with a set of numbers whose σ you have to calculate, it is much more convenient to use the following equivalent formula.

ALTERNATE DEFINITION:

$$\sigma = \sqrt{\frac{x_1^2 + x_2^2 + \cdots + x_n^2}{n} - (\bar{x})^2}$$

You can expand the formula in the definition of the σ to derive this formula. In Problem 19 we will ask you to do this for $n = 2$. Let us look at an example of how you use this alternate formula.

Example 1

Consider the set of numbers referred to earlier in the section as

Set C: 6, 2, 10, 9, 4, 1, 0, 1, 7, 10

Problem Compute the mean and standard deviation of C.

Solution We can use the alternate formula to calculate σ for this set of data.

$$x_1 = 6 \qquad x_1^2 = 36$$

$$x_2 = 2 \qquad x_2^2 = 4$$

$$x_3 = 10 \qquad x_3^2 = 100$$

$$x_4 = 9 \qquad x_4^2 = 81$$

$$x_5 = 4 \qquad x_5^2 = 16$$

$$x_6 = 1 \qquad x_6^2 = 1$$

$$x_7 = 0 \qquad x_7^2 = 0$$

$$x_8 = 1 \qquad x_8^2 = 1$$

$$x_9 = 7 \qquad x_9^2 = 49$$

$$x_{10} = \underline{10} \qquad x_{10}^2 = \underline{100}$$

Total $\qquad$ 50 $\qquad\qquad$ 388

$$\bar{x} = \frac{50}{10} = 5 \qquad n = 10$$

$$\sigma = \sqrt{\frac{388}{10} - (5)^2} = \sqrt{38.8 - 25} = \sqrt{13.8} \approx 3.7$$

There is a generalization of the alternate formula for the standard deviation that is used when the data appear repeatedly or are presented in grouped fashion, that is, in intervals. We will turn to this subject now.

Grouped Data

When the data are grouped in intervals, then the midpoints of the intervals are used as the values x_i in the formula. If each interval has a frequency of f_i, then the value x_i^2 in the formula for the σ should appear f_i times. Hence we must sum the terms $f_i x_i^2$. Therefore if the data are in grouped form, with the midpoints $x_1, \ldots, x_r$ and frequencies $f_1, \ldots, f_r$ so that the mean $\bar{x}$ is $\bar{x} = (f_1 x_1 + \cdots + f_r x_r)/(f_1 + \cdots + f_r)$, then the formula for σ becomes

$$\sigma = \sqrt{\frac{f_1 x_1^2 + f_2 x_2^2 + \cdots + f_r x_r^2}{f_i + f_2 + \cdots + f_r} - \bar{x}^2}$$

Let us look at an example with some easy numbers so you can see how to use the formula in this form. We will then consider a more complicated example.

Example 2

Suppose the data are presented as in the following table with the given values for the intervals and the frequencies. It is then necessary to compute the midpoints and the values x_i^2, $f_i x_i^2$, and $f_i x_i$.

Interval	Midpoint x_i	Frequency f_i	$f_i x_i$	x_i^2	$f_i x_i^2$
0–2	1	9	9	1	9
2–4	3	5	15	9	45
4–6	5	4	20	25	100
6–8	7	2	14	49	98
		20	58		252

$$\bar{x} = \frac{f_1 x_1 + \cdots + f_4 x_4}{f_1 + \cdots + f_4} = \frac{58}{20} = 2.9$$

$$\bar{x}^2 = (2.9)^2 = 8.41$$

$$\sigma = \sqrt{\frac{f_1 x_1^2 + \cdots + f_4 x_4^2}{f_1 + \cdots + f_4} - \bar{x}^2} = \sqrt{\frac{252}{20} - 8.41}$$

$$= \sqrt{12.6 - 8.41} = \sqrt{4.19} \approx 2.05$$

Not *the* Standard Deviation

In the prologue to this section we indicated that *the* American Revolution is really *an* American revolution, since two of them actually took place. The same is true of the standard deviation. The standard deviation presented above is called more formally the *population standard deviation.* It is used whenever the data constitute the entire population being considered.

If only a portion of the population forms the data, then another measure of dispersion, called the *sample standard deviation,* is used. Let us denote it by σ'. Then the two deviations are related by the formula

$$\sigma' = \sigma \sqrt{\frac{n}{n - 1}}$$

where n is the number of data.

In order to illustrate the difference, consider the following. If you take a survey of 10 of the wealthiest people in the United States and find that their ages are

62, 84, 47, 58, 68, 60, 62, 59, 71, 73

then you should use the sample standard deviation σ'. However, if these 10 turn out to be actually *the* 10 wealthiest persons, the data would have to be considered as a population rather than as a sample. For this data you will find that:

$$\bar{x} = 64.40 \qquad \sigma = 9.58 \qquad \sigma' = 10.10$$

where

$$\sigma' = 9.58 \quad \sqrt{\frac{10}{9}} = \frac{9.58}{3} \sqrt{10}$$

Note that σ' will always be a little greater than σ; a sample is allowed more leeway than an entire population.

We think that having to learn one standard deviation is enough for anyone who is studying statistics for the first time; so we shall not pursue the sample standard deviation any further. Thus whenever we write "standard deviation" we mean "population standard deviation." You should be aware, however, that a second type of standard deviation exists and is used frequently by pollsters and social scientists. We might point out too that if you purchase a calculator which has a standard deviation key, you should read the owner's manual to determine which one it displays. All Hewlett-Packard calculators, for instance, display the sample standard deviation, so if you are using an H-P calculator in this course you will have to convert to the population standard deviation to make your results correspond to ours.

Epilogue: "The Federalist" Paper Caper

Madison or Hamilton—who was the author of the disputed "Federalist" papers? Even though they held opposing political views, neither would claim the honor because they realized the greater good of convincing the New York electorate to vote for ratification of the new Constitution. They perfected their common writing style to such a degree that two centuries of inquiries have not been able to uncover conclusive evidence.

Then in 1964 Mosteller and Wallace cracked the Madison-Hamilton legalistic writing style by studying marker words—words that are exceptionally common to an author's writing style. Now we will present the data gathered by Mosteller and Wallace on 50 essays by Madison, 48 by Hamilton, and the 12 disputed papers. Many marker words were identified, but we will center our attention on the word "upon." Table 1 presents the distribution for this word in the 48 papers by Hamilton, Table 2 gives the distribution in the 50 Madison papers, and Table 3 gives the distribution for the 12 disputed papers. In each table the mean and σ are calculated.

TABLE 1

Frequency Distribution of Rate per 1000 Words in 48 Hamilton Papers for the Word "Upon"

Rate per 1000 Words		Number of			
Interval	Midpoint x_i	Papers f_i	$f_i x_i$	x_i^2	$f_i x_i^2$
0.5–1.5	1	5	5	1	5
1.5–2.5	2	10	20	4	40
2.5–3.5	3	12	36	9	108
3.5–4.5	4	10	40	16	160
4.5–5.5	5	8	40	25	200
5.5–6.5	6	2	12	36	72
6.5–7.5	7	1	7	49	49
		48	160	140	634

$$\bar{x} = \frac{f_1 x_1 + \cdots + f_7 x_7}{f_1 + \cdots + f_7} = \frac{160}{48} \approx 3.3 \qquad \bar{x}^2 \approx 10.89$$

$$\sigma = \sqrt{\frac{f_1 x_1^2 + \cdots + f_7 x_7^2}{f_1 + \cdots + f_7} - \bar{x}^2} = \sqrt{\frac{634}{48} - 10.89}$$

$$= \sqrt{13.21 - 10.9} = \sqrt{2.32} \approx 1.52$$

TABLE 2

Frequency Distribution of Rate per 1000 Words in 50 Madison Papers for the Word "Upon"

Rate per 1000 Words		Number of			
Interval	Midpoint x_i	Papers f_i	$f_i x_i$	x_i^2	$f_i x_i^2$
0.0–0.4	0.2	41	8.2	0.04	1.64
0.4–0.8	0.6	2	1.2	0.36	0.72
0.8–1.2	1.0	4	4.0	1.00	4.00
1.2–1.6	1.4	1	1.4	1.96	1.96
1.6–2.0	1.8	2	3.6	3.24	6.48
		50	18.4	6.60	14.80

$$\bar{x} = \frac{f_1 x_1 + \cdots + f_5 x_5}{f_1 + \cdots + f_5} = \frac{18.4}{50} \approx 0.37 \qquad \bar{x}^2 \approx 0.14$$

$$\sigma = \sqrt{\frac{14.8}{50} - 0.14} \approx \sqrt{0.30 - 0.14} = \sqrt{0.16} = 0.4$$

TABLE 3

Frequency Distribution of Rate per 1000 Words in 12 Disputed "Federalist" Papers for the Word "Upon"

Rate per 1000 Words		Number of			
Interval	Midpoint x_i	Papers f_i	$f_i x_i$	x_i^2	$f_i x_i^2$
0.0–0.4	0.2	11	2.2	0.04	0.44
0.4–0.8	0.6	0	0.0	0.36	0.00
0.8–1.2	1.0	0	0.0	1.00	0.00
1.2–1.6	1.4	1	1.4	1.96	1.96
		12	3.6	3.36	2.40

$$\bar{x} = \frac{3.6}{12} = 0.3 \qquad \bar{x}^2 = 0.09$$

$$\sigma = \sqrt{\frac{2.4}{12} - 0.09} = \sqrt{0.2 - 0.09} = \sqrt{0.11} \approx 0.33$$

The means and standard deviations are displayed in Table 4, which makes the inference clear. The distribution of the disputed papers fits the data from the Madison papers much more closely than the Hamilton papers for the word "upon."

Mosteller and Wallace looked at many other words and the same conclusions were evident. Their study shows that it is extremely likely that Madison was the author of the disputed "Federalist" papers.

TABLE 4

Mean and Standard Deviation for 48 Hamilton, 50 Madison, and 12 Disputed Papers for the Word "Upon"

	Hamilton	Madison	Disputed
Mean	3.3	0.37	0.3
σ	1.52	0.4	0.33

Summary

To measure the spread of a list of data about its mean one often first finds the range of the data, which is the largest value in the list minus the smallest value.

$$\text{Range} = \text{largest value} - \text{smallest value}$$

The range does not tell how the data are dispersed in between the extreme values.

In an intuitive sense, the mean $\bar{x}$ measures the center of the data, and the standard deviation σ gives a measure of how far from the mean the data is spread. The standard deviation provides a convenient and yet somewhat arbitrary measure of how close the various data are to the mean. Most of the entries will be within 1 σ of $\bar{x}$. Almost all the entries will be within 2 σ's of $\bar{x}$, and only rarely will an entry be farther than 3 σ's from $\bar{x}$.

The definition of σ for a list of numbers $x_1, \ldots, x_n$ and mean $\bar{x}$ is given by

$$\sigma = \sqrt{\frac{(x_1 - \bar{x})^2 + (x_2 - \bar{x})^2 + \cdots + (x_n - \bar{x})^2}{n}}$$

where the variance is defined to be the expression under the square root sign, that is, variance $= \sigma^2$, or, equivalently, $\sigma = \sqrt{\text{variance}}$. The definition of the σ can be expressed in a form that is usually more convenient to use for calculation. The alternate form of the definition is

$$\sigma = \sqrt{\frac{x_1^2 + x_2^2 + \cdots + x_n^2}{n} - \bar{x}^2}$$

If the data are grouped in intervals with midpoints $x_1, \ldots, x_n$ and frequencies $f_1, \ldots, f_n$, then the standard deviation is given by

$$\sigma = \sqrt{\frac{f_1 x_1^2 + f_2 x_2^2 + \ldots + f_n x_n^2}{f_1 + \ldots + f_n} - \bar{x}^2}$$

EXERCISES

1 Each of the following sets of data has a mean of 10. Which has a spread about 10 that is the least? the greatest?

Set 1: 15, 11, 10, 14, 7, 5, 8, 10, 16, 4

Set 2: 12, 11, 10, 10, 9, 9, 8, 11, 10, 10

Set 3: 11, 12, 10, 9, 12, 11, 12, 5, 8

2 Each of the following sets of data has a mean of 50. Which has a spread about the mean that is the least? the greatest?

Set 1: 60, 50, 40, 30, 55, 50, 45, 45, 50, 75

Set 2: 60, 50, 50, 50, 20, 25, 50, 25, 90, 80

Set 3: 60, 55, 45, 55, 55, 45, 45, 45, 45, 50

3   Find the mean for each of the sets of data. Which set has the greatest spread about the mean? the least spread?

Set 1: 0, −1, 2, −1, −2, 2

Set 2: 0, −1, 10, 10, 2, 9

Set 3: 50, 52, 54, 47, 52, 45

4   Find the mean for each of the sets of data. Which set has the greatest spread about the mean? the least spread?

Set 1: 0, 8, 10, 2, 7, 3

Set 2: 0, −5, −10, 8, −15, 12, 10

Set 3: 0, 2, 1, 0, 1, 1, 2, −1, 1, 1, 3

5   Each of the following sets of data has a mean equal to 5. Use the intuitive description of the σ to determine whether the σ is closer to 1, 5, or 10. No calculations are necessary.

Set 1: 8, 10, 11, 1, 0

Set 2: 6, 5, 4, 7, 3

Set 3: 15, 15, −5, −5, 5

6   Use the intuitive description of the σ to determine whether the σ for each set of data is closer to 1, 5, or 10. No calculations are necessary.

Set 1: 40, 45, 55, 50, 60

Set 2: 10, 25, 15, 13, 17

Set 3: 10, 8, 12, 11, 9

7   For the sketches of the distributions below match which mean and σ best describe the data.
(a)   Mean is 5 and σ is 1
(b)   Mean is 5 and σ is 5
(c)   Mean is 5 and σ is 3

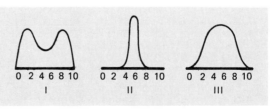

8 For the sketches of the distributions below which data have the greatest σ? smallest σ?

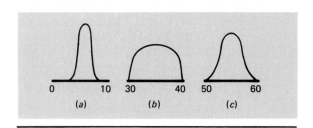

(a) (b) (c)

In Problems 9 to 14 calculate the σ for the given sets of data.

9 0, 1, 3, 1, 2, 0, 6, 2, 7, 3

10 $-1, 0, -3, 1, 4, 0, -2, 5, 1, -3$

11 10, 15, 18, 11, 12, 15, 20, 10, 12, 18

12 25, 20, 18, 15, 24, 22, 16, 20, 17, 23

13 22, 18, 32, 40, 20, 28, 38, 42, 15, 45

14 40, 48, 60, 55, 58, 65, 42, 45, 52, 35

In Problems 15 to 18 use the alternate formula for data grouped in intervals to calculate the σ for the given sets of data.

15

Interval	Frequency
0–2	3
2–4	7
4–8	5
8–10	10
10–12	5

17

Interval	Frequency
10–12	1
12–14	8
14–16	10
16–18	5
18–20	15
20–22	3

16

Interval	Frequency
0–4	15
4–8	18
8–12	20
12–16	28
16–20	20

18

Interval	Frequency
0.5–1.5	10
1.5–2.5	8
2.5–3.5	7
3.5–4.5	8
4.5–5.5	6
5.5–6.5	3

19 Verify that the formula for the σ can be simplified to the alternate formula for a data set with $n = 2$, that is, for a data set with two entries, x_1 and x_2. In other words, show that

$$\sqrt{\frac{(x_1 - \bar{x})^2 + (x_2 - \bar{x})^2}{2}} = \sqrt{\frac{x_1^2 + x_2^2}{2} - \bar{x}^2}$$

20 Calculate the σ for data set C using the first formula, and verify that your answer is the same as the one given in the text in Example 1.

21 Determine the mean and standard deviation of each of the three sets of test scores given in Figure 1.

22 J. E. Kerrich tossed a coin 1000 times and recorded the number of heads. He conducted this experiment 10 times, and obtained the following number of heads (see Section 6-2).

   502, 511, 497, 529, 504, 476, 507, 528, 504, 529

Compute the mean and standard deviation. Do you think that the coin is fair?

8-4 THE NORMAL CURVE

We have considered many events that appear to occur wholly at random. They do not seem to lend themselves to accurate prediction in any way when a single occurrence is observed. Some examples have been the toss of a coin, the roll of a die, and the color of the pea plant flower. Some regularity and predictability of these occurrences can be seen only when they are repeated very many times in exactly the same way. In our everyday language we say that one such occurrence takes place strictly "by chance." For example, if we place a bet on number 13 on a roulette wheel and 13 shows up, we say we are very lucky because the chances of 13 coming up are slim.

   We have also spent a great deal of time discussing data sets that are normally distributed, which we have taken to mean that the distribution of the data is bunched up in the middle about the mean and gradually tapers off to the left and right in a similar way. Examples of data that are usually normally distributed are heights and weights of large populations selected at random. When a bias enters the selection process, the distribution may vary a great deal from a normally distributed one. The previous examples we considered were income, heights, and weights.

   In this section we will discuss what we mean by normally distributed in greater detail. The concept is based on a geometrical entity called the *normal curve*. It was created in the early 1700s by the mathematician Abraham De Moivre in order to solve gambling problems. About a century later Karl Gauss used the normal curve to describe chance variations in common occurrences. The basic question that Gauss raised was: When we conduct an experiment or observe a phenomenon and the result is different from what was expected, when can this result be attributed to chance variation and when can it be said to be caused by some outside factor? Gauss realized that he had to fully understand the phenomenon of chance variation in order to be able to rule it out as the cause and conclude that the outside factor is the culprit.

Let us consider some anecdotes:

Anecdote 1 The Fair Coin Revisited

If one tosses a coin 100 times, one expects to observe about 50 heads. If one observes 40 heads, can this be attributed to chance variation? How about the observation of 20 heads, or 10 heads? When can we say that the coin is not fair?

Anecdote 2 Smoke Gets in Your Eyes—and Lungs

Today most people believe that smoking is dangerous to their health. This was not the prevailing attitude as few as 30 years ago, when some well-known scientists ridiculed those who warned against the use of tobacco. The government thought that the issue was important enough to fund large-scale experiments to determine the answer. In early studies the results seemed to indicate that tobacco was dangerous.

The eminent medical statistician from Johns Hopkins University, Raymond Pearl, had been keeping detailed records of the health experiences of hundreds of families in the Baltimore area. As the tobacco and health question became a serious study, Pearl decided to use his information to conduct a study of the longevity of all males in his files and compare it with their smoking habits. Pearl found that 65 percent of the non-smokers survived to age 60, whereas only 45 percent of the heavy smokers lived that long. Not many were convinced by Pearl's data, however. If Pearl's study were correct, it would be the duty of the medical profession and the government health agencies to inform the public and take drastic steps to help the public avoid the hazard. But could Pearl's results be due to chance variation? Perhaps another set of data could produce alternate conclusions. Perhaps some other factor that is unrelated to tobacco entered the picture to increase the smoker's death rate. If Pearl's results were erroneous, it would be a grave mistake for the government to interfere in the expansive tobacco industry and the private lives of the public. It was clear that better statistical methods were needed to ensure that the results of tobacco and health experiments could be judged attributable to chance variation or not. Today a new drug will not be approved by the U.S. Food and Drug Administration until it satisfactorily passes a long list of statistical experiments designed to ensure that the intended effect of the drug is significant and its side effects are minimal. These statistical tests have their roots in the concept of chance variation.

Anecdote 3 Reagan's Proudest Role

On Tuesday, November 4, 1980, Jimmy Carter ran for reelection for the U.S. Presidency, against Ronald Reagan and long shot John Anderson. The Gallup poll published on November 4, from interviews of more than

3500 potential voters, said that Reagan had the backing of 46 percent of its sample and Carter had 43 percent. Anderson drew 7 percent, 1 percent named other candidates, and 3 percent interviewed from October 30 to November 1 were undecided. By allocating the undecided vote the Gallup organization said that the final standings were Reagan, 47 percent; Carter, 44 percent; Anderson, 8 percent; and others, 1 percent. But the *margin for error* (our emphasis) was 3 percent. Thus the Gallup organization called the race dead even.

The Washington Post conducted a national survey of 1000 registered voters, and it showed Carter having the support of 42 percent; Reagan, 39 percent; and Anderson, 7 percent. The *margin for error* was 4 percent, and so again the race was viewed as dead even.

The important point here is that even though the major polls adhere to strict rules to ensure that the sample chosen accurately reflects the whole population, in this case all registered voters, they must build into their formulas a restriction for chance variation. That is, the polls recognize that the individuals they talk to may not accurately reflect the rest of the population. It is through the study of chance variation that the polls determine that they may be in error—in this case by 3 percent with the Gallup poll or 4 percent with *The Washington Post.*

Were their predictions accurate? Certainly not! The 1980 presidential race was a runaway for Ronald Reagan and the conservative Republican party. Reagan garnered 52 percent of the vote and a political earthquake shook Congress. The Republicans took control of the Senate for the first time in 25 years. "Oh God, what a mess, we've been Carterized," sighed Democratic committeewoman Billie Carr of Texas as at least 12 Democratic senators accompanied Carter in defeat. A senior aide of Senator Edward Kennedy, one of the few liberals left in Congress, said, "The implications of their victory are incredible."

How could the polls have predicted a dead heat on Tuesday morning when later in the day Americans would overwhelmingly turn out for Reagan and his Republican army? Were the polls wrong? What about a chance variation of 4 percent; perhaps it should have been 10 percent?

These are very serious questions that polling agencies will be trying to answer for years to come. Some partial answers are immediate. The polls published on Tuesday morning were conducted through Saturday, November 1, and did not take into account the dramatic events of November 2, when the Ayatollah Khomeini, Iran's religious and de facto political leader, offered for the first time in a year concrete conditions for the release of the 52 American hostages. Perhaps many Americans saw this as an overt attempt by Iran to influence the American political process and viewed it as the proverbial straw that broke the camel's back, thus prompting them to switch to Reagan. Or perhaps there were a great many who could not tell a pollster that they were going to vote for a former actor but, when reaching the polling booth, recalled what he stood for and preferred his positions to Carter's.

In any event, both Carter and Reagan admitted that their private polls conducted on Sunday and Monday, November 2 and 3, showed a landslide victory for Reagan. This was evidenced by Carter's early concession on Tuesday evening. The polls conducted before the weekend probably did reflect the mood of the country, but the unforeseen events over the weekend most likely altered the amount that the undecided vote and the chance variation should have been allocated. In other words, the Gallup poll's conclusion that Carter had 43 percent support and Reagan 46 percent was true, but over the weekend most of the 3 percent undecided leaned toward Reagan. That would make it 49 percent for Reagan to 43 percent for Carter, which is within the 3 percent margin for error.

These three anecdotes point out that it is necessary to understand the concept of chance variation when studying statistics. To understand chance variation we must study the normal curve. The two are intimately entwined.

Standard Deviation

The normal curve has an unwieldy looking equation:

$$y = \frac{1}{\sqrt{2\pi}}\, e^{-x^2/2}$$

where e is an irrational number whose decimal expansion is approximately 2.71828. . . . Do not let the equation scare you though. It is easy to graph it, easy to remember the shape and some elementary facts about it, and easy to find areas under the curve by looking in a table or using a hand calculator.

The graph of the normal curve is given in Figure 1. Study the curve for a short while before reading on and see if you can discover some elementary facts about it.

The values on the horizontal axis are the ordinary units one finds on a cartesian graph, but in statistics they have a special meaning, so we give them the formidable name of *standard units,* which we will explain shortly.

In its strictest mathematical sense, the units on the vertical axis are not frequencies or even relative frequencies, but simply numbers that are

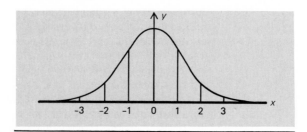

FIGURE 1

obtained by substituting the value of x into the equation for the normal curve. For our purposes, we will consider a value y on the vertical axis to be the area of a rectangle whose height is the corresponding value of x on the horizontal axis and whose width is 1 standard unit. In this way the units on the vertical axis correspond to the relative frequencies that we encountered when studying histograms. This is a natural approach because we will use the normal curve to approximate histograms of data sets.

Let us single out some of the features of the normal curve.

1   The area under the curve and above the x axis is exactly 1.

2   The curve is symmetric about the y axis. In other words, the part of the curve to the right of the y axis is the mirror image of the part of the curve to the left of the y axis.

3   The curve is always above the x axis. It never touches the x axis, but as it proceeds to the right, it gets closer and closer to the positively directed x axis. Likewise, to the left, it approaches the negatively directed x axis without ever touching it. In Figure 1 it appears that the curve touches the x axis somewhere between 3 and 4, but that is only because it gets so close to the x axis that it is impossible to indicate the distance between the curve and the x axis on the graph.

4   The units on the x axis correspond to standard deviations and the value $x = 0$ corresponds to the mean. Thus the value $x = 1$ is one standard deviation above or to the right of the mean, $x = 2$ is two standard deviations to the right of the mean, and $x = -1$ is one standard deviation to the left or below the mean. The values on the x axis of the graph of the normal curve are called *standard units*. Therefore standard units tell us how far above or below the mean a value is.

The primary use to which we will put the normal curve is finding areas under the curve. We assume that "area under the normal curve" means that the area is above the x axis. Since the total area under the normal curve is 1, or 100 percent, any other area is a decimal between 0 and 1, or a percentage between 0 and 100 percent. Examples 1 to 5 will demonstrate how some of the more important areas can be found.

Example 1
What is the area under the normal curve to the right of $x = 0$? Since this area is the same as the area to the left of zero, and since the total area is 1, the area to the right must be 0.5, or 50 percent, of the total area. Similarly, the area to the left of $x = 0$ is 0.5.

Example 2

The area under the normal curve between $x = -1$ and $x = 1$ is about 0.68. If a set of data is normally distributed, then about 68 percent of the values would lie within one standard deviation of the mean. For the normal curve $x = -1$ is 1 σ to the left of the mean, and $x = 1$ is 1 σ to the right, so the statement "within 1 σ of the mean" means between $x = -1$ and $x = 1$ for the normal curve. Thus we see that the areas under the normal curve, in this case 0.68, correspond to percentages of data, in this case 68 percent.

Example 3

The area under the normal curve between $x = 0$ and $x = 1$ is one-half the area between $x = -1$ and $x = 1$ because the curve is symmetric about the y axis. Hence the area between $x = 0$ and $x = 1$ is 0.34 ($= \frac{1}{2} \cdot 0.68$).

Example 4

The area under the normal curve between $x = -2$ and $x = 2$ is about 0.95. We say that if a set of data is normally distributed, then about 95 percent of the data lie within 2 σ's of the mean. For the normal curve, the value $x = 2$ is 2 σ's to the right of the mean and $x = -2$ is 2 σ's to the left of the mean.

Example 5

The area under the normal curve to the right of $x = 2$ can be found by noting that the area between $x = -2$ and $x = 2$ accounts for about 95 percent of the total area, so there is about 5 percent of the area remaining. There are two regions under the curve that comprise this area, the region to the right of $x = 2$ and the region to the left of $x = -2$. But these regions have equal areas because of the symmetry of the normal curve. Hence the area under the normal curve to the right of $x = 2$ is 0.025 or 2.5 percent ($= \frac{1}{2}$ of 5 percent).

It is important to understand the distinction between data sets that are normally distributed and the normal curve itself. The normal curve is a mathematical abstraction, a creation of the mind, so to speak. It is not derived from any real-world phenomena or data. It is, in a certain sense, an ideal form, one that is useful in describing a given set of data derived from natural events under appropriate circumstances.

In the same way we say that a set of data is normally distributed if it is close to the shape of the precise mathematical ideal form of the normal curve.

The fundamental difference between the normal curve and data that are normally distributed is that the normal curve is *continuous* while data sets are necessarily *discrete*. Among concrete data we must have only a finite number of observations, but the normal curve as a model must describe an infinite number of observations. In other words, a data set has a distribution that consists of a finite number of entries arranged in a finite number of cells, and a histogram can be created to graphically depict the data as a sequence of rectangles. The normal curve is defined on the entire real line, a continuous and infinite set of numbers whose graph is a continuous curve.

Yet, even in the face of all these differences between the normal curve and sets of real data, it is amazing to find how convenient and reasonably accurate the normal curve is when it is used to describe a great many distributions. Even though no data set can ever have an infinite number of entries and so no distribution can ever be exactly normally distributed, it does not matter a great deal because it is the utility of the model that we are interested in, and the utility is vast.

Using a Table to Find Areas Under the Normal Curve

At the end of this section there is a table that gives values that are areas under the normal curve between the mean $x = 0$ and a positive value for x. For example, in order to use the table to find the area under the normal curve between $x = 0$ and $x = 0.70$, you look up the value $x = 0.70$ in the third-from-the-left column headed by x. Immediately to the right of $x = 0.70$ in the column headed A, for "area," is the value 0.2580. This means that the area under the normal curve from $x = 0$ to $x = 0.70$ is 0.2580. Thus about 25.8 percent of the total area is between $x = 0$ and $x = 0.70$. (See Figure 2.) Try to apply this technique to the values in the next example.

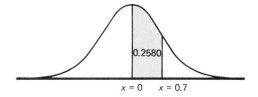

$x = 0$ $x = 0.7$

FIGURE 2   Area under the normal curve between $x = 0$ and $x = 0.7$.

Example 6

Problem   Find the area under the normal curve between $x = 0$ and the given values of x. See Figure 3 on page 540.

(a) $x = 0.48$ (b) $x = 1.00$ (c) $x = 2.95$

Solution   (a) $x = 0.48$ is found in the left-most column and it corresponds to the number $A = 0.1844$, which is the desired area. [See Figure 3(a).]

(b) $x = 1.00$ is found in the third column headed by x and it corresponds to the area $A = 0.3413$. Earlier we said that the area between $x = 0$ and $x = 1$ is "approximately 34 percent," and now we see from the table that a more exact value for the area is 34.13 percent. [See Figure 3(b).]

(c) $x = 2.95$ is found on the second page in the third column headed by x and it corresponds to the area $A = 0.4984$. Notice that this is very close to 0.5, or $\frac{1}{2}$ the total area. Recall that the entire area under the normal curve is 1, so the area to the right of $x = 0$ is exactly 0.5. [See Figure 3(c).]

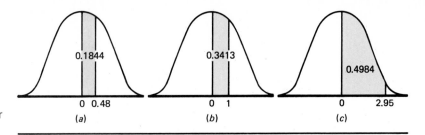

FIGURE 3 Some areas under the normal curve.

There are several ways in which we will use the table of areas under the normal curve. All the areas in the table represent areas between $x = 0$ and a positive value $x = a$. But there are other types of areas that you will want to find. They are: (I) the area to the right of a positive value $x = a$; (II) the area between a negative value $x = a$ and $x = 0$; (III) the area to the left of a negative value $x = a$; and (IV) the area between two values $x = a$ and $x = b$. We will handle each type of area separately.

(I) Now let us use the table to find the area under the normal curve that is to the right of a positive value, say $x = a$. (See Figure 4.) Recall that the total area to the right of $x = 0$ is 0.5 and the area between $x = 0$ and $x = a$ is the value A in the table. Figure 4 gives us a graphical description of the fact that the area to the right of $x = a$ is the area to the right of $x = 0$ minus the area between $x = 0$ and $x = a$. Thus:

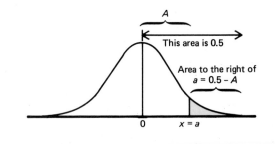

FIGURE 4 The area under the normal curve to the right of a positive value $x = a$.

The area under the normal curve to the right of a positive value $x = a$ is $0.5 - A$, where A is the area between $x = 0$ and $x = a$ found in the table corresponding to a.

Let us explain the use of this procedure in the next example.

Example 7
Problem Find the area under the normal curve to the right of the given positive values.

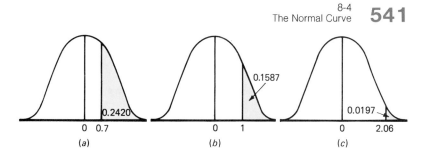

0.1587

0.2420

0.0197

0 0.7   0 1   0 2.06

(a)   (b)   (c)

FIGURE 5

(a) $x = 0.70$ (b) $x = 1.00$ (c) $x = 2.06$

Solution   (a) The value in the table corresponding to $x = 0.70$ is $A = 0.2580$, and so the area to the right of $x = 0.70$ is

$$0.5 - A = 0.5 - 0.2580 = 0.2420$$

[See Figure 5(a).]
   (b) The value in the table corresponding to $x = 1.00$ is $A = 0.3413$, so the area to the right of $x = 1.00$ is $0.5 - A = 0.5 - 0.3413 = 0.1587$. [See Figure 5(b).]
   (c) The area to the right of $x = 2.06$ is $0.5 - 0.4803 = 0.0197$. [See Figure 5(c).]

(II) We now tackle the problem of finding areas under the normal curve between a negative value $x = a$ and $x = 0$. From Figure 6 you can see that the area between $x = a$ and $x = 0$ is equal to the area between $x = 0$ and $x = -a$ [since a is assumed to be negative, say, for example -2, then $-a$ is positive, as, for example, $-(-2) = 2$]. This is because the normal curve is symmetric about $x = 0$.

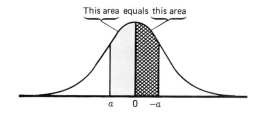

This area equals this area

a 0 $-a$

FIGURE 6

This fact makes it an easy matter to find the area under the normal curve between a negative value $x = a$ and $x = 0$. Do you see how? Simply remove the negative sign from a (by taking the absolute value and/or finding $-a$), look up that value in the table, and the value of A in the table is equal to the desired area.

The area under the normal curve between a negative value $x = a$ and $x = 0$ is the value A in the table corresponding to the positive value $x = -a$.

Let us look at an example that demonstrates the use of the rule.

Example 8

Problem   Find the area under the normal curve between the given negative value and $x = 0$.

(a) $x = -1.00$ (b) $x = -1.5$ (c) $x = -2.00$

Solution   We simply look up the areas corresponding to the values (a) $x = 1.00$, (b) $x = 1.5$, and (c) $x = 2.00$. They are

(a) $A = 0.3413$ [See Figure 7(a).]

(b) $A = 0.4332$ [See Figure 7(b).]

(c) $A = 0.4773$ [See Figure 7(c).]

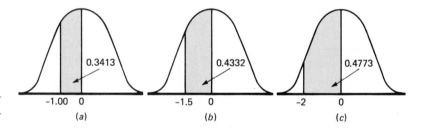

FIGURE 7   Some areas under the normal curve between a negative value and $x = 0$.

(III) In a similar manner you can see from Figure 8 that the area to the left of a negative value $x = a$ is equal to the area to the right of $x = -a$ because of the symmetry of the normal curve.

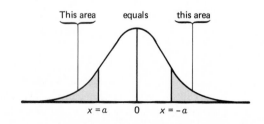

FIGURE 8

The rule is obtained directly from the rule concerning areas to the right of a positive value.

The area under the normal curve to the left of a negative value $x = a$ is equal to $0.5 - A$, where A is the area found in the table corresponding to the positive value $x = -a$.

Let us look at an example to demonstrate how to use the rule.

Example 9

Problem Find the area under the normal curve to the left of the given negative value.

$$(a) \ x = -0.50 \qquad (b) \ x = -1.48 \qquad (c) \ x = -2.50$$

Solution We must look up the areas corresponding to the positive values $(a) \ x = 0.50$, $(b) \ x = 1.48$, and $(c) \ x = 2.50$ and then subtract the number from 0.5.

(a) The area is $0.5 - A = 0.5 - 0.1915 = 0.3085$. [See Figure 9$(a)$.]

(b) The area is $0.5 - A = 0.5 - 0.4306 = 0.0694$. [See Figure 9$(b)$.]

(c) The area is $0.5 - A = 0.5 - 0.4938 = 0.0062$. [See Figure 9$(c)$.]

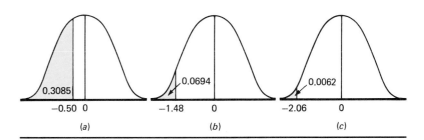

FIGURE 9

(a)   (b)   (c)

(IV) There is one more type of problem concerning areas under normal curves to consider: the area between two values $x = a$ and $x = b$. We will separate the problem into two cases: case A—a and b have the same sign, that is, they are both positive or both negative; case B—a and b have different signs, that is, one is positive and one is negative. Figure 10 on page 544 shows the possibilities.

Case A: Consider the graphical description of case A, when both a and b are positive, given in Figure 11. We are looking for the area between $x = a$ and $x = b$. Let us refer to it as area A. Figure 11(a) shows the area from $x = 0$ to $x = a$, which is obtained from the table;

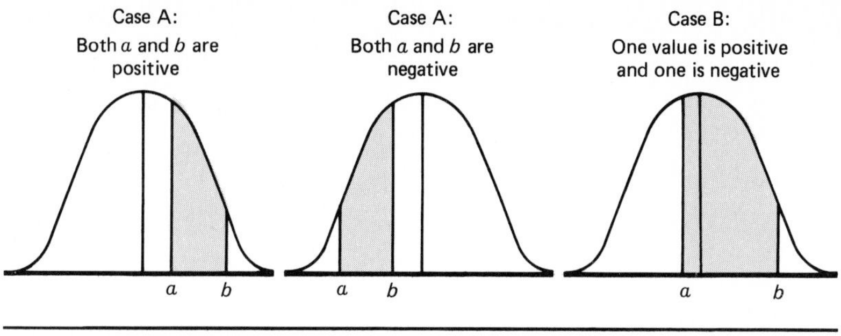

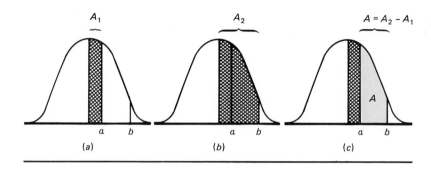

FIGURE 10

FIGURE 11

call it A_1. Figure 11(b) shows the area from $x = 0$ to $x = b$; call it A_2. Notice that A_2 consists of two distinct areas, the area from $x = 0$ to $x = a$, that is, the area A_1, and the area from $x = a$ to $x = b$, the desired area A. Therefore $A = A_2 - A_1$, as shown in Figure 11(c).

The situation is the same if both $x = a$ and $x = b$ are negative. This leads us to the following rule.

If $x = a$ and $x = b$ are both positive or both negative, look up the area between $x = 0$ and $x = a$ and the area between $x = 0$ and $x = b$, then the area under the normal curve between $x = a$ and $x = b$ is the smaller subtracted from the larger of these two areas.

This rule might look a little more complicated than the previous ones, but if you can easily look up the areas in the table, then the application of the rule is quite easy. Let us demonstrate it by an example.

Example 10
Problem Find the area under the normal curve between the two given values.

(a) $x = 0.50$ and $x = 1.50$ (b) $x = -0.78$ and $x = -1.36$

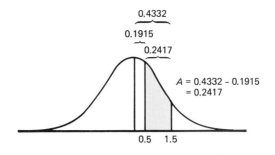

FIGURE 12

Solution (*a*) We first look up the areas in the table corresponding to the given values of x. For $x = 0.50$ the area is 0.1915, and for $x = 1.50$ the area is 0.4332. The area between $x = 0.50$ and $x = 1.50$ is then

$$0.4332 - 0.1915 = 0.2417$$

(See Figure 12.)

(*b*) We first look up the areas in the table corresponding to the positive values $x = 0.78$ and $x = 1.36$. For $x = 0.78$ the area is 0.2823, which is the area between $x = -0.78$ and $x = 0$. Similarly, the area corresponding to $x = 1.36$ is 0.4131. Hence the area between $x = -0.78$ and $x = -1.36$ is

$$0.4131 - 0.2823 = 0.1308$$

(See Figure 13.)

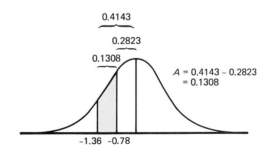

FIGURE 13

We can now consider case *B*.

Case B: $x = a$ is negative and $x = b$ is positive. As seen in Figure 14, the area between $x = a$ and $x = b$ consists of two parts: the area between $x = a$ and $x = 0$, which is the area in the table corresponding to the positive value $x = -a$; and the area between $x = 0$ and $x = b$, which is the area in the table corresponding to $x = b$. Therefore the rule is very simple.

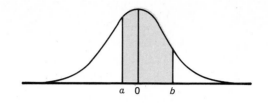

FIGURE 14

If $x = a$ is negative and $x = b$ is positive, then the area under the normal curve between $x = a$ and $x = b$ is the sum of the areas from $x = 0$ to $x = b$ and from $x = a$ to $x = 0$, which is found in the table by looking up $x = b$ and $x = -a$.

Let us demonstrate the rule with an example.

Example 11

Problem   Find the area under the normal curve between the given values of $x = a$ and $x = b$.

   (*a*) $x = -0.80$ and $x = 1.64$ (*b*) $x = -2.15$ and $x = 0.38$

Solution   (*a*) The area corresponding to $x = -0.80$ (for which we look up the value $x = 0.80$) is 0.2881, and the area corresponding to $x = 1.64$ is 0.4495. Hence the area between $x = -0.80$ and $x = 1.64$ is

   $0.2881 + 0.4495 = 0.7376$

(See Figure 15.)

   (*b*) The area corresponding to $x = -2.15$ (remember to look up $x = 2.15$) is 0.4842, and the area corresponding to $x = 0.38$ is 0.1480. Therefore the area between $x = -2.15$ and $x = 0.38$ is

   $0.4842 + 0.1480 = 0.6322$

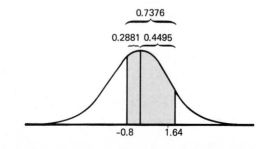

FIGURE 15

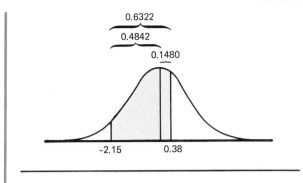

0.6322

0.4842

0.1480

−2.15   0.38

FIGURE 16

(See Figure 16.)

Summary

The normal curve is a bell-shaped curve that is symmetric about the center $x = 0$, the mean, and the total area under the curve is 1, or 100 percent. We gave an intuitive description of the connection between raw data and the abstract definition of the normal curve. It is quite remarkable how closely so many varied phenomena follow the symmetry of this aesthetically pleasing curve. Countless investigators of physical, social, and economic inquiries have demonstrated that real-world data often fit the normal distribution. Phenomena such as height, weight, IQ, national test scores, and even the wear of a carpet have been found to be distributed normally.

We learned how to find areas under the normal curve by using the table at the end of the section and the various elementary properties of symmetry of the curve.

In the next section we will investigate how to take raw data that is distributed normally and use the normal curve to approximate information about the histogram, or distribution, of the data.

EXERCISES

In Problems 1 to 4 find the area under the normal curve from $x = 0$ to the given value, without looking at the table.

1   $x = 1$ 3 $x = -1$

2   $x = 2$ 4 $x = -2$

In Problems 5 to 10 find the area under the normal curve without looking at the table.

5   Between $x = -1$ and $x = 2$ 8   To the right of $x = 0$

6   Between $x = -2$ and $x = 1$ 9   To the left of $x = 0$

7   To the right of $x = 1$ 10   To the left of $x = -1$

In Problems 11 to 16 use the table to find the area under the normal curve from $x = 0$ to the given value.

11 $x = 0.77$ **14** $x = 1.81$

12 $x = 0.32$ **15** $x = 2.09$

13 $x = 1.45$ **16** $x = 2.88$

In Problems 17 to 22 use the table to find the area under the normal curve from the given value to $x = 0$.

17 $x = -0.65$ **20** $x = -1.72$

18 $x = -0.95$ **21** $x = -2.38$

19 $x = -1.55$ **22** $x = -3.05$

In Problems 23 to 37 use the table to find the area under the normal curve.

23 To the right of $x = 0.72$

24 To the right of $x = 1.10$

25 To the right of $x = 2.08$

26 To the left of $x = -0.01$

27 To the left of $x = -1.90$

28 To the left of $x = -2.28$

29 Between $x = 0.51$ and $x = 1.82$

30 Between $x = 0.92$ and $x = 2.15$

31 Between $x = 1.23$ and $x = 1.63$

32 Between $x = -1.85$ and $x = -0.60$

33 Between $x = -2.11$ and $x = -0.25$

34 Between $x = -0.90$ and $x = -0.72$

35 Between $x = -0.48$ and $x = 1.66$

36 Between $x = -0.19$ and $x = 1.83$

37 Between $x = -2.05$ and $x = 2.05$

In Problems 38 to 43 a new type of area is described—one that includes the entire area to the right of $x = 0$ or to the left of $x = 0$, which, of course, is equal to 0.5. You must find the other part of the area described and then add 0.5 to it. Find the area under the normal curve.

38 To the right of $x = -0.50$ **41** To the left of $x = 1.18$

39 To the right of $x = -1.68$ **42** To the left of $x = 0.50$

40 To the right of $x = -2.00$ **43** To the left of $x = 1.84$

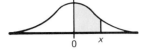

Areas Under the Standard Normal Curve

The Column Under A Gives the Proportion of the Area Under the Entire Curve Which Is Between $x = 0$ and a Positive Value of x.

x	A	x	A	x	A	x	A
.00	.0000	.49	.1879	.98	.3365	1.47	.4292
.01	.0040	.50	.1915	.99	.3389	1.48	.4306
.02	.0080	.51	.1950	1.00	.3413	1.49	.4319
.03	.0120	.52	.1985	1.01	.3438	1.50	.4332
.04	.0160	.53	.2019	1.02	.3461	1.51	.4345
.05	.0199	.54	.2054	1.03	.3485	1.52	.4357
.06	.0239	.55	.2088	1.04	.3508	1.53	.4370
.07	.0279	.56	.2123	1.05	.3531	1.54	.4382
.08	.0319	.57	.2157	1.06	.3554	1.55	.4394
.09	.0359	.58	.2190	1.07	.3577	1.56	.4406
.10	.0398	.59	.2224	1.08	.3599	1.57	.4418
.11	.0438	.60	.2258	1.09	.3621	1.58	.4430
.12	.0478	.61	.2291	1.10	.3643	1.59	.4441
.13	.0517	.62	.2324	1.11	.3665	1.60	.4452
.14	.0557	.63	.2357	1.12	.3686	1.61	.4463
.15	.0596	.64	.2389	1.13	.3708	1.62	.4474
.16	.0636	.65	.2422	1.14	.3729	1.63	.4485
.17	.0675	.66	.2454	1.15	.3749	1.64	.4495
.18	.0714	.67	.2486	1.16	.3770	1.65	.4505
.19	.0754	.68	.2518	1.17	.3790	1.66	.4515
.20	.0793	.69	.2549	1.18	.3810	1.67	.4525
.21	.0832	.70	.2580	1.19	.3830	1.68	.4535
.22	.0871	.71	.2612	1.20	.3849	1.69	.4545
.23	.0910	.72	.2642	1.21	.3869	1.70	.4554
.24	.0948	.73	.2673	1.22	.3888	1.71	.4564
.25	.0987	.74	.2704	1.23	.3907	1.72	.4573
.26	.1026	.75	.2734	1.24	.3925	1.73	.4582
.27	.1064	.76	.2764	1.25	.3944	1.74	.4591
.28	.1103	.77	.2794	1.26	.3962	1.75	.4599
.29	.1141	.78	.2823	1.27	.3980	1.76	.4608
.30	.1179	.79	.2852	1.28	.3997	1.77	.4616
.31	.1217	.80	.2881	1.29	.4015	1.78	.4625
.32	.1255	.81	.2910	1.30	.4032	1.79	.4633
.33	.1293	.82	.2939	1.31	.4049	1.80	.4641
.34	.1331	.83	.2967	1.32	.4066	1.81	.4649
.35	.1368	.84	.2996	1.33	.4082	1.82	.4656
.36	.1406	.85	.3023	1.34	.4099	1.83	.4664
.37	.1443	.86	.3051	1.35	.4115	1.84	.4671
.38	.1480	.87	.3079	1.36	.4131	1.85	.4678
.39	.1517	.88	.3106	1.37	.4147	1.86	.4686
.40	.1554	.89	.3133	1.38	.4162	1.87	.4693
.41	.1591	.90	.3159	1.39	.4177	1.88	.4700
.42	.1628	.91	.3186	1.40	.4192	1.89	.4706
.43	.1664	.92	.3212	1.41	.4207	1.90	.4713
.44	.1700	.93	.3238	1.42	.4222	1.91	.4719
.45	.1736	.94	.3264	1.43	.4236	1.92	.4726
.46	.1772	.95	.3289	1.44	.4251	1.93	.4732
.47	.1808	.96	.3315	1.45	.4265	1.94	.4738
.48	.1844	.97	.3340	1.46	.4279	1.95	.4744

Areas Under the Standard Normal Curve *(Continued)*

x	A	x	A	x	A	x	A
1.96	.4750	2.45	.4929	2.94	.4984	3.43	.4997
1.97	.4756	2.46	.4931	2.95	.4984	3.44	.4997
1.98	.4762	2.47	.4932	2.96	.4985	3.45	.4997
1.99	.4767	2.48	.4934	2.97	.4985	3.46	.4997
2.00	.4773	2.49	.4936	2.98	.4986	3.47	.4997
2.01	.4778	2.50	.4938	2.99	.4986	3.48	.4998
2.02	.4783	2.51	.4940	3.00	.4987	3.49	.4998
2.03	.4788	2.52	.4941	3.01	.4987	3.50	.4998
2.04	.4793	2.53	.4943	3.02	.4987	3.51	.4998
2.05	.4798	2.54	.4945	3.03	.4998	3.52	.4998
2.06	.4803	2.55	.4946	3.04	.4988	3.53	.4998
2.07	.4808	2.56	.4948	3.05	.4989	3.54	.4998
2.08	.4812	2.57	.4949	3.06	.4989	3.55	.4998
2.09	.4817	2.58	.4951	3.07	.4989	3.56	.4998
2.10	.4821	2.59	.4952	3.08	.4990	3.57	.4998
2.11	.4826	2.60	.4953	3.09	.4990	3.58	.4998
2.12	.4830	2.61	.4955	3.10	.4990	3.59	.4998
2.13	.4834	2.62	.4956	3.11	.4991	3.60	.4998
2.14	.4838	2.63	.4957	3.12	.4991	3.61	.4999
2.15	.4842	2.64	.4959	3.13	.4991	3.62	.4999
2.16	.4846	2.65	.4960	3.14	.4992	3.63	.4999
2.17	.4850	2.66	.4961	3.15	.4992	3.64	.4999
2.18	.4854	2.67	.4962	3.16	.4992	3.65	.4999
2.19	.4857	2.68	.4963	3.17	.4992	3.66	.4999
2.20	.4861	2.69	.4964	3.18	.4993	3.67	.4999
2.21	.4865	2.70	.4965	3.19	.4993	3.68	.4999
2.22	.4868	2.71	.4966	3.20	.4993	3.69	.4999
2.23	.4871	2.72	.4967	3.21	.4993	3.70	.4999
2.24	.4875	2.73	.4968	3.22	.4994	3.71	.4999
2.25	.4878	2.74	.4969	3.23	.4994	3.72	.4999
2.26	.4881	2.75	.4970	3.24	.4994	3.73	.4999
2.27	.4884	2.76	.4971	3.25	.4994	3.74	.4999
2.28	.4887	2.77	.4972	3.26	.4994	3.75	.4999
2.29	.4890	2.78	.4973	3.27	.4995	3.76	.4999
2.30	.4893	2.79	.4974	3.28	.4995	3.77	.4999
2.31	.4896	2.80	.4974	3.29	.4995	3.78	.4999
2.32	.4898	2.81	.4975	3.30	.4995	3.79	.4999
2.33	.4901	2.82	.4976	3.31	.4995	3.80	.4999
2.34	.4904	2.83	.4977	3.32	.4996	3.81	.4999
2.35	.4906	2.84	.4977	3.33	.4996	3.82	.4999
2.36	.4909	2.85	.4978	3.34	.4996	3.83	.4999
2.37	.4911	2.86	.4979	3.35	.4996	3.84	.4999
2.38	.4913	2.87	.4980	3.36	.4996	3.85	.4999
2.39	.4916	2.88	.4980	3.37	.4996	3.86	.4999
2.40	.4918	2.89	.4981	3.38	.4996		
2.41	.4920	2.90	.4981	3.39	.4997		
2.42	.4922	2.91	.4982	3.40	.4997		
2.43	.4925	2.92	.4983	3.41	.4997		
2.44	.4927	2.93	.4983	3.42	.4997		

8-5 THE NORMAL APPROXIMATION FOR DATA

If a list of data is assumed to be normally distributed, then the normal curve can be used to find areas under the histogram of the data. This procedure, called the *normal approximation*, is explained in this section. Basically, the procedure calls for adjusting the horizontal and vertical axes of the normal curve so that it approximates the histogram. The closer the normal curve is to the shape of the histogram, the closer the areas under the normal curve will be to the areas under the histogram.

The procedure is very handy because the investigator does not actually have to construct the histogram for the data but just use the approximation methods that we will now discuss in order to determine information about the data. Thus the normal approximation entails replacing the original histogram, or even bypassing the construction of the histogram, by the normal curve and then computing the desired areas.

Let us first consider an example in which we describe the approximation method.

Example 1

In a recent Health Examination Survey, taken from *Vital and Health Statistics,* conducted by the Department of Health and Human Services, the average height of women was 64 inches with a standard deviation of about 2.5 inches. If we assume that the heights of women in the United States are normally distributed, then we can use the normal curve to approximate areas under the histogram of the data by first drawing the horizontal axis of the histogram immediately above the horizontal axis of the normal curve, with their respective means (64 inches and 0 standard units) coinciding and the units on each axis determined so that a distance of one standard deviation on one axis is the same length on the other. In other words, on the axis of the histogram one standard deviation is 2.5 (inches), and on the normal curve axis one standard deviation is 1 (standard unit). Since 64 corresponds to 0 on the respective axes, $64 + 2.5 = 66.5$ corresponds to $0 + 1 = 1$ and $64 + 2(2.5) = 64 + 5 = 69$ corresponds to $0 + 2 \cdot 1 = 2$. The correspondence is presented in Figure 1.

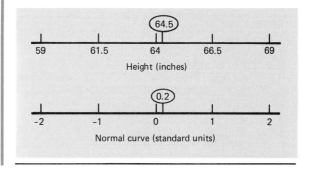

FIGURE 1

As soon as we establish that 64 (inches) corresponds to 0 (standard units) and 66.5 (inches) corresponds to 1 (standard unit), we have defined a correspondence between all the points on the axes. For example, if we wanted to determine the number to which 64.5 inches corresponds, we note that $64.5 - 64 = 0.5$, which is $1/5$ ($= 0.5/2.5$) the distance between 64 and 66.5, so the point on the standard unit axis to which it corresponds is $1/5$, which is $1/5$ the distance between 0 and 1. In other words 64.5 is $1/5$ or 0.2 standard units from the mean 64, because

$$\frac{64.5 - 64}{2.5} = \frac{0.5}{2.5} = 0.2$$

See Figure 1.

Once the correspondence is made between the horizontal axes, the correspondence between the histogram and the normal curve is completed because the vertical axes can both be considered as percentages. To approximate the area under the histogram between the value 64 and 66.5, we find the area between their corresponding values of the normal curve, 0 and 1. In the previous section this area was found to be 0.3413 or 34.13 percent. Therefore approximately 34 percent of the women in the United States are between 64 and 66.5 inches tall. (See Figure 2.)

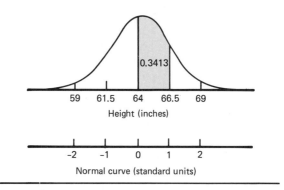

FIGURE 2 Using the normal curve to approximate the number of women whose height is between 64 and 66.5 inches. Thirty-four percent of the area under the normal curve is between 64 and 66.5.

This example demonstrates that there are two steps involved when using the normal curve to approximate areas under a histogram. First, the values on the horizontal axis must be converted to standard units. Second, the two points between which the area under the histogram is to be found are converted to standard units and the area under the normal curve is computed. This area is a percentage and therefore the approximation is this percentage.

The formula used to make the conversion from units of a set of data to standard units was described in Example 1. If x is a value on the horizontal axis of the histogram and we want to determine how many standard units it is from the mean, we first find the distance it is from the mean $\bar{x}$, namely, $x - \bar{x}$, and then divide this number by the standard deviation σ. This procedure is meant to express a value of raw data in terms of how far above or below the mean it is by computing the distance from the mean as a multiple of the standard deviation. Values above the mean have a plus sign, while values below the mean have a negative sign. These measures are usually called *standard scores* or *standard units*.

PROCEDURES FOR CONVERTING TO STANDARD UNITS: If a set of data has a mean $\bar{x}$ and standard deviation σ and if x is any value, then the value z in standard units that corresponds to x is determined by

$$z = \frac{x - \bar{x}}{\sigma}$$

Note that if z is negative then x is less than $\bar{x}$, and if z is positive x is larger than $\bar{x}$. This formula can be used to convert measurements into standard units no matter how the data are distributed. If the data are normally distributed, the values of z, called z scores, are sometimes referred to as *normal values*.

In the remainder of this section we will look at a few more examples of approximating percentages of areas under histograms of data by using the normal curve. The areas under the normal curve are percentages of the total area under the normal curve. This percentage becomes the approximation of the area under the histogram of the raw data.

Let us start with one more example from the Health Examination Survey.

Example 2

Let us assume that heights of women in the United States are normally distributed. In the Health Examination Survey mentioned in Example 1, the mean height was 64 inches with a standard deviation of 2.5 inches.

Problem   What percentage of women have a height between 61.5 and 64 inches?

Solution   The graph in Figure 2 will solve this problem. You can see intuitively that 61.5 is one standard deviation to the left of the mean, 64 inches. Hence 61.5 corresponds to -1 standard unit. We can derive this from the formula by letting $x = 61.5$, $\bar{x} = 64$, and $\sigma = 2.5$. This yields

$$z = \frac{x - \bar{x}}{\sigma} = \frac{61.5 - 64}{2.5}$$

$$= \frac{-2.5}{2.5} = -1$$

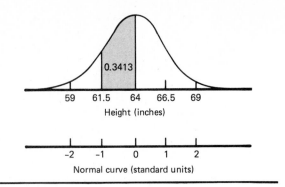

FIGURE 3   Thirty-four percent of women have a height between 61.5 and 64 inches.

From the previous section we know that the area under the normal curve between $x = -1$ and $x = 0$ is 0.3413. Hence our approximation is 34.13 percent. Therefore approximately 34 percent of the women have a height between 61.5 and 64 inches. See Figure 3.

Example 3

Many IQ (intelligence quotient) exams are designed so that the mean score is 100 and the standard deviation is 20. It is assumed that IQ scores are normally distributed.

Problem   What proportion of scores will fall (*a*) above 120, (*b*) above 148?

Solution   To convert the scores to standard units we let $\overline{x} = 100$ and $\sigma = 20$. For part (*a*) we let $x = 120$, and thus our formula for the corresponding standard unit yields

$$z = \frac{120 - 100}{20} = \frac{20}{20} = 1$$

and so 120 corresponds to 1 standard unit. In the previous section we found that the area under the normal curve to the right of 1 is 0.1587. Therefore we conclude that about 16 percent of the scores will fall above 120. This means that roughly 1 person in 6 has an IQ above 120.

For part (*b*) we let $x = 148$, and thus our formula for the corresponding standard unit yields

$$z = \frac{148 - 100}{20} = \frac{48}{20} = 2.40$$

and so 148 corresponds to 2.4 standard units. In other words, 148 is 2.4 standard deviations (2.4 · 20) from the mean (100). In the previous section it was stated that the area under the normal curve to the right of 2.4 is 0.0082. Therefore we conclude that about 0.82 percent of the scores will fall above 148. See Figure 4.

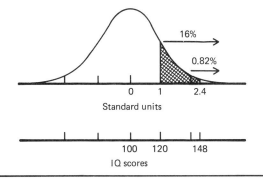

FIGURE 4

Example 4

When you decided to go to college, you probably took the SAT exam administered by the College Entrance Examinations Board. You were given your score and told how well you did compared with others who took the exam. The scale used by the College Board is a bit complicated, but we can use the normal curve to explain its meaning in part. A more detailed and thorough treatment can be found in an article by William H. Angoff.* All scores on the SAT are between 200 and 800. It is assumed that the scores are normally distributed in this range. Suppose that on your exam the mean was about 500 and a standard deviation was about 100† on the mathematics section.

Problem If your score on the mathematics section of the SAT was 600, what is the percentage of people who had a higher score than you?

Solution Since we assume that the scores are normally distributed, we can answer the question by determining the z score for 600 and then finding the area under the normal curve to the right of that z score. To compute the z score corresponding to 600 we use the formula, with $x = 600$, $\overline{x} = 500$, and $\sigma = 100$, which yields

$$z = \frac{600 - 500}{100} = \frac{100}{100} = 1$$

We previously determined that the area under the normal curve to the right of 1 is 0.1587. Therefore about 16 percent of the students scored higher than 600. See Figure 5.

---

*William H. Angoff, "How We Calibrate College Board Scores," *The College Board Review*, no. 68 (Summer), 1968.

†The average of a particular exam may differ from another given at a later date, so these numbers are approximations. They are fairly close though. The 1982–1983 mathematical SAT average was about 468.

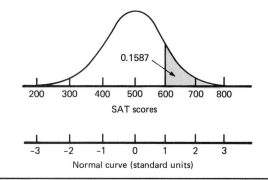

FIGURE 5   About 16 percent of the students had scores above 600.

Example 5

Problem   Suppose your score on the mathematics section of the SAT was 650. What percentage of people had a higher score than you?

Solution   Using the procedure outlined in Example 4, we compute the z score for 650, which is

$$z = \frac{650 - 500}{100} = \frac{150}{100} = 1.5$$

The percentage of people that had a higher score than you is approximately equal to the area under the normal curve to the right of 1.5, which, from the table, is 0.0688. Hence approximately 7 percent of the people scored higher than 650. See Figure 6.

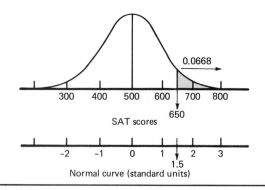

FIGURE 6   About 7 percent of the students had scores above 650.

We have prefaced each problem in this section with the caveat, "assume that the data are normally distributed." This means that the investigator who is studying the data knows from some previous experience, such as experiments done many times before, that the data are normally distributed. As we saw earlier, often data sets exhibit other distributions, such as a uniform, a skewed, or a bimodal distribution.

To further explain our procedure in this section, consider the case when a researcher generates data and assumes it is normally distributed when in fact it is not!

Example 6 (Step 1: The Error)

A researcher investigates incomes in a particular city. Data are gathered and fed into a computer, which finds the mean to be $\bar{x} = 23.75$ (in thousands of dollars) and the standard deviation to be $\sigma = 10$. In order to make some inferences about the data the researcher assumes the data are normally distributed.

Problem Calculate the percentage of the data that is between $x = 15$ and $x = 20$ (thousands of dollars) by using the approximation of the data by the normal curve, with the given mean and standard deviation.

Solution We have, for $x = 20$,

$$z = \frac{20 - 23.75}{10} = \frac{-3.75}{10} = -0.375$$

and for $x = 15$ we have

$$z = \frac{15 - 23.75}{10} = \frac{-8.75}{10} = -0.875$$

The area under the normal curve between $z = 0$ and $z = -0.38$ (where we have rounded off -0.375) is 0.1480, and between $z = 0$ and $z = -0.88$ is 0.3106. Hence the area between $z = -0.38$ and $z = -0.88$ is 0.1626 ($= 0.3106 - 0.1480$). Therefore there is about 16 percent of the data between $x = 15$ and $x = 20$. Thus the researcher concludes that about 16 percent of the residents of the city have incomes between \$15,000 and \$20,000.

How accurate is the conclusion? The accuracy depends upon how close the data distribution is to the normal distribution. We discovered earlier that income often is not distributed normally but rather has a distribution that is skewed to the right. To determine which type of distribution fits, we must construct the histogram.

Example 7 (Step 2: Realizing the Error)

The assistant of the researcher in the previous example wonders if the assumption that the data are normally distributed is correct. After data are listed in tabular form by using eight intervals as in Table 1, the mean $\bar{x}$ and the standard deviation σ are calculated and the histogram is constructed in Figure 7. Notice that it is skewed to the right.

Problem From the table determine the percentage of individuals who have an income between \$15,000 and \$20,000. Was the magnitude of the error made by the assumption that the data was normally distributed?

Solution It's easy! Directly from the table we see that 25 percent of the data is between $x = 15$ and $x = 20$. The error made by the researcher in assuming the data to be normally distributed was about 9 percent ($= 25\% - 16\%$).

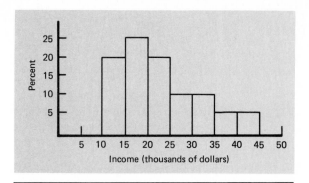

FIGURE 7

TABLE 1

Income Thousands of Dollars	Percent P	Midpoint x	Px	x^2	Px^2
10–15	20	12.5	2.50	156.25	31.25
15–20	25	17.5	4.375	306.25	76.5625
20–25	20	22.5	4.50	506.25	101.25
25–30	10	27.5	2.75	756.25	75.625
30–35	10	32.5	3.25	1056.25	105.625
35–40	5	37.5	1.875	1406.25	70.3125
40–45	5	42.5	2.125	1806.25	90.3125
45–50	5	47.5	2.375	2256.25	112.8125
			23.750		663.7500

Mean $\bar{x} = 23.75$ $\sigma = Px^2 - \bar{x}^2 = 99.6875 \approx 10$

EXERCISES

In Problems 1 to 4 convert the horizontal axis for the raw data into a horizontal axis for the normal curve in standard units.

1 $\bar{x} = 50, \sigma = 10$

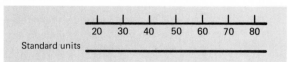

2 $\bar{x} = 100, \sigma = 10$

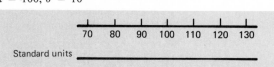

3 $\bar{x} = 15,\ \sigma = 3$

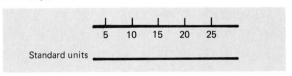

4 $\bar{x} = 22,\ \sigma = 4$

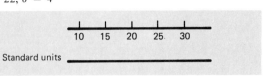

In Problems 5 to 8 determine the mean $\bar{x}$ and the standard deviation σ of the data, given that the corresponding horizontal axis for the normal curve was constructed immediately below the axis for the data.

5

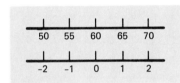

6

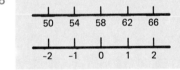

7

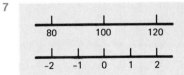

8

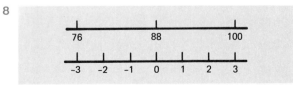

In Problems 9 to 14 construct a horizontal axis for the data showing the mean $\bar{x}$ and the points that are one and two standard deviations above and below $\bar{x}$, and then draw the horizontal axis for standard units immediately below.

9   $\bar{x} = 10, \sigma = 3$ 12 $\bar{x} = 72, \sigma = 5$

10   $\bar{x} = 10, \sigma = 5$ 13 $\bar{x} = 59, \sigma = 2.5$

11   $\bar{x} = 64, \sigma = 4$ 14 $\bar{x} = 800, \sigma = 63$

In Problems 15 to 22 use the normal approximations to determine the percentage of data between the values $x = x_1$ and $x = x_2$, where the data have the given $\bar{x}$ and σ.

15   $\bar{x} = 10, \sigma = 5, x_1 = 10, x_2 = 15$

16   $\bar{x} = 20, \sigma = 5, x_1 = 15, x_2 = 20$

17   $\bar{x} = 15, \sigma = 2, x_1 = 13, x_2 = 17$

18   $\bar{x} = 21, \sigma = 3, x_1 = 18, x_2 = 27$

19   $\bar{x} = 100, \sigma = 10, x_1 = 95, x_2 = 110$

20   $\bar{x} = 110, \sigma = 30, x_1 = 95, x_2 = 120$

21   $\bar{x} = 810, \sigma = 50, x_1 = 790, x_2 = 850$

22   $\bar{x} = 550, \sigma = 75, x_1 = 525, x_2 = 600$

In Problems 23 to 26 use the normal approximation to determine the percentage of women in the United States whose height is between the values $x = x_1$ and $x = x_2$, where $\bar{x} = 64$ and $\sigma = 2.5$.

23   $x_1 = 61.5, x_2 = 66.5$ 25 $x_1 = 64, x_2 = 65$

24   $x_1 = 59, x_2 = 64$ 26 $x_1 = 60, x_2 = 67$

In Problems 27 to 30 use the normal approximation to determine the percentage of individuals whose IQ are between the values $x = x_1$ and $x = x_2$, where $\bar{x} = 100$ and $\sigma = 20$.

27   $x_1 = 120, x_2 = 130$ 29 $x_1 = 125, x_2 = 130$

28   $x_1 = 120, x_2 = 150$ 30 $x_1 = 130, x_2 = 160$

In Problems 31 to 34 use the normal approximations to determine the percentage of individuals who had the given scores on the SAT exam, where $\bar{x} = 500$ and $\sigma = 100$.

31   Between 500 and 550   33   Over 675

32   Less than 600   34   Over 750

8-6 SAMPLE SURVEYS

One of the most important uses of the science of statistics is the sample survey. An investigator—say, a polling agency such as the Gallup poll—wants to find out information about a large set of people, called the *population.* You can find articles in almost every daily newspaper that quote a major polling firm that supports the writer's contention. This is especially true during national elections, where whole articles are based on the results of polls showing which candidate is ahead and which ones are gaining or losing ground.

The polls attempt to predict who would win the election if it were conducted at the time that the survey is given. The results of the surveys, especially when compared with previous ones, help the candidates determine their strengths and weaknesses. Thus they are watched very carefully, not only by the public but by the politicians themselves. In fact, most major candidates hire their own polling agencies to determine the preference of the electorate.

Do the pollsters interview the entire population—in the case of an election, all the eligible voters? Of course not. Such a task would not only be inordinately expensive but also too time-consuming to produce up-to-date results. Therefore it is necessary to interview a small *sample,* a subset, of the entire population. The investigator then makes a generalization about the population from the concrete facts ascertained from the sample. We say that the investigator makes an *inference* about the whole population from a relatively small part.

The investigator must ensure that the sample is chosen so that it is very much like the whole population. For example, if you wanted to determine the average height, weight, and salary of the adult population in the United States, it would be a grave mistake to choose as a sample the members of the National Football League Players Association. How about the adult population in the town where you live? You would probably get a better sample as far as height and weight goes, but perhaps you live in a relatively affluent or relatively poor neighborhood.

We mentioned earlier three numerical measurements, height, weight, and salary, of a population. Numerical facts about a population are called *parameters.* If an investigator wants to predict a presidential election, the most important parameter is the percentage of eligible voters who will vote for a particular candidate. A sample is chosen and numerical facts are measured exactly in the sample, in which case the facts are called *statistics.* The investigator then estimates the value of the parameters, using the exact value of the statistics. Therefore statistics calculated from the sample are used to estimate the parameters of the population.

How should a sample be chosen? Are there specific procedures or merely guidelines? Should names be chosen at random, say from a phone book, or should particular individuals be identified and then interviewed?

When Walter Cronkite declared Ronald Reagan the next President of the United States on November 4, 1980, he did not have in front of him a record of all the votes of the electorate. In fact he had only a very small percentage, somewhere in the neighborhood of 30 percent, of the votes counted. He took a sample of the total population, those who actually voted, and made an inference from his knowledge of the sample of the population.

There is immense pressure on the networks to be first in their prediction. The higher the ratings the bigger the advertising pot becomes. This desire to be first must be weighed against the possibility of error. If a network predicts, on the basis of returns, that Tito Schwartz will win the senatorial race, and then later, voting goes differently than expected and Tito loses, no one will remember that their prediction came first. The egg on their face could be permanent. Such stories are not rare. Perhaps the most glaring error ever committed via sampling was that of the *Literary Digest*.

The *Literary Digest*

To most observers in 1936, the outcome of the presidential election was clear. Franklin Delano Roosevelt, the incumbent Democrat, was running against Alfred Landon, the Republican governor of Kansas. The campaign issues were red-hot items and the candidates held vastly different views. The economy of the country was devastated by the great depression. Roosevelt had already started his New Deal program, using massive monetary intervention of the federal government to create jobs, whereas Landon called for a more laissez faire approach, claiming that the federal government itself was a major cause of the problem. Roosevelt's pitch appeared to reach more individuals. There were almost 10 million unemployed, and rampant inflation eroded the real income of most of those who were working. Landon's approach would have little immediate effect on the average worker. Thus most people assumed Roosevelt would win handily.

The *Literary Digest* was a very popular and influential magazine in the 1920s and 1930s. Its prestige rivaled that of *Time* magazine today. The *Digest* had correctly predicted every presidential election since 1916. The magazine conducted another survey in 1936 to determine who would win the presidential election. It was to be the largest and most extensive survey of its type ever conducted. Almost 2.4 million responses were gathered. The result of the poll was clear—it would be Landon in a landslide. The survey showed 57 percent would vote for Landon and 43 percent for Roosevelt. One month later Roosevelt won the election, 62 percent to 38 percent. Soon thereafter the *Literary Digest* went bankrupt.

In the months prior to the election, articles in the *Digest* boasted about how thorough their polling techniques were. They took great pains to ensure that their sample—10 million strong—would accurately represent the voting public. They were so sure of themselves that they maintained that not only would their poll predict the winner but also the actual percentage of victory would be accurately determined.

How could such a prestigious magazine have committed so staggering an error? The answers to this question provide the framework for the present methods of polling and sampling.

When is an investigator justified in selecting a sample rather than surveying the total population? The only reasons for drawing a sample are economics and inconvenience. It would certainly not be economically feasible, nor even possible, to poll the total voting populace in 1936.

How large should the sample be? This is a very sticky problem that always faces the statistician. Too small a sample will not give adequate results. Is there such a thing as too large a sample? Here is where inconvenience enters into the problem. The *Literary Digest* erroneously assumed that "the bigger the better." As it turned out, it was far too difficult for the *Digest* to control an initial mailing list of 10 million. They thought they could randomly select a typical cross section by choosing names from telephone books and club memberships. Today selecting names from a telephone book is not a bad way to obtain a random sample, but in 1936 it was a horrendous procedure. Do you see why? Even today the method of selecting names from club membership lists screens out the poor and those who don't feel the inclination to belong to a club, which is a significant segment of our society. The reason why the *Digest* selected these procedures is that they provided a convenient means of obtaining names, telephone numbers, and addresses. However, in 1936 many people could not afford telephones. This procedure had worked in the past (recall that the *Digest* had accurately predicted the presidential elections since 1916) because the rich and poor had voted similarly. In 1936 Landon's platform appealed to the rich, whereas Roosevelt's appealed to the poor, so the *Digest's* techniques of polling excluded a large portion of Roosevelt supporters.

The mistake that the *Digest* made is now referred to as *selection bias;* that is, the procedure used to select the members of the sample was biased in that, by its very nature, it drew upon the preponderance of individuals favoring one outcome. In this case many more rich than poor people were contacted, and they tended to favor Landon.

If a survey has selection bias, the size of the sample makes no difference. The error will be magnified as more individuals are selected. It is like a coat manufacturer whose goods are priced so low that $1 is lost on every coat sold. The manufacturer will never recoup the losses by selling more coats.

> When a sample is chosen, every effort must be made to avoid selection bias, the systematic tendency to exclude a significant segment of the population.

In Figure 1 we give a pictorial representation of how the *Digest* sample did not represent the whole population. Clearly more individuals were interviewed who favored Landon as opposed to those who favored FDR.

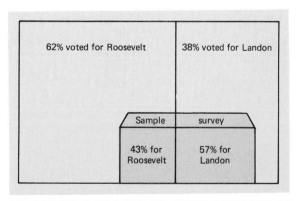

FIGURE 1 The *Literary Digest* survey polled too many people who favored Landon. The selection process was skewed toward Landon.

62% voted for Roosevelt

38% voted for Landon

Sample survey

43% for Roosevelt

57% for Landon

Even though this mistake of the *Digest* poll was significant enough to cause an error of massive proportions, their procedure had another grave error in sampling. A large number of those chosen to be interviewed refused to respond. Remember that 10 million were identified and only 2.4 million responded, so the *Digest* poll had 7.6 million nonrespondents. It has been shown that often nonrespondents differ from respondents in significant ways. It is not correct to assume that the nonrespondents are similar to those who respond. In fact, they are different in one obvious way—they did not respond! Often the reasons for not responding are significant. For instance, could it be that the majority of the nonrespondents in the *Digest* poll actually voted for Roosevelt and did not respond because they felt that the outcome was obvious? If so, the *Digest* poll would have come much closer to predicting the outcome of the election correctly if the results of the nonrespondents' views had been included. Thus even if a survey organization has identified a representative subset of the population, if the response rate is too low a serious distortion can be introduced. This is called *nonresponse bias*.

When a sample is chosen, every effort must be made to avoid nonresponse bias, the effect of too many individuals earmarked for the sample becoming nonrespondents.

Let us demonstrate the effect of nonresponse bias with a simple example.

Example 1

A survey organization is asked to determine whether Mayor Brown should run for governor, necessitating that he resign as mayor. The organization sends detailed questionnaires to a sample of 1400 individuals, selected scientifically, throughout the state in order to assess the mayor's chances. Responses from 1000 were received, with 550 in favor of the mayor's candidacy and 450 against.

Problem 1 Suppose 80 percent of those who did not respond are not in favor of his candidacy. How would the inclusion of the nonrespondents have affected the outcome of the poll, if they had responded?

Problem 2 Why might so high a percentage of the nonrespondents not favor his candidacy?

Solution 1 There are 400 nonrespondents and 320 ($= 80$ percent $\cdot$ 400) do not favor his candidacy. Hence a total of 770 ($= 450 + 320$) do not favor his candidacy, while only 630 ($= 1400 - 770$) favor his candidacy. The inclusion of the nonrespondents' views drastically affects the outcome of the poll. Without the nonrespondents included, it seems clear that he should resign and run for governor. With the nonrespondents included, he would be facing an uphill battle, which may not be worth the gamble.

Solution 2 There are many legitimate reasons why the percentage could actually be as high as assumed. Many people in another part of the state may never have heard of him and therefore would not only decline to vote for him but would also feel that filling out a questionnaire would be a waste of time. Also, those who are ardently in favor of his candidacy are much more likely to be amenable to taking time to respond.

In the *Digest* poll, each type of error tended to exclude the poor from the count, and they primarily supported Roosevelt.

We have identified two taboos to be avoided—selection bias and nonresponse bias—but we still have the question before us: How does an investigator go about choosing a representative sample while avoiding these errors? It is clear that the investigator must make sure that the sample is similar to the population in terms of the most significant characteristics. This might seem paradoxical at first because it implies that a random sample cannot be entirely random if the polling agency is allowed to earmark certain groups within the population to be interviewed. Some surveys, however, especially large ones, must be conducted in this manner because a sample chosen randomly, without bias, is impossible to achieve within the legitimate constraints of time and money.

Polls Elect Dewey, People Elect Truman

By 1948 the polls had achieved a degree of sophistication that allowed them to accurately predict any important election, whether it was a federal or state race, or so they thought. They recognized that a procedure that appears to be random may have serious flaws of design, as did the *Literary Digest* poll of 1936. It has been thoroughly substantiated that particular types of groups tend to vote certain ways. The classic study in this area is Berelson et al., "Voting" (cited in Section 7-4). They separated the voting populace into groups by three major areas: (1) occupation, income, and status; (2) religion, race, and ethnicity; (3) place of residence (regional and urban-rural divisions). They studied, in depth, the voting populace of the city of Elmira, New York. Certainly no one would attempt to suggest that one city is a valid sample for the whole country. One obvious fact that dispels such a notion is that Elmira is in the north, and northern and southern voters, even in the same political party, behave differently. Nonetheless, the primary conclusion of the Berelson study that certain group types tend to vote certain ways is an accepted fundamental assumption in forecasting elections.

In 1948 the major polls understood this concept to a certain extent. For example, they realized that they must accurately determine what percentage of each significant minority would vote for Truman, but the urban vote might differ substantially from the suburban and rural votes. Thus the sample chosen must accurately resemble as many significant group types as possible within the constraints of time and money.

By 1948 polling agencies had improved the process of drawing a sample to such an extent that individual interviewers were instructed to question a very restricted set of types of people. For example, a typical interviewer for Gallup was told to contact 15 people, of which 8 were to be male and 7 female. Of the males, 2 were to be black and 6 white. Exactly 8 were to live in the suburbs and 7 in the city. Other interviewers were given equally precise instructions even though the specific quotas may have been slightly different. The individual quotas were assigned so that the sum of persons interviewed matched the polling agency's determination of a typical cross section of the population. The percentage of representation of each category (such as sex, age, residence) in the sample was equal to the percentage of the category in the population. The specific percentage of the sample can be accurately determined from census data.

Therefore the polling agencies brimmed with confidence as the election approached—right? Right! They accurately predicted the election to within a few percentage points—right? Wrong! They blew it again!

Harry Truman was the incumbent but he was not an elected president. In 1944, when Truman was serving in the Senate, FDR selected him as his running mate. Even though many Democratic leaders felt that Roosevelt might not live through his fourth term because of his poor health, and thus the choice for the vice president might very well succeed to the Presidency, Truman was nonetheless a compromise candidate, as the majority of the party was split between incumbent Vice President Henry A. Wallace and Supreme Court Justice William O. Douglas. Late in the afternoon of April 12, 1945, Mrs. Eleanor Roosevelt summoned Truman to the White House to tell him that FDR had died.

Truman assumed the Presidency amid the chaotic atmosphere of the end of World War II. His was a stormy administration, and by 1948 it seemed impossible for him to win a full term by election. Even though he won the nomination on the first ballot, the Democratic party was sharply divided in loyalty. Two separate factions developed, one supporting Wallace and the other favoring Strom Thurmond.

On the other hand, the Republicans were united and confident. They nominated the seasoned governor of New York, Thomas Dewey, and chose the popular Governor Earl Warren of California, later to become Chief Justice of the Supreme Court, as his running mate.

After the nominations the candidates and the polls went to work. Soon it appeared that the prediction would be a cinch. Dewey's appeal was convincing—a breath of fresh air and a new start from the impending and abrupt end of the long Democratic hold on the White House since 1932. The Democratic party seemed unable and unwilling to counteract the trend. The three major polls, Gallup, Roper, and Crossley, all predicted a significant margin of victory for Dewey. Everyone was confident in a Dewey victory, except the fiery Truman. The *Chicago Tribune* even had its morning headline on election day prepared and released early in order to snatch a "scoop." The predictions are recorded in Table 1.

TABLE 1

Candidates	Predictions of the Polls		
	Gallup	Roper	Crossley
Dewey	50	53	50
Truman	44	38	45
Wallace	4	4	3
Thurmond	2	5	2

Not only did Truman win with 50 percent of the vote as opposed to 45 percent for Dewey, but one of the biggest upsets in political history was graphically represented by the electoral vote landslide for Truman, who won 28 states to only 16 for Dewey and 4 for Thurmond.

Where did the polls go wrong? It certainly seemed as though the technique of sampling by quotas had the sampling process nailed down. A great deal of study concerning sampling ensued as a result of the 1948 election. One of the most thoughtful and thorough descriptions of the shortcomings of the polls is contained in "The Pre-election Polls of 1948," by F. Mosteller and others (Social Science Research Council, New York, 1949). They pointed out that while the system of quotas itself is not faulty, it did not go far enough in reducing bias to a minimum because it allowed the interviewers too much freedom within the categories. That is, once the category as defined, say, a black male living in a medium-sized city with an average income, the interviewer could choose anyone he or she wanted. This does not seem too critical, and it was deemed totally insignificant in 1948, but note that this freedom allows the interviewer to select only those people with whom he or she feels comfortable talking. Within any specific demographic category there are those who will vote Republican and those who will vote Democratic. The assumption made by the polling agencies in 1948, and previously in those years when the quota system was practiced, was that the interviewers would randomly select individuals and thus an accurate sample would be chosen. Not so! The interviewers tended to select a predominantly large percentage of Republicans. Why? It is simple. Those who tend to vote Republican, especially since 1936, are, on the whole, better educated and wealthier than Democrats. Thus they live on nicer streets and tend to speak their mind more freely, in short, they are somewhat easier to interview. It was much more comfortable and convenient for the interviewers to single out this type of person, and all too often the respondent was Republican.

Mosteller et al. point out this tendency in graphic terms by comparing Gallup's prediction of the Republican vote with the actual percentage in the four elections from 1936 to 1948 (see Table 2). Note that in each year the Gallup prediction of the Republican vote is higher than the actual vote. This is clear evidence that too many Republicans were singled out within the somewhat strict categories. In each election until 1948 the actual Democratic plurality was large enough so that the error was not so great as to lead Gallup to predict a Republican victory. But in 1948 the election was close enough that he predicted Dewey would win; the pollsters set the stage for the miraculous Truman upset victory.

Let us review the shortcomings of the 1948 sampling procedure, called *quota sampling*. The population was divided into strict demographic categories in order to ensure that each distinct significant characteristic group was

TABLE 2

| | Republican Vote | | |
Year	Gallup's Prediction	Actual Vote	Error
1936	44	38	6
1940	48	45	3
1944	48	46	2
1948	50	45	5

represented. The percentage of those interviewed from each quota exactly matched the percentage that the group occupied in the total population of eligible voters. The concept of establishing quotas so that the sample closely match the population in its demographic character was worthwhile, but the method of filling the quotas by the interviewers proved disastrous. Too much room was left for human choice, and human choice is almost always biased.

Probability Methods: A Return to Urn Problems

Since the great debacle of 1948, virtually every survey organization has avoided the bias due to the free choice by interviewers in filling the quotas by using chance methods. The idea is to eliminate any individual choice by using a *simple random sample,* that is, a sample that is selected in very much the same way that we intuitively selected marbles from urns in the earlier sections on probability. When selecting a marble "at random," we assumed that every marble had an equally likely chance to be drawn as any other one in the urn. If there were n marbles in the urn, we assigned a probability of being selected of $1/n$ to each marble.

Let us draw an even closer parallel between the sampling problem of filling quotas and urn problems.

Problem—Voter: There are 1000 eligible voters in a particular locale that fit the description of a specific quota. A sample of 100 is to be selected. What method of selection should be used?

Problem—Urn: There are 1000 marbles in an urn; some are red and some are white. A sample of 100 is to be drawn in order to make a guess as to what percentage of the marbles are red. What method of selection should be used?

In order to solve Problem—Urn it is not enough to simply reach in and draw 100 marbles, since you would most likely be selecting only those at the top. If the white marbles were put into the urn first and then the red, your sample would have far too many red marbles and would not be a representative sample. To draw a simple random sample you must first thoroughly mix the marbles so that on the first draw every marble has an equally likely opportunity of being drawn. Once the first draw has been made, the marbles should be thoroughly mixed again and the second marble drawn, with each remaining marble having an equally likely chance of being drawn. Note that all choice has been removed from the process. In this way a simple random sample is selected. It is very likely that the percentage of red marbles in the sample is very close to the percentage in the urn.

Problem—Voter is solved in very much the same way. Whenever possible the names are placed in a hat (which is very similar to an urn) and 100 are selected at random. This guarantees that the sample is chosen at random and all bias is eliminated.

The Practical Side

In large elections, where the number of names in the hat approaches 100 million, this process is not feasible. Even if it were possible to select an appropriate sample, it would be far too costly to contact them, as they would be scattered throughout the country. Therefore polling agencies have developed alternate means to draw a sample while ensuring that it is random by employing chance methods whenever possible.

One of the most common methods, albeit one of the most complicated,* is called *multistage cluster sampling.* The idea is to select the sample in stages, and at each stage a chance mechanism is used to form a group, or cluster, of entities from which the next stage of selection is conducted.

For example, the first stage in a national survey is to divide the country into regions, usually four—the west, the south, the midwest, and the northeast. A survey team is sent into each region. Within each region, the team groups together all the geographic areas with similar population. For instance, the population centers might be grouped as follows: under 20,000; between 20,000 and 50,000; between 50,000 and 100,000; between 100,000 and 500,000; over 500,000. The names of the population centers within each group are then placed in a hat and a random sample is drawn by using chance methods. The locales selected are the only ones into which interviewers are sent.

The next stage involves dividing each locale into *wards* and drawing a random sample of wards. Then each ward is divided into *precincts,* and a random sample of precincts is selected from each of the previously selected wards. The next stage consists of drawing a random sample of households from the targeted precincts. In this way specific households are identified in a random fashion that avoids every type of bias. The households are in specified geographic regions, and thus the interviews are manageable in terms of time and cost, while the regions are selected at random.

It is interesting to note that even when an interviewer is assigned a particular household, no discretion at all is allowed in the selection of the individual to interview within the household. For instance, the interviewer might be told to interview the oldest male who is an eligible voter or the youngest female who is an eligible voter. Therefore all bias due to free choice has been eliminated with this sampling design.

The Bottom Line— Does the Method Work?

"Fine!" you say; we now have a real complicated method for drawing a sample that appears to be cost-effective, but how do we know that it works better than the other methods? How do we know that this method of selection yields a sample that is similar, with regard to voting behavior, to the whole population? In other words, if you have an urn that contains 1000 marbles and you select a simple random sample of 100 consisting of 60 red and 40 white, you still are not guaranteed that there are close to 600 red marbles in the urn. In fact, you might have reached in and selected the only 60 red marbles in the urn. Here is where the study of election polls is so rewarding—soon after the prediction is made, the results are obtained. This is often not the case in other surveys. Therefore the election polls furnish a proving ground for sampling techniques.

One of the first survey organizations to utilize probability models, that is, chance methods for choosing a sample, was the American Institute of Public Opinion, founded in 1935 by George Gallup for the "purpose of determining the public's views on the important political, social, and economic issues of the day." The surveys, referred to as the Gallup poll, have been continuous reports that distribute to the press and private industry "what people think,

---

*For a thorough and well-written description of this method, see D. Freedman et al., "Statistics," W. W. Norton, New York, 1978, chapter 22.

TABLE 3

The Gallup Poll Record in Presidential Elections Since 1948

Year	Sample Size (Approx.)	Winning Candidate	Gallup Poll Prediction for Winner, %	Election Result for Winner, %	Error, %
1952	5400	Eisenhower	51	55.4	+4.4
1956	8100	Eisenhower	59.5	57.8	−1.7
1960	8000	Kennedy	51	50.1	+0.9
1964	6600	Johnson	64	61.3	+2.7
1968	4400	Nixon	43	43.5	+0.5
1972	3700	Nixon	62	61.8	−0.2
1976	3400	Carter	49.5	51.1	−1.6

not merely what they do."* In 1936 Gallup used a sample of about 50,000 and predicted Roosevelt's victory. However, his methods were not perfect, as he predicted FDR's percentage to be 56 percent, whereas the actual count was 62 percent. Again in 1948 Gallup was off by a similar small amount, about 5 percent, but this time, as we mentioned earlier, the error led him to predict that Dewey would defeat Truman.

Since 1948 the Gallup poll has been perfecting the sampling technique of multistage cluster sampling, which relies on probability methods. The results, recorded in Table 3, are convincing.

Summary

In this section we discussed the maturing process of sampling techniques used by survey organizations in order to give you a capsule view of one aspect of the ascent of statistics—the opinion poll. Opinion polls have become commonplace in industry, and they have developed into a creative arm of our government. Their growth from ineffectual guesses to scientific instruments whose measurements survive the acid test of election night can be seen through their monumental gaffes and subsequent restructuring presented in this section.

Statistics as a mere gathering of facts dates back thousands of years, with governments taking censuses of the citizenry to assess taxes. Statistics did not emerge as a science until the 1800s. Karl Pearson, a British geneticist, is known as the founder of the science of statistics for his work in applying statistical methods to biological data. In 1892 he wrote the "Grammar of Science," which is regarded as a classic textbook on the scientific method.

While statistics is used in diverse fields such as business and industry, sciences such as biology and agriculture, and social sciences such as political science and sociology, never is it more useful than when it is applied in sampling. Often the whole population cannot be measured, so an inference must be made from a relatively small subset, a sample. It is risky to draw conclusions

---

*George Gallup, "Opinion Polling in a Democracy," in J. M. Tanur et al. (eds.), "Statistics: A Guide to the Unknown," Holden-Day, San Francisco, 1972, pp. 146–152.

from a small number of observations; yet it is usually impossible to measure all cases. Statisticians can never be completely certain that their samples accurately reflect the measurements of the population, so they must rely on a proven method of selection when drawing the sample. But if the whole population can never be measured, how does the investigator know that the method is reliable? Here is where the sample surveys applied to presidential elections enter the picture. It provides us with one of the few incidents where the prediction can be compared with the results. The reliability of the sample can be accurately judged.

The improvement of sampling techniques is clear from the results of the election polls. In 1936 the major poll had an error of 19 percent. For the seven national elections between 1936 and 1948 (including the off-year elections when some congressional seats are contested but not the Presidency), the Gallup poll had an average error of 4 percentage points. For the 11 national elections after 1948, the average error of the Gallup poll was 1.6 percentage points.

Answers to Odd-Numbered Exercises

Chapter 1

Section 1-1

1 23

3 155

5 150

7 7

9 133

11 3126

13 ||| ∩
||| ∩∩

15 |||| ∩∩ 999
||||| ∩∩∩ 9999

17 || 99
||| 99

19 (numeral symbol)

21 (numeral symbol)

23 (numeral symbol)

25 (numeral symbol)

27 (a) ∩∩∩ 99 ; (b) Exchange each numeral for the next highest value; (c) ∩ 99 (symbols)

29 Append a zero to the number.

31 (a) (numeral symbol) ; (b) the same; (c) the context

33 No. The numeral for 366 is (numeral symbol) .

35 1983

37 Year, month, day

39 (c)

Section 1-2

1 1801

3 6126

5 (numeral symbol)

7 (numeral symbol)

9 (numeral symbol)

11 (a) 10; (b) 105; (c) 1805

13 Two hundred three

15 Two thousand one

17 Hindu-Arabic: 10,018; Egyptian:

19 Hindu-Arabic: 1,200,304; Egyptian:

21 123,450

23 26,512

25 101. The two 1's have different meanings, depending upon their position.

27 (a) || $\mathcal{9}$;(b) No

29 (a) 5; (b) 5; (c) 0; (d) 0; (e) undefined

31 20 for both

33 10

35 It is positional.

37 (a) No; (b) the numbers from 13 to 19; (c) the numbers from 20 to 99; (d) 20

39 (c)

Section 1-3

1 $(8 \times 10^4) + (4 \times 10^2) + (2 \times 10) + (1 \times 10^0)$

3 1,000,000

5 81

7 729

9 $81 \times 729 = 59,049$

11 $729/81 = 9$

13 $a^m/a^n = a^{m-n}$

15 $3^{12} = 81 \times 81 \times 81 = 531,441$

17 3^5

19 1110

21 111010

23 10110100

25 10

27 74

29 277

31 101111

33 110001

35 10111110100

37 Yes

39 (a) 10^4 (b) 10^5

41 The number of zeros is equal to the exponent.

43 (c)

Section 1-4

1	Divisible by 2	19	Divisible by 4
3	Divisible by 2, 5, and 10	21	Divisible by 4 and 8
5	Divisible by 5	23	$2^4 \times 5 \times 31$
7	Not divisible by 2, 5, or 10	25	$2 \times 3^2 \times 5^2 \times 89$
9	Divisible by 2, 5, and 10	27	$2 \times 3^4 \times 5 \times 7 \times 11$
11	Divisible by 3	29	$2^3 \times 11 \times 13^2$
13	Divisible by 3	31	28
15	Divisible by 3 and 9	33	284
17	Divisible by 4 and 8	35	The first

37 Yes. The digital sum, 15, is divisible by 3.

39	Quotient	49	53
41	$2520 = 5 \times 7 \times 8 \times 9$	51	True
43	$11 + 7$ and $13 + 5$	53	False
45	175,560	55	True
47	16	57	True

Section 1-5

3 666

5 (*a*) and (*d*)

7 $T_8 = 36$

9 $T_{12} = 78$

11 $(1 + 2 + 4 + 8 + 16) \times 16 = 496$

13 The digital root of 6 is 6. The digital root of the other perfect numbers is 1.

15 $T_1 + T_2 = 4 \qquad T_2 + T_3 = 9 \qquad T_3 + T_4 = 16$
$T_4 + T_5 = 25 \qquad T_5 + T_6 = 36 \qquad T_{n-1} + T_n = n^2$

17 It works for all *n*. In general, $T_n = T_{n-1} + n$.

19

Number of points	5	6
Number of regions	16	31

This problem deals a blow to our human tendency to draw general conclusions from patterns. We expect 6 points to produce 32 regions, but only 31 are possible. If you do not believe us, try drawing 32.

21 The fifth and sixth perfect numbers end in 6.

23	Yes	29	1210
25	No	31	76,084
27	Yes	33	6232

Section 1-6

1 The 19 numbers between 888 and 906

3 (a) $1, 2, 2^2, 2^3, 2^4, 2^5$; (b) 6, 11, 100; (c) $n + 1$

5 $Z_2 = 8, Z_3 = 26, Z_5 = 242, Z_7 = 2186$. Pattern: if p is prime, then Z_p is composite. There are no Z-primes.

7 $5^3 = 15^2 - 10^2$ and $6^3 = 21^2 - 15^2$

9	True	15	True
11	True	17	False
13	True	19	100 divisors

Chapter 2

Section 2-1

1	$\frac{2}{11} = \frac{1}{6} + \frac{1}{66}$	9	$\frac{2}{15}$
3	$\frac{2}{19} = \frac{1}{12} + \frac{1}{76} + \frac{1}{114}$	11	$\frac{1}{30}$
5	$\frac{2}{7} = \frac{1}{4} + \frac{1}{28}$	13	$\frac{7}{30}$
7	$\frac{2}{13} = \frac{1}{7} + \frac{1}{91}$	15	$\frac{1}{12}$

17 The fraction $2/n$ can be reduced to a unit fraction when n is an even number.

19	$\frac{2}{13}$	29	$\frac{1}{14} + \frac{1}{a}$, where $a > 210$
21	$\frac{7}{10}$	31	(a) 1; (b) 1
23	$\frac{3}{7}$	33	9
25	(a) $\frac{3}{4} - \frac{12}{44} = \frac{21}{44}$; (b) yes	35	67
27	$\frac{9}{20}$	37	$\frac{1}{10}$ mg

39 Between $67\frac{1}{2}$ and $70\frac{1}{2}$ inches

41 Fair is $\frac{48}{5}$ minutes per mile; good is 8 minutes per mile.

43 $23\frac{1}{4}$ inches

45 Approximation; 7 feet per mile = 1960 feet/280 miles.

47 $\frac{1}{2}$

Section 2-2

1	$255/256$	17	$\frac{2}{3}$
3	$99,999,999/100,000,000$	19	0
5	9840	21	$\frac{1}{9}$
7	$\frac{15}{16}$	23	3
9	$\frac{63}{64}$	25	Diverges
11	$\frac{15}{16}$	27	Diverges
13	$\frac{15}{16}$	29	Diverges
15	$\frac{1}{9}$	31	$1/256$

33 (*a*) 5; (*b*) $\dfrac{1-1}{1-1} = \dfrac{0}{0}$ is undefined; (*c*) The formula is not true when $r = 1$.

35 About $5

37 (*a*) $r = 7, n = 5$; (*b*) 19,607

39 $1/\sqrt{2}$

Section 2-3

1	16.104	13	0.75
3	0.0281	15	$1023/10,000$
5	843 ⓪ 1 ② 7 ③	17	$\frac{7}{9}$
7	1 ① 9 ③	19	$\frac{4}{11}$
9	$0.\overline{09}$	21	22.156
11	$0.\overline{142857}$		

23 (*a*) 11 hours 51 minutes 11 seconds; (*b*) 7 minutes 7 seconds per mile (amazing!); (*c*) 44

25 20.6

27 4 minutes 55 seconds

29 2 hours 27 minutes 20 seconds

31 Montreal's pct. is correct, but Pittsburgh's is .560. Also, St. Louis is .5155 and Philadelphia .5148.

Section 2-4

1	17	7	Irrational	13	Rational
3	35	9	Irrational	15	Irrational
5	Rational	11	Irrational	17	Irrational

19 8.5

21 14.1

23 $\sqrt{8}$

27 Greater

29 2.2361

31 (*a*) $\frac{9}{4}$; (*b*) no; (*c*) no

33 $\sqrt{2}$

35 $1/\sqrt{2}$

37 $\frac{1}{2}$

39 No

41 No

43

Decimal Places	Approx. of $\sqrt{8}$	Square of Approx.
1	2.8	7.84
2	2.82	7.9524
3	2.828	7.997584
4	2.8284	7.99984656
5	2.82843	8.0000162649
6	2.828427	7.999999294329
7	2.8284271	7.99999986001441

45 0 decimal places

47 1 decimal place

49 Since 5 is prime

51 Step 7

Section 2-5

1 Not well defined

3 Well defined

5 Not well defined

7 $\{1, 2, 3, 4, 6, 12\}$

9 $\{6\}$

11 $\{-3, -2, -1, 0\}$

13 Set of natural numbers less than 8. (Other answers possible.)

15 Set of natural numbers less than 16 that are multiples of 3. (Other answers possible.)

17 Set of natural numbers between 10 and 14 inclusive. (Other answers possible.)

19 $A \not\subset S, B \subset S$

21 $A \not\subset S, B \not\subset S$

23 $A \subset S, B \subset S$

25 $\{4, 5, 6\}$

27 $\{4, 5, 6, \ldots\}$

29 $\{1, 3, 5, \ldots\}$

31 $\{1, 2, 3, 4\}$

33 $\{2, 4, 8, 16, 32, 64\}$

35 $\{-2, -1, 0, 1, 2, 3, \ldots\}$

37 Q

39 $[3, 4]$

41 $\emptyset$

43 $\{2\}$

45 $\{1, 2\}$

47 $\emptyset$

49

51

53

Section 2-6

5 Germanium. Because it was found in Germany. (See Chapter 10 of "The Ascent of Man" for a fascinating account of this discovery.)

7 We do not find the logic compelling.

9 This issue is very controversial among mathematicians. We accept the proof, but some of our colleagues have serious reservations about it and others reject it entirely.

11 Zero degree of certainty. (Jupiter has 17 moons as of September 1983.)

13 The water level will fall, a conclusion that follows from Property 4. If you do not believe us, try it. You'll shout "Eureka."

15 Let $a = 1$ and $b = 2$ in all four cases.

Chapter 3

Section 3-1

1 $64,000/81$ (≈ 790.12) cubed cubits

3 384 cubed cubits

5 1280 cubed cubits

7 9

9 9

11 1

13 3

15 11

17 -1

19 (a) $\frac{256}{81}$; (b) 3; (c) 3.1416; (d) π

21 (d)

23 $A = 5$

25 64

27 Not correct. Set $a = 3$, $b = 4$, and $c = 5$.

29 (b) $5\frac{1}{4}$ palms; (c) $5\frac{1}{2}$ palms; (d) $7b/2h$ palms

31 (a) It is doubled; (b) It is tripled. (c) It is increased by a multiple of n.

33 (a) It is doubled; (b) It is quadrupled.

35 The sum of the first 10 positive integers is 55. The sum of the first 100 positive integers is 5050.

Section 3-2

1   Step 2: Post. 3; Step 3: Post. 1; Step 5: Def. 15; Step 6: C.N. 1

3   Step 1: Post. 2; Step 2: Prop. 2; Step 5: C.N. 5

5   This is an implicit assumption.

9   Yes

11 No, because Prop. 456 does not precede Prop. 345.

13 No. Proposition 20 rules out the 1-2-3 and 1-1-4 triangles.

15 Three: 4-4-4, 2-5-5, 3-4-5.

19 To minimize the number of assumptions

21 Def. 20

23 Prop. 6

25 Two such triangles can be constructed.

27 Prop. 11 and Prop. 12

31 (e) They are equal.

33 (a) F; (b) F; (c) F; (d) T; (e) F; (f) F; (g) T

35 The length of the hairline to the crown is $\frac{1}{40}$ if $b - a$ is used and $\frac{1}{12}$ if $d - c$ is used. (Other answers are possible.)

Section 3-3

1   Not a statement

3   True statement

5   False statement

7   If it rains, then I drive to school.

9   If you are a man, then you are mortal.

11 If two lines are perpendicular to a line, then they are parallel.

13 If the results of a Rorschach test show positive correlations with the results of other personality tests, then it can be considered a valid test of personality.

15 If things are equal to the same thing, then they are equal to one another.

17 In any triangle, if one side is greater than another, then the angle opposite it is greater than the angle opposite the other side.

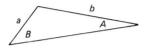

If $b > a$, then $B > A$.

19 If you must stop, then it is red.

21 If I drive to work, then it is raining.

23 If a triangle is equilateral, then all three angles are equal.

25 Both are true.

27 The statement is false; its converse is true.

29 The statement is false; its converse is false.

31 Both are false.

33 If you are not in New York State then you are not in New York City. They are true.

35 If he did not proclaim strict adherence to capitalist doctrines, then he was not a member of the Red Guard. They are false.

37 If a figure is a circle, then it consists of a set of points equidistant from a given point. If a figure consists of a set of points equidistant from a given point, then it is a circle.

39 If it is abnormal psychology, then it is the study of all forms of abnormal behavior in humans. If it is the study of all forms of abnormal behavior in humans, then it is abnormal psychology.

41 If the disease is rubella, then it is called German measles. If a disease is called German measles, then it is rubella.

43 An angle inscribed in a semicircle is a right angle.

Section 3-4

1 You live in the United States. Therefore you are an American citizen.

3 You live in Miami. Therefore you live in Florida.

5 Three noncolinear points are not given.

7 Valid

9 Invalid

11 If Mary is in Montreal, then she is in Canada.

13 If a four-sided figure has three right angles, then it is a rectangle.

15 If x and y are consecutive integers, then their product is an even integer.

17 (1), (3), (4), (2). If a and b are even numbers, then $a + b$ is an even number.

19 (2), (4), (1), (3). If a and b are odd numbers, then ab is an odd number.

21 (2), (4), (3), (5), (6), (1). The statement "If $(a - b)(a + b) = (a - b)b$, then $a + b = b$" is not true if $a = b$.

23 Promise breakers are very communicative.

25 Animals that hop well are not kangaroos.

27 Babies are saints.

29 Donkeys are not easy to swallow.

Section 3-5

1 (*a*) (3) C.N. 1; (4) Prop. 27.
(*b*) (1) Prop. 13: (3) C.N. 1; (4) C.N. 3

3 (2) C.N. 2 (implicit); (3) Prop. 13; (4) C.N. 1 (implicit); (5) Post. 5; (7) Prop. 15; (8) C.N. 1; (9) C.N. 2; (10) Prop. 13; (11) C.N. 1

5 (d, x) and (b, z)

7 L_1 and L_2

9 (1) Hypothesis; (2) Prop. 13; (3) C.N. 3; (4) Post. 27

11 $x + y = 180°$ [Hypothesis]
$x + z = 180°$ [Prop. 13]
$y = z$ [C.N. 3]
$y = w$ [Prop. 15]
$w = z$ [C.N. 1]

13 (*a*) (2) Post. 1; (3) Post. 2; (4) Prop. 2; (5) Post. 1; (6) Prop. 10; (9) Prop. 4; (10) C.N. 1; (11) C.N. 5; (12) C.N. 1 (implicit).
(*b*) (1) Prop. 13; (2) Post. 5; (3) C.N. 3

Section 3-6

1 Prop. 48 is the converse of Prop. 47. They can be combined to read, "A triangle is a right triangle if and only if the square of one side is equal to the sum of the squares of the other two sides."

3 Bisect each side of the square (Prop. 10), erect a perpendicular at each bisecting point (Prop. 11), and join the vertices of the square to the points where the perpendiculars intersect the circle.

5 A group of seven. Trio (3), quartet (4), quintet (5), sextet (6).

7 Six

9 Nine

11 The medians intersect in one point.

13 (*b*) and (*g*)

Chapter 4

Section 4-1

1 (1) Post. 1; (2) Post. 1; (3) Prop. 31; (4) Prop. 2; (5) Post. 1;
(6) Post. 1; (7) Prop. 31; (8) Prop. 30; (10) Post. 1 and Post. 2

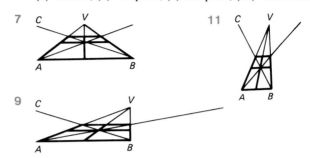

7

9

11

13 Refer to Figure 4. Draw lines *VH* and *VI*. The trapezoid *AFGB* contains the 16 squares in perspective.

15 *E* is the midpoint.

17 If *VM* ⊥ *AB*, then *VE* = *EM*. Otherwise *VE* and *EM* are not comparable.

19 6

21 Not necessarily.

23 Yes

25 Trisect *AB* at points *M* and *N* and draw lines *VM* and *VN*. The line joining *VM* ∩ *BC* to *VN* ∩ *AD* is one horizontal line, and the other is the line joining *VM* ∩ *AD* to *VN* ∩ *BC*.

27 Christ's head

29 The two central figures, Plato and Aristotle

Section 4-2

1 (*a*) and (*c*)

3 (1) ∠*A* and ∠*B* are right angles [Def. of S-quad].
(2) ∠*A* + ∠*B* = 180° [C.N. 2].
(3) *AD* is parallel to *BC* [Prop. 28].
(4) ∠*D* is a right angle [RAH].
(5) ∠*A* + ∠*D* = 180° [C.N. 2].
(6) *AB* is parallel to *DC* [Prop. 28].

5 An S-quad is not a parallelogram, so it cannot be a rectangle.

7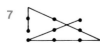

9 (1) Prop. 11; (2) Prop. 2; (3) Post. 1; (7) C.N. 2 (implicitly);
(8) Post. 4; (9) C.N. 5

Section 4-3

3 No

5 Yes

7 In spherical geometry, two lines intersect in exactly two points.

9 Triangle *ABC* has greater than 180°.

11 The exterior angle at *A* is equal to the interior angle at *B*.

13 Rectangles do not exist in spherical geometry.

Chapter 5

Section 5-1

1 *A*(1, 1), *B*(2, 0), *C*(1, −2), *D*(−3, −3), *E*(−2, 3), *F*(−5, 1)

3

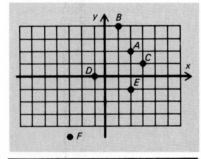

Square, 8, 4

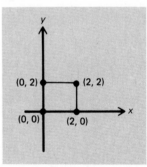

7 Triangle, 18, 12

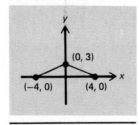

9

x	−2	−1	0	1	2	3
y	−7	−5	−3	−1	1	3
Solution	(−2,−7)	(−1,−5)	(0,−3)	(1,−1)	(2,1)	(3,3)

11

x	−2	−1	0	1	2	3
y	6	5	4	3	2	1
Solution	(−2,6)	(−1,5)	(0,4)	(1,3)	(2,2)	(3,1)

13

x	−2	−1	0	1	2	3
y	5	2	1	2	5	10
Solution	(−2,5)	(−1,2)	(0,1)	(1,2)	2,5)	(3,10)

15

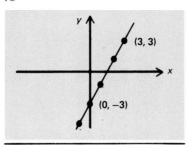

17

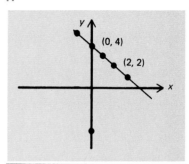

19

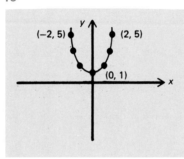

21

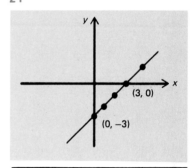

23

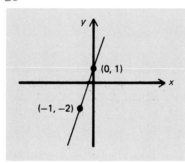

25

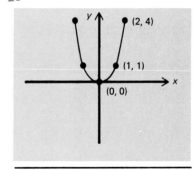

27

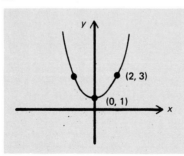

29

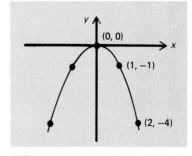

31

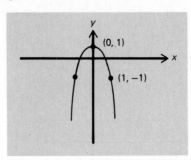

33

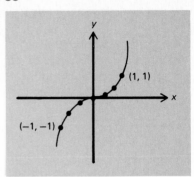

35

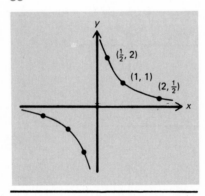

37 Think $(20 + 1)(20 - 1) = 20^2 - 1^2 = 399.$

39 Think $(20 + 2)(20 - 2) = 20^2 - 2^2 = 396.$

Section 5-2

1 3

3 $\frac{1}{3}$

5 $\frac{1}{2}$

7 2

9

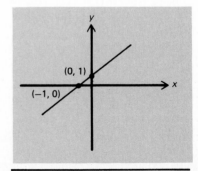

11

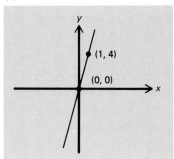

13

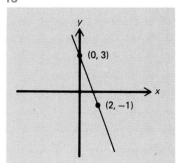

15 $2, -1$

17 $1, 0$

19 $1, 2$

21 No

23 No slope

25 0

27 (a) $y - 1 = -7(x - 2)$; (b) $y = -7x + 15$

29 (a) $y - 0.5 = 1.4(x - 2)$; (b) $y = 1.1x - 2.3$

31 $5x - y = 1$

33 $3x - y = 1$

35 $y = 2x + 7$

37 $x = -5$

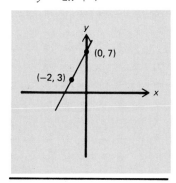

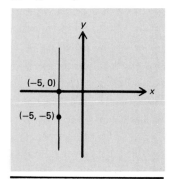

39 $y = 2x - 1$

41 m

43 l_1

45 Positive

47 0

49 Negative

51 $y = -\frac{5}{2}x + 5$

53 $y = -\frac{5}{2}x - 5$

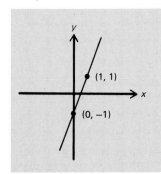

Section 5-3

1	(2, 0)	**11**	(1, −2)	**21**	(0, 0)
3	(4, 0)	**13**	$(\frac{7}{5}, \frac{3}{5})$	**23**	(100, 20)
5	(4, 2)	**15**	(1, −3)	**25**	Inconsistent
7	(2, 0)	**17**	Inconsistent	**27**	Dependent
9	(1, 1)	**19**	$(\frac{2}{3}, -\frac{5}{3})$	**29**	Dependent

Section 5-4

1 $x < \frac{3}{2}$

3 $x \le -1$

5 (1, 2)—no, (2, 2)—no

7 (−1, 1)—yes, (0, 0)—no

9 (0, 0)—below, (1, 7)—above

11 (1, −1)—below, (−1, −1)—above

13

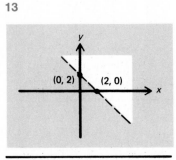

15

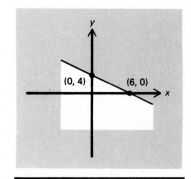

17

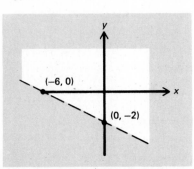

19

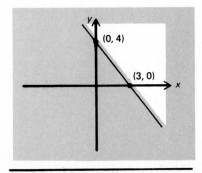

21

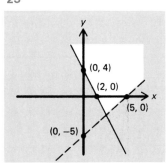

23

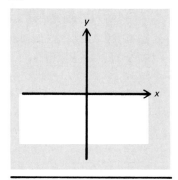

25

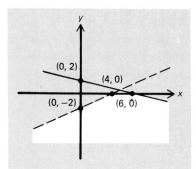

27

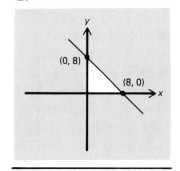

29

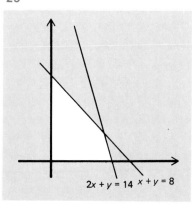

$2x + y = 14$ $x + y = 8$

Section 5-5

1

x	y	100
150	100	12,000
40	30	

Maximize $P = 40x + 30y$
subject to

$$x + y \leq 100$$
$$150x + 100y \leq 12,000$$
$$x, y \geq 0$$

3

45	9	45
9	6	15
4	2	

Minimize $F = 4x + 2y$
subject to
$$45x + 9y \geq 45$$
$$9x + 6y \geq 15$$
$$x, y \geq 0$$

5

3	6	36
3	2	24
132	80	

Minimize $C = 132x + 80y$
subject to
$$3x + 6y \leq 36$$
$$3x + 2y \geq 24$$
$$x, y \geq 0$$

7

3	4	48
100	200	2000
500	800	

Maximize $P = 500x + 800y$
subject to
$$3x + 4y \leq 48$$
$$100x + 200y \leq 2000$$
$$x, y \geq 0$$

9 Maximize $P = 0.5x + 0.7y$ subject to

$$0.3x + 0.4y \leq 120$$
$$0.2x + 0.1y \leq 60$$
$$x, y \geq 0$$

11 Maximize $R = 0.15x + 0.12y$ subject to

$$x + y \leq 15 \text{ (in millions)}$$
$$y \geq 2x$$
$$x \geq 2$$
$$y \geq 2$$

13 Maximize $R = 1000x + 1200y$ subject to

$$x + y \leq 1000$$
$$8000x + 10,000\,y \leq 9,000,000$$
$$x, y \geq 200$$

15 Minimize $C = 20,000x + 15,000y$ subject to

$$150x + 100y \geq 1000$$
$$50x + 50y \geq 450$$
$$x, y \geq 0$$

Section 5-6

1

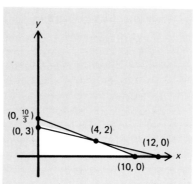

3

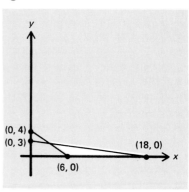

5 None	**17** 8800 at (8, 6)
7 40 at (10, 0)	**19** 210 at (0, 300)
9 88 at (2, $\frac{8}{3}$)	**21** 1.95 million at (5, 10)
11 3400 at (40, 60)	**23** 1,100,000 at (500, 500)
13 $\frac{40}{7}$ at ($\frac{5}{7}$, $\frac{10}{7}$)	**25** 145,000 at (2, 7)
15 1032 at (6, 3)	**27** 3600 at (8, 6)

Chapter 6

Section 6-1

1 They both are. One is the correct answer to the question "What is the theoretical probability?" and the other is the answer to "What is the empirical probability?"

3 There are four ways of getting two marbles into the center placeholder: by selecting the first marble from the left placeholder with the first marble from the right placeholder; or the second marble from the left with the first from the right; or the second from the left with the first from the right; or the second from the left with the second from the right. But in each case the middle placeholder will have one red and one white marble.

5 Red-red, red-white, white-red, white-white

7 Half are pure-bred and half are hybrid

9 25% both red, 25% both white, 50% red and white

11 Heads-heads, heads-tails, tails-tails. (What about tails-heads?)

13 Red-red, red-black, black-red, black-black

15 (*a*) About 800; (*b*) all; (*c*) about 400; (*d*) about 300

17 (*a*) About 1858; (*b*) all; (*c*) about 929; (*d*) about 696

19 It will be tall-stemmed.

21 About $8023 \times 2 = 16{,}046$

23 About 6018, 6022

25 (*a*) 6022/8023, (*b*) 152/580

Section 6-2

1

Outcome	red	black	equally likely
Probability	1/2	1/2	

3

Outcome	picture	no picture	not equally likely
Probability	3/13	10/13	

5

Outcome	prime	composite	neither	not equally likely
Probability	1/2	1/3	1/6	

7

Outcome	red	black	equally likely
Probability	1/2	1/2	

9 Red-red, red-black, black-red, black-black

11 $\frac{1}{6}$		19 $\frac{5}{6}$		25 12	
13 $\frac{1}{5}$		21 $\frac{1}{3}$		27 8	
15 $\frac{1}{5}$		23 $\frac{4}{7}$		29 7	
17 $\frac{4}{5}$					

Section 6-3

1 $\frac{1}{8}$	25 $\frac{1}{2}$	49 Free Parking
3 $\frac{7}{8}$	27 $\frac{7}{10}$	51 $\frac{2}{5}$
5 $\frac{1}{8}$	29 $\frac{3}{10}$	53 $\frac{1}{11}$
7 $\frac{1}{2}$	31 $\frac{5}{14}$	55 $\frac{5}{6}$
9 $\frac{2}{9}$	33 $\frac{7}{10}$	57 2 to 3
11 $\frac{1}{6}$	35 1	59 1 to 1
13 $\frac{5}{6}$	37 $\frac{1}{4}$	61 1 to 9
15 $\frac{5}{12}$	39 $\frac{1}{26}$	
17 $\frac{1}{2}$	41 $\frac{3}{4}$	
19 $\frac{5}{18}$	43 $\frac{25}{52}$	
21 $\frac{1}{6}$	45 $\frac{5}{12}$	
23 $\frac{1}{6}$	47 $\frac{7}{18}$	

Section 6-4

1 $\frac{1}{3}$

3 $\frac{1}{2}$

5 $\frac{1}{9}$

7 $\frac{5}{18}$

9 $\frac{1}{6}$

11 $\frac{25}{51}$

13 $\frac{25}{102}$

15 $\frac{1}{17}$

17 $\frac{1}{221}$

19 $\frac{2}{9}$

21 $\frac{1}{45}$

23 $\frac{7}{15}$

25 $\frac{1}{3}$

27 $\frac{14}{45}$

29 $\frac{51}{100}$

31 Without

33 $\frac{17}{45}$

35 0.54

37 (a) 0.56; (b) 0.06

39 No

41 Yes

43 $\frac{38}{97}$

45 $\frac{36}{97}$

47 $\frac{8}{97}$

49 $\frac{52}{268}$

51 $\frac{1}{54,145}$

53 $\frac{94}{54,145}$

Section 6-5

1 Yes

3 Yes

5 No

7 0.1

9 0.72

11 $\frac{1}{16}, \frac{1}{16}, \frac{15}{16}$

13 $\frac{1}{256}, \frac{1}{256}, \frac{255}{256}$

15 $\frac{1}{16}$

17 $\frac{1}{6}$

19 $\frac{1}{125}$

21 $\frac{11}{75}$

23 $\frac{1}{16}$

25 $\frac{9}{169}$

27 0.0005

29 2

31 Here is one way; there are others. Suppose E and F are independent. Then
$\Pr(E \cap F) = \Pr(E) \times \Pr(F) = \Pr(E) + \Pr(F)$ since E and F are
mutually exclusive. Divide by $\Pr(F)$ and get $\Pr(E) = \Pr(E)/\Pr(F) + 1$,
which implies that $\Pr(E) > 0$, a contradiction.

33 $\frac{1}{16,384}$

Section 6-6

1 676

3 6,760,000

5 12

7 1000

9 27,000

11 20

13 30

15 11,880

17 10

19 45

21 15,504

23 Permutations, 90

25 Permutations, 15,600

27 Permutations, 8

29 Permutations, 720

31 Be careful! Use the multiplication principle, 48

33 Combinations, 15,504

35 (a) Combinations, 21; (b) permutations, 42

Section 6-7

1 $\frac{1}{100}$ 7 $\frac{1}{56}$

3 $\frac{1}{6}$ 9 $\frac{1}{55}$

5 $\frac{1}{2}$

11

Outcome	0	1	2	3	4
Probability	$\frac{1}{16}$	$\frac{1}{4}$	$\frac{3}{8}$	$\frac{1}{4}$	$\frac{1}{16}$

2 has the largest probability

13 (a) $\frac{252}{1024} = \frac{63}{256}$; (b) $\frac{210}{1024} = \frac{105}{512}$; (c) $\frac{210}{1024} = \frac{105}{512}$; (d) $\frac{11}{1024}$; (e) $\frac{7}{128}$

15 $\frac{924}{4096} = \frac{231}{1024}$

17 $\frac{2}{7}$ 23 $\frac{1}{3}$

19 $\frac{1}{260}$ 25 $\frac{5}{33}$

21 $\frac{5}{12}$ 27 $\frac{35}{99}$

29 (a) $\frac{1}{6}$; (b) $\frac{1}{30}$; (c) $\frac{29}{30}$

31 (a) $364 \times 363 \times 362/(356)^4 = \approx 0.98$; (b) 0.02

Chapter 7

Section 7-1

1 It should be close.

3 It should be close.   5   About half the time

7 We found them to be very different.

9 From Example 1, about 14

11 From a list of random numbers, select them two at a time. Let it be a hit if the first number is greater than or equal to 50 or a miss if not. If the next number is greater than or equal to 75, then it is a hit. The next number corresponds to the first player, and so on. The probability is $\frac{4}{7}$.

13 (a) About 0.24; (b) $\frac{33}{150}$, or about 0.22

15 (a) 0.76; (b) $(0.76)^{44}$

17 Use Table 1. Let 0-3 correspond to a hit and 4-9 correspond to an out: 0.420.

19 (a) 12,843/20,763 $\approx$ 0.62; (b) 738/1729; (c) $1 - (12!/2)/2^{10}$

Section 7-2

A 1. $\frac{229}{576}$

B 1. Major E against women, major A against men
   3. (a) 0.82; (b) 0.68; (c) 0.07

C 1. (a) 7%; (b) 8%; (c) 17%; (d) 30%
   3. Users: (a) 53%; (b) 48%; (c) 56%; (d) 66%
      Nonusers: (a) 40%; (b) 36%; (c) 41%; (d) 57%

D 1. Children in grades 1 and 3 might be significantly different from those in grade 2. The groups are too homogeneous. The expense of this experiment was less because the groups were in the same location rather than spread out, so it was easier to administer.

E 1. $\frac{243}{686}$

3. 3488

F 1. $\frac{69}{121}$

G 1. 0.4913

H 1. (a) $\frac{6}{11}$; (b) $\frac{4}{9}$; (c) not exactly, but it is close: suppose 99 pups were born early and 99 late—there would be $(\frac{6}{11})$ 99 = 54 males born early and $(\frac{4}{9})$ 99 = 44 born late, a total of 98, not 99.

I 1. Assuming independence, 0.0004

Section 7-3

1 There are many difficulties, among which are: some people do not have phones; some people have unlisted numbers; some people are listed more than once.

3 Yes. Nonrespondents are very often a very significant population.

9 No, because the rate of contraction of polio for those who did not get vaccinated is higher.

11 This study is just as faulty from a statistical point of view as the Wangensteen study.

Section 7-4

1 Descriptive statistics is the set of numbers that measure or describe data. Inferential statistics is the process of making judgments based upon the descriptive statistics of a set of data.

3 The Admiralty's measure did not take into account the severity of the damage. The OR group used ship-months lost. They took into account how long it took to return the ship to action or to replace it.

5 Deleting the lower part of the vertical axis tends to exaggerate the difference between the yearly totals.

7 To indicate when the new law went into effect

9 (a) 0.35; (b) 0.19; (c) 0.31; (d) 0.08; (e) 0.74

Chapter 8

Section 8-1

1 Not enough spread to get an overall picture of the data: results are too specific

3 (a) 2; (b) 3.7, so choose 4; (c) 4.2, so choose 4; (d) 5.6, so choose 6; (e) 2.1, so choose 2; (f) 4.36, so choose 4; (g) 1.73, so choose 2; (h) 4.99, so choose 5

5

Number	18	19	20	21	22	23	24	25	26	27	28	29	30
Frequency	7	6	5	0	2	1	3	1	1	1	0	0	1

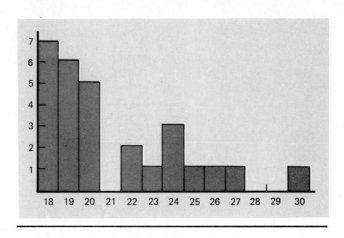

7

Interval { to	−10	−4	1	6	11	16	21	26	31	36	
from	− 5	0	5	10	15	20	25	30	35	40	+
Frequency	1	1	11	11	9	9	2	3	1	2	1

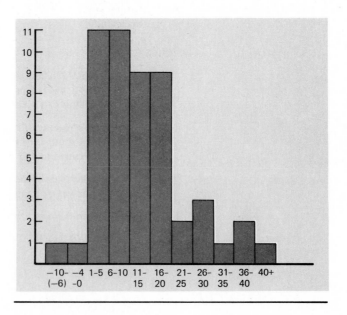

9

Wins	10	11	12	13	14	15	16	17	18	19	20+
Frequency	4	3	4	8	9	2	4	1	2	2	1

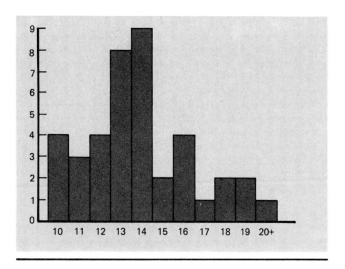

11

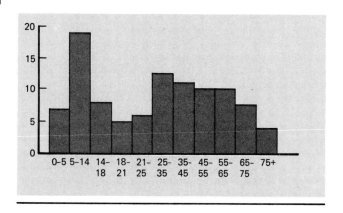

13 Many people round off their GPA when they report it.

15

Inches	70	71	72	73	74	75	76	77	78	79
Frequency	5	6	4	5	11	6	7	1	3	1

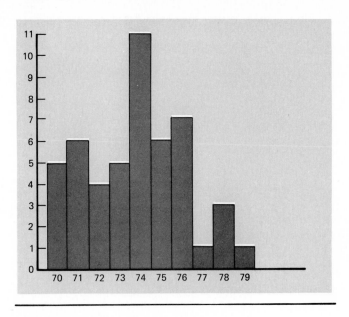

17 We apologize; authors have to have some fun too.

19 Decline

21 (*a*) The peanut graph is a frequency polygon; (*b*) about 200,000,000 pounds; (*c*) yes

Section 8-2

1	6	**11**	1.64	**21**	3.31
3	0	**13**	6.71	**23**	9.46
5	2.6	**15**	4, 4	**25**	2
7	16.1	**17**	4, 3.5	**27**	91
9	.05	**19**	12, 12.5	**29**	Both near 10

31 Mean near 10; median near 5

33 About 62 inches

35 *n* will always be greater than or equal to *r*.

37 Mean—508.7; median—505.5

39 It is not exact because not all teams have the same number of players on their rosters, but it is close. Be careful though; a very significant error can be made by calculating the average this way.

Section 8-3

1 Least—set 2; greatest—set 1

3 Set 1—0; set 2—5; set 3—50; least—set 1; greatest—set 2

5 Set 1—5; set 2—1; set 3—10

7 a—II; b—I; c—III

9 2.37

11 3.36

13 10.36

15 6.63, 3.43

17 16.62, 2.73

19 Expand the parentheses and simplify, using the formula for the mean.

21 (a) 6, 0; (b) 6, 1.4; (c) 6, 5.1

Section 8-4

1 0.5

3 0.34

5 0.81

7 0.16

9 0.5

11 0.2794

13 0.4265

15 0.4817

17 0.2422

19 0.4394

21 0.4913

23 0.2358

25 0.0188

27 0.0287

29 0.2706

31 0.0678

33 0.3839

35 0.6359

37 0.9596

39 0.9535

41 0.8810

43 0.9671

Section 8-5

1
20	30	40	50	60	70	80
−3	−2	−1	0	1	2	3

3
6	9	12	15	18	21	24
−3	−2	−1	0	1	2	3

5 $x = 60, \sigma = 5$

7 $x = 100, \sigma = 10$

9
4	7	10	13	16
−2	−1	0	1	2

11
56	60	64	68	72
−2	−1	0	1	2

13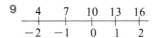
54	56.5	59	61.5	64
−2	−1	0	1	2

15 0.3413

17 0.6826

19 0.5328

21 0.4435

23 0.6826

25 0.1554

27 0.0919

29 0.0388

31 0.1915

33 0.0401

Figure Credits

Index

Absolute geometry, 242
Accu-Weather, 457, 461
Achilles, 174−175
Acute angle, 225
Addition rule, 378
"Adventures of a Mathematician" (Ulam),
   446−447
AE (algebraic entry), 108
"Africa Counts" (Zaslavsky), 3, 6
Age of Pericles (see Pericles)
Age of Reason, 266
Ahmes, 82, 87, 89−90, 154ff.
Alberti, Leon Battista, 232−236, 238−239
Alexander the Great, 220
Alexandria, 164
Algebra, 89, 267ff.
Algebraic entry (AE), 108
Algebraic geometry, 271
"Alice in Wonderland" (Carroll), 199
al-Khowarizmi, 21
Alphabetic system, 57
AMA (American Medical Association),
   461, 471n.
Amenemhet, 152
Amenhotep, 7−8
American Health Foundation, 321
"American Jury, The" (Kalvern and
   Zeisel), 391n.
American Mathematical Society, 44
American Naturalist, The, 355
American Revolution, 519
Amicable numbers, 68
Analytic geometry, 266, 271
Anaxagoras, 219
Anderson, John, 534−536
Angle, 225
Angoff, William H., 555
Antecedent, 177, 187
Anthropology, 2, 3, 42, 136
Antiquity, three problems of, 219
Apollo, 415
Apollonius, 214, 220
Approximations:
   decimal, 125−126
   normal, 551ff.

Arabs, 90
Archeology, 20, 105
Archimedes, 147−148, 220
Architecture, 230
   Egyptian, 153
   Greek, 165
Area, 156
Argument, 187
Aristotle, 120, 186, 346
Arithmetic, 56
Arithmetic average, 506
Arithmetic mean (see Mean)
Art, 105
   Egyptian, 153, 230
   Greek, 231
   Renaissance, 231−237, 240
"Art of Conjecturing, The" (Bernoulli),
   438
Artifacts, 3, 136
Arusha Maasai, 6
"Ascent of Man, The" (Bronowski), 58
Asimov, Isaac, 42
Athena Parthenos, 165
Atomism, 45
Australopithecus, 3
Automobile fatalities, 481−483, 487
Average, 477
Axioms, 166

Babylonian number system, 20, 28
Babylonians, 213
Bar graph, 483
Bardi, 19
Base:
   of a number, 34
   of a number system, 5, 20, 24, 29, 33
Baseball, 118, 458−459, 466, 478, 486
Beast, number of the, 43, 425, 433
Bell, Eric Temple, 44, 266
Bell Telephone Laboratories, 33
Beni-Hasan, 152
Berelson, B., 483−486, 565
Bernoulli, James, 437−438
Berresford, Geoffrey C., 455

Best system, 42
Bhaskhara, 29
Bias, 471, 476
Binary system, 33, 37–39
Biology, 349–350
"Biometry" (Sokal and Rohlf), 355
Blind experiment, 475
Bloom, Norman, 148
Bolyai, Janos, 240, 251, 253, 258–261
Bolyai, Wolfgang, 240, 252
Boole, George, 186
Boundary, 225
Bowa, Larry, 458
Bowmar Brain, 107, 108
Boyer, Carl, 278
Brahmagupta, 136
Brahmi letters, 20
Brahmi numerals, 22
Brauer, Richard, 54
Braun, Martin, 105
"Brethren, The" (Woodward and
    Armstrong), 147
Brett, George, 454
"Briefe and True Report of the New-Found
    Land of Virginia, A," 305
British Exchequer, 5
"Broca's Brain" (Sagan), 130, 148
Bronowski, Jacob, 58, 174
Brunelleschi, Filippo, 230–232
Bugilai, 5

Caffeine, 140
Calculators, 33, 108–112, 125–126, 449
Calculus, 5, 62, 240, 255
Canon, 230
    Egyptian, 153
    Greek, 173
Cantor, Georg, 131, 135–136, 139
Card problems, 353, 361, 362, 364,
    385–386, 402–405
Cardano, Jerome, 136, 347
Carlton, Steve, 458
Carpenter's Problem, 295
Carr, Billie, 535
Carroll, Lewis (Charles Lutwidge
    Dodgson), 199
Carter, Jimmy, 534–536
Cartesian coordinate system, 270
Casting out nines, 47
"Catch-22" (Heller), 43, 66
Central tendency, 505ff.
Chace, A. B., 82
Chemistry, 141
Chicago Tribune, 566

China, 20
Chinese mathematics, 136, 294
Chip revolution, 33
Chromosome, 349
Chuquet, Nicholas, 136
Circle, 225
Circumstantial evidence, 196
Classical Greeks, 45, 231, 240, 347
Classification of finite simple groups, 54
Clever Hans (horse), 130
Cole, Frank Nelson, 44, 74, 112
College Entrance Examination Board, 555
Collins, Janet, 196
Collins, Malcolm, 196
Columbia (space shuttle), 415
Combination, 428, 432, 435, 438–441
Combined Shipping Adjustment Board, 318
Common notions, 166
Compass, 167
Compatible events, 377
Complement, 133, 382
Complementary probability, 382
Composite, 46
Computer, 33, 39, 414, 423, 449, 450
Conclusion, 187
Conditional probability, 390, 393–405, 457
Conditional statements, 176
Confounding effects, 483
Congruence, 183
Conic sections, 220
"Conics" (Apollonius), 214
Conjecture, 73, 146
    Fermat's, 143
    Mersenne's, 74
Consequent, 177, 187
Constraints, 322
Constructible figures, 215
Contrapositive, 180–181
Control group, 471, 475–476
Converge, 102
Converse, 179–180
Coordinate system, 269
Corner points, 335
Count, 3
Counterexample, 75
Counting numbers, 45
    (See also Natural numbers)
Crossley poll, 566
Cube, 217
Cubits, 154
Cultural universal, 42
Cuneiform, 28
Current Population Survey (CPS), 490ff.
Cylinder, 154
Cylindrical granary, 154

Dantzig, George, 145, 265, 319–320, 332
Darwin, Charles, 351, 489
Death of St. Francis, The (Giotto), 231
Deathday, 468
Decagon, 217
Decimal approximations, 125–126
Decimal numbers, 105ff.
Decimal numeration system, 33
Decimal representation, 112–116, 123–124, 126
Decimals:
  repeating, 113, 123–124
  terminating, 113, 123–124
Dedekind, Richard, 131
Deduction, 166ff.
Delian problem, 219
"Della Pictura" (Alberti), 232
de Méré (*see* Méré)
Demographics, 93
De Moivre, Abraham, 437–438
De Morgan, Augustus, 122
Dependent, 303
Dependent events, 411
Descartes, René, 136, 266ff.
Descriptive statistics, 477
Deviation, standard, 520ff.
  population, 526
  sample, 526
Dewey, Thomas, 566
Diameter, 225
Dice, 359, 374
Didion, Joan, 94
Digit, 5
Digital root, 47
Digital sum, 46
Digital watches, 33
"Discourse on the Method of Reasoning Well and Seeking Truth in the Sciences" (Descartes), 267
Discus thrower, 165
Disjoint sets, 135
Disraeli, Benjamin, 477
Diverge, 102
Divisible, 46
"Doctrine of Chances" (De Moivre), 438
Dodecahedron, 217
Dodgson, Charles Lutwidge (Lewis Carroll), 199
Dominant trait, 349
Double-blind test, 459, 472, 473, 476
Douglas, William O., 171–172, 566
Dozenal Society of America, 42
Drugs, 463, 465
Dürer, Albrecht, 22

e, 122
Ecology, 105
Eddim, Nasir, 248
Education, 94
Educational psychology, 2
Egan, George, 424
Egan, Lillian, 424
Egyptian art, 153, 230
Egyptian civilization, 7, 19, 152, 346
Egyptian numeration system, 7–11, 25–26, 42, 136
Eiffel Tower, 147
Einstein, Albert, 171, 251, 261, 456, 469
Electrons, 46
"Elements" (Euclid), 47, 63, 66, 72, 165ff., 213ff., 240–241, 259–260
Elements, 45
Elimination, method of, 303
Ellipse, 220
Elmira, 484–486
Empirical probability, 455ff.
Empty set, 135
Endpoint convention, 491
Energy, nuclear, 94
ENIAC, 107
Epicureans, 169, 171, 246, 251
Equally likely outcomes, 359, 362–364, 367, 373, 394–395, 438–441
Equilateral triangle, 167
Equivalent statements, 182
Euclid, 47, 63, 66, 72, 164, 240–241, 250, 258, 263, 266
"Euclid Freed of Every Flaw" (Saccheri), 241
Eudoxus, 214
Euler, Leonhard, 76
Event, 371
Expanded version of a number, 35
Expected number, 448
Experimentation in the sciences, 158
Exponential notation, 34
Exponents, 34–37, 96
Extremities, 225

Factorial, 430
Farming, 93
"Federalist, The," papers, 519, 527–530
Fermat, Pierre de, 76, 277, 346–347, 437
Fermat's conjecture, 143
Fibonacci, 21, 34
Field hockey, 400
Fifth postulate, 203ff., 240–241, 252–253
Figure, 225
Finger gesturing, 5

Finite set, 132
"Finnegan's Wake" (Joyce), 172
Fitness, 94
   (*See also* Running)
Food and Drug Administration, 140, 146
Foreshortening, 231
Form of a linear equation, 286ff.
Four-color problem, 145−146
Fractions, 84ff., 90
Freedman, D., 569*n.*
Frequency distribution, 491
Frequency polygon, 493
Friendly numbers, 68
Function:
   of one variable, 157
   of two variables, 155
Fundamental theorem of arithmetic, 47

Galileo, 59, 456
Gallup, George, 569
Gallup poll, 536ff.
Galois, Evariste, 54
Galton, Sir Francis, 489
Garden of Eden, 19
Gauss, Karl Friedrich, 251−252, 259−261,
   295, 533−534
Geiberger, Al, 119
Gematria, 57, 59
General form, 288ff.
Generation of plants, 348
Genes, 349, 356
Genetic code, 42
Genetics, 346−356
Genotype, 351
Geography, 94
Geometric series, 102
"Geometry" (Descartes), 278
Germanium, 33
Giotto, 231
Gödel, Kurt, 46
Golden Age of Greece, 164
Golding, Evelyn, 222
Golf, 118
Gombaud, Antoine (*see* Méré)
Goodall, Jane, 128
Googol, 44
Googolplex, 44
Grand unification, 46
Graph of an equation, 272ff.
Graphical solution, 297
Graunt, John, 489, 505−506
Great Britain, 462, 479−481
Great circle, 257, 262
Greece, Golden Age of, 164

Greek art, 231
Greeks, classical (*see* Classical Greeks)
Ground line, 234
Group, 54
Gutenberg, Johann, 21, 22

Half-life, 95
Hamilton, Alexander, 519, 527−530
Hans, Clever (horse), 130
Hanukkah, 66
Harmonic series, 102
Harper, Chandler, 118
Harriot, Thomas, 305
Health Examination Survey, 551, 553
Heinlein, Robert, 43
Heller, Joseph, 43, 66
Heredity, 346−347
Heron of Alexandria, 129−130
Hesiod, 219
Hevesy, Paul de, 104
Hexagon, 216
Hexagram, 223
Heyerdahl, Thor, 19ff.
Hieratic, 7
Hieroglyphics, 7, 82
Hill, Andrew, 2
Hindu-Arabic numeration system, 6, 8,
   19ff., 29
Hindus, 19, 90
Hippasus, 119−120, 122, 133, 175
Hippopotamus hunt, 165
Histograms, 489−505
Home computers, 33
Homer, 219
Homo habilis, 3
Homo sapiens, 3
Horizon, 234
Horizontal line, 284
House of Commons, 5
"How We Calibrate College Board Scores"
   (Angoff), 555
Huygens, Christian, 147
Hybrid, 348
Hyperbola, 220
Hypothesis, 187

Icosahedron, 217
Ideal form, 165
If-then statements, 176
Iff, 48, 50
Inconsistent system, 303
Independent events, 409−414, 418−421
Independent trials, 413

Indus Valley, 19
Inequalities, 306
Inference, 561
"Inference and Disputed Authorship: The
    Federalist" (Mosteller and Wallace),
    520*n*.
Inferential statistics, 477–479
Infinite set, 133
Infinite sum, 95, 100–102
Integers, 132
Integrated circuits, 33
Intersection, 135
"Introduction to Plane and Solid Loci"
    (Fermat), 278
Invalid arguments, 194–196
Iodine 129, 104
Iodine 131, 95–100, 105
Iodine 133, 104
Ionization, 95
Irrational numbers, 119ff., 135, 144, 175
Isaiah 44:13, 172
Ishango tally bone, 4, 18, 32, 53
Isosceles triangle, 173, 182

Jackson, Reggie, 409–410, 478
Jacobs, Franklin, 94
Janko group, 54
Japan, 396
Jay, John, 519, 527–530
Jericho, 3
Jevons, W. Stanley, 54
Johanson, Donald, 3
Jordan, Camille, 295
*Journal of the American Medical
    Association*, 471*n*.
Joyce, James, 172
Jupiter, 147, 219
Jury, 390–392

Kalvern, H., Jr., 391
Kamikaze attacks, 396–398
Kantorovich, L. V., 320
Kasner, Edward, 44
Kennedy, Edward, 535
Kepler, Johann, 107, 218, 456
Kerrich, J. E., 360, 455, 466
Khachian, L. G., 144–145, 320
Khalix, 62
Khan, Genghis, 248
Khet, 156
Khomeini, Ayatollah, 535
Killer innings, 466
Kimball, G. E., 396*n*.

Kline, Morris, 94
Koopmans, T. C., 320
Kroeber, A. L., 28
Kubrick, Stanley, 23

Laetoli Hominid No. 18, 3
Lagrange, Joseph Louis, 240
Landon, Alfred, 562–565
Langcor, Willard W., 360, 466
Last Supper, The (Leonardo), 233, 239
Lazarsfeld, P., 483*n*.
Leakey, Mary, 2, 3
Leakey, Richard, 3
Least common multiple, 85
Leibniz, Gottfried, 186
Leonardo da Vinci, 233
Leontief, Wassily, 265, 319
Lincoln, Abraham, 32
Line, 166, 225, 279ff.
Linear equations, 279ff.
Linear inequalities, 306, 309
Linear programming, 145, 265, 319ff.
Literary Digest, 562–565
Lobachevsky, Nicholas, 250–253,
    258–261, 263
Lobachevsky's tenth axiom, 251
Logic, 168ff., 187ff.
Logical variant, 179
Logistica, 56
London, 462
Lottery, 424–427, 433–435, 437, 438,
    441
Lucas, Eduard, 74
Luzinski, Greg, 458
Lyons-Sims group, 54

McEnroe, John, 422
McPhee, W., 483*n*.
Madison, James, 519, 527–530
Magen David, 223
Magic square, 294
Mahavira, 29
Mail, 33
Mantegna, Andrea, 233
Marijuana, 465–466
Marker words, 520, 527–530
Masaccio (Tommaso di Giovanni Guidi),
    231
Mathieu group, 54
Matte, Tom, 390
Maxwell, James Clerk, 142–143
Mayan numeration system, 6–7, 11–16,
    25–26, 42, 136

Mayer, Jean, 321
Mean, 477, 506ff.
Median, 508ff.
Menaechmus, 219
Mendel, Johann Gregor, 346−356, 358−359, 363, 366, 489ff.
Mendeleev, Dmitri Ivanovich, 141−143, 147
Menes, 7
Menorah, 66
Méré, Antoine Gombaud, Chevalier de, 408−409, 421−422
Mermin, N. D., 141
Mersenne, Marin, 74
Mersenne numbers, 63, 111−112
Mersenne's conjecture, 74
Meteorology, 456
"Methods of Operations Research" (Morse and Kimball), 396*n*.
Mill, John Stuart, 266
Mixed-base system, 25
Mode, 508
Modus ponens, 144, 187−188, 190
Modus tollens, 187−188, 190
Monopoly, 369, 382, 387, 389
Monte Carlo method, 446ff., 455
Morse, P. M., 396*n*.
Morse code, 12
Mosteller, Frederick, 360, 520, 527−530, 567
Motorola, Inc., 33
Multilateral, 225
Multiple, 46
Multiplication principle, 425−427
Multistage cluster sampling, 569
Murphy, Calvin, 410, 424
Museum of Alexandria, 164
Mutually exclusive events, 377, 420
Myers, Joel, 456−457, 461
Myriad, 131

Napier, John, 57, 107
National Geographic, 128
Natural numbers, 45, 132
Natural sciences, 158
Nature, 3
Negative numbers, 136
Neugebauer, Otto, 20
Neutrons, 46
New England Journal of Medicine, 472
New York Public Library, 34
New York Times, The, 130, 140, 467
Newton, Sir Isaac, 251, 437
"Nine Chapters on the Mathematical Art," 294

Nine dots, 247, 249
Nobel Prize, 320
Nonagon, 217
Non-Euclidean geometry, 229, 260
Nonrespondents, 564
Nonresponse bias, 564
Normal approximation, 551ff..
Normal curve, 533ff.
Normal distribution, 492
Normal values, 553
Nuclear energy, 94
Nucleus, 46
Number of the beast, 43, 425, 433
Number words, 5
Numerals, 6
    Brahmi, 22
    Roman, 18, 21, 32
Numeration system, 6−7
    decimal, 33
    Egyptian (*see* Egyptian numeration system)
    Hindu-Arabic, 6, 8, 19ff., 29
    Mayan (*see* Mayan numeration system)
    vertical, 12
Nursing, 94

Objective function, 322
Oblong, 225
Obtuse angle, 225
Obtuse angle hypothesis, 243, 246
Octagon, 217
Octahedron, 217
Octal notation, 42
Octonary system, 42
Odds, 389
Ontogeny, 135−137
Operations research (OR), 396, 479, 487
Optimum solution, 335
Order of a group, 54
Ordered pair, 269
Origin, 269
Ouden, 28
Ozark, Danny, 458−459

Painting:
    Egyptian, 153
    Greek, 165
Papyri, Greek, 92
Parabola, 220
Parallel straight lines, 225, 283
Parallel systems, 414−418
Parameters, 561
Parliament, 5
Parthenon, 165

Pascal, Blaise, 346–347, 421–422, 437–438
Pattern recognition, 56ff.
Pauling, Linus, 473–474
Pea plants, 347–356
Pearl, Raymond, 534
Pearson, Karl, 489
Pediatrics, 93
Pentagon, 216
People v. Collins, 195
Perfect number, 63, 93
Perfect square, 60
   (*See also* Square numbers)
Pericles, 164, 219
Permutation, 428, 432, 438–440
Perpendicular, 225
Pharmacy, 93
Pharoah, 152
Phenotype, 351
Phillips, David, 468–471
Phylogeny, 135–137
Physical sciences, 158
Physics, 142
Pi, 122, 159, 162
Piero della Francesca, 232
Pittsburgh Steelers, 497, 512–513
Place value, 26, 29
Plane angle, 225
Plato, 45, 217, 220
Platonic instruments, 167, 217
Platonic solids, 217
Playfair, John, 241
Playfair's axiom, 241
Poincaré, Henri, 137
Point, 166, 225
Point-slope form, 286ff.
Political science, 93
Polyclitis, 173
Polygon, 216
   frequency, 493
Polyhedra, 216
Polynomial, 71
Population, 561
Position of a number system, 20, 25–29
Positional system, 26, 29
Postulates, 166
Powers (*see* Exponents)
PR (personal record), 454
"Precious Mirror of the Four Elements," 294
"Pre-election Polls of 1948, The" (Mosteller), 567
Pregnancy test, 460–461
Presidential elections, 486, 487
Prime number, 4, 46, 63, 70ff., 83
Primes, 106

"Principia Mathematica" (Russell and Whitehead), 186
Printing press, 21, 22
Probability, 361
Probability experiment, 359, 367, 370
Product rule, 399–400, 411, 413
Projective geometry, 236
Proof, 140ff.
Propositions, 167
Protons, 46
Psychology, 2, 147
Publius, 519
Pyramids, 20, 152
   Cheops, 163
   Gizeh, 153, 163
Pythagoras, 57ff., 69, 174, 175, 214
Pythagorean theorem, 58
Pythagorean triple, 60
Pythagoreans, 119–120, 122–123, 133, 137

Quadrilateral, 159, 225
Quarks, 46
Quetelet, Adolph, 489
Quipu, 18, 107
Quota sampling, 567

Racecar driving, 118
Radioactivity, 94–100
Raleigh, Sir Walter, 305
Rameses II, temple of, 153, 165
Random, 364
Random numbers, 449
Raphael, 233
Rational number, 90, 112–116, 132, 136
Reagan, Ronald, 422–423, 468, 534–536
Real line, 269
Real number system, 135
Real numbers, 134
Recessive trait, 349
Rectilinear, 225
Reductio ad absurdum, 72
Regnier, Edme, 473–475
Regular polygon, 216
Regular soiids, 217
Reinhold, Robert, 140
Relativity, 171, 237, 261
Reliability, 414
Rem, 95
Renaissance art, 231–237, 240
Repeating decimal, 113, 123–124
Replacement:
   with, 366, 429
   without, 366, 401–405, 429
Resurrection (Piero), 232, 239

Revelations, 32, 57
Reverse Polish notation (RPN), 108
Review of Allergy, 473
Rhind papyrus, 82–91, 104, 136, 152, 154–164, 214
Rhomboid, 225
Rhombus, 225
Ribicoff, Abraham, 481–482, 486
Riemann, Bernhard, 252–253, 258–261, 263
Riemann's tenth axiom, 253
Right angle hypothesis, 243
Rise, 280–283
Ritchie, Don, 117–118
Ritzel, G., 474
Rohlf, F. J., 355
Roman Empire, 231
Roman numerals, 18, 21, 32
Roosevelt, Eleanor, 566
Roosevelt, Franklin Delano, 562–566
Roper poll, 566
Rose, Pete, 449–451, 454
Rosetta Stone, 20, 82
Roulette, 368
Round-off error, 109–110, 126
Rozin, Paul, 147
RPN (reverse Polish notation), 108
Ruffin, 472
Run, 280–283
Running, 94, 117–118
Russell, Bertrand, 186

S-quad, 242
Saccheri, Girolamo, 229, 241–248, 250–251, 253, 261
Safire, William, 130–131
Sagan, Carl, 130, 148
St. James Led to Martyrdom (Mantegna), 233, 239
St. Peter Resuscitating the Son of the King of Antioch (Masaccio), 231
Salazar, Alberto, 94
Sales, 94
Salk, Jonas, 459, 461
Salk vaccine, 456, 459–460, 464–465, 467, 476
Sample, 561
Sample space, 371, 37
Sample surveys, 473–475
Schmidt, Mike, 118
Scholastic Aptitude Test (SAT), 555
School of Athens, The (Raphael), 233, 239
Science, 467, 476
Scientific American, 93
Scientific method, 157–159

Score, 32
Sculpture:
    Egyptian, 153
    Greek, 165
Seconds, 106
Selection bias, 563–564
Semicircle, 225
Septagon, 216
Series, systems in, 414–418
Setat, 156
Sets, 132ff.
Sex, 466
Sex discrimination, 462
Shammos, 66
Ship maneuvering, 396
Ship-months, 480
Sifr, 28
Silicon, 33
Simplex method, 320
Skewed distribution, 493
Slope, 280ff.
Slope-intercept form, 286ff.
Sneva, Tom, 118
Soda-pop example, 448–449, 451
Sokal, R. R., 355
Solution:
    of an equation, 272ff.
    of a system, 297ff.
Souchak, Mike, 118
Space shuttle, 415
Speed limits, 481–482, 486
Spherical geometry, 250ff.
Sporadic groups, 54
Sports, 94
Springfield Joggers Club, 111
Square, 217
Square numbers, 60, 123
Square root, 120ff.
Standard deviation, 520ff.
    population, 526
    sample, 526
Standard scores, 553
Standard units, 536ff.
Star of David, 223
Statement, 175
"Statistics" (Freedman et al.), 569*n*.
"Statistics: A Guide to the Unknown" (Tanur), 468*n*.
Stevin, Simon, 105ff., 123–124
Stifel, Michael, 57
Stock market quotations, 42
Straight line, 225
Straightedge, 167
Strange Geometry, 243–244, 247
Streck, Ron, 119
Subset, 133

Substitution, method of, 301
Suds, 94
Sumer, 19, 152, 153
Summit angle, 242
Sunya, 28
Surface, 225
Swetz, Frank, 21
Symbolic logic, 175
System of two linear equations, 296
Systems in series, 414−418

Tally, 4, 7, 17, 24
Tally sticks, 5
Tanur, J. M., 468*n.*
Tanzania, 2, 3
Telephone, 33
Television, 33
"Tenth, The" (Stevin), 106
Terminating decimal, 113, 123−124
Tetrahedron, 217
Texas Instruments, Inc., 107
Thales, 174, 185
Thales' theorem, 185
Theorem, 73, 146
Thompson group, 54
Thorp, Edward O., 408
Three Mile Island, 94
Three problems of antiquity, 219
Thurmond, Strom, 566
Tigris (Heyerdahl's boat), 19−20
Transversal, 204
Trapezia, 225
Trapezoid, 160, 235
Treatment group, 471−475−476
Tree diagram, 366, 398−404, 418, 428, 440
Triangular numbers, 61
Trilateral, 225
Triskaidekaphobia, 65
Truman, Harry, 565ff.
TWA (Trans World Airlines), 250, 256−258
"2001: A Space Odyssey" (film), 23
Tymoczko, Thomas, 147

U 234 (isotope of uranium), 103
Ulam, Stanislaw, 446−447
Ulcers, 471−472
Unger, Franz, 347
Uniform distribution, 496
Union, 134
Unit fraction, 83, 106
Unitas, Johnny, 390

U.S. Air Force, 265, 319
U.S. Census Bureau, 501−502
U.S. Department of Justice, 320
U.S. Food and Drug Administration, 534
U.S. recommended daily allowance (U.S. RDA), 321ff.
Units, standard, 536ff.
Universal set, 134
Urn problems, 364−367, 384−385, 401−402, 414, 440−441

Vanishing point, 234, 239
Variance, 523
Variant, 179
Venn diagram, 133, 408
Verne, Jules, 42
Vertex, 222
Vertical line, 284
Vertical numeration system, 12
Viète, Francois, 136
Vigesimal system, 25
"Vital and Health Statistics" (Department of Health and Human Services), 551
Vitamin C, 473−475
Vitruvius, 174
Volume, 154ff.
Voting, 483−486
"Voting: A Study of Opinion Formation in a Presidential Election" (Berelson, Lazarsfeld, and McPhee), 483*n.,* 565

Wallace, D. L., 520, 527−530
Wallace, Henry, 566
Wangansteen, 471−472, 477
War Production Board, 319
Warner-Lambert Company, 460, 467
Warren, Earl, 566
Washington Post, The, 535
Water-meter readings, 31
Whitehead, Alfred North, 186
Wynder, Ernest, 321

y intercept, 282ff.
Yin and yang, 59
Yossarian (of "Catch-22"), 43, 66

Zaslavsky, Claudia, 3
Zeisel, H., 391
Zeno, 174, 175, 214
Zephirum, 28
Zero, 16, 20, 26−29, 136

DATE DUE			

GAYLORD   PRINTED IN U.S.A.